Springer Monographs in Mathematics

This series publishes advanced monographs giving well-written presentations of the "state-of-the-art" in fields of mathematical research that have acquired the maturity needed for such a treatment. They are sufficiently self-contained to be accessible to more than just the intimate specialists of the subject, and sufficiently comprehensive to remain valuable references for many years. Besides the current state of knowledge in its field, an SMM volume should ideally describe its relevance to and interaction with neighbouring fields of mathematics, and give pointers to future directions of research.

Tullio Ceccherini-Silberstein • Fabio Scarabotti •
Filippo Tolli

Representation Theory of Finite Group Extensions

Clifford Theory, Mackey Obstruction, and the Orbit Method

Springer

Tullio Ceccherini-Silberstein
Dipartimento di Ingegneria
Università degli Studi del Sannio
Benevento, Italy

Fabio Scarabotti
Dipartimento SBAI
Sapienza University of Rome
Roma, Italy

Filippo Tolli
Dipartimento di Matematica e Fisica
Roma Tre University
Roma, Italy

ISSN 1439-7382 ISSN 2196-9922 (electronic)
Springer Monographs in Mathematics
ISBN 978-3-031-13872-0 ISBN 978-3-031-13873-7 (eBook)
https://doi.org/10.1007/978-3-031-13873-7

This Springer imprint is published by the registered company Springer Nature Switzerland AG
The registered company address is: Gewerbestrasse 11, 6330 Cham, Switzerland

To the memory of my beloved "aunties" zia Cornelia Trevisani and zia Marcella Treves, in primis, and Adriana Mondolfi, zia Alma Enriques Ghiron, Augusta and zia Gina Castelnuovo, suor Cesarina, Césarine Poiget, zia Eda Berger Vidale, Elisa Frontali, Fiorenza Almagià Maranelli, Giacometta Limentani, zie Giovanna e Regina Franchetti, zia Giuliana Treves Artom, Ida Sacchetti, zia Laura Zelasio, Leah Loewenwirth Reuveni, Livia e Sidda Garofolini, Maria Danelon, Nella Mortara, Nella Volterra, Paola Franchetti, tata Rita Alverà, zia Silvia Treves Levi Vidale, Teresa Tagliaferri, Mlle Yvonne Bâtard, and, last but not least, Anna Palaggi

To Cristina and Virginia

To my mom, my sister Rossella, my wife Stefania, and my niece Valentina

Preface

Given two groups N and H, the problem of classifying the groups G such that we have a short exact sequence

$$1 \longrightarrow N \stackrel{\iota}{\longrightarrow} G \stackrel{\pi}{\longrightarrow} H \longrightarrow 1$$

is called the *extension problem*. This short exact sequence (and, by abuse of language, the group G) is called an extension of H by N. In the framework of representation theory, a natural question arises, namely:

Is it possible to describe the irreducible representations of G by means of the structure of the group extension?

This problem was faced by many mathematicians, including I. Schur, A.H. Clifford, and G. Mackey (who actually worked on locally compact groups). More recently, contributions and expositions were given by M. Isaacs, B. Huppert, Y.G. Berkovich and E.M. Zhmud, and by J.M.G. Fell and R.S. Doran.

The aim of the present monograph is to give an exposition of some of the many facets of this problem, at an elementary level and pointing to concrete applications. Indeed, the representation theory of several classes of groups is determined by the study of such group extensions: this is the case, for instance, for wreath products and semidirect products with Abelian normal subgroups; we have devoted an entire monograph [28] (and several papers) to the first case and Chapter 12 of [30] to remarkable applications of the second case (which is also widely used in the Lie groups setting). Actually, both cases are particular examples of a general principle, which we call the little group method (after Mackey and E. Wigner). We give a precise formulation of this principle and ask two fundamental questions:

When is it possible to apply this method?

When it is not possible, is there an alternative way to construct the irreducible representations of the group extension?

The present book aims at developing powerful methods and tools, ranging from low-dimensional group cohomology to Clifford theory and to Mackey-Wigner theory, providing, via a thorough analysis of all such related intertwined concepts, a complete answer to the above questions. Our general approach—with an emphasis on the fundamental contributions of Mackey, from Mackey's lemma (on the decomposition of $\mathrm{Res}^G_K \mathrm{Ind}^G_H$, where $H, K \leq G$) to Mackey's obstruction (whose terminology we borrowed from Fell and Doran)—has a functional flavor and therefore it is more suitable for applications in a harmonic analytical framework. We clearly have in mind that an alternative approach, based on character theory as Isaacs' [67], on the one hand gives rise, sometimes, to shorter proofs and elegant applications to group theory, but, on the other hand, seems to be not sufficient in order to develop all the powerful tools for harmonic analysis and applications.

This monograph is definitively not suitable for a general course on representation theory. Indeed, the topics treated are too advanced and possibly too monothematic for such a general target. However, we believe that it might be successfully used as a valid support for such a general course, for, say, a subsequent course on "topics in representation theory" as well as for some parallel "reading courses" and/or an "advanced seminar." Last but not least, the mature researcher may find it useful as a reference to Clifford theory, Mackey's machinery (originally developed for locally compact groups), and Kirillov's orbit method.

It is structured as follows.

In Chap. 1, we give some preliminary results on representation theory and group extensions: we prove just the results which are not standard.

Chapter 2 is an exposition of Clifford theory which relates the representation theory of a (finite) group G to that of its normal subgroups $N \leq G$ through the operations of restriction, induction, and extension. We study the conjugation action of G (and of $H = G/N$) on $\widehat{N}$, the (unitary) dual of N. Given $\sigma \in \widehat{N}$, we denote by $I_G(\sigma)$ the stabilizer of σ (usually called the inertia group). We study several particular cases and applications of Clifford theory, in particular we give a precise formulation of the little group method: it may be applied when every $\sigma \in \widehat{N}$ has an extension $\widetilde{\sigma}$ to $I_G(\sigma)$ (i.e., $\widetilde{\sigma}$ is an irreducible representation of $I_G(\sigma)$ and its restriction to N is σ). In this case, every irreducible representation of G is of the form $\mathrm{Ind}^G_{I_G(\sigma)}(\widetilde{\sigma} \otimes \overset{\vartriangle}{\psi})$, where ψ is an irreducible representation of $H_G(\sigma) := I_G(\sigma)/N$ and $\overset{\vartriangle}{\psi} \in \widehat{I_G(\sigma)}$ denotes its inflation to $I_G(\sigma)$, that is, $\overset{\vartriangle}{\psi}(h) := \psi(hN)$ for all $h \in I_G(\sigma)$.

Chapters 3, 4, and 5 are devoted to the study of the case when $A := G/N$ is Abelian. Again, there are several remarkable cases where this condition is satisfied, for instance the case when N is an index two subgroup of G (which includes $G = \mathfrak{S}_n$, the symmetric group, and $N = \mathfrak{A}_n$, the alternating group). In the general case (Chap. 3), we examine the dual action of $\widehat{A}$ on $\widehat{G}$ given by tensoring two representations and its relationships with the conjugating action. In Chap. 4, we examine the particular case when A is Abelian and every $\sigma \in \widehat{N}$ has an extension to $I_G(\sigma)$, so that the little group method may be applied. In particular, we examine the case when A is of prime order. In Chap. 5, we discuss the representation theory of the alternating group, of metacyclic, quaternionic, and dihedral groups. We also examine the case when A is of order p^2.

Chapter 6 is devoted to the orbit method for step-2 nilpotent groups of odd order. When this condition is satisfied, we have a natural and powerful correspondence between Lie rings and nilpotent groups, and this correspondence can be used to describe the irreducible representations of the nilpotent groups. The main tool is an action of the group on the ring and the coadjoint action, namely the corresponding action on the dual of the Lie ring (regarded as an Abelian group). We describe the behavior of this construction with respect to induction and restriction and we apply it to the finite Heisenberg group. In this chapter, we also introduce some cohomological machinery that is also used in the subsequent chapters.

In Chap. 7, we answer the questions raised at the beginning of this preface: we show that the little group method may be applied if and only if a suitable unitary 2-cocycle (the Mackey obstruction) is trivial. We then introduce (unitary) projective representations and use them to give a complete answer to the problem of the construction of the irreducible representations of a finite group extension. We also examine the commutant of an induced representation and the relative Hecke algebra.

Chapters 8 and 9 are devoted to the study of induction for (unitary) projective representations: this includes both Mackey and Clifford theories. With respect to our main source (Mackey's paper on unitary representations of locally compact group extension) we give more direct proofs, involving detailed cocycle computations.

In Chap. 10, we expose the theory of projective representations for finite Abelian groups. Some results are given under the assumption that the group is of odd order. We then apply this theory in order to describe the irreducible representations of finite metabelian (i.e., step-2 solvable) as well as step-2 nilpotent groups.

Finally, in the Appendix, we review some historical aspects and discuss some additional topics.

We express our deep gratitude to Pierre de la Harpe, Slava Grigorchuk, Marty Isaacs, Daniel Rockmore, and Jean-Pierre Serre for useful remarks and suggestions, as well as for their interest and encouragement. Finally, we warmly thank Elena Griniari from Springer Verlag for her continuous care and most precious help at various stages of the preparation of the manuscript.

Rome, Italy
April 2022

Tullio Ceccherini-Silberstein
Fabio Scarabotti
Filippo Tolli

Contents

Chapter 1
Preliminaries

In this chapter, we fix notation and recall some basic facts of linear algebra, of finite groups and their representation theory. For more details on representation theory we refer to the monographs by Alperin and Bell [3], Berkovich and Zhmud [16], Bump [21], Curtis and Reiner [38, 39], Dornhoff [43], Fell and Doran [48], Fulton and Harris [51], Gallagher [52], Grove [54], Huppert [65], Isaacs [67], James and Kerber [70], Mackey [95], Serre [121], Simon [125], as well as ours [24, 27, 30]. As for the theory of finite groups, we also refer to the books by Grove [55], M. Hall [57], Huppert and Blackburn [66], Kurosh [84], Machì [90], Robinson [108], Rotman [114], Serre [122], and Sternberg [126].

1.1 Representations of Finite Groups

All vector spaces considered here are complex. Moreover, we shall equip every finite dimensional vector space V with a scalar product denoted by $\langle \cdot, \cdot \rangle_V$ and associated norm $\|\cdot\|_V$; we usually omit the subscript if the vector space we are referring to is clear from the context. Given two finite dimensional vector spaces W and U, we denote by $\mathrm{Hom}(W, U)$ the vector space of all linear maps from W to U. When $U = W$ we write $\mathrm{End}(W) := \mathrm{Hom}(W, W)$ and denote by $\mathrm{GL}(W) \subseteq \mathrm{End}(W)$ the general linear group of W consisting of all bijective endomorphisms of W. Also, for $T \in \mathrm{Hom}(W, U)$ we denote by $T^* \in \mathrm{Hom}(U, W)$ the adjoint of T.

We define a (normalized) Hilbert-Schmidt scalar product on $\mathrm{Hom}(W, U)$ by setting

$$\langle T_1, T_2 \rangle_{\mathrm{Hom}(W,U)} := \frac{1}{\dim W} \mathrm{tr}(T_2^* T_1) \tag{1.1}$$

T. Ceccherini-Silberstein et al., *Representation Theory of Finite Group Extensions*, Springer Monographs in Mathematics,
https://doi.org/10.1007/978-3-031-13873-7_1

for all $T_1, T_2 \in \mathrm{Hom}(W, U)$, where $\mathrm{tr}(\cdot)$ denotes the trace of endomorphisms; note that,

$$\langle T_1, T_2\rangle_{\mathrm{Hom}(W,U)} = \frac{\dim U}{\dim W}\langle T_2^*, T_1^*\rangle_{\mathrm{Hom}(U,W)}. \tag{1.2}$$

In particular, the map $T \mapsto \sqrt{\dim U/\dim W}\ T^*$ is an (antilinear) isometry from $\mathrm{Hom}(W, U)$ onto $\mathrm{Hom}(U, W)$. Finally, denoting by $I_W : W \to W$ the identity operator, we have $\|I_W\|_{\mathrm{End}(W)} = 1$.

We now recall some basic facts on the representation theory of finite groups.

Let G be a finite group. A *unitary representation* of G is a pair (θ, W), where W is a finite dimensional vector space and $\theta : G \to \mathrm{GL}(W)$ is a group homomorphism such that $\theta(g)$ is unitary (that is, $\theta(g)^*\theta(g) = I_W$) for all $g \in G$. In the sequel, the term "unitary" will be omitted. We denote by $d_\theta := \dim(W)$ the dimension of the representation (θ, W) and by $(\iota_G, \mathbb{C})$ the *trivial representation* of G, that is, the one-dimensional G-representation defined by $\iota_G(g) := \mathrm{id}_{\mathbb{C}}$ for all $g \in G$.

The *kernel* of a representation (θ, W) of G is $\ker\theta := \{g \in G : \theta(g) = I_W\}$ and one says that the representation is *faithful* if $\ker\theta = \{1_G\}$.

Let (θ, W) be a G-representation. A subspace $V \leq W$ is said to be *G-invariant* provided $\theta(g)V \subseteq V$ for all $g \in G$. Writing $\theta|_V(g) = \theta(g)|_V$ for all $g \in G$, we have that $(\theta|_V, V)$ is a G-representation, called a *subrepresentation* of θ. One says that θ is *irreducible* provided the only G-invariant subspaces are trivial (equivalently, θ admits no proper subrepresentations).

A finite group A is Abelian if and only if all of its irreducible representations are one-dimensional (see Exercise 10.3.6 in [30]). If this is the case, an irreducible representation of A is just a map $\chi : A \to \mathbb{T} := \{z \in \mathbb{C} : |z| = 1\}$ such that $\chi(ab) = \chi(a)\chi(b)$ for all $a, b \in A$. Usually, we shall use multiplicative notation also in the Abelian setting.

Let (θ, W) and (ρ, U) be two G-representations. We denote by

$$\mathrm{Hom}_G(W, U) := \{T \in \mathrm{Hom}(W, U) : T\theta(g) = \rho(g)T, \forall g \in G\},$$

the space of all *intertwining operators*. When $U = W$ we write $\mathrm{End}_G(W) := \mathrm{Hom}_G(W, W)$. We equip $\mathrm{Hom}_G(W, U)$ with a scalar product by restricting the Hilbert-Schmidt scalar product (1.1). Finally, observe that if T belongs to $\mathrm{Hom}_G(W, U)$ then T^* belongs to $\mathrm{Hom}_G(U, W)$. Indeed, for all $g \in G$ we have

$$T^*\rho(g) = T^*\rho(g^{-1})^* = (\rho(g^{-1})T)^* = (T\theta(g^{-1}))^* = \theta(g^{-1})^*T^* = \theta(g)T. \tag{1.3}$$

Moreover, the map $T \mapsto T^*$ is an antilinear isomorphism between the vector spaces $\mathrm{Hom}_G(W, U)$ and $\mathrm{Hom}_G(U, W)$.

One says that (θ, W) and (ρ, U) are *equivalent*, and we shall write $(\theta, W) \sim (\rho, U)$, or, simply, $\theta \sim \rho$ or $W \cong U$ (see Remark 1.2), if there exists a bijective intertwining operator $T \in \mathrm{Hom}_G(W, U)$.

The vector space $\mathrm{End}_G(W)$ of all intertwining operators of (θ, W) with itself, when equipped with the multiplication given by the composition of maps and the adjoint operation, is a $*$-algebra (see [27, Chapter 7], [30, Sections 10.3 and 10.6]), called the *commutant* of (θ, W). We can thus express the well known *Schur lemma* as follows: (θ, W) is irreducible if and only if its commutant is one-dimensional (as a vector space), that is, it reduces to the scalars (the scalar multiples of the identity I_W).

We denote by $\widehat{G}$ a (fixed, once and for all) complete set of pairwise-inequivalent irreducible representations of G. This is called the *dual* of G. It is well known (cf. [24, Theorem 3.9.10] or [30, Theorem 10.3.13.(ii)]) that there is a bijection between $\widehat{G}$ and the set of conjugacy classes of elements in G so that, in particular, $\widehat{G}$ is finite. Moreover, if $\theta, \rho \in \widehat{G}$ we set $\delta_{\theta,\rho} := 1$ (resp. $:= 0$) if $\theta = \rho$ (resp. otherwise).

Let (θ, W) and (ρ, U) be two G-representations.

The *direct sum* of θ and ρ is the G-representation $(\theta \oplus \rho, W \oplus U)$ defined by

$$[(\theta \oplus \rho)(g)](w, u) := (\theta(g)w, \rho(g)u)$$

for all $g \in G$, $w \in W$ and $u \in U$. The *internal tensor product* $\theta \otimes \rho$ is the G-representation on $W \otimes U$ defined by setting

$$[(\theta \otimes \rho)(g)](w \otimes u) := \theta(g)w \otimes \rho(g)u$$

for all $g \in G$, $w \in W$ and $u \in U$.

Moreover, if $W \subseteq U$ is a G-invariant subspace, then $W^{\perp} := \{u \in U : \langle u, w\rangle_U = 0, \forall w \in W\}$, the orthogonal complement of W in U, is also a G-invariant subspace, and $\rho = \rho|_W \oplus \rho|_{W^{\perp}}$. From this, one deduces that every representation ρ decomposes as a (finite) direct sum of irreducible subrepresentations.

More generally, when θ is equivalent to a subrepresentation of ρ, we say that θ is *contained* in ρ and write $\theta \preceq \rho$.

Suppose that (θ, W) is irreducible. Then the nonnegative number

$$m = m_\theta^\rho := \mathrm{dimHom}_G(W, U) \tag{1.4}$$

is the *multiplicity* of θ in ρ. This means that one may decompose

$$U = U_1 \oplus U_2 \oplus \cdots \oplus U_m \oplus U_{m+1}$$

with $(\rho|_{U_i}, U_i) \sim (\theta, W)$ for all $i = 1, 2, \ldots, m$ and θ is not contained in $\rho|_{U_{m+1}}$. The G-invariant subspace $U_1 \oplus U_2 \oplus \cdots \oplus U_m \le U$ is called the *W-isotypic component* of U and is denoted by mW. One also says that ρ (or, equivalently, U) contains m copies of θ (resp. of W). If this is the case, we say that $T_1, T_2, \ldots, T_m \in \mathrm{Hom}_G(W, U)$ yield an *isometric orthogonal decomposition* of mW if $T_iW \leq U \ominus U_{m+1}$, and, in addition,

$$\langle T_i w_1, T_j w_2\rangle_U = \langle w_1, w_2\rangle_W \delta_{i,j} \tag{1.5}$$

for all $w_1, w_2 \in W$ and $i, j = 1, 2, \ldots, m$. This implies that the subrepresentation $mW = U_1 \oplus U_2 \oplus \cdots \oplus U_m$ is equal to the *orthogonal* direct sum $T_1W \oplus T_2W \oplus \cdots \oplus T_mW$, and each operator T_j is a isometry from W onto $U_j \equiv TW_j$. For a quite detailed analysis of this decomposition, we refer to [30, Section 10.6]. In particular, the following isomorphism is quite useful. Suppose that $\theta \sim \oplus_{\eta\in J} m_\eta \eta$ and $\rho \sim \oplus_{\eta\in K} n_\eta \eta$ are the decompositions into irreducible representations. This means that $J, K \subseteq \widehat{G}$ and $m_\eta > 0$ for $\eta \in J$ (resp. $n_\eta > 0$ for $\eta \in K$). Then we have a vector space isomorphism

$$\mathrm{Hom}_G(\theta, \rho) \cong \bigoplus_{\eta\in J\cap K} M_{m_\eta, n_\eta}(\mathbb{C}), \tag{1.6}$$

where $M_{m,n}(\mathbb{C})$ denotes the space of all $m \times n$ matrices with complex coefficients. If $\rho = \theta$ then (1.6) is a $*$-algebra isomorphism (cf. (1.38) for an important particular case and [30, Section 10.6] for a quite detailed analysis). More generally, if $\theta \sim \oplus_{i=1}^n \psi_i$ and $\rho \sim \oplus_{j=1}^m \mu_j$ are decompositions into arbitrary (not necessarily irreducible) representations then

$$\mathrm{dimHom}_G(\theta, \rho) = \sum_{i=1}^n \sum_{j=1}^m \mathrm{dimHom}_G(\psi_i, \mu_j) \equiv \mathrm{dimHom}_G(\rho, \theta). \tag{1.7}$$

One says that a representation σ is *multiplicity-free* if it decomposes without multiplicities, that is, one may decompose $\sigma \sim \bigoplus_{i=1}^n \sigma_i$ with $\sigma_i \in \widehat{G}$ and $\sigma_i \not\sim \sigma_j$ for all $i \neq j$.

1.2 The Group Algebra and the Left-Regular Representation

We denote by $L(G)$ the *group algebra* of G. This is the vector space of all functions $f : G \to \mathbb{C}$ equipped with the *convolution product* $*$ defined by setting

$$[f_1 * f_2](g) := \sum_{h\in G} f_1(h) f_2(h^{-1}g) = \sum_{h\in G} f_1(gh) f_2(h^{-1}),$$

for all $f_1, f_2 \in L(G)$ and $g \in G$. We shall endow $L(G)$ with the scalar product $\langle\cdot,\cdot\rangle_{L(G)}$ defined by setting

$$\langle f_1, f_2\rangle_{L(G)} := \sum_{g\in G} f_1(g)\overline{f_2(g)} \tag{1.8}$$

for all $f_1, f_2 \in L(G)$. The *Dirac functions* $\{\delta_g : g \in G\}$, where δ_g is defined by $\delta_g(g) := 1$ and $\delta_g(h) := 0$ if $h \neq g$, constitute a natural orthonormal basis for $L(G)$. We shall also equip $L(G)$ with the *involution* $f \mapsto f^*$, where $f^*(g) := \overline{f(g^{-1})}$, for all $f \in L(G)$ and $g \in G$. It is straightforward to check that $(f_1 * f_2)^* = f_2^* * f_1^*$, for all $f_1, f_2 \in L(G)$. We shall thus regard $L(G)$ as a $*$-algebra.

The *left-regular representation* of G is the G-representation $(\lambda_G, L(G))$ defined by setting $[\lambda_G(h)f](g) := f(h^{-1}g)$, for all $f \in L(G)$ and $h, g \in G$. Similarly, the *right-regular representation* of G is the G-representation $(\rho_G, L(G))$ defined by setting $[\rho_G(h)f](g) := f(gh)$, for all $f \in L(G)$ and $h, g \in G$. Note that the left-regular and right-regular representations commute, that is,

$$\lambda_G(g_1)\rho_G(g_2) = \rho_G(g_2)\lambda_G(g_1) \tag{1.9}$$

for all $g_1, g_2 \in G$.

Given $f \in L(G)$, the (right) *convolution operator* with *kernel* f is the linear map $T_f : L(G) \to L(G)$ defined by

$$T_f f' := f' * f \tag{1.10}$$

for all $f' \in L(G)$. We have

$$T_{f_1 * f_2} = T_{f_2} T_{f_1} \tag{1.11}$$

and

$$T_{f^*} = (T_f)^*, \tag{1.12}$$

for all f_1, f_2 and f in $L(G)$. Moreover, $T_f \in \mathrm{End}_G(L(G))$, the commutant of the left-regular representation (this is a consequence of (1.9)) and the map

$$\begin{array}{ccc} L(G) & \longrightarrow & \mathrm{End}_G(L(G)) \\ f & \longmapsto & T_f \end{array} \tag{1.13}$$

is a $*$-anti-isomorphism of $*$-algebras (see [27, Proposition 1.5.2] or [30, Proposition 10.3.5]).

It is easy to check that $T_f \delta_g = \lambda_G(g) f$ and $\mathrm{tr}(T_f) = |G| f(1_G)$. We deduce that

$$\langle T_{f_1}, T_{f_2} \rangle_{\mathrm{End}(L(G))} = \frac{1}{|G|}\mathrm{tr}\left[(T_{f_2})^* T_{f_1}\right] = \frac{1}{|G|}\mathrm{tr}\left[T_{f_1 * f_2^*}\right] = [f_1 * f_2^*](1_G) = \langle f_1, f_2 \rangle_{L(G)} \tag{1.14}$$

for all $f_1, f_2 \in L(G)$. This shows that the map (1.13) is an isometry.

Let (θ, W) be a representation of G and let $\{w_1, w_2, \ldots, w_{d_\theta}\}$ be an orthonormal basis of W. The corresponding *matrix coefficients* $u_{i,j}^\theta \in L(G)$ are defined by setting

$$u_{i,j}^\theta(g) := \langle \theta(g) w_j, w_i \rangle \tag{1.15}$$

for all $i, j = 1, 2, \ldots, d_\theta$ and $g \in G$.

Proposition 1.1 *Let $\theta, \rho \in \widehat{G}$. Then*

$$\langle u_{i,j}^\theta, u_{h,k}^\rho \rangle = \frac{|G|}{d_\theta}\delta_{\theta,\rho}\delta_{i,h}\delta_{j,k} \quad \textit{(orthogonality relations)}, \tag{1.16}$$

$$u_{i,j}^\theta * u_{h,k}^\rho = \frac{|G|}{d_\theta}\delta_{\theta,\rho}\delta_{j,h}u_{i,k}^\theta \quad \textit{(convolution properties)}, \tag{1.17}$$

and

$$u_{i,j}^\theta(g_1g_2) = \sum_{\ell=1}^{d_\theta} u_{i,\ell}^\theta(g_1)u_{\ell,j}^\theta(g_2) \tag{1.18}$$

for all $i, j = 1, 2, \ldots, d_\theta$, $h, k = 1, 2, \ldots, d_\rho$, and $g_1, g_2 \in G$.

Proof See [24, Lemma 3.6.3 and Lemma 3.9.14] or [30, Lemma 10.2.10, Lemma 10.2.13, and Proposition 10.3.6]. ■

The sum $\chi^\theta := \sum_{i=1}^{d_\theta} u_{i,i}^\theta \in L(G)$ of the diagonal entries of the matrix coefficients of the G-representation (θ, W) is called the *character* of θ. Note that $\chi^\theta(g) = \mathrm{tr}(\theta(g))$ for all $g \in G$. Finally, the celebrated *Peter-Weyl theorem* states that

$$\lambda_G \sim \bigoplus_{\theta \in \widehat{G}} d_\theta \theta, \tag{1.19}$$

is the decomposition of the left regular representation into irreducible representations (or, rather, isotypic components) and

$$\sum_{\theta \in \widehat{G}} d_\theta^2 = |G|. \tag{1.20}$$

Given two G-representations θ and ρ, with θ irreducible, the multiplicity $m = m_\theta^\rho$ of θ in ρ (cf. 1.4) may be expressed as the normalized scalar product of the corresponding characters χ^θ and χ^ρ, that is,

$$m = \frac{1}{|G|}\langle \chi^\theta, \chi^\rho \rangle_{L(G)}. \tag{1.21}$$

(see, for instance [30, Proposition 10.2.18]).

We recall that a *central function* on a group G is a function $f \in L(G)$ that is constant on the conjugacy classes of G, equivalently $f(g_1g_2) = f(g_2g_1)$ for all $g_1, g_2 \in G$. The central functions form a commutative subalgebra of the convolution algebra. A vector space basis for this subalgebra is made up of the characters of all the irreducible representations of G.

The characters satisfy a set of fundamental orthogonality relations. Denote by $C(G)$ the conjugacy classes of G and for $\theta \in \widehat{G}$ and $\mathscr{C} \in C(G)$ denote by $\chi^\theta(\mathscr{C}) := \chi^\theta(g)$ the (common) value of χ^θ on the elements $g \in \mathscr{C}$. If also $\tau \in \widehat{G}$ then

$$\sum_{\mathscr{C}\in C(G)} \chi^\theta(\mathscr{C})\overline{\chi^\tau(\mathscr{C})}|\mathscr{C}| = |G|\delta_{\theta,\tau}. \tag{1.22}$$

The dual orthogonality relations are (cf. [30, Proposition 10.2.17 and Corollary 10.3.14]):

$$\sum_{\theta\in\widehat{G}} \chi^\theta(\mathscr{C})\overline{\chi^\theta(\mathscr{D})}\sqrt{|\mathscr{C}|\cdot|\mathscr{D}|} = |G|\delta_{\mathscr{C},\mathscr{D}} \tag{1.23}$$

for all $\mathscr{C}, \mathscr{D} \in C(G)$.

The *derived* subgroup (also called the *commutator* subgroup) of G is the subgroup G' (also denoted by $[G, G]$) generated by all the *commutators* $[g, h] := g^{-1}h^{-1}gh$, where $g, h \in G$. We recall that G' is normal (even characteristic) and that if $N \trianglelefteq G$ then the quotient group G/N is Abelian if and only if $G' \leq N$; see [3, 90, 108, 114]. Moreover, G has exactly $|G|/|G'|$ one-dimensional (and therefore irreducible) representations; these are given by the injective map

$$\widehat{G/G'} \ni \omega \longmapsto \overset{\triangle}{\omega} \in \widehat{G} \tag{1.24}$$

defined by setting $\overset{\triangle}{\omega}(g) := \omega(gG')$ for all $g \in G$; see [24, Proposition 9.2.5] and Definition 2.2.

1.3 Induced Representations

Let now $K \leq G$ be a subgroup. We denote by $(\mathrm{Res}^G_K\theta, W)$ the *restriction* of the G-representation (θ, W) to K, that is, the K-representation defined by $[\mathrm{Res}^G_K\theta](k) := \theta(k)$ for all $k \in K$.

Given a K-representation (σ, V) of K, denote by $\lambda = \mathrm{Ind}^G_K\sigma$ the *induced representation* (see, for instance, [21, 25, 27, 28, 30, 48, 125, 126]). We recall that the representation space of λ is given by

$$\mathrm{Ind}^G_K V := \{f : G \to V \text{ such that } f(gk) = \sigma(k^{-1})f(g), \text{ for all } g \in G, k \in K\} \tag{1.25}$$

and the action is given by setting

$$[\lambda(g)f](g') := f(g^{-1}g'), \tag{1.26}$$

for all $f \in \mathrm{Ind}_K^G V$ and $g, g' \in G$.

Let $\mathcal{T} \subseteq G$ be a *left-transversal* for K, that is, a complete set of representatives for the left-cosets gK of K in G. Then we have the decomposition

$$G = \bigsqcup_{t \in \mathcal{T}} tK, \tag{1.27}$$

where, from now on, $\bigsqcup$ denotes a disjoint union. For $v \in V$ we define $f_v \in \mathrm{Ind}_K^G V$ by setting

$$f_v(g) := \begin{cases} \sigma(g^{-1})v & \text{if } g \in K \\ 0 & \text{otherwise.} \end{cases} \tag{1.28}$$

The induced representation $\mathrm{Ind}_K^G \sigma$ is unitary with respect to the scalar product $\langle \cdot, \cdot \rangle_{\mathrm{Ind}_K^G V}$ defined by

$$\langle f_1, f_2 \rangle_{\mathrm{Ind}_K^G V} := \frac{1}{|K|} \sum_{g \in G} \langle f_1(g), f_2(g) \rangle_V = \sum_{t \in \mathcal{T}} \langle f_1(t), f_2(t) \rangle_V \tag{1.29}$$

for all $f_1, f_2 \in \mathrm{Ind}_K^G V$. We have $d_\lambda = [G : K] d_\sigma = |\mathcal{T}| d_\sigma$: indeed, if $\{v_j : j = 1, 2, \ldots, d_\sigma\}$ is an orthonormal basis in V then the set

$$\{\lambda(t) f_{v_j} : t \in \mathcal{T}, j = 1, 2, \ldots, d_\sigma\} \tag{1.30}$$

is an orthonormal basis in $\mathrm{Ind}_K^G V$ (see [25, Theorem 2.1] and [30, Theorem 11.1.11]).

Suppose now that $U(k) \in M_{d_\sigma}(\mathbb{C})$ is a unitary matrix realization of $\sigma(k)$ for all $k \in K$ (where $M_m(\mathbb{C})$, $m \in \mathbb{N}$, denotes the $*$-algebra of all $m \times m$ complex matrices) and let $\mathcal{T} = \{t_1, t_2, \ldots, t_m\}$. Then the block matrix

$$B(g) := \begin{pmatrix} U(t_1^{-1} g t_1) & U(t_1^{-1} g t_2) & \cdots & U(t_1^{-1} g t_m) \\ U(t_2^{-1} g t_1) & U(t_2^{-1} g t_2) & \cdots & U(t_2^{-1} g t_m) \\ \vdots & \vdots & \cdots & \vdots \\ U(t_m^{-1} g t_1) & U(t_m^{-1} g t_2) & \cdots & U(t_m^{-1} g t_m) \end{pmatrix} \in M_{d_\lambda}(\mathbb{C}) \tag{1.31}$$

where $U(t_i^{-1} g t_j) := 0$ whenever $t_i^{-1} g t_j \notin K$, is a unitary matrix realization of $\lambda(g) = \mathrm{Ind}_K^G \sigma(g)$ for all $g \in G$ (see [25] and [30, Theorem 11.1.11]).

By taking traces, we immediately get the *Frobenius character formula* for induced representations: the character of $\mathrm{Ind}_K^G\sigma$ is given by

$$\chi^{\mathrm{Ind}_K^G\sigma}(g) = \sum_{t\in\mathcal{T}:t^{-1}gt\in K} \chi^\sigma(t^{-1}gt). \tag{1.32}$$

The *transitivity of induction* is expressed by the isomorphism

$$\mathrm{Ind}_H^G[\mathrm{Ind}_K^H V] \cong \mathrm{Ind}_K^G V, \tag{1.33}$$

where H is an intermediate subgroup: $K \leq H \leq G$. We recall the construction of this isomorphism: the left hand side may be seen as the vector space of all maps $F\colon G \times H \to V$ such that $F(gh, h_0k) = \sigma(k^{-1})F(g, hh_0)$, for all $g \in G$, $h, h_0 \in H$, and $k \in K$. Moreover, the right hand side is as in (1.25). Then the isomorphism in (1.33) is given by the map

$$F \mapsto f$$

where $f(g) := F(g, 1_G)$ for all $g \in G$ (note that F is uniquely determined by f, because $F(g, h) = F(gh, 1_G) = f(gh)$ for all $g \in G$ and $h \in H$). See [27, Proposition 1.6.6], [28, Proposition 1.1.10], or [30, Proposition 11.1.5].

A well known relation between the induction of a K-representation (σ, V) and a G-representation (θ, W) is expressed by the so called *Frobenius reciprocity* (cf. [27, Theorem 1.6.11], [28, Theorem 1.1.19], [30, Theorem 11.2.1], or [116]) which establishes the vector space isomorphism

$$\mathrm{Hom}_G(W, \mathrm{Ind}_K^G V) \cong \mathrm{Hom}_K(\mathrm{Res}_K^G W, V). \tag{1.34}$$

In particular, if for $L \in \mathrm{Hom}_K(\mathrm{Res}_K^G W, V)$ we set

$$[\check{L}w](g) := \frac{1}{\sqrt{|G/K|}} L\theta(g^{-1})w, \tag{1.35}$$

for all $w \in W$ and $g \in G$, then the map $L \mapsto \check{L}$ yields a linear isometric isomorphism between $\mathrm{Hom}_K(\mathrm{Res}_K^G W, V)$ and $\mathrm{Hom}_G(W, \mathrm{Ind}_K^G V)$.

Another useful isomorphism (as G-representation spaces) is the following:

$$W \otimes \mathrm{Ind}_K^G V \cong \mathrm{Ind}_K^G \left[\left(\mathrm{Res}_K^G W\right) \otimes V\right], \tag{1.36}$$

which is realized by the isometric isomorphism

$$\xi\colon W \otimes \mathrm{Ind}_K^G V \to \mathrm{Ind}_K^G \left[\left(\mathrm{Res}_K^G W\right) \otimes V\right]$$

defined by setting $[\xi(w \otimes f)](g) := \theta(g^{-1})w \otimes f(g)$, for all $w \in W$, $f \in \mathrm{Ind}_K^G V$, and $g \in G$. See [30, Theorem 11.1.16]).

Let $J := \{\theta \in \widehat{G} : \theta \preceq \mathrm{Ind}_K^G \sigma\}$ denote a complete set of pairwise inequivalent irreducible G-representations contained in $\mathrm{Ind}_K^G \sigma$. For $\theta \in J$ we denote by W_θ its representation space and by $m_\theta := \mathrm{dimHom}_G(\theta, \mathrm{Ind}_K^G \sigma) \geq 1$ its multiplicity in $\mathrm{Ind}_K^G \sigma$. Then

$$\mathrm{Ind}_K^G V \cong \bigoplus_{\theta \in J} m_\theta W_\theta \tag{1.37}$$

is the decomposition of $\mathrm{Ind}_K^G V$ into irreducible G-representations and we have the $*$-isomorphism of $*$-algebras

$$\mathrm{End}_G(\mathrm{Ind}_K^G V) \cong \bigoplus_{\theta \in J} M_{m_\theta}(\mathbb{C}), \tag{1.38}$$

(cf. [30, Theorem 10.6.3]).

Remark 1.2 In (1.36), (1.37), and (1.38) we have used the symbol $\cong$ to denote an *isomorphism* (with respect to the corresponding algebraic structure). We will use the equality symbol $=$ to denote an *explicit decomposition*. For instance, in the present setting, this corresponds to the choice of a set $\{I_{\theta,j} : \theta \in J, j = 1, 2, \ldots, m_\theta\} \subseteq \mathrm{Hom}_G(W_\theta, \mathrm{Ind}_K^G V)$ such that

$$\mathrm{Ind}_K^G V = \bigoplus_{\theta \in J} \bigoplus_{j=1}^{m_\theta} I_{\theta,j} W_\theta$$

is an orthogonal decomposition.

Recall that a representation θ is said to be *real* if all of its matrix coefficients are real-valued. If it is not real, one says that θ is *quaternionic* if its character χ^θ is real-valued, and *complex* otherwise. The well known *Frobenius-Schur theorem* (cf. [24, Theorem 9.7.7]) states as follows:

$$\frac{1}{|G|} \sum_{g \in G} \chi^\theta(g^2) = \begin{cases} 1 & \text{if } \theta \text{ is real} \\ 0 & \text{if } \theta \text{ is complex} \\ -1 & \text{if } \theta \text{ is quaternionic.} \end{cases} \tag{1.39}$$

We now give a first look at the relation between the real/complex/quaternionic types of $\sigma \in \widehat{K}$ and $\mathrm{Ind}_K^G \sigma$, when the latter representation is irreducible as well. We refer to [24] and [125] for the general theory. However, we shall use a different terminology: the *adjoint* representation of σ is the representation σ' whose matrix coefficients are the complex conjugate of those corresponding to σ, and we shall say that σ is *selfadjoint* when $\sigma \sim \sigma'$ (the term "conjugate" will be used in Clifford theory with a different meaning, see Chap. 2). Recall also that $\chi^{\sigma'} = \overline{\chi^{\sigma}}$, that is, the character of the adjoint σ' is the complex conjugate of the character of σ, so that σ is selfadjoint if and only if its character is real. Moreover, if V' is the dual space of V then σ' is the representation of K on V' defined by setting

$$[\sigma'(k)\varphi](v) := \varphi[\sigma(k^{-1})v], \tag{1.40}$$

for all $k \in K$, $v \in V$, and $\varphi \in V'$.

Denote by $\xi \colon V' \to V$ the *Riesz map*, so that $f(v) = \langle v, \xi(f)\rangle_V$ for all $v \in V$ and $f \in V'$. Recall that ξ is antilinear, that is, $\xi(\alpha_1 f_1 + \alpha_2 f_2) = \overline{\alpha}_1\xi(f_1) + \overline{\alpha}_2\xi(f_2)$ for all $\alpha_i \in \mathbb{C}$ and $f_i \in V'$, $i = 1, 2$. Then, for all $k \in K$ we have

$$\sigma'(k) = \xi^{-1}\sigma(k)\xi. \tag{1.41}$$

Suppose now that $K \ni k \mapsto U(k)$ is a matrix realization of $\sigma \in \widehat{K}$. When σ is selfadjoint, there exists a unitary matrix W such that

$$\overline{U(k)} = WU(k)W^*, \tag{1.42}$$

for all $k \in K$, and

$$W\overline{W} = \overline{W}W = \pm I \qquad \text{according to } \sigma \text{ being real (resp. quaternionic)}, \tag{1.43}$$

where $\overline{W}$ is the conjugate matrix and W^* the adjoint (see [24, Lemma 9.7.6]).

Proposition 1.3 *Suppose that* $\mathrm{Ind}_K^G\sigma$ *is irreducible. If* σ *is real (resp. quaternionic) then* $\mathrm{Ind}_K^G\sigma$ *is real (resp. quaternionic).*

Proof If σ is real/quaternionic then we can use W in (1.42) to define the unitary matrix $W_1 := \begin{pmatrix} W & 0 & 0 & \cdots & 0 \\ 0 & W & 0 & \cdots & 0 \\ \vdots & \vdots & & \ddots & \vdots \\ 0 & 0 & & \cdots & W \end{pmatrix}$, where W is repeated $|\mathcal{T}|$-times. If $B(g)$ is as in (1.31), it is easy to check that $W_1 B(g) W_1^* = \overline{B(g)}$; moreover, $W_1\overline{W_1} = \overline{W_1}W_1 = \pm I$, according to $W\overline{W} = \pm I$. ■

If σ is complex then $\mathrm{Ind}_K^G\sigma$ may be real/complex/quaternionic. We will study this question in Theorem 2.50 when K is a normal subgroup.

1.4 Further Results on Induced Representations

Let G be a finite group, $K \leq G$, and let (σ, V) be an irreducible K-representation. Choose $v \in V$ with $\|v\| = 1$ and an orthonormal basis $\{v_j : j = 1, 2, \ldots, d_\sigma\}$ in V with $v_1 = v$. We will use the following version of the orthogonality relations, adapted to the restriction on a subgroup.

Lemma 1.4 *For all $w \in V$ and $1 \le i, j \le d_\sigma$, we have:*

$$\frac{d_\sigma}{|K|}\sum_{k\in K}\langle\sigma(k)v_i, v_j\rangle\sigma(k^{-1})w = \langle w, v_j\rangle v_i. \tag{1.44}$$

Proof Let $w := \sum_{r=1}^{d_\sigma}\alpha_r v_r$. Then, for all $1 \le s \le d_\sigma$, we have

$$\begin{aligned}\left\langle\frac{d_\sigma}{|K|}\sum_{k\in K}\langle\sigma(k)v_i, v_j\rangle\sigma(k^{-1})w, v_s\right\rangle &= \sum_{r=1}^{d_\sigma}\alpha_r\frac{d_\sigma}{|K|}\sum_{k\in K}\langle\sigma(k)v_i, v_j\rangle\langle v_r, \sigma(k)v_s\rangle\\ \text{(by (1.16))} &= \sum_{r=1}^{d_\sigma}\alpha_r\delta_{i,s}\delta_{j,r}\\ &= \langle w, v_j\rangle\delta_{i,s}\end{aligned}$$

so that the left hand side of (1.44) is equal to

$$\sum_{s=1}^{d_\sigma}\langle w, v_j\rangle\delta_{i,s}v_s = \langle w, v_j\rangle v_i.$$

■

In particular, for $w = v_j$ we get:

$$v_i = \frac{d_\sigma}{|K|}\sum_{k\in K}\langle\sigma(k)v_i, v_j\rangle\sigma(k^{-1})v_j. \tag{1.45}$$

Define $\varphi \in L(K) := \{f : K \to \mathbb{C}\}$ by setting

$$\varphi(k) = \frac{d_\sigma}{|K|}\langle v, \sigma(k)v\rangle \tag{1.46}$$

for all $k \in K$, where $v := v_1$ is as above. From now on, we identify $L(K)$ as the subalgebra of $L(G)$ consisting of all functions supported on K. Thus, we regard φ also as an element in $L(G)$. We then define the convolution operator $P : L(G) \to L(G)$ by setting

$$Pf := f * \varphi \tag{1.47}$$

for all $f \in L(G)$. We also define the operator

$$T_v : \mathrm{Ind}_K^G V \to L(G) \tag{1.48}$$

by setting

$$[T_v f](g) := \sqrt{d_\sigma/|K|}\langle f(g), v\rangle \tag{1.49}$$

for all $f \in \mathrm{Ind}_K^G V$ and $g \in G$ and denote its image by

$$\mathscr{I}(G, K, \varphi) := T_v\left(\mathrm{Ind}_K^G V\right) \subseteq L(G).$$

The following theorem is taken from [116] (see also [31, Theorem 2.2]). We reproduce the proof for the sake of completeness.

Theorem 1.5

(1) *The operator T_v belongs to $\mathrm{Hom}_G(\mathrm{Ind}_K^G V, L(G))$ and it is an isometry; in particular, $\mathscr{I}(G, K, \varphi)$ is a λ_G-invariant subspace of $L(G)$ equivalent to $\mathrm{Ind}_K^G \sigma$ as a G-representation.*
(2) *The function φ satisfies the identities*

$$\varphi * \varphi = \varphi \ \text{ and } \ \varphi^* = \varphi; \tag{1.50}$$

moreover P is the orthogonal projection of $L(G)$ onto $\mathscr{I}(G, K, \varphi)$. In other words,

$$\mathscr{I}(G, K, \varphi) = \{f * \varphi : f \in L(G)\} \equiv \{f \in L(G) : f * \varphi = f\}. \tag{1.51}$$

Proof

(1) It is immediate to check that

$$T_v\lambda(g) f = \lambda_G(g) T_v f$$

for all $g \in G$ and $f \in \mathrm{Ind}_K^G V$, showing that $T_v \in \mathrm{Hom}_G(\mathrm{Ind}_K^G V, L(G))$. We now prove that T_v is an isometry by using the basis in (1.30). For $t_1, t_2 \in \mathscr{T}$ and $i, j = 1, 2, \ldots, d_\sigma$, we have

$$\begin{aligned}
\langle T_v\lambda(t_1) f_{v_i}, T_v\lambda(t_2) f_{v_j}\rangle_{L(G)} &= \frac{d_\sigma}{|K|}\sum_{g\in G}\langle f_{v_i}(t_1^{-1}g), v\rangle\overline{\langle f_{v_j}(t_2^{-1}g), v\rangle} \\
\text{(by (1.28))} \quad &= \frac{d_\sigma}{|K|}\delta_{t_1,t_2}\sum_{k\in K}\langle \sigma(k^{-1})v_i, v_1\rangle\overline{\langle \sigma(k^{-1})v_j, v_1\rangle} \\
\text{(by (1.16))} \quad &= \delta_{t_1,t_2}\delta_{i,j} \\
&= \langle \lambda(t_1) f_{v_i}, \lambda(t_2) f_{v_j}\rangle_{\mathrm{Ind}_K^G(V)}.
\end{aligned}$$

(2) The first identity in (1.50) follows from the convolution properties of matrix coefficients ((1.17) with $\rho = \theta = \sigma$ and $i = j = h = k = 1$) and ensures that P is an idempotent; the second identity in (1.50) follows from the fact that φ is a diagonal matrix coefficient and, together with (1.17), ensures that P is self-adjoint. As a consequence, $P = P^2 = P^*$ is an orthogonal projection. Moreover, for all $f \in \mathrm{Ind}_K^G V$ and $g \in G$ we have

$$\begin{aligned}
[(T_v f) * \varphi](g) &= \left(\frac{d_\sigma}{|K|}\right)^{3/2} \sum_{k \in K} \langle f(gk^{-1}), v\rangle \langle v, \sigma(k)v\rangle \\
\text{(by (1.25))} \quad &= \left(\frac{d_\sigma}{|K|}\right)^{3/2} \sum_{k \in K} \langle \sigma(k) f(g), v\rangle \langle v, \sigma(k)v\rangle \\
&= \left(\frac{d_\sigma}{|K|}\right)^{3/2} \left\langle f(g), \sum_{k \in K} \langle \sigma(k)v, v\rangle \sigma(k^{-1})v \right\rangle \\
\text{(by (1.45))} \quad &= \sqrt{\frac{d_\sigma}{|K|}} \langle f(g), v\rangle \\
&= (T_v f)(g),
\end{aligned}$$

that is, $PT_v f = T_v f$. We deduce that the range $\mathrm{ran}P$ of P contains $T_v(\mathrm{Ind}_K^G V) = \mathscr{I}(G, K, \varphi)$. Let now show that, in fact, the range of P is contained in (and therefore equals) $\mathscr{I}(G, K, \varphi)$. Indeed, for all $f \in L(G)$ and $g \in G$ we have

$$\begin{aligned}
[Pf](g) &= \sum_{k \in K} f(gk)\varphi(k^{-1}) \\
\text{(by (1.46))} \quad &= \frac{d_\sigma}{|K|} \left\langle \sum_{k \in K} f(gk)\sigma(k)v, v \right\rangle \\
&= [T_v F](g)
\end{aligned}$$

where $F \in V^G$ is defined by $F(g) := \sqrt{\frac{d_\sigma}{|K|}} \sum_{k \in K} f(gk)\sigma(k)v$. Since F belongs to $\mathrm{Ind}_K^G V$, as one immediately checks, we conclude that $\mathrm{ran}P = T_v(\mathrm{Ind}_K^G V)$.

■

1.5 Semidirect Products, Wreath Products, and Group Extensions

This section is devoted to two basic constructions in group theory that involve normal subgroups. We begin with a well known notion; for more details we refer to [3, 90, 108, 114, 125].

Definition 1.6 Let G be a finite group and let $N, H \leq G$ be two subgroups of G. Then G is the (internal) *semidirect product* of N by H and we write $G = H \ltimes N$, if

(i) N is a normal subgroup: $N \trianglelefteq G$,
(ii) $G = HN$,
(iii) $N \cap H = \{1_G\}$.

Proposition 1.7 *Suppose that G is a semidirect product of N by H. Then*

(1) $G/N \cong H$;
(2) *every $g \in G$ has a* unique *expression of the form $g = hn$ with $h \in H$ and $n \in N$;*
(3) *for $h \in H$ define $h^\flat \in Aut(N)$ by setting $n^{h^\flat} := h^{-1}nh$ for all $n \in N$. Then the map $H \ni h \mapsto h^\flat \in Aut(N)$ is a group homomorphism (*conjugation homomorphism*);*
(4) *if $h_1n_1, h_2n_2 \in G$ as in (2), then their product is given by*

$$h_1n_1 \cdot h_2n_2 = h_1h_2[h_2^{-1}n_1h_2 \cdot n_2] = h_1h_2n_1^{h_2^\flat}n_2. \tag{1.52}$$

Conversely, suppose that H and N are two (finite) groups with a homomorphism $H \ni h \mapsto h^\flat \in Aut(N)$, where the action of $h^\flat$ on $n \in N$ is denoted by $n^{h^\flat}$. Set $G := \{(h, n) : h \in H, n \in N\}$ and endow it with the operation

$$(h_1, n_1) \cdot (h_2, n_2) := (h_1h_2, n_1^{h_2^\flat}n_2)$$

(cf. Eq. (1.52)). Then G is a group isomorphic to the (internal) semidirect product of $\widetilde{N} := \{(1_H, n) : n \in N\} \cong N$ by $\widetilde{H} := \{(h, 1_N) : h \in H\} \cong H$. Also, G is called the external semidirect product *of N by H with respect to the homomorphism $h \mapsto h^\flat$ and it is denoted by $G = H \ltimes_\flat N$. Moreover, the following conditions are equivalent:*

(a) *G is isomorphic to the* direct *product $\widetilde{H} \times \widetilde{N}$;*
(b) *$\widetilde{H}$ is normal in G;*
(c) *$h^\flat = id_N$, the trivial automorphism of N, for all $h \in H$.*

Proof The proof is just an easy exercise and it is left to the reader. ∎

Clearly, the internal and external semidirect products are equivalent constructions and we shall make no distinction between them.

We now consider the most general situation.

Definition 1.8 Let N and H be two groups. An *extension* of H by N is a triple $(G; \iota, \pi)$ where: G is a group, $\iota\colon N \to G$ is an injective homomorphism, and $\pi\colon G \to H$ is a surjective homomorphism such that $\ker(\pi) = \iota(N)$. In other words, an extension of H by N is a *short exact sequence*

$$1 \longrightarrow N \stackrel{\iota}{\longrightarrow} G \stackrel{\pi}{\longrightarrow} H \longrightarrow 1.$$

In particular,

$$G/N \cong H.$$

Two extensions $(G_1; \iota_1, \pi_1)$ and $(G_2; \iota_2, \pi_2)$ of H by N are said to be *equivalent* (or *congruent*) if there exists a group isomorphism $\eta\colon G_1 \to G_2$ such that the diagram

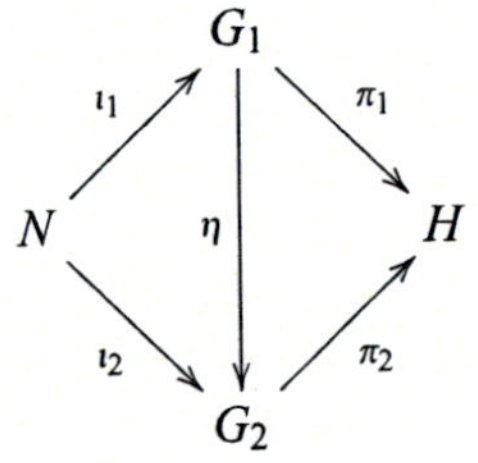

is commutative, that is,

$$\iota_2 = \eta\iota_1 \qquad \text{and} \qquad \pi_1 = \pi_2\eta. \tag{1.53}$$

The extension is *Abelian* when H is Abelian.

In the following example, we present two non-equivalent extensions $(G_1; \iota_1, \pi_1)$ and $(G_2; \iota_2, \pi_2)$ of $H = \mathbb{Z}_3$ by $N = \mathbb{Z}_3$ with $G_1 = G_2 = \mathbb{Z}_9$. This shows that in the definition above the commutativity of the corresponding diagram cannot be omitted.

Example 1.9 Let $N = H := \mathbb{Z}_3$ and $G_1 = G_2 := \mathbb{Z}_9$. For $j = 1, 2$ define the injective homomorphism $\iota_j\colon N \to G_j$ (resp. the surjective homomorphism $\pi_j\colon G_j \to H$) by setting $\iota_j([n]_3) := [3jn]_9$ (resp. $\pi_j([n]_9) := [n]_3$) for all $n \in \mathbb{Z}$. As $\iota_j(N) = \{[0]_9, [3]_9, [6]_9\} = \ker(\pi_j)$ for $j = 1, 2$, we deduce that $(G_1; \iota_1, \pi_1)$ and $(G_2; \iota_2, \pi_2)$ are two extensions of H by N. Let us show that they are not equivalent. Recall (cf. [30, Corollary 1.5.3]) that the automorphism group $\mathrm{Aut}(\mathbb{Z}_n)$ (of the additive group $\mathbb{Z}_n$) is isomorphic to $\mathscr{U}(\mathbb{Z}_n)$ the (multiplicative) group of units of the ring $\mathbb{Z}_n$. Thus, every $\eta \in \mathrm{Aut}(G)$ is of the form $\eta([n]_9) := [un]_9$ with $u \in \mathscr{U}(\mathbb{Z}_9)$. Suppose, by contradiction, that (1.53) holds. Then, condition $\iota_2 = \eta\iota_1$

implies that $u \equiv 2 \bmod 3$, while condition $\pi_1 = \pi_2\eta$ implies $u \equiv 1 \bmod 3$. This is clearly impossible. Thus, $(G_1; \iota_1, \pi_1)$ and $(G_2; \iota_2, \pi_2)$ are not equivalent.

Let $(G; \iota, \pi)$ be an extension of H by N. To simplify the expressions, we shall write $\overline{n}$ (resp. $\overline{N}$) instead of $\iota(n)$ for $n \in N$ (resp. $\iota(N)$), and, given $g \in \iota(N) \subseteq G$, we shall denote by $\iota^{-1}(g)$ the unique element $n \in N$ such that $\overline{n} = g$. A *section* of the extension is a map $t\colon H \to G$ which is a right-inverse of π, that is,

$$\pi(t(h)) = h$$

for all $h \in H$. Clearly, the set $\{t(h) : h \in H\}$ is a system of representatives for the $\overline{N}$-cosets, that is,

$$G = \sqcup_{h\in H} t(h)\overline{N} \equiv \sqcup_{h\in H} \overline{N} t(h), \tag{1.54}$$

where $\sqcup$ denotes a disjoint union. In other words, $t(h)\overline{N} = \{g \in G : \pi(g) = h\}$ and every $g \in G$ may be *uniquely* written in the form $g = t(h)\overline{n}$, with $h = \pi(g) \in H$ and $\overline{n} = t(h)^{-1}g$, that is, $n = \iota^{-1}\left[t(h)^{-1}g\right] \in N$. Clearly, the map t is not uniquely defined and, in general, it is *not* a homomorphism; in the case when there exists a section t which is also a homomorphism, one says that $(G; \iota, \pi)$ is a *split* extension (cf. Exercise 1.14). A section $t\colon H \to G$ is *normalized* if $t(1_H) = 1_G$.

Moreover, we shall use the following notation: for $g \in G$ we define $g^\flat \in \mathrm{Aut}(N)$ by setting

$$n^{g^\flat} := \iota^{-1}(g^{-1}\overline{n}g), \tag{1.55}$$

for all $n \in N$. Clearly, the map $G \ni g \mapsto g^\flat \in \mathrm{Aut}(N)$ is a group homomorphism, that is, $(g_1g_2)^\flat = g_1^\flat g_2^\flat$ for all $g_1, g_2 \in G$, and its kernel is the centralizer $C_G(\overline{N}) := \{g \in G : g\overline{n} = \overline{n}g \text{ for all } n \in N\}$ of $\overline{N}$. The map $n \mapsto \overline{n}^\flat$ is nothing but the canonical homomorphism of N onto $\mathrm{Inn}(N) \le \mathrm{Aut}(N)$, the group of all *inner* automorphisms of N, whose kernel is $Z(N)$, the center of N.

For $h, k \in H$ we have

$$\pi[t(h)t(k)] = \pi(t(h))\pi(t(k)) = hk = \pi[t(hk)]$$

and therefore there exists a unique $n(h, k) \in N$ such that

$$t(h)t(k) = t(hk)\overline{n}(h, k). \tag{1.56}$$

Moreover, for each $h \in H$ we set $h^\sharp = t(h)^\flat \in \mathrm{Aut}(N)$, that is,

$$n^{h^\sharp} := \iota^{-1}(t(h)^{-1}\overline{n}t(h)), \tag{1.57}$$

for all $n \in N$. The map $H \ni h \mapsto h^\sharp \in \mathrm{Aut}(N)$ clearly depends on the section t (as well as on ι) and, in general, is not a group homomorphism (cf. (1.58) and Exercise 1.14).

Proposition 1.10 *For all $h, k, j \in H$ and $m, \ell \in N$ the following hold:*

$$h^\sharp k^\sharp = (hk)^\sharp n(h,k)^\flat, \tag{1.58}$$

$$n(h, kj)n(k, j) = n(hk, j)n(h,k)^{j^\sharp} \text{ (cocycle identity)} \tag{1.59}$$

and

$$t(h)\overline{m} \cdot t(k)\overline{\ell} = t(hk)\left[\overline{n}(h,k)\overline{m^{k^\sharp}\ell}\right] \text{ (multiplication in } G\text{)}. \tag{1.60}$$

Moreover, if t is normalized,

$$n(h, 1_H) = n(1_H, h) = 1_N \quad \text{and} \quad 1_H^\sharp = id_N, \tag{1.61}$$

and

$$t(h^{-1}) = t(h)^{-1}\overline{n}(h, h^{-1}) = \overline{n}(h^{-1}, h)t(h)^{-1}. \tag{1.62}$$

Proof We have:

$$\begin{aligned} m^{h^\sharp k^\sharp} &= \iota^{-1}(t(k)^{-1}t(h)^{-1}\overline{m}t(h)t(k)) \\ \text{(by (1.56))} \quad &= \iota^{-1}(\overline{n}(n,k)^{-1}t(hk)^{-1}\overline{m}t(hk)\overline{n}(h,k)) \\ &= m^{(hk)^\sharp n(h,k)^\flat}, \end{aligned}$$

and this proves (1.58). Moreover,

$$t(h)[t(k)t(j)] = t(h)t(kj)\overline{n}(k,j) = t(hkj)\overline{n}(h,kj)\overline{n}(k,j)$$

and

$$[t(h)t(k)]t(j) = t(hk)\overline{n}(h,k)t(j) = t(hk)t(j)\overline{n(h,k)^{j^\sharp}} = t(hkj)\overline{n}(hk,j)\overline{n(h,k)^{j^\sharp}}.$$

By the associativity law in G, the two quantities above are equal, and (1.59) follows. Then

$$t(h)\overline{m} \cdot t(k)\overline{\ell} = t(h)t(k)\overline{m^{k^\sharp}\ell} = t(hk)\left[\overline{n}(h,k)\overline{m^{k^\sharp}\ell}\right]$$

yields (1.60) Finally, if t is normalized, from $t(h) = t(h)t(1_H) = t(h)\overline{n}(h, 1_H)$ we deduce the first identity (1.61) (the second one is analogous), while, from $t(h)t(h^{-1}) = t(1_H)\overline{n}(h, h^{-1}) = \overline{n}(h, h^{-1})$, we deduce the first identity in (1.62) (the second one is analogous). ■

Corollary 1.11 *If t is normalized then, for all $h \in H$ and $m \in N$,*

$$t(h) = \overline{n}(h, h^{-1})t(h^{-1})^{-1} = t(h^{-1})^{-1}\overline{n}(h^{-1}, h) \tag{1.63}$$

and

$$t(h)\overline{m}t(h)^{-1} = \overline{m^{(h^{-1})^\sharp[\overline{n}(h,h^{-1})^{-1}]^\flat}}.$$

Proof The equalities in (1.63) follow immediately from (1.62). Moreover, from (1.62) and (1.63) we deduce

$$t(h)\overline{m}t(h)^{-1} = \overline{n}(h, h^{-1})t(h^{-1})^{-1}\overline{m}t(h^{-1})\overline{n}(h, h^{-1})^{-1} = \overline{m^{(h^{-1})^\sharp[n(h,h^{-1})^{-1}]^\flat}}.$$

■

Another useful identity is the following: for $h, k \in H$ we have

$$t(k)^{-1}t(h) = t(k^{-1}h)\overline{n}(k, k^{-1}h)^{-1}. \tag{1.64}$$

Indeed, from (1.56) we deduce immediately that

$$\overline{n}(k, k^{-1}h)^{-1} = \left[t(h)^{-1}t(k)t(k^{-1}h)\right]^{-1} = t(k^{-1}h)^{-1}t(k)^{-1}t(h).$$

Definition 1.12 Let H and N be two finite groups. A *Schreier factor system* for H and N is a pair

$$n\colon H \times H \to N, \qquad H \ni h \mapsto h^\sharp \in \mathrm{Aut}(N)$$

of maps satisfying the identities (1.58) and (1.59). The factor system is *normalized* if also the conditions in (1.61) are satisfied.

Note that Definition 1.12 does not require the existence of the group G (nor the associated section t). On the other hand, if the factor system is constructed as above by means of a group extension, then (1.60) may be used to compute the product of two elements of G. In the following exercise we ask to externalize this construction, that is, to define a group extension starting from a given factor system.

Exercise 1.13 Suppose that we are given a normalized factor system as in Definition 1.12. Equip $G = H \times N$ with a product by setting $(h, m) \cdot (k, \ell) :=$

$\left(hk, n(h,k)m^{k^\sharp}\ell\right)$, for all $h, k \in H$ and $m, \ell \in N$ (cf. Eq. (1.60)). Prove that this yields a group structure on G, and that (G, ι, π), where $\iota(n) := (1_H, n)$ and $\pi(h, n) := h$ for all $h \in H$ and $n \in N$, is a group extension of H by N. Also show that with a suitable choice of a section $t: H \to G$, the associated factor system coincides with the given one.

Exercise 1.14 Suppose that G is an extension of H by N. Prove that the following statements are equivalent:

(a) there exists a section $t: H \to G$ which is also a homomorphism;
(b) G is the semidirect product of N by H.

Moreover, if this is the case, the associated factor system satisfies the conditions $n(h, k) = 1_N$ and $(hk)^\sharp = h^\sharp k^\sharp$, for all $h, k \in H$ (the latter replaces (1.58)). As mentioned above, in this case we say that G is a *split extension*.

We refer to the monographs by Alperin and Bell [3], Knapp [80], and Kurosh [84] for more extensive expositions on Schreier's factor systems and to the books by Kurosh [84], Machì [90], Robinson [108], and Rotman [114] for introductory expositions on the theory of group extensions and the associated cohomological machinery. See also Chap. 6 for the particular case of *central* group extensions.

1.6 Regular Wreath Products and the Kaloujnine-Krasner Theorem

In this final section, we give the definition of a regular wreath product of two groups (a more general group theoretical construction will be presented in Sect. 2.10) and discuss the Kaloujnine-Krasner theorem which, with a short exact sequence $1 \longrightarrow N \xrightarrow{\iota} G \xrightarrow{\pi} H \longrightarrow 1$ of groups (that is, with a group extension $(G; \iota, \pi)$ of H by N) and a section $s: H \to G$, associates an embedding $\Phi_s: G \to N \wr H$ of G into the wreath product of N and H. Also, we shall present a new result from [32] relating equivalence of group extensions (cf. Definition 1.8) with a suitable equivalence for the associated embeddings (see below).

Let N and H be two finite groups. Denote by N^H the set of all maps $f: H \to N$; it is a group under pointwise multiplication: $(ff')(h) := f(h) \cdot f'(h)$, for all $f, f' \in N^H$ and $h \in H$. Then H acts on N^H in a natural way by setting

$$(hf)(h') := f(h^{-1}h'), \tag{1.65}$$

for all $h, h' \in H$ and $f \in N^H$. Moreover, $h(ff') = (hf) \cdot (hf')$ and $(hf)^{-1} = hf^{-1}$, that is, H acts on N^H as a group of automorphisms. Note that

$$(f \cdot hf')(h') = f(h)f'(h^{-1}h'), \tag{1.66}$$

for all $h, h' \in H$ and $f, f' \in N^H$. On the set

$$N \wr H := \{(f, h) : f \in N^H, h \in H\} \equiv N^H \times H$$

we define a multiplication law by setting:

$$(f, h)(f', h') := (f \cdot hf', hh') \tag{1.67}$$

for all $(f, h), (f', h') \in N \wr H$.

Lemma 1.15 *The set $N \wr H$ with the multiplication law (1.67) is a group. Moreover, the identity element is $(\mathbf{1}_N, 1_H)$, where $\mathbf{1}_N(h) := 1_N$ for all $h \in H$, and the inverse of $(f, h) \in N \wr H$ is given by*

$$(f, h)^{-1} = (h^{-1} f^{-1}, h^{-1}). \tag{1.68}$$

Finally, the conjugation action of $N \wr H$ on itself is given by:

$$(f, h)(f_0, h_0)(f, h)^{-1} = \left(f \cdot hf_0 \cdot hh_0h^{-1} f^{-1}, hh_0h^{-1}\right), \tag{1.69}$$

for all $(f, h), (f_0, h_0) \in N \wr H$.

Proof We leave it as an exercise. The reader may find a more general result with a complete proof in Lemma 2.56. ■

The group $N \wr H$ is called the *regular wreath product* of N by H. The subgroup

$$\{(f, 1_H) : f \in N^H\} \cong N^H$$

is called the *base group* and we will identify it with N^H. By taking $h_0 = 1_H$ in (1.69), we deduce that the base group is a normal subgroup. Moreover, if we identify $\{(\mathbf{1}_N, h) : h \in H\}$ with H, then the wreath product may be written as a semidirect product, namely, $N \wr H = N^H \rtimes H$. Indeed, by setting $h_0 = 1_H$ and $f = \mathbf{1}_N$ in (1.69) we recover the action of H of N^H as the restriction to H of the conjugation action on N^H (note that, with respect to the notation in Sect. 1.5, we are writing the action of the automorphism on the left).

We are now in a position to state and prove the following well known result due to Leo Kaloujnine and Marc Krasner [82, 83].

Theorem 1.16 (Kaloujnine-Krasner) *Let N and H be two groups and let $(G; \iota, \pi)$ be an extension of H by N. Then there exists an injective homomorphism*

$$\Phi : G \longrightarrow N \wr H$$

of G into the regular wreath product of N by H such that

$$\Phi(\iota(N)) = \Phi(G) \cap N^H \quad \text{and} \quad \Phi(G)N^H = N \wr H. \tag{1.70}$$

Proof Let $s\colon H \to G$ be a section (i.e., $\pi(s(h)) = h$ for all $h \in H$) such that $s(1_H) = 1_G$. To simplify notation, we write $s_h := s(h)$ for $h \in H$. We then have

$$G = \sqcup_{h\in H} s_h \iota(N).$$

As $\pi(s_h^{-1} g s_{\pi(g)^{-1}h}) = h^{-1}\cdot\pi(g)\cdot\pi(g)^{-1}h = 1_H$, we have $s_h^{-1} g s_{\pi(g)^{-1}h} \in \ker\pi = \iota(N)$, for all $g \in G$ and $h \in H$. Therefore, for each $g \in G$ we may define a map $f_g\colon H \longrightarrow N$ by setting

$$f_g(h) := \iota^{-1}(s_h^{-1} g s_{\pi(g)^{-1}h}), \tag{1.71}$$

for all $h \in H$. We claim that the map $\Phi_s\colon G \to N \wr H$ defined by setting

$$\Phi_s(g) := (f_g, \pi(g)) \tag{1.72}$$

for all $g \in G$, is the desired injective homomorphism. First of all, for $g, g' \in G$ and $h \in H$ we have

$$\begin{aligned}
\left\{f_g \cdot [\pi(g) f_{g'}]\right\}(h) &= f_g(h)\cdot f_{g'}\left[\pi(g)^{-1}h\right] \quad \text{(by (1.66))}\\
&= \iota^{-1}(s_h^{-1} g s_{\pi(g)^{-1}h})\cdot\iota^{-1}(s_{\pi(g)^{-1}h}^{-1} g' s_{\pi(g')^{-1}\pi(g)^{-1}h}) \quad \text{(by (1.71))}\\
&= \iota^{-1}(s_h^{-1} g g' s_{\pi((gg')^{-1})h}) \quad \text{(as } \iota \text{ and } \pi \text{ are homomorphisms)}\\
&= f_{gg'}(h) \quad \text{(again by (1.71))}
\end{aligned}$$

that is,

$$f_g \cdot [\pi(g) f_{g'}] = f_{gg'}. \tag{1.73}$$

Then, for all $g, g' \in G$,

$$\begin{aligned}
\Phi_s(g)\Phi_s(g') &= (f_g, \pi(g))(f_{g'}, \pi(g')) && \text{(by (1.72))}\\
&= (f_g \cdot \pi(g) f_{g'}, \pi(g)\pi(g')) && \text{(by (1.67))}\\
&= (f_{gg'}, \pi(gg')) && \text{(by (1.73))}\\
&= \Phi_s(gg') && \text{(by (1.72))}
\end{aligned}$$

showing that Φ_s is a homomorphism. Let now $g \in \ker\Phi_s$. Then $f_g = \mathbf{1}_N$ and $\pi(g) = 1_H$ and therefore, recalling that $s_{1_H} = 1_G$,

$$1_G = \iota(1_N) = \iota(\mathbf{1}_N(1_H)) = \iota(f_g(1_H)) = s_{1_H}^{-1} g s_{\pi(g)^{-1}1_H} = g.$$

This shows that Φ_s is injective.

We are only left to show (1.70). The first equality immediately follows from the fact that for $g \in G$ one has $\Phi_s(g) \in N^H$ if and only if $\pi(g) = 1_H$, that is, if and only if $g \in \ker \pi = \iota(N)$.

Let now $(f, h) \in N \wr H$. Set $g := s_h \in G$ and define $f' \in N^H$ by setting $f' := h^{-1}(f_g^{-1} \cdot f)$ (cf. (1.66), (1.67), and (1.71)). We then have

$$\Phi_s(g)(f', 1_H) = (f_g, h)(f', 1_H) = (f_g \cdot hf', h) = (f, h).$$

This shows that $\Phi_s(G)N^H = N \wr H$. ■

The embedding $\Phi_s \colon G \to N \wr H$ defined by (1.72) is called the *Frobenius embedding* associated with the section $s \colon H \to G$ (cf. [72, 6.2.8]).

Let N and H be two finite groups and let $(G_1; \iota_1, \pi_1)$ and $(G_2; \iota_2, \pi_2)$ be two extensions of H by N. We say that two embeddings $\Phi_1 \colon G_1 \to N \wr H$ and $\Phi_2 \colon G_2 \to N \wr H$ are *equivalent*, and we write $\Phi_1 \approx \Phi_2$, if there exists a group isomorphism $\eta \colon G_1 \to G_2$ such that

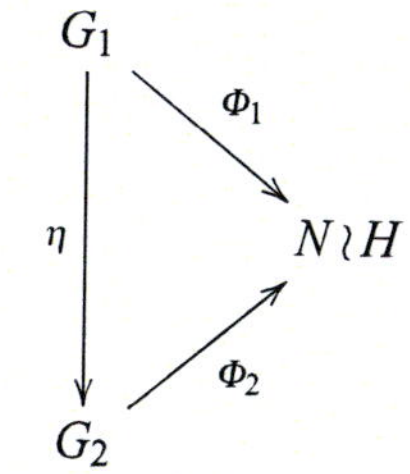

is commutative, that is, $\Phi_1 = \Phi_2 \circ \eta$.

The remaining part of the section is devoted to relating the equivalence of two extensions $(G_1; \iota_1, \pi_1)$ and $(G_2; \iota_2, \pi_2)$ and that of the corresponding Frobenius embeddings Φ_1 and Φ_2 (associated with two sections $s_1 \colon H \to G_1$ and $s_2 \colon H \to G_2$, respectively).

We start by analyzing the dependence of the Frobenius embedding $\Phi \colon G \to N \wr H$ on the choice of the section $s \colon H \to G$.

Lemma 1.17 *Let $(G; \iota, \pi)$ be a group extension of H by N and let $s \colon H \to G$ and $s' \colon H \to G$ be two sections such that $s_{1_H} = 1_G = s'_{1_H}$. Then the corresponding embeddings $\Phi = \Phi_s \colon G \to N \wr H$ and $\Phi' = \Phi_{s'} \colon G \to N \wr H$ are equivalent.*

Proof We have $\Phi(g) = (f_g, \pi(g))$ where $f_g(h) = \iota^{-1}(s_h^{-1} g s_{\pi(g)^{-1}h})$ and $\Phi'(g) = (f'_g, \pi(g))$ where $f'_g(h) = \iota^{-1}({s'}_h^{-1} g s'_{\pi(g)^{-1}h})$ for all $g \in G$ and $h \in H$. For each element $k \in H$ there exists a unique element $n_k \in N$ such that $s'_k = s_k\iota(n_k)$. This way, for all $g \in G$ and $h \in H$, we have

$$\begin{aligned} f'_g(h) &= \iota^{-1}(s'^{-1}_h g s'_{\pi(g)^{-1}h}) \\ &= \iota^{-1}((s_h\iota(n_h))^{-1} g s_{\pi(g)^{-1}h}\iota(n_{\pi(g)^{-1}h})) \\ &= \iota^{-1}(\iota(n_h^{-1})(s_h^{-1} g s_{\pi(g)^{-1}h})\iota(n_{\pi(g)^{-1}h})) \\ &= n_h^{-1}\iota^{-1}(s_h^{-1} g s_{\pi(g)^{-1}h}) n_{\pi(g)^{-1}h} \\ &= n_h^{-1} f_g(h) n_{\pi(g)^{-1}h}. \end{aligned} \tag{1.74}$$

Consider the map $f\colon H \to N$ defined by setting $f(h) := n_h{}^{-1}$ for all $h \in H$. Using (1.69) we get, for all $g \in G$,

$$(f, 1_H)(f_g, \pi(g))(f, 1_H)^{-1} = \left(f \cdot f_g \cdot \pi(g) f^{-1}, \pi(g)\right)$$

with

$$\left(f \cdot f_g \cdot \pi(g) f^{-1}\right)(h) = f(h) f_g(h) f^{-1}(\pi(g)^{-1}h) = n_h^{-1} f_g(h) n_{\pi(g)^{-1}h} \tag{1.75}$$

for all $h \in H$. Comparing (1.74) and (1.75), we deduce that conjugation by $(f, 1_H) \in N \wr H$ yields and automorphism α of $N \wr H$ mapping $\Phi(G)$ onto $\Phi'(G)$. Then, setting

$$\eta := (\Phi')^{-1} \circ \alpha \circ \Phi$$

yields an automorphism of $(G; \iota, \pi)$ such that $\Phi = \Phi' \circ \eta$. In other words, the embeddings Φ and Φ' are equivalent. ∎

More generally, we have the following:

Theorem 1.18 *Let N and H be two finite groups and let $(G_1; \iota_1, \pi_1)$ and $(G_2; \iota_2, \pi_2)$ be two extensions of H by N. Let Φ_1 and Φ_2 denote the corresponding Frobenius embeddings into $N \wr H$ (associated with sections $s_1\colon H \to G_1$ and $s_2\colon H \to G_2$, respectively). Then $(G_1; \iota_1, \pi) \sim (G_2; \iota_2, \pi)$ if and only if $\Phi_1 \approx \Phi_2$.*

Proof For $j = 1, 2$, let $s^j\colon H \to G_j$ be a section such that $s^j_{1_H} = 1_{G_j}$ and set $\Phi_i = \Phi_{s^j}$. Suppose first that $\Phi_1 \approx \Phi_2$ and let $\eta\colon G_1 \to G_2$ be a group isomorphism such that $\Phi_1 = \Phi_2 \circ \eta$, that is,

$$(f^{(2)}_{\eta(g)}, \pi_2(\eta(g))) = (f^{(1)}_g, \pi_1(g)) \tag{1.76}$$

for all $g \in G_1$. Looking at the second component in (1.76), we get $\pi_1 = \pi_2 \circ \eta$. Let now $n \in N$ and set $g := \iota_1(n) \in G_1$. Then (1.76) gives $f^{(2)}_{\eta(g)}(h) = f^{(1)}_g(h)$ for all

$h \in H$. Taking $h = 1_H$, so that $s^j_{\pi_j(g)^{-1}h} = s^j_h = s^j_{1_H} = 1_{G_j}$ for $j = 1, 2$, we get, on the one hand

$$f^{(1)}_g(1_H) = f^{(1)}_{\iota_1(n)}(1_H) = \iota_1^{-1}(\iota_1(n)) = n$$

and, on the other hand,

$$f^{(2)}_{\eta(g)}(1_H) = f^{(2)}_{\eta(\iota_1(n))}(1_H) = \iota_2^{-1}(\eta(\iota_1(n))).$$

We deduce that $n = \iota_2^{-1}(\eta(\iota_1(n)))$, equivalently, $\iota_2(n) = \eta(\iota_1(n))$, and, as $n \in N$ was arbitrary, $\iota_2 = \eta \circ \iota_1$. This shows that $(G_1; \iota_1, \pi_1) \sim (G_2; \iota_2, \pi_2)$.

Vice versa, suppose that the extensions $(G_1; \iota_1, \pi_1)$ and $(G_2; \iota_2, \pi_2)$ are equivalent and let $\eta\colon G_1 \to G_2$ be a group isomorphism yielding the equivalence of extensions. Let us show that the same automorphism η implements the equivalence between the embeddings Φ_1 and Φ_2, that is, $\Phi_1 = \Phi_2 \circ \eta$. Let $s\colon H \to G_1$ be a section such that $s_{1_H} = 1_{G_1}$. Then $\Phi_1(g) = (f^{(1)}_g, \pi_1(g))$ with

$$f^{(1)}_g(h) := \iota_1^{-1}(s_h^{-1} g s_{\pi_1(g)^{-1}h})$$

for all $g \in G_1$ and $h \in H$. By virtue of the lemma, we may suppose that the section $t := \eta \circ s\colon H \to G_2$, implements the embeddings Φ_2, that is, $\Phi_2(g) = (f^{(2)}_g, \pi_2(g))$ with

$$f^{(2)}_g(h) = \iota_2^{-1}(t_h^{-1} g t_{\pi_2(g)^{-1}h}) \tag{1.77}$$

for all $g \in G_2$ and $h \in H$.

Let $g \in G_1$. Keeping in mind that $\iota_2 = \eta \circ \iota_1$, we have, for all $h \in H$,

$$\begin{aligned} f^{(1)}_g(h) &= \iota_1^{-1}(s_h^{-1} g s_{\pi_1(g)^{-1}h}) \\ &= (\iota_2^{-1} \circ \eta)\left(s_h^{-1} g s_{\pi_1(g)^{-1}h}\right) \\ &= \iota_2^{-1}\left(\eta(s_h)^{-1}\eta(g)\eta(s_{\pi_1(g)^{-1}h})\right) \\ &= \iota_2^{-1}\left(t_h^{-1}\eta(g)t_{\pi_1(g)^{-1}h}\right) \\ &= f^{(2)}_{\eta(g)}(h) \quad \text{(by (1.77))}, \end{aligned}$$

thus showing that $f^{(1)}_g = f^{(2)}_{\eta(g)}$. Finally, keeping in mind that $\pi_1(g) = \pi_2(\eta(g))$, we deduce that $\Phi_1(g) = \Phi_2(\eta(g))$. This shows that $\Phi_1 \approx \Phi_2$. ∎

In the following example we describe two non-equivalent embeddings corresponding to two non-equivalent extensions of two finite groups.

Example 1.19 Let $N = H := \mathbb{Z}_3$ and $G := \mathbb{Z}_9$, and consider the (non-equivalent) extensions $(G; \iota_1, \pi_1)$ and $(G; \iota_2, \pi_2)$ of H by N described in Example 1.9. As $\pi_1 = \pi_2$, we can use the common section $s_{[n]_3} := [n]_9$ for $n = 0, 1, 2$. For $j = 1, 2$, the corresponding embeddings $\Phi_j : G \to N \wr H$ are then given by $\Phi_j([m]_9) := (f^{(j)}_{[m]_9}, \pi_j([m]_9)) = (f^{(j)}_{[m]_9}, [m]_3)$, for $m = 0, 1, \ldots, 8$, where

$$
\begin{aligned}
f^{(1)}_{[0]_9} &= ([0]_3, [0]_3, [0]_3) & f^{(2)}_{[0]_9} &= ([0]_3, [0]_3, [0]_3)\\
f^{(1)}_{[1]_9} &= ([1]_3, [0]_3, [0]_3) & f^{(2)}_{[1]_9} &= ([2]_3, [0]_3, [0]_3)\\
f^{(1)}_{[2]_9} &= ([1]_3, [1]_3, [0]_3) & f^{(2)}_{[2]_9} &= ([2]_3, [2]_3, [0]_3)\\
f^{(1)}_{[3]_9} &= ([1]_3, [1]_3, [1]_3) & f^{(2)}_{[3]_9} &= ([2]_3, [2]_3, [2]_3)\\
f^{(1)}_{[4]_9} &= ([2]_3, [1]_3, [1]_3) & f^{(2)}_{[4]_9} &= ([1]_3, [2]_3, [2]_3)\\
f^{(1)}_{[5]_9} &= ([2]_3, [2]_3, [1]_3) & f^{(2)}_{[5]_9} &= ([1]_3, [1]_3, [2]_3)\\
f^{(1)}_{[6]_9} &= ([2]_3, [2]_3, [2]_3) & f^{(2)}_{[6]_9} &= ([1]_3, [1]_3, [1]_3)\\
f^{(1)}_{[7]_9} &= ([0]_3, [2]_3, [2]_3) & f^{(2)}_{[7]_9} &= ([0]_3, [1]_3, [1]_3)\\
f^{(1)}_{[8]_9} &= ([0]_3, [0]_3, [0]_3) & f^{(2)}_{[8]_9} &= ([0]_3, [0]_3, [0]_3).
\end{aligned}
$$

As remarked in Example 1.9, the automorphism group $\mathrm{Aut}(G) = \mathrm{Aut}(\mathbb{Z}_9)$ is isomorphic to $\mathscr{U}(\mathbb{Z}_9)$ the (multiplicative) group of units of the ring $\mathbb{Z}_9$. Thus, every $\eta \in \mathrm{Aut}(G)$ is of the form $\eta([n]_9) = [un]_9$ with $u \in \mathscr{U}(\mathbb{Z}_9) = \{[1]_9, [2]_9, [4]_9, [5]_9, [7]_9, [8]_9\}$. Suppose that there exists $\eta \in \mathrm{Aut}(G)$ such that $\Phi_1 = \Phi_2 \circ \eta$. As $\pi_2(g) = \pi_1(g) = \pi_2(\eta(g))$ for all $g \in G$, we necessarily have $u = [1 + k3]_9$ with $k \in \{0, 1, 2\}$. As $f^{(1)}_{[1]_9} \neq f^{(2)}_{[1]_9}$ (resp. $f^{(1)}_{[4]_9} \neq f^{(2)}_{[1]_9}$, resp. $f^{(1)}_{[7]_9} \neq f^{(2)}_{[1]_9}$) we see that no such $\eta \in \mathrm{Aut}(G)$ exists. We deduce that Φ_1 and Φ_2 are non-equivalent.

Remark 1.20 In [32, Example 4] we presented a "dual" version of Example 1.19. The involved groups are the same, namely, $H = N := \mathbb{Z}_3$ and $G := \mathbb{Z}_9$ but, differently, we consider the extensions $(G; \iota_1, \pi_1)$ and $(G; \iota_2, \pi_2)$ where, for $j = 1, 2$, the injective homomorphism $\iota_j : N \to G_j$ (resp. the surjective homomorphism $\pi_j : G_j \to H$) is defined by setting $\iota_j([m]_3) := [3m]_9$ for all $m = 0, 1, 2$ (resp. $\pi_j([m]_9) := [jm]_3$ for all $m = 0, 1, 2, \ldots, 8$). We then showed that, once again, these two extensions are non-equivalent. Moreover, defining, for $j = 1, 2$, the section $s^{(j)} : H \to G_j$ by setting $s^{(j)}_{[m]_3} := [jm]_9$ for $m = 0, 1, 2$, and denoting by $\Phi_j = \Phi_{s^{(j)}} : G_j \to N \wr H$ the corresponding Frobenius embedding, so that $\Phi_j([m]_9) = (f^{(j)}_{[m]_9}, [jm]_3)$ for all $m = 0, 1, 2, \ldots, 8$, we showed that $\Phi_1 \not\approx \Phi_2$, in accordance with Theorem 1.18.

Chapter 2
Clifford Theory

In this chapter, we establish some basic facts on Clifford theory for induced representations from a normal subgroup and we give the first applications of the theory.

2.1 Preliminaries and Notation

First of all, we establish all the notation used in this section. Let G be a finite group and let $N \trianglelefteq G$ be a normal subgroup of G (we will use the symbol B to denote the normal subgroup when it is Abelian). Set $H := G/N$ (we will use the symbol A to denote this quotient when it is Abelian). Here we regard $\widehat{G}$ (resp. $\widehat{N}$) as the set of all irreducible G-representations (resp. N-representations) modulo the equivalence $\sim$: by abuse of notation we shall not distinguish a representation and its equivalence class.

We will use the notation on group extensions in Sect. 1.5 but we think of N as a subgroup of G so that we avoid the use of the notation $\iota(n)$ or $\overline{n}$ to indicate $n \in N$ as an element of G.

Definition 2.1 Let $\sigma \in \widehat{N}$. We set

$$\widehat{G}(\sigma) := \{\theta \in \widehat{G} : \sigma \preceq \mathrm{Res}^G_N(\theta)\} \equiv \{\theta \in \widehat{G} : \theta \preceq \mathrm{Ind}^G_N(\sigma)\}.$$

For all $g \in G$ the *g-conjugate* of σ is the representation ${}^g\sigma \in \widehat{N}$ defined by

$$ {}^g\sigma(n) := \sigma(g^{-1}ng) \tag{2.1}$$

T. Ceccherini-Silberstein et al., *Representation Theory of Finite Group Extensions*, Springer Monographs in Mathematics,
https://doi.org/10.1007/978-3-031-13873-7_2

for all $n \in N$. The subgroup

$$I_G(\sigma) := \{g \in G : {}^g\sigma \sim \sigma\}$$

is called the *inertia group* of $\sigma \in \widehat{N}$. Finally, we set $\widehat{I_G}(\sigma) := \{\psi \in \widehat{I_G(\sigma)} : \psi \preceq \mathrm{Ind}_N^{I_G(\sigma)}\sigma\}$ and $H_G(\sigma) := I_G(\sigma)/N$; this is a subgroup of H.

It is easy to see that (2.1) defines an action of G on $\widehat{N}$, i.e. ${}^{g_1g_2}\sigma = {}^{g_1}({}^{g_2}\sigma)$ for all $g_1, g_2 \in G$. Note also that ${}^g\sigma(n) = \sigma\left(n^{g^\flat}\right)$, where $g \mapsto g^\flat$ is the action of G on N defined in (1.55). Clearly, $I_G(\sigma)$ is the stabilizer of σ in G and it contains the subgroup N: ${}^m\sigma(n) = \sigma(m)^{-1}\sigma(n)\sigma(m)$ if $m, n \in N$ and therefore ${}^m\sigma \sim \sigma$. In particular, $\widehat{G}({}^g\sigma) = \widehat{G}(\sigma)$ because ${}^g\theta \sim \theta$ for all $\theta \in \widehat{G}$ and $g \in G$.

Fix a normalized section $t\colon H \to G$ and recall the decomposition (1.54). As in (2.1), we may define an action of H on $\widehat{N}$ by setting ${}^h\sigma := {}^{t(h)}\sigma$, that is,

$$ {}^h\sigma(n) := \sigma(t(h)^{-1}nt(h)) \equiv \sigma\left(n^{h^\sharp}\right), \tag{2.2}$$

for all $h \in H$ and $n \in N$; cf. (1.57). Note that it does not depend on the choice of the section t: if $g = t(h)n \in t(h)N$ and $m \in N$, then

$$ {}^g\sigma(m) = \sigma(g^{-1}mg) = \sigma(n)^*\sigma\big(t(h)^{-1}mt(h)\big)\sigma(n) = \sigma(n)^*\, {}^h\sigma(m)\sigma(n)$$

and therefore ${}^h\sigma \sim {}^g\sigma$. Moreover, we have indeed an action of H: from (1.56) it follows that

$$\begin{aligned}\left[{}^h\left({}^k\sigma\right)\right](m) &= \sigma\left(t(k)^{-1}t(h)^{-1}nt(h)t(k)\right) = \sigma\left(n(h,k)^{-1}t(hk)^{-1}nt(hk)n(h,k)\right)\\ &= \sigma(n(h,k))^*\sigma\left(t(hk)^{-1}mt(hk)\right)\sigma(n(h,k))\\ &= \sigma(n(h,k))^* \cdot {}^{hk}\sigma(m)\cdot\sigma(n(h,k)),\end{aligned} \tag{2.3}$$

for all $m \in N$, that is, ${}^h\left({}^k\sigma\right) \sim {}^{hk}\sigma$. Alternatively, since N is contained in the kernel of the action of G on $\widehat{N}$, the above is nothing but the natural action of $H = G/N$ on $\widehat{N}$.

Suppose that $\mathscr{R} = \mathscr{R}(\sigma)$ is a set of representatives for the left $H_G(\sigma)$-cosets in H, i.e. $H = \sqcup_{r\in\mathscr{R}} rH_G(\sigma)$. Then $\{t(r) : r \in \mathscr{R}(\sigma)\}$ is a set of representatives for the left $I_G(\sigma)$-cosets in G:

$$G = \sqcup_{r\in\mathscr{R}} t(r)I_G(\sigma). \tag{2.4}$$

Moreover, $\{{}^g\sigma : g \in G\} \equiv \{{}^r\sigma : r \in \mathscr{R}\}$ and the representations ${}^r\sigma$'s are pairwise non–equivalent. In $\widehat{N}$ we define an equivalence relation $\approx$ by setting $\sigma_1 \approx \sigma_2$ if there exists $g \in G$ such that ${}^g\sigma_1 \sim \sigma_2$. Let Σ be a set of representatives of the corresponding quotient space $\widehat{N}/\approx$. Then

$$\widehat{N} = \sqcup_{\sigma \in \Sigma}\{{}^r\sigma : r \in \mathscr{R}(\sigma)\} \tag{2.5}$$

is the partition of $\widehat{N}$ into the orbits of G.

Note also that if $N \leq L \leq G$ is any intermediate subgroup then, for all $g \in G$, one has the equivalence

$$\mathrm{Ind}_N^L\, {}^g\sigma \sim {}^g\mathrm{Ind}_N^L\sigma. \tag{2.6}$$

Indeed, if V_σ (resp. $V_{{}^g\sigma}$) is the representation space of σ (resp. ${}^g\sigma$) then (cf. (1.25)) the map $\mathrm{Ind}_N^L V_\sigma \in f \to {}^gf \in \mathrm{Ind}_N^L V_{{}^g\sigma}$, where ${}^gf(h) = f(g^{-1}hg)$ for all $h \in L$, is a vector space isomorphism because

$${}^gf(hn) = f(g^{-1}hng) = f(g^{-1}hgg^{-1}ng) = \sigma(g^{-1}n^{-1}g)f(g^{-1}hg) = {}^g\sigma(n^{-1})\,{}^gf(h),$$

for all $n \in N, h \in L$. Moreover, setting $\lambda_\sigma := \mathrm{Ind}_N^L\sigma$ (resp. $\lambda_{{}^g\sigma} := \mathrm{Ind}_N^L{}^g\sigma$), cf. (1.26), we have, for all $h, k \in L$,

$$\begin{aligned}
{}^g\left({}^g\lambda_\sigma(h)f\right)(k) &= \left(\lambda_\sigma(g^{-1}hg)f\right)(g^{-1}kg) \\
&= f(g^{-1}h^{-1}gg^{-1}kg) \\
&= f(g^{-1}h^{-1}kg) \\
&= {}^gf(h^{-1}k) \\
&= \left(\lambda_{{}^g\sigma}(h){}^gf\right)(k),
\end{aligned}$$

thus showing that the above map also establishes the equivalence $\lambda_{{}^g\sigma} \sim {}^g\lambda_\sigma$, which is nothing but (2.6).

By the classical correspondence theorem in group theory (see [90, Theorem 2.5], [108, 1.4.6] and [114]), the restriction of t to $H_G(\sigma)$, namely $t|_{H_G(\sigma)}\colon H_G(\sigma) \to I_G(\sigma)$, is a normalized section of the extension $I_G(\sigma)$ of $H_G(\sigma)$ by N, so that

$$I_G(\sigma) = \sqcup_{h \in H_G(\sigma)} t(h)N \tag{2.7}$$

is the coset decomposition of $I_G(\sigma)$ over N. Then from (2.4), (2.7) and (1.56) we may deduce a more refined form of (1.54), namely:

$$G = \sqcup_{r \in \mathscr{R}} t(r) I_G(\sigma) = \sqcup_{r \in \mathscr{R}} \sqcup_{h \in H_G(\sigma)} t(rh)N. \tag{2.8}$$

Finally, we set

$$c = c_\sigma := [I_G(\sigma) : N]. \tag{2.9}$$

We end this section with two important notions in the representation theory of group extensions.

Definition 2.2 If ψ is an H-representation its *inflation* is the G-representation $\overset{\vartriangle}{\psi}$ defined by setting

$$\overset{\vartriangle}{\psi}(g) := \psi(gN) \quad \forall g \in G. \tag{2.10}$$

In other words, $\overset{\vartriangle}{\psi}(g) = \psi(\pi(g))$, where $\pi : G \to H$ is the projection map. As an example, if λ is the left-regular representation of H then $\overset{\vartriangle}{\lambda}$ is equivalent to the so-called *permutation representation* of G over H, that is,

$$\overset{\vartriangle}{\lambda} \sim \mathrm{Ind}_N^G \iota_N, \tag{2.11}$$

where ι_N is the trivial representation of N. Moreover, if ψ is irreducible then $\overset{\vartriangle}{\psi}$ is irreducible as well and if $\psi \sim \psi_1 \oplus \psi_2$ then $\overset{\vartriangle}{\psi} \sim \overset{\vartriangle}{\psi}_1 \oplus \overset{\vartriangle}{\psi}_2$. A first example of inflation is in (1.24).

Definition 2.3 Let K be a subgroup of G and let $\sigma \in \widehat{K}$. An *extension* of σ to G is a representation $\widetilde{\sigma} \in \widehat{G}$ such that $\mathrm{Res}_K^G \widetilde{\sigma} = \sigma$.

Clearly, the extension of a representation has the same dimension, moreover it may be not unique or it may not even exist. In the case of a normal subgroup, existence will be studied in Sect. 7.1 while, assuming existence, a characterization of all such extensions will be presented in Corollary 2.20.

We end these preliminaries by the following elementary but remarkable fact.

Proposition 2.4 *Suppose that $N \trianglelefteq G$ and that $\sigma \in \widehat{N}$ has an extension τ to an intermediate subgroup $L \leq G$. Then $L \leq I_G(\sigma)$.*

Proof Indeed, if $h \in L$ then for all $n \in N$ we have

$${}^h\sigma(n) = \sigma(h^{-1}nh) = \tau(h^{-1}nh) = \tau(h^{-1})\tau(n)\tau(h) = \tau(h)^{-1}\sigma(n)\tau(h),$$

showing that ${}^h\sigma \sim \sigma$. ■

2.2 Basic Clifford Theory

In the first result of this section, we specify the classical *Mackey lemma* ([30, Theorem 11.5.1], see also Theorem 8.23 for a generalization) in the present setting.

Theorem 2.5 *Let $\sigma \in \widehat{N}$ and $\theta \in \widehat{G}(\sigma)$. Denote by $\ell = \ell_\theta$ the multiplicity of σ in $\mathrm{Res}^G_N\theta$ and let $c = c_\sigma := [I_G(\sigma) : N]$ (cf. (2.9)). Then we have:*

(1) $\mathrm{Res}^G_N\mathrm{Ind}^G_N\sigma \sim \bigoplus_{h\in H} {}^h\sigma \sim c\bigoplus_{r\in\mathscr{R}} {}^r\sigma$ *is the decomposition of* $\mathrm{Res}^G_N\mathrm{Ind}^G_N\sigma$ *into irreducible representations.*
(2) $\dim\mathrm{End}_G(\mathrm{Ind}^G_N\sigma) = c$.
(3) $\mathrm{Res}^G_N\theta \sim \ell\bigoplus_{r\in\mathscr{R}} {}^r\sigma$.
(4) $d_\theta = \ell|G/I_G(\sigma)|d_\sigma$ *and there is a $*$-algebra isomorphism*

$$\mathrm{End}_N(\mathrm{Res}^G_N\theta) \cong \underbrace{M_c(\mathbb{C}) \oplus \cdots \oplus M_c(\mathbb{C})}_{\ell \text{ times}}.$$

Proof

(1) Let V denote the representation space of σ. For all $h \in H$, set

$$Z_h := \{f : G \to V \text{ s.t. } f(t(k)n) = \delta_{h,k}\sigma(n^{-1})f(t(h)), \forall n \in N, k \in H\}. \tag{2.12}$$

From (1.54) it follows that

$$\mathrm{Ind}^G_N V = \bigoplus_{h\in H} Z_h. \tag{2.13}$$

In other words, Z_h is made up of those $f \in \mathrm{Ind}^G_N V$ whose support lies in $t(h)N$. Moreover, the map $\widetilde{L}_h : V \to Z_h$, given by $[\widetilde{L}_h v](t(k)n) := \delta_{h,k}\sigma(n^{-1})v$ for all $v \in V$ $k \in H$ and $n \in N$, is a linear isomorphism of vector spaces.

Setting $\lambda := \mathrm{Ind}^G_N\sigma$ one has

$$\begin{aligned}
[\lambda(n)\widetilde{L}_h v](t(k)m) &= [\widetilde{L}_h v](n^{-1}t(k)m)\\
\text{(by (1.57))}\quad &= [\widetilde{L}_h v]\left(t(k)\left(n^{-1}\right)^{k^\sharp} m\right)\\
&= \delta_{h,k}\sigma\left(m^{-1}n^{k^\sharp}\right)v\\
\text{(by (2.2))}\quad &= \delta_{h,k}\sigma(m^{-1})[{}^k\sigma(n)v]\\
&= \left[\widetilde{L}_h\, {}^h\sigma(n)v\right](t(k)m)
\end{aligned} \tag{2.14}$$

for all $v \in V$, $k \in H$ and $n, m \in N$. In other words, Z_h is $\lambda(n)$-invariant and $\lambda(n)\widetilde{L}_h = \widetilde{L}_h\, {}^h\sigma(n)$, that is $\widetilde{L}_h$ is an intertwiner inducing an equivalence between $(\mathrm{Res}^G_N\lambda, Z_h)$ and $({}^h\sigma, V)$. Therefore $\mathrm{Res}^G_N(\mathrm{Ind}^G_N\sigma)$ is equivalent to $\bigoplus_{h\in H} {}^h\sigma$ and (1) follows as, by (2.8) and (2.3),

$$\bigoplus_{h\in H} {}^h\sigma \sim \bigoplus_{r\in\mathscr{R}} \bigoplus_{h\in H_G(\sigma)} {}^{rh}\sigma \sim c \bigoplus_{r\in\mathscr{R}} {}^r\sigma.$$

(2) Since, by (1), the multiplicity of the irreducible N-representation σ in $\mathrm{Res}^G_N\mathrm{Ind}^G_N\sigma$ is equal to c, we have

$$\mathrm{Hom}_N(\sigma, \mathrm{Res}^G_N\mathrm{Ind}^G_N\sigma) \cong \mathbb{C}^c$$

and therefore, by Frobenius' reciprocity (cf. (1.34)) and (1.7),

$$\mathrm{dimEnd}_G(\mathrm{Ind}^G_N\sigma) = \mathrm{dimHom}_N(\mathrm{Res}^G_N\mathrm{Ind}^G_N\sigma, \sigma) = c.$$

(3) One easily checks that, for $g \in G$, the map

$$\begin{array}{ccc} \mathrm{Hom}_N(\sigma, \mathrm{Res}^G_N\theta) & \longrightarrow & \mathrm{Hom}_N({}^g\sigma, \mathrm{Res}^G_N\theta) \\ T & \longmapsto & \theta(g)T \end{array}$$

is an isometric isomorphism. Therefore every ${}^g\sigma$ has multiplicity ℓ in $\mathrm{Res}^G_N\theta$. Moreover, by Frobenius' reciprocity, we have that $\mathrm{Ind}^G_N\sigma$ contains exactly ℓ copies of θ. Therefore, every irreducible sub-representation in $\mathrm{Res}^G_N\theta$ is also a sub-representation of $\mathrm{Res}^G_N\mathrm{Ind}^G_N\sigma$. But, by (1), the latter contains only irreducible sub-representations of the form ${}^r\sigma$, and this ends the proof.

(4) The first identity follows immediately from (3), while the isomorphism just requires the use of (1.6).

■

Remark 2.6

(1) It is easy to check that each operator $\widetilde{L}_h$ is an isometric immersion of V into the space of $\mathrm{Ind}^G_N\sigma$ and that

$$\mathrm{Ind}^G_N V = \bigoplus_{h\in H} \widetilde{L}_h V$$

is an explicit form of the isomorphism in (1) (cf. Remark 1.2). Moreover, if $T_1, T_2, \dots, T_\ell$ is an orthonormal basis of $\mathrm{Hom}_N(\sigma, \mathrm{Res}^G_N\theta)$ and W is the space of θ then it is easy to check that

$$W = \bigoplus_{r\in\mathscr{R}} \bigoplus_{j=1}^{\ell} \theta(t(r))T_j V,$$

is an explicit form of the decomposition in Theorem 2.5.(3). Also, for $r \in \mathscr{R}$ and $1 \le j \le \ell$ we define $S_{r,j}\colon W \to \mathrm{Ind}_N^G V$ (now we think of V as the space of ${}^r\sigma$) by means of (1.35), that is, by setting

$$\left[S_{r,j}w\right](g) := \frac{1}{\sqrt{|G/N|}} T_j^* \theta(t(r)^{-1}g^{-1})w,$$

for all $g \in G$, $w \in W$. Then $\{S_{r,j} : j = 1, 2, \ldots, \ell\}$ is an orthonormal basis of $\mathrm{Hom}_G(\theta, \mathrm{Ind}_N^G {}^r\sigma)$ (take also into account (1.3)).

(2) By means of (1.64) we can generalize the calculations in the proof of Theorem 2.5.(1). Namely, it is immediate to check that for all $f \in \mathrm{Ind}_N^G V$, $s, k \in H$ and $n, m \in N$ we have:

$$[\lambda(t(s)n)f](t(k)m) = \sigma(m)^*\sigma(n(s, s^{-1}k))\, {}^{s^{-1}k}\sigma(n) f(t(s^{-1}k)),$$

while if $v \in V$, $h \in H$, the action on $\lambda(t(h))f_v$ (cf. (1.30)) is given by:

$$\lambda(t(s)n)[\lambda(t(h))f_v] = \lambda(t(sh)) f_{\sigma(n(s,h))\, {}^h\sigma(n)v}.$$

The last formula, for $s = 1_H$, reduces to (2.14).

A more detailed analysis of $\mathrm{End}_G(\mathrm{Ind}_N^G\sigma)$ will be given in Sect. 7.5.

Definition 2.7 The number $\ell_\theta := \dim\mathrm{Hom}_N(\sigma, \mathrm{Res}_N^G\theta)$ (i.e. the multiplicity of σ in $\mathrm{Res}_N^G\theta$) is called the *inertia index* of $\theta \in \widehat{G}(\sigma)$ with respect to N.

By Frobenius reciprocity, the inertia index is also equal to the multiplicity of θ in $\mathrm{Ind}_N^G\sigma$ and by Theorem 2.5.(3), it only depends on θ and N, that is $\ell_\theta = \dim\mathrm{Hom}_N({}^r\sigma, \mathrm{Res}_N^G\theta)$ for all $r \in \mathscr{R}$. Note also that $\{{}^r\sigma : r \in \mathscr{R}\}$ are (up to equivalence) the pairwise inequivalent irreducible N-subrepresentations of $\mathrm{Res}_N^G\theta$.

In particular,

$$\mathrm{Ind}_N^G\sigma \sim \mathrm{Ind}_N^G {}^r\sigma \sim \bigoplus_{\theta \in \widehat{G}(\sigma)} \ell_\theta\theta. \tag{2.15}$$

Remark 2.8 Clearly, $I_{I_G(\sigma)}(\sigma) = I_G(\sigma)$, and Theorem 2.5 with G replaced by $I_G(\sigma)$ yields:

$$\mathrm{Res}_N^{I_G(\sigma)}\mathrm{Ind}_N^{I_G(\sigma)}\sigma \sim \underbrace{\sigma \oplus \sigma \oplus \cdots \oplus \sigma}_{c \text{ times}} \quad \text{and} \quad \mathrm{Res}_N^{I_G(\sigma)}\eta \sim \underbrace{\sigma \oplus \sigma \oplus \cdots \oplus \sigma}_{\ell_\eta \text{ times}}, \tag{2.16}$$

for all $\eta \in \widehat{I_G}(\sigma)$. In particular,

$$d_\eta = \ell_\eta d_\sigma. \tag{2.17}$$

The following proposition collects some immediate but quite important consequences of Theorem 2.5.

Proposition 2.9

(1) $\sum_{\theta\in\widehat{G}(\sigma)}\ell_\theta^2 = c\ (\equiv [I_G(\sigma) : N])$. *In particular,* $|\widehat{G}(\sigma)| = 1$ *if and only if* $\ell_\theta^2 = c$ *for some* $\theta \in \widehat{G}(\sigma)$.

(2)
$$\widehat{G} = \sqcup_{\sigma\in\Sigma}\widehat{G}(\sigma) \tag{2.18}$$

and the map

$$\{{}^r\sigma : r \in \mathscr{R}(\sigma)\} \longmapsto \widehat{G}(\sigma) \tag{2.19}$$

yields a bijection between the parts of the partitions (2.5) *and* (2.18).

(3) *If* λ_H *is the left regular representation of* H *then for all* $\theta \in \widehat{G}(\sigma)$ *we have (cf.* (2.10)*):*

$$\theta\otimes\overset{\triangle}{\lambda_H} \sim \ell_\theta\bigoplus_{r\in\mathscr{R}}\mathrm{Ind}_N^G({}^r\sigma) \sim |\mathscr{R}|\ell_\theta\bigoplus_{\rho\in\widehat{G}(\sigma)}\ell_\rho\rho. \tag{2.20}$$

(4) *The character of* $\mathrm{Ind}_N^G\sigma$ *is given by:*

$$\chi^{\mathrm{Ind}_N^G\sigma}(t(h)n) = \begin{cases} 0 & \text{if } h \neq 1_H,\\ c\sum_{r\in\mathscr{R}}\chi^{{}^r\sigma}(n) & \text{if } h = 1_H,\end{cases} \tag{2.21}$$

for all $h \in H$, $n \in N$.

(5) *Let* $\theta \in \widehat{G}$ *and suppose that* $\mathrm{Res}_N^G\theta$ *contains* ι_N. *Then* $\mathrm{Res}_N^G\theta \sim d_\theta\iota_N$ *and* $\ker\theta \geq N$.

Proof

(1) From (2.15) and Theorem 2.5.(3) we get immediately

$$\mathrm{Res}_N^G\mathrm{Ind}_N^G\sigma \sim \bigoplus_{\theta\in\widehat{G}(\sigma)}\ell_\theta\mathrm{Res}_N^G\theta \sim \left[\sum_{\theta\in\widehat{G}(\sigma)}\ell_\theta^2\right]\left(\bigoplus_{r\in\mathscr{R}}{}^r\sigma\right),$$

and therefore Theorem 2.5.(1) forces the identity $\sum_{\theta\in\widehat{G}(\sigma)}\ell_\theta^2 = c$. Alternatively, we may use (2.15), the structure of the commutant in (1.38), and Theorem 2.5.(2).

(2) This is an immediate consequence of Theorem 2.5.(3). Just note that if $\tau \in \widehat{G}$, then taking $\gamma \in \widehat{N}$ contained in $\mathrm{Res}_N^G\tau$ and denoting by $\sigma \in \Sigma$ the representative of the G-orbit containing γ, we have $\tau \in \widehat{G}(\sigma)$.

(3) Keeping in mind (2.11), we have:

$$\begin{aligned}
\theta \otimes \overset{\triangle}{\lambda}_H &\sim \theta \otimes \mathrm{Ind}_N^G \iota_N \\
\text{(by (1.36))} \qquad &\sim \mathrm{Ind}_N^G \left[\left(\mathrm{Res}_N^G \theta \right) \otimes \iota_N \right] \\
\text{(by Theorem 2.5.(3))} \qquad &\sim \ell_\theta \bigoplus_{r \in \mathscr{R}} \mathrm{Ind}_N^G ({}^r\sigma) \\
\text{(by (2.15))} \qquad &\sim |\mathscr{R}| \ell_\theta \bigoplus_{\rho \in \widehat{G}(\sigma)} \ell_\rho \rho.
\end{aligned}$$

(4) By Frobenius' character formula (1.32), we have:

$$\chi^{\mathrm{Ind}_N^G \sigma}(t(h)n) = \sum_{\substack{k \in H: \\ t(k)^{-1} t(h) n t(k) \in N}} \chi^\sigma \left[t(k)^{-1} t(h) n t(k) \right].$$

Since $\pi \left[t(k)^{-1} t(h) n t(k) \right] = k^{-1} h k$, we have $t(k)^{-1} t(h) n t(k) \in N$ if and only if $k^{-1} h k = 1_H$, that is, if and only if $h = 1_H$ and $k \in H$. This gives (2.21) for $h \neq 1_H$. Finally, taking into account Theorem 2.5.(1), we complete the proof of formula (2.21).

(5) In this case, ${}^g\iota_N \sim \iota_N$ for all $g \in G$, so that by Theorem 2.5.(3), $\mathrm{Res}_N^G \theta$ only contains ι_N.

■

From (2.18), (2.19), and (2.5) it follows that the left hand side of (1.20) may be written in the forms:

$$\sum_{\theta \in \widehat{G}} d_\theta^2 = \sum_{\sigma \in \Sigma} \sum_{\theta \in \widehat{G}(\sigma)} d_\theta^2 = \sum_{\sigma \in \widehat{N}} \frac{1}{|\mathscr{R}(\sigma)|} \sum_{\theta \in \widehat{G}(\sigma)} d_\theta^2. \tag{2.22}$$

Remark 2.10

(1) It is worthwhile to check the dimensions in (2.20):

$$\begin{aligned}
\dim \left(\theta \otimes \overset{\triangle}{\lambda}_H \right) &= d_\theta \, |H| \\
\text{(by Theorem 2.5.(4))} \quad &= \ell_\theta \, |\mathscr{R}| \, d_\sigma \, |H| \\
&= \ell_\theta \, |\mathscr{R}| \, \dim \mathrm{Ind}_N^G \sigma \\
\text{(by (2.15))} \quad &= \ell_\theta \, |\mathscr{R}| \sum_{\rho \in \widehat{G}(\sigma)} \ell_\rho d_\rho.
\end{aligned}$$

(2) Note that, when $N = \{1_G\}$, Proposition 2.9.(1) reduces to (1.20).
(3) The first equivalence in (2.20) is a particular case (namely, the case of induction from a normal subgroup) of the so called *Mackey tensor product theorem*; see [30, Theorem 11.5.3] for the general case. Two other particular cases, that may be proved in the same way, are:

$$\theta \otimes \mathrm{Ind}_N^G \sigma \sim \ell_\theta \bigoplus_{r \in \mathcal{R}} \mathrm{Ind}_N^G \left({}^r\sigma \otimes \sigma\right), \qquad \text{for all } \theta \in \widehat{G}(\sigma),$$

and

$$\mathrm{Ind}_N^G \sigma \bigotimes \mathrm{Ind}_N^G \tau \sim c \bigoplus_{r \in \mathcal{R}} \mathrm{Ind}_N^G \left({}^r\sigma \otimes \tau\right), \qquad \text{for all } \tau \in \widehat{N}.$$

We now study an intermediate step: the induction from N to the inertia group $I_G(\sigma)$.

Lemma 2.11 *Let $\sigma \in \widehat{N}$ and let*

$$\mathrm{Ind}_N^{I_G(\sigma)} \sigma \sim \bigoplus_{\eta \in \widehat{I_G}(\sigma)} \ell_\eta \eta \tag{2.23}$$

be the decomposition of $\mathrm{Ind}_N^{I_G(\sigma)} \sigma$ into $I_G(\sigma)$-irreducible representations ($\ell_\eta > 0$ is the multiplicity of η and therefore its inertia index with respect to N). Then

(1)

$$\mathrm{Ind}_N^G \sigma \sim \bigoplus_{\eta \in \widehat{I_G}(\sigma)} \ell_\eta \mathrm{Ind}_{I_G(\sigma)}^G \eta \tag{2.24}$$

is the decomposition of $\mathrm{Ind}_N^G \sigma$ into its G-irreducible components (i.e., the $\mathrm{Ind}_{I_G(\sigma)}^G \eta$, with $\eta \in \widehat{I_G}(\sigma)$, are G-irreducible and pairwise inequivalent).

(2) *If $\theta \in \widehat{G}(\sigma)$ then $\theta \sim \mathrm{Ind}_{I_G(\sigma)}^G \eta$ for some (unique) $\eta \in \widehat{I_G}(\sigma)$.*

Proof

(1) First note that, from Theorem 2.5.(2) and with the same notation therein, we get that

$$\dim\mathrm{End}_{I_G(\sigma)} \left(\mathrm{Ind}_N^{I_G(\sigma)} \sigma\right) = c = \dim\mathrm{End}_G \left(\mathrm{Ind}_N^G \sigma\right).$$

Indeed, $I_G(\sigma)$ coincides with the inertia group of σ with respect to $I_G(\sigma)$ (cf. (2.9) and Remark 2.8). Moreover, the decomposition (2.24) follows from (2.23) and transitivity of induction (cf. (1.33)). Then, by applying (1.38) to (2.23) and (1.7) to (2.24), we deduce that

$$\sum_{\eta\in\widehat{I_G}(\sigma)} \ell_\eta^2 = c = \sum_{\psi,\eta\in\widehat{I_G}(\sigma)} \ell_\psi \ell_\eta \mathrm{dimHom}_G(\mathrm{Ind}_{I_G(\sigma)}^G\psi, \mathrm{Ind}_{I_G(\sigma)}^G\eta).$$

Therefore $\mathrm{dimHom}_G\left(\mathrm{Ind}_{I_G(\sigma)}^G\psi, \mathrm{Ind}_{I_G(\sigma)}^G\eta\right) = \delta_{\psi,\eta}$, and irreducibility of the representations $\mathrm{Ind}_{I_G(\sigma)}^G\eta$'s follows.

(2) This is an immediate consequence of (1): if $\theta \preceq \mathrm{Ind}_N^G\sigma$ then it must be equal to one of the $\mathrm{Ind}_{I_G(\sigma)}^G\eta$ with $\eta \in \widehat{I_G}(\sigma)$ in (2.24).

■

At the end of Sect. 2.5 we show how one deduces the irreducibility of $\mathrm{Ind}_{I_G(\sigma)}^G\eta$ from Mackey's irreducibility criterion.

Theorem 2.12 (Clifford Correspondence) *Let $\sigma \in \widehat{N}$. Then the map*

$$\begin{array}{ccc}\widehat{I_G}(\sigma) & \longrightarrow & \widehat{G}(\sigma) \\ \eta & \longmapsto & \mathrm{Ind}_{I_G(\sigma)}^G\eta\end{array}$$

is a bijection. Moreover the inertia index of $\eta \in \widehat{I_G}(\sigma)$ with respect to N coincides with the inertia index of $\mathrm{Ind}_{I_G(\sigma)}^G\eta$ with respect to N, that is,

$$\ell_\eta = \ell_{\mathrm{Ind}_{I_G(\sigma)}^G\eta}. \tag{2.25}$$

Finally,

$$\mathrm{Res}_N^{I_G(\sigma)}\eta \sim \ell_\eta\sigma. \tag{2.26}$$

Proof The first part is just a reformulation of Lemma 2.11. Moreover, from (2.23) and (2.24) it follows that the inertia index of η with respect to N coincides with the inertia index of $\mathrm{Ind}_{I_G(\sigma)}^G\eta$ with respect to N. Finally, an application of Theorem 2.5.(3) with G replaced by $I_G(\sigma)$ (cf. Remark 2.8) yields (2.26). ■

Corollary 2.13 *Let $\sigma \in \widehat{N}$ and $\theta \in \widehat{G}(\sigma)$. Suppose that $I_G(\sigma) \trianglelefteq G$, let $\eta \in \widehat{I_G}(\sigma)$ such that $\theta \sim \mathrm{Ind}_{I_G(\sigma)}^G\eta$. Then*

$$\mathrm{Res}_{I_G(\sigma)}^G\theta \sim \bigoplus_{r\in\mathscr{R}(\sigma)} {}^r\eta, \tag{2.27}$$

where the $I_G(\sigma)$-representations in the right hand side are (irreducible and) pairwise inequivalent.

Proof From Frobenius reciprocity, the equivalence ${}^r\theta \sim \theta$, and normality of $I_G(\sigma)$ in G, we deduce that $\mathrm{Res}^G_{I_G(\sigma)}\theta$ contains ${}^r\eta$ with multiplicity 1 for all $r \in \mathscr{R}(\sigma)$. Since

$$\dim\mathrm{Res}^G_{I_G(\sigma)}\theta = d_\theta = \dim\mathrm{Ind}^G_{I_G(\sigma)}\eta = [G : I_G(\sigma)]d_\eta = |\mathscr{R}(\sigma)|d_\eta,$$

and $d_{{}^r\eta} = d_\eta$ for all $r \in \mathscr{R}(\sigma)$, the decomposition (2.27) follows. ■

Note that, conversely, from (2.27) and $\mathrm{Res}^{I_G(\sigma)}_N {}^r\eta \sim \ell_\theta {}^r\sigma$ (cf. Theorem 2.12), one may immediately deduce Theorem 2.5.(3).

Corollary 2.14 *Let $\sigma \in \widehat{N}$ and $\theta \in \widehat{G}(\sigma)$. Then $\ell_\theta = 1$ if and only if there exists $\eta \in \widehat{I_G}(\sigma)$ such that*

$$\mathrm{Res}^{I_G(\sigma)}_N\eta \sim \sigma \qquad \textit{and} \qquad \mathrm{Ind}^G_{I_G(\sigma)}\eta \sim \theta.$$

Proof Indeed, by Clifford's correspondence we have $\theta \sim \mathrm{Ind}^G_{I_G(\sigma)}\eta$ for some $\eta \in \widehat{I_G}(\sigma)$ and $\ell_\theta = \ell_\eta$ (cf. Eq. (2.25)). Then, we conclude by using the second equivalence in (2.16). ■

In Corollary 7.22, we will show that, in general, the inertia index of σ is equal to the dimension of an irreducible cocycle representation of the quotient $I_G(\sigma)/N$.

Remark 2.15 The Clifford correspondence provides a method for determining $\widehat{G}$, starting from $\widehat{N}$. The steps are the following:

- Determine a set Σ as in (2.5).
- For each $\sigma \in \Sigma$ determine $I_G(\sigma)$.
- Compute the decomposition $\mathrm{Ind}^{I_G(\sigma)}_N\sigma = \oplus_{\eta\in\widehat{I_G}(\sigma)}\ell_\eta\eta$, in order to determine $\widehat{I_G}(\sigma)$.
- Then, $\widehat{G}(\sigma) = \left\{\mathrm{Ind}^G_{I_G(\sigma)}\eta : \eta \in \widehat{I_G}(\sigma)\right\}$.
- Finally, $\widehat{G} = \sqcup_{\sigma\in\Sigma}\widehat{G}(\sigma)$.

In the next section, we will show how to simplify this procedure in some particular, important cases. Another procedure, which determines $\widehat{I_G}(\sigma)$ and therefore also $\widehat{G}(\sigma)$, in a different way, is presented in Sect. 7.2.

In Table 2.1 we summarize the results obtained so far.

2.3 First Applications and the Little Group Method

In this section, we examine several particular cases of Clifford theory, in particular the following extremal cases: $I_G(\sigma) = G$, $I_G(\sigma) = N$, $H_G(\sigma)$ Abelian, and $\ell_\theta = 1$. We deduce several important results which are quite useful in the applications. We begin with a simple remark and an elementary proposition.

Table 2.1

Induction:
$\mathrm{Ind}_N^G\sigma = \bigoplus_{\eta\in\widehat{I_G(\sigma)}} \ell_\eta \mathrm{Ind}_{I_G(\sigma)}^G \eta$
$\uparrow$
$\mathrm{Ind}_N^I\sigma = \bigoplus_{\eta\in\widehat{I_G(\sigma)}} \ell_\eta \eta$
$\uparrow$
$\sigma \in \widehat{N}$
Restriction from $I_G(\sigma)$:
$\eta \in \widehat{I_G}(\sigma)$
$\downarrow$
$\mathrm{Res}_N^{I_G(\sigma)}\eta = \ell_\eta\sigma$
Restriction from G:
$\mathrm{Ind}_{I_G(\sigma)}^G\eta \in \widehat{G}(\sigma)$
$\downarrow$
$\mathrm{Res}_N^G(\mathrm{Ind}_{I_G(\sigma)}^G\eta) = \ell_\eta \bigoplus_{r\in\mathscr{R}} {}^r\sigma$

Remark 2.16

(1) If $I_G(\sigma) = G$, that is ${}^g\sigma \sim \sigma$ for all $g \in G$, then $H_G(\sigma) \equiv H$, $\mathscr{R} = \{1_G\}$, $c = |H|$, and we have:

$$\mathrm{Res}_N^G\mathrm{Ind}_N^G\sigma \sim \underbrace{\sigma\oplus\sigma\oplus\cdots\oplus\sigma}_{c \text{ times}} \qquad \text{and} \qquad \mathrm{Res}_N^G\theta \sim \underbrace{\sigma\oplus\sigma\oplus\cdots\oplus\sigma}_{\ell_\theta \text{ times}},$$

and $d_\theta = \ell_\theta d_\sigma$, for all $\theta \in \widehat{G}(\sigma)$, cf. Remark 2.8.
In particular, if $I_G(\sigma) = G$ for all $\sigma \in \widehat{N}$ then $\Sigma \equiv \widehat{N}$ and the decomposition of the left regular representation of G (1.19) becomes

$$\lambda_G \sim \bigoplus_{\sigma\in\widehat{N}} d_\sigma \bigoplus_{\theta\in\widehat{G}(\sigma)} \ell_\theta\theta, \tag{2.28}$$

cf. (2.17) and (2.18). Alternatively, start from $\lambda_G \sim \mathrm{Ind}_{\{1_G\}}^G\iota$ and use transitivity of induction. For instance, if $N = B$ is a central subgroup (that is, B is contained in the center of G; see also Chap. 6), then for all $\sigma = \xi \in \widehat{B}$ we have $I_G(\xi) = G$ because $g^{-1}bg = b$, for all $g \in G$ and $b \in B$. In particular, $\mathrm{Res}_N^G\theta = \ell_\theta\xi$ and $d_\xi = 1$ implies that $d_\theta = \ell_\theta$.

(2) It is easy to check that the following facts are equivalent:

(a) $I_G(\sigma) = G$ and $\widehat{G}(\sigma) = \{\theta\}$;
(b) $\mathrm{Res}_N^G\theta \sim \ell_\theta\sigma$ and $\ell_\theta^2 = [G : N]$;
(c) $I_G(\sigma) = G$ and $\chi^\theta(t(h)n) = 0$ for $h \neq 1_H$.

For instance, to show that (c) $\Longrightarrow$ (a), just note that from (c) and Theorem 2.5(3) it follows that $\mathrm{Res}^G_N\theta \sim \ell_\theta\sigma$. Then, from Proposition 2.9.(4) we deduce that $\chi^{\mathrm{Ind}^G_N\sigma} \sim \frac{c}{\ell_\theta}\chi^\theta$, and therefore $\widehat{G}(\sigma) = \{\theta\}$. Note also that we have $c = [G : N]$ so that (cf. Proposition 2.9.(1)) $\ell^2_\theta = c$.

If the above conditions are satisfied, one says that θ and σ are *fully ramified*; cf. [67, Exercise (6.3)].

Proposition 2.17 *Suppose that G/N is Abelian and let $N \leq M \leq G$. Let $\psi \in \widehat{M/N}$ and denote by $\overset{\vartriangle}{\psi}$ is its inflation to M. Then $I_G(\overset{\vartriangle}{\psi}) = G$.*

Proof Let $g \in G$ and $m \in M$. Then

$$^g\overset{\vartriangle}{\psi}(m) = \overset{\vartriangle}{\psi}(g^{-1}mg) = \psi(g^{-1}mgN) = \psi(mN) = \overset{\vartriangle}{\psi}(m).$$

■

In the following theorem, which is quite important, we examine the case in which $I_G(\sigma) = G$ (cf. Remark 2.16) and the inertia index of $\theta \in \widehat{G}(\sigma)$ with respect to N is equal to 1. By Theorem 2.5.(3), these conditions are equivalent to $\mathrm{Res}^G_N\theta \sim \sigma$.

Theorem 2.18 (Gallagher [52]) *Let $\sigma \in \widehat{N}$ and let $\theta \in \widehat{G}(\sigma)$. Suppose that θ is an extension of σ. Then*

$$\mathrm{Ind}^G_N\sigma \sim \bigoplus_{\psi\in\widehat{H}} d_\psi(\theta \otimes \overset{\vartriangle}{\psi}), \tag{2.29}$$

where the $\theta \otimes \overset{\vartriangle}{\psi}$'s are irreducible and pairwise non-equivalent G-representations,

$$\widehat{G}(\sigma) = \{\theta \otimes \overset{\vartriangle}{\psi} : \psi \in \widehat{H}\} \quad \text{and} \quad \ell_{\theta\otimes\overset{\vartriangle}{\psi}} = d_\psi.$$

Moreover, if $\tau \in \widehat{G}$, then $\mathrm{Res}^G_N\tau \sim \sigma$ if and only if $\tau \sim \theta \otimes \overset{\vartriangle}{\omega}$ for some $\omega \in \widehat{H}$ with $d_\omega = 1$.

Proof As remarked above, we have $I_G(\sigma) = G$ and $\ell_\theta = 1$, so that the first equivalence in (2.20) now becomes:

$$\mathrm{Ind}^G_N\sigma \sim \theta \otimes \overset{\vartriangle}{\lambda}_H. \tag{2.30}$$

Since $\lambda_H \sim \oplus_{\psi\in\widehat{H}} d_\psi \psi$ (cf. (1.19)), we have $\overset{\vartriangle}{\lambda}_H \sim \oplus_{\psi\in\widehat{H}} d_\psi \overset{\vartriangle}{\psi}$, and (2.29) follows.

Applying Theorem 2.5.(2) and observing that $c = |H|$, we get that

$$\mathrm{dimEnd}_G(\mathrm{Ind}^G_N\sigma) = |H|.$$

From (1.20), (2.29), and (1.7) it follows that

$$\sum_{\psi\in\widehat{H}} d_\psi^2 = |H| = \dim\mathrm{End}_G(\mathrm{Ind}_N^G\sigma) = \sum_{\psi_1,\psi_2\in\widehat{H}} d_{\psi_1}d_{\psi_2}\mathrm{Hom}_G(\theta\otimes\overset{\vartriangle}{\psi_1},\theta\otimes\overset{\vartriangle}{\psi_2}).$$

This yields $\dim\mathrm{Hom}_G(\theta\otimes\overset{\vartriangle}{\psi_1},\theta\otimes\overset{\vartriangle}{\psi_2}) = \delta_{\psi_1,\psi_2}$, that is, the irreducibility and pairwise inequivalence of the $\theta\otimes\overset{\vartriangle}{\psi}$'s. Then (2.29) is a decomposition into irreducible representations and d_ψ is equal to the inertia index of $\theta\otimes\overset{\vartriangle}{\psi}$ with respect to N.

Finally, if $\tau\in\widehat{G}$ and $\mathrm{Res}_N^G\tau\sim\sigma$, then $\tau\in\widehat{G}(\sigma)$ and therefore it must coincide with one of the $\theta\otimes\overset{\vartriangle}{\psi}$'s with inertia index equal to 1, that is, with a $\theta\otimes\overset{\vartriangle}{\psi}$ with $d_\psi=1$. The converse is obvious. ■

Remark 2.19 Under the above hypotheses, using the map ξ that yields (1.36), one can immediately deduce an explicit isometric form of the isomorphism (2.30): it is given by the map $\widetilde{\xi}\colon V\otimes L(H)\to\mathrm{Ind}_N^G V$, where $[\widetilde{\xi}(v\otimes f)](g):=f(gN)\theta(g^{-1})v$, for all $v\in V$, $f\in L(H)$, and $g\in G$ (now V is the space of both σ and θ).

The following corollary is just a reformulation of the last statement in Theorem 2.18.

Corollary 2.20 *Suppose that $\sigma\in\widehat{N}$ admits an extension $\widetilde{\sigma}\in\widehat{G}$. Then the set of all such extensions is given by $\{\widetilde{\sigma}\otimes\overset{\vartriangle}{\omega}:\omega\in\widehat{H}, d_\omega=1\}$.*

Corollary 2.21 *Suppose that $A:=G/N$ is Abelian, so that $G'\le N$.*

(1) *Let $\gamma\in\widehat{G}$ and suppose that $d_\gamma=1$. Set $\xi:=\mathrm{Res}_N^G\gamma$. Then the following holds:*

- *$\xi\in\widehat{N}$;*
- *γ is an extension of ξ;*
- *$I_G(\xi)=G$;*
- *$\widehat{G}(\xi)=\{\gamma\otimes\overset{\vartriangle}{\omega}:\omega\in\widehat{A}\}$;*
- *$\mathrm{Ind}_N^G\xi\sim\bigoplus_{\omega\in\widehat{A}}(\gamma\otimes\overset{\vartriangle}{\omega})$;*
- *$\ell_{\gamma\otimes\overset{\vartriangle}{\omega}}=1$ for all $\omega\in\widehat{A}$.*

(2) *There are exactly $[N:G']$ one-dimensional $\xi\in\widehat{N}$ admitting an extension to G and such that $I_G(\xi)=G$. In particular, there are at least $[N:G']$ one-dimensional fixed points for the action of G on $\widehat{N}$.*

Proof (1) It is obvious that $\xi\in\widehat{N}$ and that γ is an extension of ξ. All the other statements are particular cases of Theorem 2.18. (2) By the correspondence (1.24), there are $[G:G']$ such (one-dimensional) $\gamma\in\widehat{G}$. Moreover, each $\xi=\mathrm{Res}_N^G\gamma\in\widehat{N}$

is the restriction of $|A| = [G : N]$ such γ's. Therefore the number of these ξ's is $[G : G']/[G : N] = [N : G']$. ■

We now prove a partial converse to Gallagher's theorem (Theorem 2.18); it is taken from [111, 112].

Proposition 2.22 *Let $\theta \in \widehat{G}$ and suppose that the representations $\theta \otimes \overset{\triangle}{\psi}$, $\psi \in \widehat{H}$, are irreducible and pairwise non-equivalent. Then $\sigma := \mathrm{Res}^G_N \theta$ is irreducible and its inertia group coincides with the whole of G.*

Proof From Theorem 2.5.(3) it follows that, for $\psi \in \widehat{H}$,

$$\mathrm{Res}^G_N \left(\theta \otimes \overset{\triangle}{\psi}\right) \sim d_\psi \mathrm{Res}^G_N \theta \sim d_\psi \ell_\theta \bigoplus_{r \in \mathscr{R}(\sigma)} {}^r\sigma.$$

From Frobenius reciprocity we deduce that the multiplicity of $\theta \otimes \overset{\triangle}{\psi}$ in $\mathrm{Ind}^G_N \sigma$ is equal to the multiplicity of σ in $\mathrm{Res}^G_N \left(\theta \otimes \overset{\triangle}{\psi}\right)$, and therefore it is equal to $d_\psi \ell_\theta$.

Since the $\theta \otimes \overset{\triangle}{\psi}$'s are pairwise non-equivalent, $\mathrm{Ind}^G_N \sigma$ contains $|\widehat{H}|$ irreducible pairwise non-equivalent representations of dimension $d_\theta d_\psi$, each of them with multiplicity $d_\psi \ell_\theta$. Thus,

$$\begin{aligned} [G : N] d_\sigma &= \dim \mathrm{Ind}^G_N \sigma \\ &\geq \sum_{\psi \in \widehat{H}} d_\psi^2 \ell_\theta d_\theta \\ \text{(by (1.20))} \quad &= |H| \ell_\theta d_\theta \\ &= [G : N] \ell_\theta d_\theta \\ (d_\theta \geq d_\sigma \text{ and } \ell_\theta \geq 1) \quad &\geq [G : N] d_\sigma. \end{aligned}$$

This forces $\ell_\theta = 1$ and $d_\sigma = d_\theta$. We deduce that $\mathrm{Res}^G_N \theta \sim \sigma$ and $I_G(\sigma) = G$. ■

In the following proposition, we establish a particular criterion for the extendibility of a representation from N to G. It is taken from [65]. It is important in view of Theorem 2.18.

Proposition 2.23 *Let $\sigma \in \widehat{N}$. Suppose that G/N is cyclic and $I_G(\sigma) = G$. Then σ has an extension θ to G.*

Proof Let s be the order of G/N and let $g \in G$ such that gN generates G/N. Since ${}^g\sigma \sim \sigma$, there exists a unitary matrix $U \in \mathrm{GL}(V_\sigma)$ such that

$$ {}^g\sigma(n) = U^* \sigma(n) U \tag{2.31}$$

for all $n \in N$. As $g \mapsto {}^g\sigma$ is an action, we deduce that

$$ {}^{g^s}\sigma(n) = (U^s)^*\sigma(n)U^s, $$

while, as $g^s \in N$,

$$ {}^{g^s}\sigma(n) = \sigma(g^{-s}ng^s) = \sigma(g^{-s})\sigma(n)\sigma(g^s), $$

so that

$$ \sigma(n) = \sigma(g^s)(U^s)^*\sigma(n)U^s\sigma(g^s)^*. $$

By Schur's Lemma, $\sigma(g^s)(U^s)^*$ is a multiple of the identity: there exists $\alpha \in \mathbb{C}$, $|\alpha| = 1$, such that

$$ \sigma(g^s) = \alpha U^s. \tag{2.32} $$

Let β be an s-root of α and set $V := \beta U$. Then (2.32) becomes

$$ \sigma(g^s) = V^s \tag{2.33} $$

while (2.31) becomes

$$ {}^g\sigma(n) = V^*\sigma(n)V. \tag{2.34} $$

We are now in a position to define the required extension of σ. Set

$$ \theta(g^k n) := V^k\sigma(n) \tag{2.35} $$

for all $n \in N$ and $0 \le k \le s-1$ (observe that $G = \sqcup_{k=0}^{s-1} g^k N$) and let us verify that θ is indeed a G-representation. First of all, note that if $k \in \mathbb{Z}$, say, $k = r + \ell s$, with $0 \le r < s-1$ and $\ell \in \mathbb{Z}$, then

$$ \begin{aligned} \theta(g^k n) &= \theta(g^r g^{\ell s} n) \\ \text{(by (2.35) and } g^{\ell s} \in N) \quad &= V^r\sigma(g^{\ell s})\sigma(n) \\ \text{(by (2.33))} \quad &= V^k\sigma(n). \end{aligned} \tag{2.36} $$

Moreover, if $k, \ell \in \mathbb{Z}$ and $n_1, n_2 \in \mathbb{N}$, then, taking into account (1.57), we have

$$ \begin{aligned} \theta(g^k n_1 g^\ell n_2) &= \theta\left(g^{k+\ell} n_1^{(g^\ell)^\sharp} n_2\right) \\ \text{(by (2.2), (2.34) and (2.36))} \quad &= V^{k+\ell}V^{-\ell}\sigma(n_1)V^\ell\sigma(n_2) \\ \text{(by (2.36) again)} \quad &= \theta(g^k n_1)\theta(g^\ell n_2). \end{aligned} $$

This shows that θ is a representation. Finally, it is obvious that $\mathrm{Res}^G_N\theta = \sigma$. ■

In Remark 10.19 we shall give another proof of Proposition 2.23.

We now present a second simple criterion for extendibility, based on the following remark. It is again taken from [65].

Remark 2.24 We recall (cf. (1.24)) that there is a natural correspondence between the one-dimensional representations of G and the representations in $\widehat{G/G'}$. Moreover, it is easy to prove that if $B \leq A$ are Abelian and $\xi \in \widehat{B}$, then ξ has an extension to A: in fact, $\mathrm{Ind}^A_B\xi$ is made up of all $\gamma \in \widehat{A}$ such that $\mathrm{Res}^A_B\gamma = \xi$ (this may be proved directly, but it is also an immediate consequence of Frobenius reciprocity), and therefore any such γ yields an extension.

Proposition 2.25 *Let ξ be a one-dimensional representation of N. Then ξ has an extension to G if and only if*

$$G' \cap N \leq \ker \xi. \tag{2.37}$$

If this is the case, then $I_G(\xi) = G$.

Moreover, if $G = N \rtimes H$ is a semidirect product and $I_G(\xi) = G$, then (2.37) *is satisfied.*

Proof If ξ has an extension γ to G, then, recalling that $d_\gamma = d_\xi = 1$, we have $G' \leq \ker \gamma$ so that $G' \cap N \leq \ker \mathrm{Res}^G_N \gamma = \ker \xi$. Conversely, suppose that $G' \cap N \leq \ker \xi$. Then by setting

$$\widetilde{\xi}(nG') := \xi(n) \tag{2.38}$$

for all $n \in N$, we get a character of $G'N/G'$. Indeed, first note that $\widetilde{\xi}$ in (2.38) is well defined: if $n, m \in N$ and $nG' = mG'$ then $n^{-1}m \in G' \cap N \leq \ker \xi$, so that $\xi(n) = \xi(m)$. Moreover,

$$\widetilde{\xi}(nG'mG') = \widetilde{\xi}(nmG') = \xi(nm) = \xi(n)\xi(m) = \widetilde{\xi}(nG')\widetilde{\xi}(mG')$$

for all $n, m \in N$. Then, $\widetilde{\xi}$ has an extension $\widetilde{\widetilde{\xi}}$ from $G'N/G'$ to G/G' because these are Abelian groups (cf. Remark 2.24). Then the required extension of ξ to G is given by applying the bijection (1.24) to $\widetilde{\widetilde{\xi}}$.

Note also that (2.37) directly implies that $I_G(\xi) = G$. Indeed, if $g \in G, n \in N$ then $g^{-1}ngn^{-1} \in G' \cap N \leq \ker \xi$ so that ${}^g\xi(n) = \xi(n)$.

Suppose now that $G = N \rtimes H$ and $I_G(\xi) = G$. Set $K := \ker \xi \leq N$. Then, for all $g \in G$ and $n \in N$,

$$\xi(g^{-1}ng) = {}^g\xi(n) = \xi(n)$$

so that $n^{-1}g^{-1}ng \in K$ and therefore

$$ngK = gnK. \tag{2.39}$$

In other words, N/K is in the center of G/K. In particular, N/K is Abelian, that is, $N' \leq K$. If now $g_1 = n_1h_1, g_2 = n_2h_2 \in G$, then taking into account (2.39) we get

$$g_1^{-1}g_2^{-1}g_1g_2 = h_1^{-1}n_1^{-1}h_2^{-1}n_2^{-1}n_1h_1n_2h_2 \in h_1^{-1}h_2^{-1}h_1h_2K \subseteq H'K$$

since $n_1^{-1}n_2^{-1}n_1n_2 \in N' \leq K$. This implies that $G' \leq H'K$ and therefore

$$\begin{aligned} & G' \cap N \leq H'K \cap N \\ (HN = NH \text{ and } K \trianglelefteq N) \quad & = K(H' \cap N) \\ (G = N \rtimes H \Rightarrow H \cap N = \{1_G\} \Rightarrow H' \cap N = \{1_G\}) \quad & = K = \ker \xi. \end{aligned}$$

■

From Theorem 2.18 we can also deduce a very important result, namely the method of little group of Mackey and Wigner. It is a very useful way to get a complete list of irreducible representations for a wide class of groups.

Theorem 2.26 (The Little Group Method)

(1) *Let $\sigma \in \widehat{N}$ and suppose it has an extension $\widetilde{\sigma}$ to $I_G(\sigma)$. Then*

$$\widehat{G}(\sigma) = \left\{ \mathrm{Ind}_{I_G(\sigma)}^G (\widetilde{\sigma} \otimes \overset{\vartriangle}{\psi}) : \psi \in \widehat{H_G(\sigma)} \right\}, \tag{2.40}$$

where $\overset{\vartriangle}{\psi}$ denotes the inflation of ψ to $I_G(\sigma)$.

(2) *Suppose now that* every *$\sigma \in \widehat{N}$ has an extension $\widetilde{\sigma}$ to its inertia group $I_G(\sigma)$. Let $\Sigma \subseteq \widehat{N}$ be as in (2.5). Then*

$$\widehat{G} = \left\{ \mathrm{Ind}_{I_G(\sigma)}^G (\widetilde{\sigma} \otimes \overset{\vartriangle}{\psi}) : \sigma \in \Sigma, \psi \in \widehat{H_G(\sigma)} \right\}. \tag{2.41}$$

More precisely, the right hand side of (2.41) is a list of all irreducible G-representations and, for different values of σ and ψ, the corresponding representations are pairwise non-equivalent.

Proof

(1) From Theorem 2.18 applied with $I_G(\sigma)$ in place of G and $\widetilde{\sigma}$ in place of θ, we have

$$\mathrm{Ind}_N^{I_G(\sigma)} \sigma = \bigoplus_{\psi \in \widehat{H_G(\sigma)}} d_\psi (\widetilde{\sigma} \otimes \overset{\vartriangle}{\psi})$$

and $\widehat{I_G}(\sigma) = \{\widetilde{\sigma} \otimes \overset{\triangle}{\psi} : \psi \in \widehat{H_G(\sigma)}\}$. Then from the Clifford correspondence (Theorem 2.12) we deduce (2.40).

(2) This follows from (2.18) and (2.40).

■

In Sect. 7.1 we will investigate a cohomological condition for the applicability of the Little Group Method; see Remark 7.5.

The following proposition is the normal subgroup version of *Mackey's irreducibility criterion*; see ([30, Corollary 11.4.6]).

Proposition 2.27 *Let* $\sigma, \sigma_1 \in \widehat{N}$. *Then the induced representation* $\mathrm{Ind}_N^G\sigma$ *is irreducible if and only if* $I_G(\sigma) = N$. *Moreover, if* $I_G(\sigma) = I_G(\sigma_1) = N$, *then* $\mathrm{Ind}_N^G(\sigma) \sim \mathrm{Ind}_N^G(\sigma_1)$ *if and only if* σ *and* σ_1 *are* conjugate *(i.e., there exists* $g \in G$ *such that* $\sigma_1 = {}^g\sigma$*).*

Proof The first statement follows from Theorem 2.5.(2) combined with Schur's lemma. Moreover, if $\theta := \mathrm{Ind}_N^G\sigma$ is irreducible then Theorem 2.5.(1) yields $\mathrm{Res}_N^G\theta \sim \oplus_{h\in H}\, {}^h\sigma$ and therefore, by Frobenius' reciprocity, $\mathrm{Ind}_N^G\sigma_1 \sim \theta$ if and only if $\sigma_1 \sim {}^h\sigma$ for some $h \in H$. ■

Proposition 2.28 *Let* $\sigma \in \widehat{N}$ *and suppose that* $I_G(\sigma) = N$. *Then*

(1) $\theta := \mathrm{Ind}_N^G\sigma$ *is the unique element in* $\widehat{G}(\sigma)$.
(2) $\theta \otimes \overset{\triangle}{\omega} \sim \theta$, *for all one-dimensional* $\omega \in \widehat{H}$.
(3) $\mathrm{Res}_N^G\theta \sim \oplus_{h\in H}\, {}^h\sigma$, *with the* ${}^h\sigma$*'s pairwise non-equivalent.*
(4) *The image of* $\{{}^h\sigma : h \in H\}$ *via the map* (2.19) *is the singleton* $\{\mathrm{Ind}_N^G\sigma\}$.

Proof

(1) This follows from Proposition 2.27 and Definition 2.1.
(2) Now $\theta \otimes \overset{\triangle}{\omega} \in \widehat{G}$ and $\overset{\triangle}{\omega}(n) = 1$ for all $n \in N$, so that $\mathrm{Res}_N^G(\theta \otimes \overset{\triangle}{\omega}) \equiv \mathrm{Res}_N^G\theta$. Therefore $\theta \otimes \overset{\triangle}{\omega} \in \widehat{G}(\sigma)$ and, by (1), it must be equivalent to θ.
(3) This is just Theorem 2.5.(1), with $c = 1$ and $\mathscr{R} \equiv H$.
(4) This follows immediately from (1).

■

Corollary 2.29 *Let* $\sigma \in \widehat{N}$ *and suppose that* $I_G(\sigma) \trianglelefteq G$. *Then* $I_G(\psi) = I_G(\sigma)$ *for all* $\psi \in \widehat{I_G}(\sigma)$.

Proof Let $\psi \in \widehat{I_G}(\sigma)$. Then, by Clifford correspondence (Theorem 2.12), $\mathrm{Ind}_{I_G(\sigma)}^G\psi$ is irreducible. Thus, by Proposition 2.27 we have that $I_G(\psi) = I_G(\sigma)$. ■

2.4 The Case Where $A_G(\sigma) := I_G(\sigma)/N$ is Abelian

We first present some elementary facts on the duality of finite Abelian groups.

- Let A be a finite Abelian group, let $\Omega \leq \widehat{A}$ be a subgroup of its dual, and set $C := \bigcap_{\omega \in \Omega} \ker \omega$. Then the inflation

$$\begin{aligned} \widehat{A/C} &\to \widehat{A} \\ \epsilon &\mapsto \overset{\triangle}{\epsilon} \end{aligned} \tag{2.42}$$

establishes an isomorphism between $\widehat{A/C}$ and Ω. In particular,

$$\Omega = \left\{ \overset{\triangle}{\epsilon} : \epsilon \in \widehat{A/C} \right\}. \tag{2.43}$$

To see this, first observe that the image of (2.42) equals $\{\xi \in \widehat{A} : \ker \xi \geq C\}$. Indeed, if $\xi \in \widehat{A}$ satisfies that $\ker \xi \geq C$, then setting $\epsilon(aC) := \xi(a)$, for all $a \in A$, yields an element $\epsilon \in \widehat{A/C}$ such that $\overset{\triangle}{\epsilon} = \xi$. Conversely, if $\epsilon \in \widehat{A/C}$, then $\xi := \overset{\triangle}{\epsilon} \in \widehat{A}$ satisfies that $\xi(c) = \epsilon(C) = \epsilon(1_{A/C}) = 1$, for all $c \in C$, that is, $\ker \xi \geq C$. Now, if $\xi \in \Omega$, then $\ker \xi \supseteq C$, so that $\Omega \subseteq \{\xi \in \widehat{A} : \ker \xi \geq C\}$. We claim that indeed an equality holds. We first observe that the map (2.42) is injective. On the other hand, the map

$$\begin{aligned} A &\to \widehat{\Omega} \\ a &\mapsto \psi_a, \end{aligned}$$

where $\psi_a(\omega) := \omega(a)$ for all $\omega \in \Omega$ and $a \in A$, is a surjective group homomorphism and its kernel is precisely C: for, $\psi_a = 1$ if and only if $\omega(a) = 1$ for all $\omega \in \Omega$, that is, $a \in \bigcap_{\omega \in \Omega} \ker \omega = C$ (see also [86, pp. 48–49]). We deduce that $|\widehat{A/C}| = |A/C| = |\widehat{\Omega}| = |\Omega|$ and the claim follows. This completes the argument showing that (2.42) establishes the required isomorphism.
- Let A be a finite Abelian group, let $C \leq A$ be a subgroup, and set $\Omega := \{\psi \in \widehat{A} : C \leq \ker \psi\}$. Then

$$C = \bigcap_{\psi \in \Omega} \ker \psi. \tag{2.44}$$

Indeed, on the one hand it is clear that $C \subseteq \bigcap_{\psi \in \Omega} \ker \psi$, and on the other hand, given $a \in A \setminus C$ we can find $\omega \in \widehat{A/C}$ with $\omega(aC) \neq 1$. Then $\psi := \overset{\triangle}{\omega} \in \Omega$ satisfies that $\psi(a) = \omega(aC) \neq 1$, that is, $a \notin \bigcap_{\psi \in \Omega} \ker \psi$.

Moreover, the map

$$\begin{aligned} \widehat{A} &\to \quad \widehat{C} \\ \psi &\mapsto \mathrm{Res}^A_C \psi, \end{aligned} \tag{2.45}$$

establishes a group isomorphism $\widehat{C} \cong \widehat{A}/\Omega$. Indeed, one the one hand, the kernel of this map is Ω, and, on the other hand, the map is surjective. Indeed, if $\psi \in \widehat{A}$ is contained in $\mathrm{Ind}^A_C \chi$ for some $\chi \in \widehat{C}$, then, by Frobenius' reciprocity, $\mathrm{Res}^A_C \psi = \chi$.

Note that from (2.44) and (2.43) we also have

$$C = \bigcap_{\omega \in \widehat{A/C}} \ker \overset{\vartriangle}{\omega} . \tag{2.46}$$

In the following, we suppose that $A_G(\sigma) := I_G(\sigma)/N$ is Abelian. Then its dual $\widehat{A_G(\sigma)}$ is a group and it acts on $\widehat{I_G}(\sigma)$: this action is given by the map $(\omega, \eta) \mapsto \eta \otimes \overset{\vartriangle}{\omega}$, for all $\omega \in \widehat{A_G(\sigma)}$ and $\eta \in \widehat{I_G}(\sigma)$; here $\overset{\vartriangle}{\omega}$ is the inflation of ω to $I_G(\sigma)$.

Theorem 2.30 (Roth [111]) *Let $\sigma \in \widehat{N}$ and suppose that $A_G(\sigma) = I_G(\sigma)/N$ is Abelian.*

(1) *Then the action of $\widehat{A_G(\sigma)}$ on $\widehat{I_G}(\sigma)$ is transitive. Moreover, denoting by*

$$\Omega_\eta := \mathrm{Stab}_{\widehat{A_G(\sigma)}}(\eta) = \{\omega \in \widehat{A_G}(\sigma) : \eta \otimes \overset{\vartriangle}{\omega} \sim \eta\}$$

the stabilizer of $\eta \in \widehat{I_G}(\sigma)$, we have

$$\ell_\eta^2 = |\Omega_\eta| = \frac{|A_G(\sigma)|}{|\widehat{I_G}(\sigma)|}. \tag{2.47}$$

In particular, the inertia index is the same for all $\eta \in \widehat{I_G}(\sigma)$.

(2) *Let $\eta \in \widehat{I_G}(\sigma)$ and set*

$$M_\eta := \bigcap_{\omega \in \Omega_\eta} \ker \overset{\vartriangle}{\omega} \equiv \{h \in I_G(\sigma) : \overset{\vartriangle}{\omega}(h) = 1, \forall \omega \in \Omega_\eta\}.$$

Then $N \leq M_\eta$ and the inflation to $A_G(\sigma)$ establishes an isomorphism

$$\widehat{I_G(\sigma)/M_\eta} \cong \Omega_\eta. \tag{2.48}$$

(3) *Let $\eta \in \widehat{I_G}(\sigma)$. Then there exists $\tau \in \widehat{M_\eta}$ such that: $\mathrm{Res}^{M_\eta}_N \tau \sim \sigma$ (i.e. τ is an extension of σ to M_η), $\mathrm{Res}^{I_G(\sigma)}_{M_\eta} \eta \sim \ell_\eta \tau$, $\mathrm{Ind}^{I_G(\sigma)}_{M_\eta} \tau \sim \ell_\eta \eta$, and $I_{I_G(\sigma)}(\tau) = I_G(\sigma)$. In particular, ℓ_η is also the inertia index of τ (with respect to N).*

Proof

(1) Let $\lambda = \lambda_{A_G(\sigma)}$ be the regular representation of $A_G(\sigma)$ and denote by $\overset{\Delta}{\lambda}$ its inflation to $I_G(\sigma)$. Then $\overset{\Delta}{\lambda} = \oplus_{\omega \in \widehat{A_G(\sigma)}} \overset{\Delta}{\omega}$ and (2.20), with G replaced by $I_G(\sigma)$, becomes:

$$\bigoplus_{\omega \in \widehat{A_G(\sigma)}} (\eta \otimes \overset{\Delta}{\omega}) = \eta \otimes \overset{\Delta}{\lambda} \sim \ell_\eta \bigoplus_{\rho \in \widehat{I_G}(\sigma)} \ell_\rho \rho,$$

for each $\eta \in \widehat{I_G}(\sigma)$. Comparing the left hand side and the right hand side, we deduce that the action is transitive: each $\rho \in \widehat{I_G}(\sigma)$ is equivalent to $\eta \otimes \overset{\Delta}{\omega}$ for exactly $\ell_\eta \ell_\rho$ characters ω. In particular, the cardinality of the stabilizer of η is equal to ℓ_η^2. Since the former equals $\frac{|\widehat{A_G(\sigma)}|}{|\widehat{I_G}(\sigma)|} \equiv \frac{|A_G(\sigma)|}{|\widehat{I_G}(\sigma)|}$, formula (2.47) follows.

(2) First of all we have $N \leq M_\eta$, by definition of $\overset{\Delta}{\omega}$. The remaining is an immediate consequence of the isomorphism in (2.42) (see also (2.43)), taking into account that $\bigcap_{\omega \in \Omega_\eta} \ker \omega = M_\eta / N$ and

$$\frac{I_G(\sigma)/N}{M_\eta/N} \cong \frac{I_G(\sigma)}{M_\eta}.$$

(3) Suppose that

$$\mathrm{Res}^{I_G(\sigma)}_{M_\eta} \eta \sim \bigoplus_{j=1}^{t} m_j \tau_j \tag{2.49}$$

where $\tau_1, \tau_2, \ldots, \tau_t$ are pairwise inequivalent irreducible representations of M_η (and the m_j's are their multiplicities). From Remark 2.8 and transitivity of restriction, it follows that

$$\mathrm{Res}^{M_\eta}_N \tau_j \sim s_j \sigma, \quad j = 1, 2, \ldots, t \tag{2.50}$$

for some integers s_j (note that $s_j \geq 1$ since $\eta \in \widehat{I_G}(\sigma)$). Moreover,

$$\ell_\eta \sigma \sim \mathrm{Res}^{I_G(\sigma)}_N \eta \sim \left(\sum_{j=1}^{t} m_j s_j \right) \sigma, \tag{2.51}$$

where the last equivalence follows from (2.49) and (2.50).

Now, on the one hand

$$\mathrm{Ind}_{M_\eta}^{I_G(\sigma)}\mathrm{Res}_{M_\eta}^{I_G(\sigma)}\eta \sim \eta \otimes \mathrm{Ind}_{M_\eta}^{I_G(\sigma)}\iota_{M_\eta} \sim \eta \otimes \left(\bigoplus_{\omega \in \widehat{I_G(\sigma)/M_\eta}} \overset{\Delta}{\omega} \right) \sim \ell_\eta^2 \eta \tag{2.52}$$

where the first equivalence follows from (1.36), the second one from the fact that the permutation representation (with respect to a normal subgroup) is equivalent to the inflation of the regular representation of the quotient group, and the last one follows from (1) and (2.48). On the other hand, from (2.49), it follows that

$$\mathrm{Ind}_{M_\eta}^{I_G(\sigma)}\mathrm{Res}_{M_\eta}^{I_G(\sigma)}\eta \sim \bigoplus_{j=1}^{t} m_j \mathrm{Ind}_{M_\eta}^{I_G(\sigma)}\tau_j. \tag{2.53}$$

By Frobenius reciprocity, the multiplicity of η in $\mathrm{Ind}_{M_\eta}^{I_G(\sigma)}\tau_j$ is equal to m_j for all $j = 1, 2 \dots, t$. Then, from (2.52) and (2.53) we deduce that the multiplicity of η in $\mathrm{Ind}_{M_\eta}^{I_G(\sigma)}\mathrm{Res}_{M_\eta}^{I_G(\sigma)}\eta$ is equal to $\sum_{j=1}^{t} m_j^2 = \ell_\eta^2$.

Let us show that $t = 1$ in (2.49). Indeed, we have

$$\begin{aligned}
\sum_{j=1}^{t} m_j^2 &= \ell_\eta^2 \\
\text{(by (2.51))} &= \left(\sum_{j=1}^{t} m_j s_j \right)^2 \\
&\geq \sum_{j=1}^{t} m_j^2 s_j^2 \\
(s_j \geq 1) &\geq \sum_{j=1}^{t} m_j^2.
\end{aligned}$$

Since $s_j \geq 1$ for all j, it follows that necessarily $t = 1$, $s_1 = 1$, and therefore $m_1 = \ell_\eta$. Then by taking $\tau = \tau_1$, (2.49) becomes

$$\mathrm{Res}_{M_\eta}^{I_G(\sigma)}\eta \sim \ell_\eta \tau, \tag{2.54}$$

while (2.50) becomes $\mathrm{Res}_N^{M_\eta}\tau \sim \sigma$. Moreover, (2.54) implies that $d_\eta = \ell_\eta d_\tau$ so that

$$\ell_\eta d_\eta = \ell_\eta^2 d_\tau = \left| \frac{I_G(\sigma)}{M_\eta} \right| d_\tau = \dim \mathrm{Ind}_{M_\eta}^{I_G(\sigma)}\tau$$

(where the second equality follows from (2.47) and (2.48)). Since, by Frobenius reciprocity, the multiplicity of η in $\mathrm{Ind}_{M_\eta}^{I_G(\sigma)}\tau$ is equal to the multiplicity of τ in $\mathrm{Res}_{M_\eta}^{I_G(\sigma)}\eta$, which in turn is equal to ℓ_η by (2.54), we conclude that

$$\mathrm{Ind}_{M_\eta}^{I_G(\sigma)}\tau \sim \ell_\eta \eta.$$

Finally, from (2.54) and Theorem 2.5.(3) we deduce that $I_{I_G(\sigma)}(\tau) \equiv I_G(\sigma)$. ■

Corollary 2.31 *Suppose that $A := G/N$ is Abelian and that $I_G(\sigma) = G$ for all $\sigma \in \widehat{N}$ (cf. Remark 2.16.(2)). Then $\ell = \ell_\theta = \sqrt{\frac{|A|}{|\widehat{G}(\sigma)|}}$ for all $\theta \in \widehat{G}(\sigma)$ and $\mathrm{Ind}_N^G\sigma \sim \ell \bigoplus_{\theta\in\widehat{G}(\sigma)}\theta$. Moreover, (cf. (2.28))*

$$\lambda_G \sim \bigoplus_{\sigma\in\widehat{N}} d_\sigma \sqrt{\frac{|A|}{|\widehat{G}(\sigma)|}} \bigoplus_{\theta\in\widehat{G}(\sigma)} \theta.$$

We finally examine the multiplicity free case.

Proposition 2.32 *Let $\sigma \in \widehat{N}$. Then the following conditions are equivalent:*

(a) *$\ell_\theta = 1$ for all $\theta \in \widehat{G}(\sigma)$;*
(b) *$H_G(\sigma) = I_G(\sigma)/N$ is Abelian and σ has an extension to $I_G(\sigma)$.*

Moreover, if this is the case, then $\mathrm{Res}_N^{I_G(\sigma)}\eta \sim \sigma$ for all $\eta \in \widehat{I_G}(\sigma)$.

Proof Assume (a). From (2.25) in Clifford correspondence, we deduce that $\ell_\eta = 1$ for all $\eta \in \widehat{I_G}(\sigma)$. Therefore, by Remark 2.8, $\mathrm{Res}_N^{I_G(\sigma)}\eta \sim \sigma$, showing that every $\eta \in \widehat{I_G}(\sigma)$ is an extension of σ to $I_G(\sigma)$. Then we may apply Theorem 2.18 with G replaced by $I_G(\sigma)$, and get $d_\psi = 1$ for all $\psi \in \widehat{H_G(\sigma)}$. This forces $H_G(\sigma)$ being Abelian.

Conversely, assume (b). Again from Theorem 2.18 with G replaced by $I_G(\sigma)$ (and H replaced by $H_G(\sigma)$), we deduce that $\ell_\eta = 1$ for all $\eta \in \widehat{I_G}(\sigma)$. Then (a) follows from Theorem 2.12 (Clifford correspondence).

Finally, in either case for all $\eta \in \widehat{I_G}(\sigma)$ we have $\ell_\eta = 1$ and Theorem 2.18 ensures that $\mathrm{Res}_N^{I_G(\sigma)}\eta \sim \sigma$. ■

Definition 2.33 An action is *regular* if it is transitive and all stabilizers are trivial (so that, in particular, there are no fixed points). Also, given a (not necessarily transitive) action, an orbit Ω is *regular* if the restriction of the action to Ω is regular.

With the notation and hypotheses as in Theorem 2.30, we deduce the following:

Corollary 2.34 *The action of $\widehat{A_G(\sigma)}$ on $\widehat{I_G}(\sigma)$ is regular if and only if σ has an extension to $I_G(\sigma)$.*

Corollary 2.35 *Suppose that $I_G(\sigma) = G$ and that $\ell_\theta = 1$ for all $\theta \in \widehat{G}(\sigma)$. Then $A := G/N$ is Abelian and for each $\theta \in \widehat{G}(\sigma)$ we have*

$$\mathrm{Ind}_N^G \sigma \sim \bigoplus_{\omega \in \widehat{A}} \left(\theta \otimes \overset{\Delta}{\omega}\right) \text{ and } \mathrm{Res}_N^G \theta \sim \sigma,$$

where the $\theta \otimes \overset{\Delta}{\omega}$'s are pairwise inequivalent.

In particular, in the present setting correspondence (2.19) *becomes:*

$$\{\sigma\} \mapsto \widehat{G}(\sigma) = \{\theta \otimes \overset{\Delta}{\omega} : \omega \in \widehat{A}\}. \tag{2.55}$$

Proof This is just Proposition 2.32 in the case $I_G(\sigma) = G$, taking also into account (2.29) (or Proposition 2.30). ■

We end this section with a useful but quite technical result of Isaacs. First, let us introduce the following definition.

Definition 2.36 Let G be a group and let $N \leq M$ be normal subgroups of G. Suppose that no $L \trianglelefteq G$ exists such that $N < L < M$. One then says that M/N is a *chief factor* for G.

Theorem 2.37 (Isaacs' Going Down Theorem) *Let G be a group and let $N \leq M$ be normal subgroups of G such that M/N is an Abelian chief factor for G. Suppose that*

$$\psi \in \widehat{M} \text{ and } I_G(\psi) = G;$$

$$\sigma \in \widehat{N} \text{ and } \sigma \preceq \mathrm{Res}_N^M \psi.$$

Then one (and only one) of the following possibilities occurs:

(1) $I_M(\sigma) = M$, $\mathrm{Res}_N^M \psi$ *is irreducible, and* $\mathrm{Res}_N^M \psi \sim \sigma$.

(2) $I_M(\sigma) = M$, $\mathrm{Res}_N^M \psi \sim \underbrace{\sigma \oplus \sigma \oplus \cdots \oplus \sigma}_{\ell \text{ times}}$, *and* $\ell^2 = [M : N]$. *Moreover,* $\widehat{I_M}(\sigma) = \{\psi\}$ *and* $\mathrm{Ind}_N^M \sigma \sim \underbrace{\psi \oplus \psi \oplus \cdots \oplus \psi}_{\ell \text{ times}}$.

(3) $I_M(\sigma) = N$ *and* $\mathrm{Res}_N^M \psi \sim \bigoplus_{i=1}^s \sigma_j$, *where* $\sigma_1, \sigma_2, \ldots, \sigma_s \in \widehat{N}$ *are the (pairwise inequivalent) G-conjugates of σ, with $\sigma_1 \sim \sigma$, and $s = [M : N]$.*

Proof From $I_G(\psi) = G$ it follows that ${}^g\sigma \preceq \mathrm{Res}^M_N \psi$ for all $g \in G$. Then from Theorem 2.5.(3) with M in place of G, we have

$$ {}^g\sigma \preceq \mathrm{Res}^M_N \psi \sim \ell \bigoplus_{r \in \mathscr{R}} {}^r\sigma, $$

where $\mathscr{R} = M/I_M(\sigma)$. We deduce that the set of G-conjugates of σ coincides with the set of its M-conjugates, and therefore their cardinalities are equal, that is,

$$ [G : I_G(\sigma)] = [M : I_M(\sigma)] = [M : M \cap I_G(\sigma)], \tag{2.56} $$

where the latter equality follows from $I_M(\sigma) = M \cap I_G(\sigma)$. From elementary group theory, it follows that

$$ |M I_G(\sigma)| = \frac{|M| \cdot |I_G(\sigma)|}{|M \cap I_G(\sigma)|} = |G| $$

which implies that

$$ G = M I_G(\sigma). \tag{2.57} $$

Since $N \leq I_M(\sigma) \leq M$ and M/N is Abelian we have that $I_M(\sigma) \trianglelefteq M$. Moreover, $M \trianglelefteq G$ implies that $I_M(\sigma) = M \cap I_G(\sigma) \trianglelefteq I_G(\sigma)$. Then, the normalizer of $I_M(\sigma)$ in G contains both M and $I_G(\sigma)$, so that by (2.57), $I_M(\sigma) \trianglelefteq G$. By the assumption that M/N is an Abelian chief factor for G, this implies that either $I_M(\sigma) = N$ or $I_M(\sigma) = M$.

Assume first that $I_M(\sigma) = N$. Then, Proposition 2.28.(3) (with G replaced by M and H by M/N) yields condition (3) in the statement (recall that the M-conjugates and the G-conjugates of σ coincide).

Assume now

$$ I_M(\sigma) = M. \tag{2.58} $$

Using Remark 2.16, with G replaced by M and θ by ψ, we get

$$ \mathrm{Res}^M_N \psi \sim \underbrace{\sigma \oplus \sigma \oplus \cdots \oplus \sigma}_{\ell \text{ times}}. $$

If the action of $\widehat{M/N}$ on $\widehat{I_M}(\sigma)$ is regular, then by Corollary 2.34 and (2.47) we have $\ell = 1$ and ψ is an extension of σ; this yields condition (1) in the statement.

Suppose now that the action of $\widehat{M/N}$ on $\widehat{I_M}(\sigma)$ is not regular and set

$$ \Omega_\psi := \{\omega \in \widehat{M/N} : \psi \otimes \overset{\vartriangle}{\omega} \sim \psi\} \text{ and } M_\psi := \bigcap_{\omega \in \Omega_\psi} \ker \overset{\vartriangle}{\omega} $$

(cf. Theorem 2.30.(2); recall that $\overset{\vartriangle}{\omega}$ denotes the inflation to M). Clearly,

$$N \leq M_\psi \lneqq M, \tag{2.59}$$

where the second inclusion is proper because Ω_ψ contains a nontrivial $\omega \in \widehat{M/N}$. Since, by hypothesis, $I_G(\psi) = G$, for $\omega \in \Omega_\psi$ and $g \in G$ we have that

$$\psi \otimes {}^g\overset{\vartriangle}{\omega} \sim {}^g\psi \otimes {}^g\overset{\vartriangle}{\omega} \sim {}^g\left(\psi \otimes \overset{\vartriangle}{\omega}\right) \sim {}^g\psi \sim \psi,$$

so that Ω_ψ is invariant under conjugation by G. Then M_ψ is normal in G and, by our hypotheses, (2.59) forces

$$M_\psi = N. \tag{2.60}$$

From Theorem 2.30 it follows that

$$\frac{[M : N]}{|\widehat{I_M}(\sigma)|} = \ell^2 = |\Omega_\psi| = \frac{|I_M(\sigma)|}{|M_\psi|} = \frac{|M|}{|N|},$$

where the last equality follows from (2.58) and (2.60). This yields $|\widehat{I_M}(\sigma)| = 1$ and condition (2) in the statement follows. ■

Definition 2.38 Let $N \trianglelefteq G$ and suppose that $\sigma \in \widehat{N}$ and $\theta \in \widehat{G}$. We say that a group M satisfying $N \leq M \leq G$ is an *intermediary subgroup* for σ and θ if there exists $\tau \in \widehat{M}$ (the *intermediary representation*) such that $\mathrm{Res}^M_N \tau \sim \sigma$ and $\mathrm{Ind}^G_M \tau \sim \theta$.

Observe that a necessary condition for the existence of an intermediary subgroup (for $\sigma \in \widehat{N}$ and $\theta \in \widehat{G}$) is that $\theta \in \widehat{G}(\sigma)$.

Definition 2.39 Let $N \trianglelefteq G$ and suppose that for all $\sigma \in \widehat{N}$ and $\theta \in \widehat{G}(\sigma)$ there exists an intermediary subgroup. We then say that G is a *relative M-group* with respect to N.

Theorem 2.40 *Let G be a finite group. Let $N \trianglelefteq G$ and suppose that every chief factor for G/N is Abelian of nonsquare order. Then G is a relative M-group with respect to N.*

Proof The proof is by induction on $[G : N]$. Let $\sigma \in \widehat{N}$ and $\theta \in \widehat{G}(\sigma)$.

If $[G : N] = 1$, that is, $G = N$, then $\sigma = \theta$ and we can take $M = G = N$ (and $\tau = \theta = \sigma$). Note that there are no chief factors for G/N, in this case.

Suppose now that for an integer $k \geq 1$ we have proved the assertion for all groups $\widetilde{G}$ with $N \leq \widetilde{G} \leq G$ with $[\widetilde{G} : N] \leq k$, and such that every chief factor for $\widetilde{G}/N$

is Abelian of nonsquare order. If $\mathrm{Res}^G_N\theta$ is irreducible, then $\sigma = \mathrm{Res}^G_N\theta$ and we can take $M = G$ (and $\tau = \theta$).

Assume now that $\mathrm{Res}^G_N\theta$ is reducible. Choose H with $N \lneq H \trianglelefteq G$ and minimal with respect to the condition that $\mathrm{Res}^G_H\theta$ is irreducible. Let also L with $N \leq L \trianglelefteq G$ and maximal with respect to the condition $L \lneq H$, so that $H/L \cong (H/N)/(L/N)$ is a chief factor for G/N. Set $\psi := \mathrm{Res}^G_H\theta \in \widehat{H}$ and observe that, by Theorem 2.5.(3), $I_G(\psi) = G$. Thus the hypotheses of Isaacs's theorem (Theorem 2.37) are satisfied with N replaced by L and M by H. Now, on the one hand $\mathrm{Res}^H_L\psi(= \mathrm{Res}^G_L\theta)$ is not irreducible, by minimality of H, and this excludes possibility (1). On the other hand, $[H : L]$ is not a square, and this excludes possibility (2). We thus are in case (3), so that there exist $\sigma_1, \sigma_2, \ldots, \sigma_s \in \widehat{L}$, pairwise non-equivalent, such that $\mathrm{Res}^H_L\psi \sim \bigoplus_{i=1}^s \sigma_i$ and $s = [H : L]$. By Frobenius reciprocity, $\theta \in \widehat{G}(\sigma_1)$ so that, by Clifford correspondence (Theorem 2.12), there exists $\eta \in \widehat{I_G}(\sigma_1)$ such that $\mathrm{Ind}^G_{I_G(\sigma_1)}\eta \sim \theta$. Note that $\widetilde{G} := I_G(\sigma_1) \lneq G$ by Theorem 2.5.(3). Thus, since $[\widetilde{G} : N] < [G : N]$, we may apply induction deriving the existence of a subgroup M with $N \leq M \leq \widetilde{G}$ and of a representation $\tau \in \widehat{M}$ such that $\mathrm{Res}^M_N\tau \sim \sigma$ and $\mathrm{Ind}^{\widetilde{G}}_M\tau \sim \eta$. By transitivity of induction, $\mathrm{Ind}^G_M\tau \sim \mathrm{Ind}^G_{\widetilde{G}}\eta \sim \theta$. ■

Remark 2.41 We now indicate an important case where the condition that "every chief factor is Abelian" in Theorem 2.40 is satisfied.

Recall that a group H is *solvable* if there exists a finite chain of subgroups $\{1_H\} = L_0 \leq L_1 \leq L_{k-1} \leq L_k = H$ such that, L_{j-1} is normal in L_j and L_j/L_{j-1} is Abelian, for all $j = 1, 2, \ldots, k$. Also, it is easy to see that if H is solvable, so are all of its subgroups and quotient groups (see [3, 90, 108, 114]). Now, if M is a minimal nontrivial normal subgroup of a solvable group H, then M is Abelian: indeed, its derived subgroup M' is strictly contained in M (because M is solvable), it is normal in H (because it is characteristic in M), and therefore $M' = \{1_H\}$. It follows that every chief factor for a solvable group H is Abelian (cf. Theorem 2.37): if $N \leq M$ are normal in H and no $L \trianglelefteq H$ exists such that $N < L < M$, then M/N is a minimal normal subgroup of the solvable group G/N.

When G is Abelian, the whole condition "every chief factor of G/N is Abelian of nonsquare order" is always satisfied. Indeed, every quotient of an Abelian group is Abelian and, as it follows for instance from the Cauchy Theorem for Abelian groups (cf. e.g., [30, Corollary 1.3.3]), chief factoriality forces G/N having prime order.

More generally, the above is true for all supersolvable groups (and hence for nilpotent groups); see [67].

2.5 Some Applications of Mackey Theory to Clifford Theory

In this section we treat some aspects of the Mackey theory of induced representations which are strictly connected with Clifford theory. Recall that, given a (finite) group G and two subgroups $H, K \leq G$, Mackey's theory investigates relations between induction (resp. restriction) to G (resp. to K and H) of irreducible representations of H and K (resp. of these inductions). For an exposition of the general Mackey theory we refer to [30, Chapter 11]. In this section we focus our treatment in the particular case when one of the two subgroups involved is normal.

Let $K, N \leq G$, with N normal in G, and assume the notation in Sect. 2.2 (note that K is *not* necessarily normal in G). Fix a complete system $\mathscr{S}$ of representatives for the double N-K cosets in G, that is,

$$G = \sqcup_{s\in\mathscr{S}} NsK \equiv \sqcup_{s\in\mathscr{S}} sNK = \sqcup_{s\in\mathscr{S}^{-1}} KsN \equiv \sqcup_{s\in\mathscr{S}^{-1}} KNs, \tag{2.61}$$

where $\mathscr{S}^{-1} := \{s^{-1} : s \in \mathscr{S}\}$; we assume that $1_G \in \mathscr{S}$.

Let $\gamma \in \widehat{K}$ and $s \in \mathscr{S}$. We set $K_s := N \cap sKs^{-1} \equiv s(N \cap K)s^{-1}$ and define the K_s-representation γ_s by setting $\gamma_s(x) := \gamma(s^{-1}xs)$ for all $x \in K_s$, that is, $\gamma_s \equiv {}^{s}(\mathrm{Res}^{K}_{K_s}\gamma)$. Analogously, recalling that $\mathscr{S}^{-1}$ is a complete system of representatives for the double K-N cosets in G (cf.(2.61)), given $\sigma \in \widehat{N}$ and $s \in \mathscr{S}^{-1}$, and observing that $N_s := K \cap sNs^{-1} \equiv K \cap N$ is independent of s, we define the $N \cap K$-representation σ_s by setting $\sigma_s(y) := \sigma(s^{-1}ys)$ for all $y \in N \cap K$, that is, $\sigma_s \equiv \mathrm{Res}^{N}_{N\cap K}{}^{s}\sigma$.

In the present setting, the classical *Mackey lemma* (cf. [30, Theorem 11.5.1]) has the following two formulations:

$$\mathrm{Res}^{G}_{K}\mathrm{Ind}^{G}_{N}\sigma \sim \bigoplus_{s\in\mathscr{S}^{-1}} \mathrm{Ind}^{K}_{N\cap K}\sigma_s, \tag{2.62}$$

and

$$\mathrm{Res}^{G}_{N}\mathrm{Ind}^{G}_{K}\gamma \sim \bigoplus_{s\in\mathscr{S}} \mathrm{Ind}^{N}_{K_s}\gamma_s. \tag{2.63}$$

Note also that $N \cap K \trianglelefteq K$, so that in (2.62) both inductions are from normal subgroups, while in (2.63) the restriction is to a normal subgroup.

As a first application, we prove an interesting result on the existence of intermediary subgroups, yielding a solution to [67, Exercise 6.11].

Proposition 2.42 (Isaacs) *Let G be a finite group and let $K, B \leq G$ be two subgroups with B Abelian and normal in G. Let $\gamma \in \widehat{K}$ with* $\dim\gamma = 1$, *and suppose that $\theta := \mathrm{Ind}^{G}_{K}\gamma$ is irreducible. Then there exist $\psi \in \widehat{B}$ and $\eta \in \widehat{I_{BK}(\psi)}$ such that*

$$\mathrm{Res}^{I_{BK}(\psi)}_{B}\eta = \psi \quad \textit{and} \quad \mathrm{Ind}^{G}_{I_{BK}(\psi)} \sim \theta.$$

In other words, $I_{BK}(\psi)$ is an intermediary subgroup for θ and ψ, with intermediary representation η (cf. Definition 2.38).

Proof First of all, note that BK is a subgroup of G and that, by transitivity of induction, also $\mathrm{Ind}_K^{BK}\gamma$ is irreducible. Choose $\psi \in \widehat{B}$ which is contained in $\mathrm{Res}_B^{BK}\mathrm{Ind}_K^{BK}\gamma$.

From (2.63) with BK in place of G (so that $\mathscr{S}$ reduces to $\{1_G\}$ and therefore $K_s = K_{1_G} = B \cap K$ and $\gamma_s = \gamma_{1_G} = \mathrm{Res}_{N\cap K}^K\gamma$) we deduce that

$$\mathrm{Res}_B^{BK}\mathrm{Ind}_K^{BK}\gamma \sim \mathrm{Ind}_{B\cap K}^B\mathrm{Res}_{B\cap K}^K\gamma,$$

and the right hand side decomposes without multiplicity, because it is the induction to the Abelian group B of the character $\mathrm{Res}_{B\cap K}^K\gamma$ of the subgroup $B\cap K$. It follows that the corresponding decomposition in Theorem 2.5.(3) (with G and θ therein now replaced by BK and $\mathrm{Ind}_K^{BK}\gamma$, respectively) is also multiplicity-free, so that $\ell_{\mathrm{Ind}_K^{BK}\gamma} = 1$. Using Corollary 2.14, we deduce the existence of $\eta \in \widehat{I_{BK}(\psi)}$ such that $\mathrm{Res}_B^{I_{BK}(\psi)}\eta \sim \psi$ and $\mathrm{Ind}_{I_{BK}(\psi)}^{BK}\eta \sim \mathrm{Ind}_K^{BK}\gamma$. Finally, by transitivity of induction,

$$\theta = \mathrm{Ind}_K^G\gamma \sim \mathrm{Ind}_{BK}^G\mathrm{Ind}_K^{BK}\gamma \sim \mathrm{Ind}_{BK}^G\mathrm{Ind}_{I_{BK}(\psi)}^{BK}\eta \sim \mathrm{Ind}_{I_{BK}(\psi)}^G\eta.$$

■

Now suppose that $N \subseteq K$; for instance, $K = I_G(\sigma)$. Then (cf. (2.61)),

$$N\cap K = N, \quad G = \sqcup_{s\in\mathscr{S}} sK, \quad \sigma_s \equiv {}^s\sigma, \quad K_s = N \quad \text{and} \quad \gamma_s = {}^s\left(\mathrm{Res}_N^K\gamma\right),$$

for all $s \in S$. Therefore (2.62) and (2.63) become, respectively,

$$\mathrm{Res}_K^G\mathrm{Ind}_N^G\sigma \sim \bigoplus_{s\in\mathscr{S}^{-1}} \mathrm{Ind}_N^K\, {}^s\sigma, \tag{2.64}$$

and

$$\mathrm{Res}_N^G\mathrm{Ind}_K^G\gamma \sim \bigoplus_{s\in\mathscr{S}} {}^s\left(\mathrm{Res}_N^K\gamma\right). \tag{2.65}$$

Suppose now that $\mathscr{T}$ is a complete system of representatives for the double K-K cosets in G, that is,

$$G = \sqcup_{t\in\mathscr{T}} KtK;$$

as usually, we suppose that $1_G \in \mathscr{T}$. For each $t \in \mathscr{T}$ set $K_t = K \cap (tKt^{-1})$ and choose a complete set $\mathscr{U}_t$ of representatives for the left K_t cosets in K, that is,

$$K = \sqcup_{u \in \mathscr{U}_t} uK_t.$$

Proposition 2.43 *We have*

$$G = \sqcup_{\substack{t \in \mathscr{T} \\ u \in \mathscr{U}_t}} utK,$$

i.e., $\{ut : u \in \mathscr{U}_t, t \in \mathscr{T}\}$ *is a complete set of representatives for the left* K*-cosets in* G.

Proof Let $g \in G$. Then we can write $g = k_1 t k_2$ with $k_1, k_2 \in K$ and $t \in \mathscr{T}$, and $k_1 = uh$ with $u \in \mathscr{U}_t$ and $h \in K_t$. Thus,

$$g = k_1 t k_2 = uhtk_2 = ut(t^{-1}ht)k_2 \in utK,$$

since $(t^{-1}ht)k_2 \in K$.

On the other hand, if $t_1, t_2 \in \mathscr{T}$, $u_1 \in \mathscr{U}_{t_1}$, and $u_2 \in \mathscr{U}_{t_2}$, satisfy that $u_1 t_1 = u_2 t_2 k$ with $k \in K$, then $t_1 = u_1^{-1} u_2 t_2 k \in K t_2 K$ which forces $t_1 = t_2$. Therefore, $u_1 = u_2 t_1 k t_1^{-1}$ implies that $u_1 \in u_2 K_{t_1}$ so that, necessarily, $u_1 = u_2$. ■

Note that the above proposition does not depend on N.

We now apply Clifford theory (using the fact that N is normal in K) to the decomposition (2.64).

Proposition 2.44 *With the notation above,*

$$\bigoplus_{\tau \in \widehat{K}(\sigma)} \ell_\tau \bigoplus_{t \in \mathscr{T}} \mathrm{Ind}_{K_t}^K \tau_t \sim \bigoplus_{s \in \mathscr{S}^{-1}} \mathrm{Ind}_N^K {}^s\sigma.$$

Proof First note that with K in place of G, (2.15) becomes $\mathrm{Ind}_N^K \sigma \sim \bigoplus_{\tau \in \widehat{K}(\sigma)} \ell_\tau \tau$. Also, by Mackey's lemma (cf. (2.64) and [30, Theorem 11.5.1]) we have

$$\mathrm{Res}_K^G \mathrm{Ind}_N^G \sigma \sim \bigoplus_{s \in \mathscr{S}^{-1}} \mathrm{Ind}_N^K {}^s\sigma \quad \text{and} \quad \mathrm{Res}_K^G \mathrm{Ind}_K^G \tau \sim \bigoplus_{t \in \mathscr{T}} \mathrm{Ind}_{K_t}^K \tau_t$$

for all $\tau \in \widehat{K}$. We thus have

$$\begin{aligned}
\bigoplus_{s\in\mathscr{S}^{-1}} \mathrm{Ind}_N^K {}^s\sigma &\sim \mathrm{Res}_K^G \mathrm{Ind}_N^G \sigma \\
\text{(by transitivity of induction)} &\sim \mathrm{Res}_K^G \mathrm{Ind}_K^G \mathrm{Ind}_N^K \sigma \\
&\sim \mathrm{Res}_K^G \mathrm{Ind}_K^G \left(\bigoplus_{\tau\in\widehat{K}(\sigma)} \ell_\tau \tau \right) \\
&\sim \bigoplus_{\tau\in\widehat{K}(\sigma)} \ell_\tau \mathrm{Res}_K^G \mathrm{Ind}_K^G \tau \\
&\sim \bigoplus_{\tau\in\widehat{K}(\sigma)} \ell_\tau \bigoplus_{t\in\mathscr{T}} \mathrm{Ind}_{K_t}^K \tau_t.
\end{aligned}$$

■

Remark 2.45 Note that an application of Clifford theory to (2.65) does not provide any new results. Indeed, since $N \trianglelefteq K_t$, we may apply (2.65) also with K in place of G, K_t in place of K, and $\mathscr{U}_t$ in place of $\mathscr{S}$, so that the left hand side of (2.65) may be written in the form:

$$\mathrm{Res}_N^G \mathrm{Ind}_K^G \gamma \sim \mathrm{Res}_N^K \mathrm{Res}_K^G \mathrm{Ind}_K^G \gamma \sim \mathrm{Res}_N^K \bigoplus_{t\in\mathscr{T}} \mathrm{Ind}_{K_t}^K \gamma_t \sim \bigoplus_{t\in\mathscr{T}} \bigoplus_{u\in\mathscr{U}_t} {}^u\!\left(\mathrm{Res}_N^{K_t} \gamma_t\right).$$

Since ${}^u\!\left(\mathrm{Res}_N^{K_t} \gamma_t\right) = {}^u({}^t(\mathrm{Res}_N^K \gamma)) = {}^{ut}(\mathrm{Res}_N^K \gamma)$, by virtue of Proposition 2.43 the above is just a slightly different way to write the right hand side of (2.65).

We now recall the important Mackey criterion for the irreducibility of induced representations (see, for instance, [30, Corollary 11.4.6]).

Theorem 2.46 (Mackey's Irreducibility Criterion) *Let $\gamma \in \widehat{K}$. Then $\mathrm{Ind}_K^G \gamma$ is irreducible if and only if the following condition holds:*

(∗) *for each $t \in \mathscr{T} \setminus \{1_G\}$ the K_t-representations $\mathrm{Res}_{K_t}^K \gamma$ and $\gamma_t = {}^t(\mathrm{Res}_{K_t}^K \gamma)$ do not have common irreducible subrepresentations.*

In particular, if $\sigma \in \widehat{N}$, $K = I_G(\sigma)$, and $\gamma = \eta \in \widehat{I_G}(\sigma)$, then, by virtue of Clifford's correspondence (cf. Theorem 2.12) $\mathrm{Ind}_K^G \gamma$ is irreducible (and therefore (∗) is satisfied). Note that, in fact, the following condition, stronger than (∗), holds in this case:

(∗∗) for each $t \in \mathscr{T} \setminus \{1_G\}$ the N-representations $\mathrm{Res}_N^K \gamma$ and $\gamma_t = {}^t(\mathrm{Res}_N^K \gamma)$ do not have common irreducible subrepresentations.

Indeed, we now have $\mathscr{S} \equiv \mathscr{R}$ (cf. (2.4)) and $\mathrm{Res}^K_N\gamma \sim \ell\sigma$ (cf. Remark 2.8). Moreover, since $\mathscr{T} \subseteq \mathscr{S}$, we have $\sigma \nsim {}^t\sigma$ if $t \neq 1_G$.

2.6 The G-Action on the N-Conjugacy Classes

This section is adapted from [67, Chapter 6], where more abstract constructions are considered. Suppose again that $N \trianglelefteq G$ and denote by $C(N)$ the conjugacy classes of N. If $\mathscr{C} \in C(N)$ and $g \in G$ then ${}^g\mathscr{C} := \{g^{-1}ng : n \in \mathscr{C}\}$ is again an N-conjugacy class:

$$m^{-1}(g^{-1}ng)m = g^{-1}[(gm^{-1}g^{-1})n(gmg^{-1})]g = g^{-1}[(gmg^{-1})^{-1}n(gmg^{-1})]g \in {}^g\mathscr{C},$$

for all $m, n \in N$. This way, we have defined the *conjugacy action* of G on $C(N)$.

Let ξ (resp. γ) denote the permutation character associated with the conjugacy action of G on $\widehat{N}$ (resp. $C(N)$). Denote by Ξ (resp. Γ) the corresponding permutation matrix, that is, for all $g \in G$ we have

$$\Xi(g)_{\sigma,\tau} = \delta_{\sigma,{}^g\tau} \equiv \begin{cases} 1 & \text{if } \sigma \sim {}^g\tau, \\ 0 & \text{otherwise}, \end{cases} \tag{2.66}$$

for all $\sigma, \tau \in \widehat{N}$, and

$$\Gamma(g)_{\mathscr{B},\mathscr{C}} = \delta_{\mathscr{B},{}^g\mathscr{C}} \equiv \begin{cases} 1 & \text{if } \mathscr{B} = {}^g\mathscr{C}, \\ 0 & \text{otherwise}, \end{cases} \tag{2.67}$$

for all $\mathscr{B}, \mathscr{C} \in C(N)$, respectively (compare with (1.15)).

If χ^σ is the character of $\sigma \in \widehat{N}$, then for $\mathscr{C} \in C(N)$ we denote by $\chi^\sigma(\mathscr{C}) := \chi^\sigma(g)$ the (common) value of χ^σ on the elements $g \in \mathscr{C}$. Then, clearly

$$\chi^{{}^g\sigma}(\mathscr{C}) = \chi^\sigma({}^g\mathscr{C}). \tag{2.68}$$

Note that $|\widehat{N}| = |C(N)|$ so that matrices Ξ and Γ have the same dimension. Actually, the following stronger result holds.

Proposition 2.47 *We have* $\xi = \gamma$.

Proof First of all, we need a preliminary computation that expresses the entries of the $\Xi(g)$'s in terms of the entries of the $\Gamma(g)$'s:

$$\begin{aligned}\sum_{\mathscr{B},\mathscr{C}\in C(N)} \chi^{\sigma}(\mathscr{B})\overline{\chi^{\varphi}(\mathscr{C})}\frac{|\mathscr{C}|}{|G|}\Gamma(g)_{\mathscr{B},\mathscr{C}} &= \sum_{\mathscr{C}\in C(N)} \chi^{\sigma}\left({}^{g}\mathscr{C}\right)\overline{\chi^{\varphi}(\mathscr{C})}\frac{|\mathscr{C}|}{|G|} && \text{by (2.67)}\\ &= \sum_{\mathscr{C}\in C(N)} \chi^{{}^{g}\sigma}(\mathscr{C})\overline{\chi^{\varphi}(\mathscr{C})}\frac{|\mathscr{C}|}{|G|} && \text{by (2.68)}\\ &= \delta_{{}^{g}\sigma,\varphi} && \text{by (1.22)}\\ &= \Xi(g)_{\varphi,\sigma} && \text{by (2.66).}\end{aligned}$$

Therefore,

$$\begin{aligned}\xi(g) &= \sum_{\sigma\in\widehat{N}} \Xi(g)_{\sigma,\sigma}\\ &= \sum_{\sigma\in\widehat{N}}\left(\sum_{\mathscr{B},\mathscr{C}\in C(N)} \chi^{\sigma}(\mathscr{B})\overline{\chi^{\sigma}(\mathscr{C})}\frac{|\mathscr{C}|}{|G|}\Gamma(g)_{\mathscr{B},\mathscr{C}}\right)\\ &= \sum_{\mathscr{B},\mathscr{C}\in C(N)}\sqrt{\frac{|\mathscr{C}|}{|\mathscr{B}|}}\left[\sum_{\sigma\in\widehat{N}}\chi^{\sigma}(\mathscr{B})\overline{\chi^{\sigma}(\mathscr{C})}\frac{\sqrt{|\mathscr{B}|\cdot|\mathscr{C}|}}{|G|}\right]\Gamma(g)_{\mathscr{B},\mathscr{C}}\\ \text{(by (1.23))}\quad &= \sum_{\mathscr{C}\in C(N)}\Gamma(g)_{\mathscr{C},\mathscr{C}}\\ &= \gamma(g).\end{aligned}$$

■

Corollary 2.48 *The number of orbits of the conjugacy actions of G on $\widehat{N}$ and $C(N)$ coincide.*

Proof These numbers equal the multiplicity of the trivial representation in the corresponding permutation representations: this is the *Burnside lemma* (cf. [24, Lemma 3.11.1]). By virtue of (1.21), this, in turn, equals the normalized scalar product of the trivial character with ξ and γ, respectively. ■

Corollary 2.49 *For every $g \in G$ we have $|\{\sigma \in \widehat{N} : {}^{g}\sigma \sim \sigma\}| = |\{\mathscr{C} \in C(N) : {}^{g}\mathscr{C} = \mathscr{C}\}|$.*

Proof This is just the equality $\xi(g) = \gamma(g)$: see the fixed point character formula in [30, Proposition 10.4.6] (see also [24, formula (3.22)]). ■

2.7 Real, Complex, and Quaternionic Representations and Clifford Theory

In this section we examine the relation between the real/complex/quaternionic types of $\sigma \in \widehat{N}$ and $\theta \in \widehat{G}(\sigma)$ (cf. the last part in Sect. 1.3) in two particular but quite important cases, namely, when $\mathrm{Ind}_N^G\sigma$ is irreducible and when σ has an extension to G. These results will be applied in Chap. 4.

In the first case, we thus assume that $\mathrm{Ind}_N^G\sigma$ is irreducible and therefore, by Proposition 2.27, $I_G(\sigma) = N$. As a consequence, the correspondence (2.19) is expressed as in Proposition 2.28.(4), namely, with $\{{}^r\sigma : r \in \mathscr{R}(\sigma)\} \equiv \{{}^h\sigma : h \in H\}$, where $H = G/N$ (cf. (2.2)), one associates the singleton $\{\mathrm{Ind}_N^G\sigma\}$.

Let $t\colon H \to G$ be a section with $t(1_H) = 1_G$, so that $G = Nt(H)$, and let $N \ni n \mapsto A(n) \in M_{d_\sigma}(\mathbb{C})$ be an explicit matrix realization of σ. Then the matrix realization of $\mathrm{Ind}_N^G\sigma$ is expressed as follows (cf. (1.31)):

$$B(nt(k)) = \left[A(t(j)^{-1}nt(k)t(s))\right]_{j,s\in H} \tag{2.69}$$

for all $n \in N$ and $k \in H$. Note that we write B in an $(H \times H)$-block form *without enumerating the elements of* H. Indeed, such an enumeration would be unnatural and not useful.

Theorem 2.50 *Suppose that $I_G(\sigma) = N$, let σ' be the adjoint of σ, and set $\theta := \mathrm{Ind}_N^G\sigma$.*

(1) *If σ is real (resp. quaternionic) then θ is real (resp. quaternionic).*
(2) *If σ is complex and σ' is not conjugate to σ then θ is complex.*
(3) *Suppose now that σ is complex and σ' is conjugate to σ, so that there exists $h \in H$ such that $\sigma' \sim {}^h\sigma$. Then θ is selfadjoint. Moreover, $h^2 = 1_H$ and $t(h)^2 \in N$, and if $U \in M_{d_\sigma}(\mathbb{C})$ is a unitary implementing the above equivalence, then $A(t(h)^2) = +\overline{U}U/-\overline{U}U$ and, correspondingly, θ is real/quaternionic.*

The above criterion for the type of σ, in the case where σ' is conjugate to σ, is particularly useful when $\overline{U}U = I$; for instance, this holds when σ is one-dimensional. See also Remark 2.53.

***Proof of Theorem* 2.50**

(1) This is just a particular case of Proposition 1.3.
(2) Now (2.21) yields $\chi^\theta(n) = \sum_{h\in H}\chi^{{}^h\sigma}(n)$, for $n \in N$, that is $\chi^{\mathrm{Res}_N^G\theta} = \sum_{h\in H}\chi^{{}^h\sigma}$. Then

$$\overline{\chi^{\mathrm{Res}_N^G\theta}} = \sum_{h\in H}\overline{\chi^{{}^h\sigma}} = \sum_{h\in H}\chi^{{}^h\sigma'} \neq \sum_{h\in H}\chi^{{}^h\sigma} = \chi^{\mathrm{Res}_N^G\theta},$$

because σ and σ' belong to different H-orbits. Then $\chi^{\theta'} = \overline{\chi^\theta} \neq \chi^\theta$, and θ is complex.

(3) Now σ and σ' belongs to the same H-orbits and therefore, again from (2.21),

$$\overline{\chi^\theta(n)} = \sum_{h\in H} \overline{\chi^{{}^h\sigma}(n)} = \sum_{h\in H} \chi^{{}^h\sigma'}(n) = \sum_{h\in H} \chi^{{}^h\sigma}(n) = \chi^\theta(n)$$

for all $n \in N$. Since $\chi^\theta(g) = 0$ if $g \notin N$, we conclude that $\overline{\chi^\theta} = \chi^\theta$, and θ is selfadjoint.

A matrix realization of ${}^h\sigma = {}^{t(h)}\sigma$ (cf. (2.2)) is given by

$$ {}^hA(n) := A(t(h)^{-1}nt(h)) \tag{2.70}$$

for all $n \in N$. Since $\sigma' \sim {}^h\sigma$, there exists a unitary matrix U such that

$$U^*\overline{A(n)}U = {}^hA(n) \tag{2.71}$$

for all $n \in N$. Then, for all $k \in H$ and $m \in N$ we have:

$$\begin{aligned}
{}^kA(m) &= A\left(t(h)^{-1}\cdot t(h)t(k)^{-1}mt(k)t(h)^{-1}\cdot t(h)\right)\\
\text{(by (2.70))} \quad &= {}^hA\left(t(h)t(k)^{-1}mt(k)t(h)^{-1}\right)\\
\text{(by (1.63))} \quad &= {}^hA\Big(n(h,h^{-1})\left(t(k)t(h^{-1})\right)^{-1}\cdot m\cdot\left(t(k)t(h^{-1})\right)n(h,h^{-1})^{-1}\Big)\\
\text{(by (1.56))} \quad &= {}^hA\Big(\left[n(k,h^{-1})n(h,h^{-1})^{-1}\right]^{-1} t(kh^{-1})^{-1}\cdot m\cdot\\
&\qquad\cdot t(kh^{-1})\left[n(k,h^{-1})n(h,h^{-1})^{-1}\right]\Big)\\
\text{(by (2.71))} \quad &= U^*\overline{A\Big(\left[n(k,h^{-1})n(h,h^{-1})^{-1}\right]^{-1} t(kh^{-1})^{-1}\cdot m\cdot}\\
&\qquad\overline{\cdot t(kh^{-1})\left[n(k,h^{-1})n(h,h^{-1})^{-1}\right]\Big)}U
\end{aligned}$$

and therefore

$$\overline{{}^kA(m)} = \overline{U}^*\cdot A\left(n(k,h^{-1})n(h,h^{-1})^{-1}\right)^*\cdot {}^{kh^{-1}}A(m)\cdot A\left(n(k,h^{-1})n(h,h^{-1})^{-1}\right)\cdot\overline{U}. \tag{2.72}$$

Moreover, since, as we have just seen, θ is selfadjoint, we can find a unitary matrix W such that

$$WB(g)W^* = \overline{B(g)}, \tag{2.73}$$

where B is as in (2.69). From Theorem 2.5.(1) we deduce that $\mathrm{Res}^G_N \theta \sim \oplus_{j\in H} {}^j\sigma$ (with the ${}^j\sigma$'s pairwise inequivalent) which implies that

$$B(m) = \left[\delta_{j,s}\, {}^jA(m)\right]_{j,s\in H} \tag{2.74}$$

for all $m \in N$: just note that the decomposition (2.13) is the same used in the proof of the block form (2.69). Fix $k \in H$ and let $m \in N$. Then, taking into account (2.74), we have:

$$\begin{aligned}
\sum_{j\in H} W_{k,j}\, {}^jA(m)(W_{k,j})^* &= \sum_{j,s\in H} W_{k,j}\, {}^jA(m)\delta_{j,s}(W_{k,s})^* \\
&= \left[W B(m) W^*\right]_{k,k} \\
\text{(by (2.73))} \qquad &= \overline{{}^kA(m)} \\
\text{(by (2.72))} \qquad &= \overline{U}^* A\left(n(k,h^{-1})n(h,h^{-1})^{-1}\right)^* \cdot {}^{kh^{-1}}A(m) \cdot \\
&\quad \cdot A\left(n(k,h^{-1})n(h,h^{-1})^{-1}\right)\overline{U}.
\end{aligned}$$

By taking traces we get

$$\begin{aligned}
\sum_{\substack{j\in H:\\ W_{k,j}\neq 0}} \chi^{{}^j\sigma}(m) &= \sum_{\substack{j\in H:\\ W_{k,j}\neq 0}} \mathrm{tr}\left({}^jA(m)\right) \\
&= \sum_{\substack{j\in H:\\ W_{k,j}\neq 0}} \mathrm{tr}\left(W_{k,j}\, {}^jA(m)(W_{k,j})^*\right) \\
&= \sum_{j\in H} \mathrm{tr}\left(W_{k,j}\, {}^jA(m)(W_{k,j})^*\right) \\
&= \mathrm{tr}\left({}^{kh^{-1}}A(m)\right) \\
&= \chi^{{}^{kh^{-1}}\sigma}(m)
\end{aligned}$$

for all $m \in N$. Since the ${}^j\sigma$'s are pairwise nonequivalent, by taking scalar products with distinct $\chi^{{}^j\sigma}$'s, we deduce that the sum in the left hand side consists only of one summand corresponding to $j = kh^{-1}$, equivalently,

$$W_{k,j} = 0 \quad \text{if} \quad j \neq kh^{-1}. \tag{2.75}$$

Moreover, from Schur's lemma we deduce that, up to a renormalization of W by a complex scalar of modulus 1,

$$W_{k,kh^{-1}} = \overline{U}^* A\left(n(k, h^{-1})n(h, h^{-1})^{-1}\right)^*. \tag{2.76}$$

From (1.43) we know that $W\overline{W}$ is a non zero scalar matrix (the proof of this fact is an easy application of Schur's lemma; see [24, Lemma 9.7.6]). But from (2.75) we deduce that

$$0 \neq \left[W\overline{W}\right]_{s,s} = \sum_{k \in H} W_{s,k}\overline{W}_{k,s} = W_{s,sh^{-1}}\overline{W}_{sh^{-1},sh^{-2}}\delta_{s,sh^{-2}} \tag{2.77}$$

for all $s \in H$, forcing $h^2 = 1$. Then, (1.56), (1.62), and (1.63) become (note that $h = h^{-1}$):

$$t(h)^2 = n(h, h) \in N \text{ and } t(h)^{-1} = n(h, h)^{-1}t(h). \tag{2.78}$$

Moreover,

$$\begin{aligned} A(t(h)^{-2})A(n)A(t(h)^2) &= {}^hA(t(h)^{-1}nt(h)) \\ \text{(by (2.71))} \quad &= U^*\overline{{}^hA(n)}U \\ \text{(again by (2.71))} \quad &= U^*\overline{U^*\overline{A(n)}U}U \\ &= (\overline{U}U)^*A(n)\overline{U}U. \end{aligned}$$

Hence, by Schur's lemma,

$$A(t(h)^2) = \lambda\overline{U}U \tag{2.79}$$

for some $\lambda \in \mathbb{C}$ and $|\lambda| = 1$ ($A(n)$ is unitary). Keeping in mind (2.78) and setting $n := t(h)^2 \in N$ in (2.71), we get $U^*\overline{A(t(h)^2)}U = A(t(h)^2)$. From (2.79) we also get $U^*\overline{A(t(h)^2)} = \overline{\lambda U}$ so that, altogether,

$$A(t(h)^2) = \overline{\lambda}\overline{U}U. \tag{2.80}$$

From (2.79) and (2.80) one deduces that $\lambda = \overline{\lambda}$ so that $\lambda = \pm 1$. In other words

$$A(t(h)^2) = \pm\overline{U}U. \tag{2.81}$$

Finally, (2.77) yields

$$\begin{aligned}
\left[W\overline{W}\right]_{s,s} &= W_{s,sh}\overline{W}_{sh,s}\\
\text{(by (2.76))}\quad &= \overline{U}^* A(n(s,h)n(h,h)^{-1})^* U^* \overline{A(n(sh,h)n(h,h)^{-1})}^*\\
&= \left[\overline{A(n(sh,h)n(h,h)^{-1})}U A(n(s,h)n(h,h)^{-1})\overline{U}\right]^*\\
\text{(by (2.71))}\quad &= \left[U\, {}^h\!A(n(sh,h)n(h,h)^{-1})A(n(s,h)n(h,h)^{-1})\overline{U}\right]^*\\
&= \left[UA\left(t(h)^{-1}n(sh,h)n(h,h)^{-1}t(h)n(s,h)n(h,h)^{-1}\right)\overline{U}\right]^*\\
\text{(by (1.56), (2.78))}\quad &= \left[UA\big(t(h)^{-1}\cdot t(s)^{-1}t(sh)t(h)\cdot t(h)^{-1}\cdot t(sh)^{-1}t(s)t(h)\cdot t(h)^{-2}\big)\overline{U}\right]^*\\
&= \left[UA\left(t(h)^{-2}\right)\overline{U}\right]^*\\
\text{(by (2.81))}\quad &= \pm I,
\end{aligned}$$

that is, $W\overline{W} = \pm I$ accordingly to the sign in (2.81). We then conclude by invoking (1.43). ■

Remark 2.51 More generally, if $\sigma \in \widehat{N}$ is complex with arbitrary $I_G(\sigma)$, and σ and its adjoint σ' are not conjugate, then any $\theta \in \widehat{G}(\sigma)$ is complex. The proof is the same as for Theorem 2.50.(2) by using Theorem 2.5.(3) in order to compute $\chi^{\mathrm{Res}^G_N\theta}$.

In the second case, we assume the hypotheses in Gallagher's Theorem 2.18, that is, there exists $\theta \in \widehat{G}(\sigma)$ which is an extension of σ (equivalently, $I_G(\sigma) = G$ and $\ell_\theta = 1$). The corresponding analysis that we shall develop will find particularly useful in the particular setting of Corollary 2.35 and the related correspondence (2.55).

If $C(g)$, $g \in G$, is a unitary matrix realization of θ, then $A(n) := C(n)$, $n \in N$, is a matrix realization of σ. Moreover, if $h \in H$ and $V := C(t(h))$ then, for all $n \in N$,

$$ {}^h\!A(n) = V^*A(n)V \tag{2.82}$$

is a matrix form of the equivalence ${}^h\sigma \sim \sigma$, and

$$C(nt(h)) = A(n)V. \tag{2.83}$$

Theorem 2.52 *Assume that $\theta \in \widehat{G}$ is an extension of σ. Then the following hold.*

(1) *If σ is complex then θ is complex as well.*
(2) *If σ is selfadjoint and W and V are as in (1.42) and (2.82), respectively, then $W^*\overline{V}W = \lambda V$, with $\lambda = \lambda(t(h)) \in \mathbb{C}$ and $\lambda^s = 1$, where s is the order of h. Moreover, the following hold.*

(a) *Suppose that* σ *is real. Then* θ *is real (resp. complex) if* $\lambda(t(h)) = 1$ *for all* $h \in H$ *(resp. otherwise).*
(b) *Suppose* σ *is quaternionic. Then* θ *is quaternionic (resp. complex) if* $\lambda(t(h)) = 1$ *for all* $h \in H$ *(resp. otherwise).*

Proof

(1) We have $\chi^\sigma = \chi^{\mathrm{Res}^G_N \theta}$ and therefore if σ is complex so is θ.
(2) Now suppose that σ is selfadjoint. From (1.42) and (2.82) we deduce

$$\begin{aligned} W^*\overline{V}WV^*A(n) &= W^*\overline{V}W\,{}^hA(n)V^* \\ &= W^*\overline{V}\,\overline{{}^hA(n)}WV^* \\ &= W^*\overline{A(n)V}WV^* \\ &= A(n)W^*\overline{V}WV^* \end{aligned}$$

and therefore, by Schur's lemma, $W^*\overline{V}WV^* = \lambda I$, with $\lambda = \lambda(t(h)) \in \mathbb{C}$ and $|\lambda| = 1$. Hence

$$W^*\overline{V}W = \lambda V. \tag{2.84}$$

Since $h^s = 1$ then $t(h)^s \in N$ and $V^s = C(t(h)^s) = A(t(h)^s)$. Then from (1.42) we get

$$WV^sW^* = WA(t(h)^s)W^* = \overline{A(t(h)^s)} = \overline{V}^s. \tag{2.85}$$

From (2.84) and (2.85) it follows that

$$\lambda^s V^s = W^*\overline{V}^s W = V^s,$$

so that $\lambda^s = 1$. In particular, since σ is selfadjoint one has $WA(n)W^* = \overline{A(n)}$ and (by (2.83), (1.42), and (2.84))

$$WC(nt(h))W^* = WA(n)VW^* = \overline{\lambda(t(h))A(n)V} = \overline{\lambda(t(h))C(nt(h))} \tag{2.86}$$

for all $n \in N$.

Now assume that σ is real. If $\lambda(t(h)) = 1$ for all $h \in H$ then (2.86) yields $WC(g)W^* = \overline{C(g)}$ for all $g \in G$; since $W\overline{W} = \overline{W}W = I$ then (1.43) ensures that θ is also real. Now suppose that there exists $h_0 \in H$ such that $\lambda(t(h_0)) \neq 1$ and let us show that θ is complex. By contradiction, suppose that there exists a unitary W_1 such that $W_1C(g)W_1^* = \overline{C(g)}$, for all $g \in G$. Then, restricting to N and applying Schur's lemma, we necessarily have $W_1 = \alpha W$, with $|\alpha| = 1$. This contradicts (2.86) because $\lambda(t(h_0)) \neq 1$, and (a) is proved.

One proves (b) exactly in the same way: note that, in this case, $W\overline{W} = \overline{W}W = -I$. ■

Remark 2.53 The criterion in Theorem 2.52.(2) is particularly useful when G/N is cyclic. Indeed, under this assumption, as in Proposition 2.23, we can choose $h \in G$ in such a way that Nh generates of G/N and then take $\{h^j : j = 0, 1, \ldots, s-1\}$, where $s = |G/N|$, as a complete set of representatives of the N-cosets in G. Therefore, $G = \sqcup_{j=0}^{s-1} Nh^j$ and the corresponding section is $Nh^j \mapsto h^j$, $j = 0, 1, \ldots, s-1$. Then, assuming (2.83) with this particular choice of h (note that here, for simplicity, $t(h)$ has been replaced by h), we have $C(nh^j) = A(n)V^j$ for $j = 0, 1, \ldots, s-1$. Since (2.84) yields

$$W^*\overline{V}^j W = \left[W^*\overline{V}W\right]^j = \lambda(h)^j V^j$$

it follows that $\lambda(h^j) = \lambda(h)^j$, $j = 0, 1, \ldots, s-1$. Therefore, when applying the criterion, it suffices to check whether or not $\lambda(h) = 1$ *just for the "generator"* h. Finally, note also that with this notation, when G/N is cyclic of order s, Theorem 2.50.(3) may be formulated in the following way:

Suppose that σ is complex and σ' is conjugate to σ, so that there exists $0 \leq j \leq s-1$ such that $\sigma' \sim {}^{h^j}\sigma$. Then θ is selfadjoint. Moreover, $h^{2j} \in N$, so that $j = \frac{s}{2}$, and s is even. Finally, if $\overline{A(n)} = U\,{}^{h^j}A(n)U^*$, with U unitary, then $A(h^{2j}) = +\overline{U}U / -\overline{U}U$ and, correspondingly, θ is real/quaternionic.

2.8 Semidirect Products with an Abelian Normal Subgroup

We now apply the method of little group (Theorem 2.26) to an important class of semidirect products. This particular case was developed by Frobenius for finite groups and then by Wigner for groups arising in physics such as the Poincaré group (see [126, Sections 3.9 and 3.20]). Finally, Mackey extended this method to topological groups [95].

A more general result, in the setting of finite groups, was proved in Theorem 2.26. We now particularize it in the setting of a semidirect product with an Abelian normal subgroups.

Theorem 2.54 *Let $G = B \rtimes H$ with B Abelian. Given $\xi \in \widehat{B}$, its inertia group $I_G(\xi)$ coincides with $B \rtimes H_\xi$, where $H_\xi := \{h \in H : {}^h\xi = \xi\}$. Then any $\xi \in \widehat{B}$ may be extended to a one–dimensional representation $\widetilde{\xi} \in \widehat{B \rtimes H_\xi}$ by setting*

$$\widetilde{\xi}(ah) := \xi(a) \tag{2.87}$$

for all $a \in B$ and $h \in H_\xi$. Moreover, with the notation of Theorem 2.26, one has

$$\widehat{G} = \left\{ \mathrm{Ind}_{B \rtimes H_\xi}^{G} (\widetilde{\xi} \otimes \overset{\triangle}{\psi}) : \xi \in \Sigma, \psi \in \widehat{H_\xi} \right\}.$$

Proof We only have to prove that (2.87) is an extension, but this is trivial. Indeed, ξ is invariant by conjugation with elements in H_ξ so that, if $a, b \in B$ and $h, k \in H_\xi$, one has

$$\widetilde{\xi}(ah \cdot bk) = \widetilde{\xi}(ahbh^{-1} \cdot hk) = \xi(a \cdot hbh^{-1}) = \xi(a)\xi(hbh^{-1}) = \xi(a)\xi(b) = \widetilde{\xi}(ah)\widetilde{\xi}(bk).$$

■

If $B = N$ is not Abelian and $\sigma \in \widehat{N}$ then (2.87) does not lead, in general, to an extension: this is explained and analyzed in Sect. 7.1.

A Fourier Analysis for semidirect products $G = N \rtimes H$, with H acting on an Abelian normal subgroup N, was developed by D. Rockmore in [109, Section 6]. In particular, he showed that the number $T(G)$ of operations needed to compute such Fourier transforms is $O(|N|(|G|+T(H)))$ (cf. [109, Proposition 6]) or $O(|G|(|H|+\log(|N|)))$ (cf. [109, Proposition 7]).

2.9 Semidirect Products of Abelian Groups

Let $G = B \rtimes A$, with A and B Abelian subgroups, with multiplicative notation. This case has been extensively studied by means of the little group method in Theorem 2.54: [30, chapter 12] treats four important examples of such groups, namely, the Heisenberg group and the affine groups over finite fields or over the rings $\mathbb{Z}/n\mathbb{Z}$. We now illustrate a quite elementary but powerful approach to this case. A similar approach is developed in [38, Section 47] for a class of metacyclic groups. We introduce the following notation (see also Sect. 1.5): for $b \in B$ and $a \in A$ we set ${}^{a}b := aba^{-1}$. Then for $a, a_1 \in A$ and $b, b_1 \in B$ we have

$$ba \cdot b_1 a_1 = b\,{}^{a}b_1 \cdot aa_1 \tag{2.88}$$

and

$$(ba)^{-1} = a^{-1}b^{-1}aa^{-1} = ({}^{a^{-1}}b)^{-1}a^{-1} = {}^{a^{-1}}(b^{-1})a^{-1}. \tag{2.89}$$

For $\psi \in \widehat{A}$ and $\xi \in \widehat{B}$ we define their *tensor product* $\xi \otimes \psi \in L(G)$ by setting

$$(\xi \otimes \psi)(ba) := \xi(b)\psi(a) \tag{2.90}$$

for all $a \in A$ and $b \in B$. Clearly, the set $\{\xi \otimes \psi : \xi \in \widehat{B}, \psi \in \widehat{A}\}$ is an orthogonal basis for $L(G)$ and it is worthwhile to describe the irreducible representations of G in terms of these tensor products. Note that

$$^{a}\xi(b) = \xi(^{a^{-1}}b), \tag{2.91}$$

for all $a \in A$ and $b \in B$.

Denote by $\lambda = \lambda_G$ the left regular representation of G. Then, for all $a, a_0 \in A$ and $b, b_0 \in B$, we have

$$\begin{aligned}
[\lambda(ba)(\xi \otimes \psi)](b_0a_0) &= [\xi \otimes \psi]\left((ba)^{-1} \cdot b_0a_0)\right) \\
\text{(by (2.88) and (2.89))} \ &= [\xi \otimes \psi]\left(^{a^{-1}}(b^{-1}b_0) \cdot a^{-1}a_0)\right) \\
\text{(by (2.90) and (2.91))} \ &= {}^{a}\xi(b^{-1}b_0)\psi(a^{-1}a_0) \\
&= {}^{a}\xi(b^{-1})\psi(a^{-1})\left[^{a}\xi \otimes \psi\right](b_0a_0)
\end{aligned}$$

that is,

$$\lambda(ba)(\xi \otimes \psi) = \overline{^{a}\xi(b)\psi(a)}(^{a}\xi \otimes \psi). \tag{2.92}$$

Given $\xi \in \widehat{B}$, set $A_\xi := \{a \in A : {}^{a}\xi = \xi\}$ so that, cf. Theorem 2.54, $I_G(\xi) = B \rtimes A_\xi$. Let $\mathscr{R}_\xi$ be a complete set of representatives for the cosets of A_ξ in A, so that

$$A = \sqcup_{r \in \mathscr{R}_\xi} rA_\xi. \tag{2.93}$$

We also set

$$W_{\xi,\psi} := \text{span}\{^{a}\xi \otimes \psi : a \in A\} = \text{span}\{^{r}\xi \otimes \psi : r \in \mathscr{R}_\xi\}.$$

By (2.92), the subspace $W_{\xi,\psi} \subseteq L(G)$ is λ-invariant and, by (2.93), $\dim W_{\xi,\psi} = |\mathscr{R}_\xi| = \frac{|A|}{|A_\xi|}$. Moreover, the set

$$\left\{\frac{1}{\sqrt{|G|}}\, {}^{r}\xi \otimes \psi : r \in \mathscr{R}_\xi\right\}$$

is an orthonormal basis for $W_{\xi,\psi}$. We can use it to compute the matrix coefficients and the character of the G-representation $\lambda|_{W_{\xi,\psi}}$ (we will use the orthogonality relations for characters of A and B): for $r, s \in \mathscr{R}_\xi$, $a \in A$, and $b \in B$,

$$\begin{aligned} u_{r,s}^{\xi,\psi}(ba) &= \frac{1}{|G|}\langle\lambda(ba)({}^s\xi \otimes \psi), {}^r\xi \otimes \psi\rangle_{L(G)} \\ \text{(by (2.92))} \quad &= \frac{1}{|A|\cdot|B|}\overline{{}^{as}\xi(b)\psi(a)}\,\langle{}^{as}\xi \otimes \psi, {}^r\xi \otimes \psi\rangle_{L(G)} \\ &= \overline{{}^{as}\xi(b)\psi(a)}\,\frac{1}{|B|}\langle{}^{as}\xi, {}^r\xi\rangle_{L(B)} \\ &= \begin{cases} \overline{{}^{as}\xi(b)\psi(a)} & \text{if } as \in rA_\xi \\ 0 & \text{if } as \notin rA_\xi \end{cases} \end{aligned}$$

because ${}^{as}\xi = {}^r\xi$ if and only if $as \in rA_\xi$. Moreover, for $s = r$ this condition becomes $a \in A_\xi$ so that the character $\chi^{\xi,\psi}$ of $\lambda|_{W_{\xi,\psi}}$ is given by

$$\chi^{\xi,\psi}(ba) = \sum_{r\in\mathscr{R}_\xi} u_{r,r}^{\xi,\psi}(ba) = \begin{cases} \overline{\psi(a)}\,\sum_{r\in\mathscr{R}_\xi}\overline{{}^r\xi(b)} & \text{if } a \in A_\xi \\ 0 & \text{if } a \notin A_\xi. \end{cases} \tag{2.94}$$

It follows that

$$\begin{aligned} \frac{1}{|G|}\langle\chi^{\xi,\psi}, \chi^{\xi,\psi}\rangle_{L(G)} &= \frac{1}{|A|\cdot|B|}\sum_{\substack{a\in A_\xi\\ b\in B}} |\psi(a)|^2 \cdot \left(\sum_{r\in\mathscr{R}_\xi}\overline{{}^r\xi(b)}\right)\cdot\left(\sum_{s\in\mathscr{R}_\xi}\overline{{}^s\xi(b)}\right) \\ &= \frac{|A_\xi|}{|A|\cdot|B|}\sum_{r,s\in\mathscr{R}_\xi}\langle{}^s\xi, {}^r\xi\rangle_{L(B)} \\ &= \frac{|A_\xi|\cdot|B|\cdot|\mathscr{R}_\xi|}{|A|\cdot|B|} = 1, \end{aligned}$$

so that $\lambda|_{W_{\xi,\psi}}$ is irreducible (cf. [30, Corollary 10.2.23]). Moreover, from (2.94) it also follows that $\lambda|_{W_{\xi_1,\psi_1}}$ and $\lambda|_{W_{\xi_2,\psi_2}}$ are equivalent if and only if ξ_1 and ξ_2 belong to the same A-orbit (so that $A_{\xi_1} = A_{\xi_2}$) and $\mathrm{Res}^A_{A_{\xi_1}}\psi_1 = \mathrm{Res}^A_{A_{\xi_2}}\psi_2$.

We are now in a position to give a quite satisfactory description of the representation theory of G. Let Ξ be a complete set of representatives for the A-orbits on $\widehat{B}$ (this is denoted by Σ in Theorem 2.54). Then

$$L(G) = \bigoplus_{\substack{\xi\in\Xi\\ \psi\in\widehat{A}}} W_{\xi,\psi} \tag{2.95}$$

is a very explicit decomposition of $L(G)$ into irreducible G-representations. We now regroup the irreducible representations in (2.95) into isotypic components.

First note that, given $\xi \in \Xi$, the inflation map $\widehat{A/A_\xi} \to \widehat{A}$ is an injective group homomorphism, so that we may regard $\widehat{A/A_\xi}$ as a subgroup of $\widehat{A}$. Indeed, recalling that the inflation of $\epsilon \in \widehat{A/A_\xi}$ is the representation $\overset{\triangle}{\epsilon} \in \widehat{A}$ defined by $\overset{\triangle}{\epsilon}(a) := \epsilon(aA_\xi)$ for all $a \in A$, then $\widehat{A/A_\xi}$ is identified with the kernel of the restriction map

$$\begin{aligned} \widehat{A} &\to \widehat{A_\xi} \\ \psi &\mapsto \operatorname{Res}^A_{A_\xi} \psi, \end{aligned}$$

see (2.45). Thus, for each $\xi \in \Xi$ we select a complete set $\Psi_\xi \subseteq \widehat{A}$ of representatives for the $\widehat{A/A_\xi}$ cosets in $\widehat{A}$. Note that Ψ_ξ has the same cardinality of $\widehat{A_\xi}$. Then, for each $\xi \in \Xi$

$$\widehat{A} = \left\{ \psi \otimes \overset{\triangle}{\epsilon} : \psi \in \Psi_\xi, \epsilon \in \widehat{A/A_\xi} \right\}$$

and therefore (2.95) yields

$$L(G) = \bigoplus_{\xi \in \Xi} \bigoplus_{\psi \in \Psi_\xi} \left(\bigoplus_{\epsilon \in \widehat{A/A_\xi}} W_{\xi, \psi \otimes \overset{\triangle}{\epsilon}} \right) \tag{2.96}$$

is the decomposition of $L(G)$ into isotypic components.

The proofs of all results obtained in this section so far are quite elementary: they just require very basic representation theory (character theory of Abelian groups and the first notions of representation theory of noncommutative groups, say [30, Sections 10.1 and 10.2]). We now recover the same results as an application of Theorem 2.54.

Just note that if we extend $\xi \in \widehat{B}$ to its inertia group $B \ltimes A_\xi$, by setting $\widetilde{\xi}(ba) := \xi(b)$ for all $b \in B$ and $a \in A_\xi$, then, in the notation of Theorem 2.54, we have

$$\widehat{G} = \left\{ \operatorname{Ind}^G_{B \ltimes A_\xi} (\widetilde{\xi} \otimes \overset{\triangle}{\psi}) : \xi \in \Xi, \psi \in \widehat{A_\xi} \right\}.$$

The character of $\theta := \operatorname{Ind}^G_{B \ltimes A_\xi} (\widetilde{\xi} \otimes \overset{\triangle}{\psi})$ is given by (1.32), but in the notation in (2.90):

$$\chi^\theta(g) = \sum_{\substack{r \in \mathscr{R}_\xi: \\ r^{-1}gr \in B \ltimes A_\xi}} (\xi \otimes \psi)(r^{-1}gr)$$

for all $g \in G$. Since

$$r^{-1}bar = {}^{r^{-1}}b \cdot r^{-1}ar \in B \rtimes A_\xi \Leftrightarrow r^{-1}ar \in A_\xi \Leftrightarrow a \in A_\xi,$$

we have

$$\chi^\theta(ba) = \begin{cases} \psi(a)\sum_{r\in\mathscr{R}_\xi} {}^r\xi(b) & \text{if } a \in A_\xi \\ 0 & \text{if } a \notin A_\xi, \end{cases} \tag{2.97}$$

for all $a \in A$ and $b \in B$. Comparing (2.94) and (2.97), it follows that

$$\lambda|_{W_{\overline{\xi},\overline{\psi}}} \sim \mathrm{Ind}_{B\rtimes A_\xi}^G(\widetilde{\xi} \otimes \overset{\triangle}{\psi}).$$

Remark 2.55 Let $\psi \in \widehat{A}$ and $\xi \in \widehat{B}$. The definition of the tensor product $\psi \otimes \xi$ in (2.90), based on the factorization $G = BA$, is well adapted to the left action of G via $\lambda = \lambda_G$. Since one also has $G = AB$, we may alternatively define it by setting

$$[\psi \otimes \xi](ab) := \psi(a)\xi(b)$$

for all $a \in A$ and $b \in B$. Arguing as in the proof of (2.92), it is then easy to prove that the representation space of $\mathrm{Ind}_{B\rtimes A_\xi}^G(\overline{\psi} \otimes \overline{\xi})$ is

$$\mathrm{span}\{\varphi \otimes \xi : \varphi \in \widehat{A} \text{ such that } \mathrm{Res}_{A_\xi}^A \varphi = \mathrm{Res}_{A_\xi}^A \psi\}. \tag{2.98}$$

Moreover, it is easy to check that the connection from the tensor products in (2.98) and those in (2.90) is given by

$$\varphi \otimes \xi = \frac{1}{|\mathscr{R}_\xi|} \sum_{s\in\mathscr{R}_\xi} \sum_{\substack{\varphi' \in \widehat{A}: \\ \mathrm{Res}_{A_\xi}^A \varphi' = \mathrm{Res}_{A_\xi}^A \varphi}} \varphi(s)\overline{\varphi'(s)}({}^s\xi \otimes \varphi').$$

This shows, in particular, that the representation space in (2.98) is contained in one of the isotypic components in (2.96), but *does not* coincide with one of the representation spaces in (2.95).

2.10 Representation Theory of Wreath Products of Finite Groups

We now give a second, quite important application of the little group method (Theorem 2.26), namely, a construction of all the irreducible representations of the wreath product of two finite groups (cf. Sect. 1.6). We have devoted an entire monograph to this topic (cf. [28]); we refer to it for further results, more examples, and several applications, especially to the harmonic analysis of remarkable permutation representations.

Let H be a finite group acting on a finite set X and let F be another finite group. Denote by F^X the set of all maps $f\colon X \to F$; it is a group under pointwise multiplication: $(ff')(x) := f(x) \cdot f'(x)$, for all $f, f' \in F^X$ and $x \in X$. Set theoretically, define

$$F \wr H := \{(f, h) : f \in F^X, h \in H\} \equiv F^X \times H.$$

Then H acts on F^X in a natural way, by setting

$$(hf)(x) := f(h^{-1}x), \tag{2.99}$$

for all $h \in H$, $f \in F^X$, and $x \in X$. Moreover, $h(ff') = hf \cdot hf'$ and $(hf)^{-1} = hf^{-1}$, that is, H acts on F^X as a group of automorphisms. In $F \wr H$ we then define a multiplication law by setting:

$$(f, h)(f', h') := (f \cdot hf', hh') \tag{2.100}$$

for all $(f, h), (f', h') \in F \wr H$. Clearly, $(f \cdot hf')(x) = f(x)f'(h^{-1}x)$, for all $x \in X$.

Lemma 2.56 *The set $F \wr H$ with the multiplication law (2.100) is a group. Moreover, the identity element is $(\mathbf{1}_F, 1_H)$, where $\mathbf{1}_F(x) := 1_F$ for all $x \in X$, and the inverse of $(f, h) \in F \wr H$ is given by*

$$(f, h)^{-1} = (h^{-1}f^{-1}, h^{-1}). \tag{2.101}$$

Finally, the conjugation action of $F \wr H$ on itself is given by:

$$(f, h)(f_0, h_0)(f, h)^{-1} = \left(f \cdot hf_0 \cdot hh_0h^{-1}f^{-1}, hh_0h^{-1}\right), \tag{2.102}$$

for all $(f, h), (f_0, h_0) \in F \wr H$.

Proof It easy to show that $(\mathbf{1}_F, 1_H)$ is the identity. Moreover,

$$(h^{-1}f^{-1}, h^{-1})(f, h) = (h^{-1}f^{-1} \cdot h^{-1}f, 1_H) = (h^{-1}(f^{-1}f), 1_H) = (\mathbf{1}_F, 1_H)$$

and, similarly, $(f, h)(h^{-1}f^{-1}, h^{-1}) = (\mathbf{1}_F, 1_H)$, showing (2.101). Now we check the associativity: if $(f, h), (f', h'), (f'', h'') \in F \wr H$ then

$$[(f, h)(f', h')](f'', h'') = (f \cdot hf' \cdot hg'f'', hh'h'') = (f, h)[(f', h')(f'', h'')]$$

simply because $h(f' \cdot h'f'') = hf' \cdot hh'f''$. Finally,

$$(f, h)(f_0, h_0)(f, h)^{-1} = (f \cdot hf_0, hh_0)(h^{-1}f^{-1}, h^{-1}) = \left(f \cdot hf_0 \cdot hh_0h^{-1}f^{-1}, hh_0h^{-1}\right).$$

■

The group $F \wr H$ is called the *wreath product* of F by (the permutation group) H. The subgroup

$$\{(f, 1_H) : f \in F^X\} \cong F^X$$

is called the *base group* and we will identify it with F^X. By taking $h_0 = 1_H$ in (2.102) we deduce that the base group is a normal subgroup. Moreover, if we identify $\{(\mathbf{1}_F, h) : h \in H\}$ with H, then the wreath product may be written as a semidirect product: $F \wr H = F^X \rtimes H$. Indeed, by setting $h_0 = 1_H$ and $f = \mathbf{1}_F$ in (2.102) we recover the action of H of F^X as the restriction to H of the conjugation action on F^X (note that, with respect to the notation in Sect. 1.5, we are writing the action of the automorphism on the left).

We now describe the irreducible representations of the wreath product $F \wr H$. By virtue of [30, Corollary 10.5.17], every irreducible representation of the base group F^X is of the form

$$\bigotimes_{x \in X} \sigma_x$$

where

$$\begin{aligned} X &\longrightarrow \widehat{F} \\ x &\longmapsto \sigma_x \end{aligned} \tag{2.103}$$

is any map from X to $\widehat{F}$, the dual of F (actually, for typographical reasons, with respect to [30] we use $\otimes$ in place of $\boxtimes$ to denote the *external* tensor product).

In other words, if $f_0 \in F^X$ then

$$\left(\bigotimes_{x \in X} \sigma_x\right)(f_0, 1_H) := \bigotimes_{x \in X} \sigma_x(f_0(x)) \tag{2.104}$$

that is, if $\bigotimes_{x\in X} v_x \in \bigotimes_{x\in X} V_{\sigma_x}$, where V_{σ_x} is the representation space of σ_x, then

$$\left[\left(\bigotimes_{x\in X}\sigma_x\right)(f_0, 1_H)\right]\left(\bigotimes_{x\in X} v_x\right) = \bigotimes_{x\in X}\sigma_x(f_0(x))v_x.$$

We now study the conjugation action (2.1) in the present setting.

Lemma 2.57 *The (f, h)-conjugate of $\bigotimes_{x\in X}\sigma_x$ is given by*

$$^{(f,h)}\left(\bigotimes_{x\in X}\sigma_x\right) = \bigotimes_{x\in X} {}^{f(x)}\sigma_{h^{-1}x} \sim \bigotimes_{x\in X}\sigma_{h^{-1}x}.$$

Proof For all $f \in F^X$ we have

$$\begin{aligned}
^{(f,h)}\left(\bigotimes_{x\in X}\sigma_x\right)(f_0, 1_H) &= \left(\bigotimes_{x\in X}\sigma_x\right)[(f,h)^{-1}(f_0, 1_H)(f,h)] \\
\text{(by (2.100) and (2.101))} \quad &= \left(\bigotimes_{x\in X}\sigma_x\right)(h^{-1}f^{-1}\cdot h^{-1}(f_0 f), 1_H) \\
\text{(by (2.99) and (2.104))} \quad &= \bigotimes_{x\in X}\sigma_x(f(hx)^{-1}f_0(hx)f(hx)) \\
\text{(replace } x \text{ with } h^{-1}x) \quad &= \bigotimes_{x\in X}\sigma_{h^{-1}x}(f(x)^{-1}f_0(x)f(x)) \\
\text{(by (2.1))} \quad &= \bigotimes_{x\in X}\left[{}^{f(x)}\sigma_{h^{-1}x}\right](f_0(x)) \\
\text{(by (2.104))} \quad &= \left[\bigotimes_{x\in X} {}^{f(x)}\sigma_{h^{-1}x}\right](f_0, 1_H).
\end{aligned}$$

But from $f(x) \in F$ it follows that

$$^{f(x)}\sigma_{h^{-1}x} \sim \sigma_{h^{-1}x},$$

and therefore

$$^{(f,h)}\left(\bigotimes_{x\in X}\sigma_x\right) \sim \bigotimes_{x\in X}\sigma_{h^{-1}x}.$$

■

Lemma 2.58 *Let* $\sigma = \left(\bigotimes_{x\in X} \sigma_x\right) \in \widehat{F^X}$. *Then the inertia group of* σ *with respect to* $F \wr H$ *is given by*

$$I_{F\wr H}(\sigma) = F \wr H_\sigma,$$

where $H_\sigma := \{h \in H : \sigma_{hx} \sim \sigma_x \; \forall x \in X\}$.

Proof From Lemma 2.57 it follows that

$$I_{F\wr H}(\sigma) = \{(f, h) : \sigma_{hx} \sim \sigma_x, \; \forall x \in X\}$$

and this is isomorphic to

$$F^X \rtimes H_\sigma = F \wr H_\sigma.$$

■

Note that we actually have $H_\sigma = \{h \in H : \sigma_{hx} = \sigma_x\}$.

Lemma 2.59 *Each representation* $\sigma = \left(\bigotimes_{x\in X} \sigma_x\right) \in \widehat{F^X}$ *has an extension* $\widetilde{\sigma}$ *to the whole* $I_{F\wr H}(\sigma)$ *and it is given by setting*

$$\widetilde{\sigma}(f, h)\left(\bigotimes_{x\in X} v_x\right) := \bigotimes_{x\in X} \sigma_{h^{-1}x}(f(x))v_{h^{-1}x}$$

for all $(f, h) \in F \wr H_\sigma$ *and* $\bigotimes_{x\in X} v_x \in \bigotimes_{x\in X} V_{\sigma_x}$.

Proof From the definition of $\tilde{\sigma}$, we actually have

$$\widetilde{\sigma}(f, h)\left(\bigotimes_{x\in X} v_x\right) = \bigotimes_{x\in X} \sigma_{h^{-1}x}(f(x))v_{h^{-1}x} = \bigotimes_{x\in X} \sigma_x(f(x))v_{h^{-1}x}$$

where the last equality follows from the definition of H_σ. In particular, $\mathrm{Res}^{F\wr H_\sigma}_{F^X} \widetilde{\sigma} = \sigma$. Moreover, for all $(f_1, h_1), (f_2, h_2) \in F \wr H_\sigma$ we have

$$\begin{aligned}\widetilde{\sigma}\left((f_1, h_1) \cdot (f_2, h_2)\right)\left(\bigotimes_{x\in X} v_x\right) &= \widetilde{\sigma}\left(f_1 \cdot (h_1 f_2), h_1h_2\right)\left(\bigotimes_{x\in X} v_x\right)\\ &= \bigotimes_{x\in X} \sigma_x(f_1(x) f_2(h_1^{-1}x))v_{h_2^{-1}h_1^{-1}x}.\end{aligned}$$

On the other hand

$$\begin{aligned}\widetilde{\sigma}(f_1,h_1)\left(\widetilde{\sigma}(f_2,h_2)\left(\bigotimes_{x\in X} v_x\right)\right) &= \widetilde{\sigma}(f_1,h_1)\left(\bigotimes_{x\in X}\sigma_x(f_2(x))v_{h_2^{-1}x}\right)\\ &= \bigotimes_{x\in X}\sigma_x(f_1(x))\sigma_{h_1^{-1}x}(f_2(h_1^{-1}x))v_{h_2^{-1}h_1^{-1}x}\\ &= \bigotimes_{x\in X}\sigma_x(f_1(x))\sigma_x(f_2(h_1^{-1}x))v_{h_2^{-1}h_1^{-1}x}\\ &= \bigotimes_{x\in X}\sigma_x(f_1(x)f_2(h_1^{-1}x))v_{h_2^{-1}h_1^{-1}x}.\end{aligned}$$

Therefore $\widetilde{\sigma}(f_1,h_1)\widetilde{\sigma}(f_2,h_2) = \widetilde{\sigma}((f_1,h_1)\cdot(f_2,h_2))$, showing that $\widetilde{\sigma}$ is a representation. ■

Let $\Sigma \subseteq \widehat{F^X}$ be a complete system of representatives for the $F \wr H$-conjugacy classes. Given $\sigma \in \Sigma$, denote by $\widetilde{\sigma}$ its extension to $I_{F\wr H}(\sigma)$ as shown in Lemma 2.59. For each $\psi \in \widehat{H_\sigma}$ denote by $\overset{\vartriangle}{\psi}$ its inflation to $I_{F\wr H}(\sigma)$ (cf. 2.10). Then, as an immediate consequence of Theorem 2.26.(2), we have:

Theorem 2.60

$$\widehat{F\wr H} = \left\{\mathrm{Ind}_{I_{F\wr H}(\sigma)}^{F\wr H}(\widetilde{\sigma}\otimes\overset{\vartriangle}{\psi}) : \sigma\in\Sigma, \psi\in\widehat{H_\sigma}\right\},$$

that is, the above is a list of all irreducible representations of $F\wr H$ and, for different values of σ and ψ, the corresponding representations are pairwise inequivalent.

Example 2.61 We can use the representation theory of wreath products to present examples where several of the conditions examined in the previous sections are effectively satisfied.

(1) Suppose that the action of H on X is regular (i.e., it is transitive and all stabilizers are trivial; cf. Definition 2.33); for instance, set $X = H$ with the left Cayley action. Take any nontrivial group F, choose $x_0 \in X$ and then define (2.103) by choosing a nontrivial $\sigma_{x_0} \in \widehat{F}$ and setting $\sigma_x := \iota_F$ for all $x \in X \setminus \{x_0\}$. Then, $I_{F\wr H}(\sigma) = F^X$, and $\mathrm{Ind}_{F^X}^{F\wr H}\sigma$ is irreducible. Indeed it is the unique element in $\widehat{F\wr H}(\sigma)$; cf. Proposition 9.18.
(2) Suppose now that $\{1_H\} \lneqq K \lneqq H$ and, again, that the action of H on X is regular. Fix $x_0 \in X$ and define (2.103) in such a way that it equals a fixed nontrivial F-representation on the orbit Kx_0, while it is equal to ι_F on $X \setminus Kx_0$. Then $H_\sigma = K$ and $I_{F\wr H}(\sigma) = K \wr F$.

(3) Finally, suppose that the action of H on X is arbitrary and take (2.103) to be constant. Then $H_\sigma = H$, $I_{F \wr H}(\sigma) = F \wr H$, and σ has an extension to the whole $F \wr H$. We are in the setting of Theorem 2.18.

Remark 2.62 Let F, X, and H be as before. Denote by $G \leq F^X$ the subgroup of all constant maps. Clearly, $G \cong F$. Then, the subgroup of $F \wr H$ made up of all $(f, h) \in F \wr H$ such that $f \in G$ is normal and we shall denote it by (its isomorphic group) $F \times H$. Assume all the notation in Theorem 2.60.

(1) $(F \times H) I_{F \wr H}(\sigma) = F \wr H$ and $(F \times H) \cap I_{F \wr H}(\sigma) = F \times H_\sigma$, for all $\sigma \in \Sigma$.
(2) From Mackey's lemma [30, Theorem 11.5.1], one deduces that

$$\mathrm{Res}^{F \wr H}_{F \times H} \mathrm{Ind}^{F \wr H}_{I_{F \wr H}(\sigma)} (\widetilde{\sigma} \otimes \overset{\triangle}{\psi}) \sim \mathrm{Ind}^{F \times H}_{F \times H_\sigma} \left[\left(\mathrm{Res}^{F^X}_F \sigma \right) \otimes \psi \right].$$

(3) From both sides of Frobenius reciprocity (cf. [30, Section 11.] and [116]) one deduces that if $\xi \otimes \theta \in \widehat{F \times H}$ then

$$\begin{aligned} &\text{multiplicity of } \mathrm{Ind}^{F \wr H}_{I_{F \wr H}(\sigma)} (\widetilde{\sigma} \otimes \overset{\triangle}{\psi}) \text{ in } \mathrm{Ind}^{F \wr H}_{F \times H} (\xi \otimes \theta) \\ &\quad = \mathrm{dimHom}_{F \times H_\sigma} \left(\left(\mathrm{Res}^{F^X}_F \sigma \right) \otimes \psi, \mathrm{Res}^{F \times H}_{F \times H_\sigma} (\xi \otimes \theta) \right) \\ &\quad = \left(\text{multiplicity of } \xi \text{ in } \mathrm{Res}^{F^X}_F \sigma \right) \times \left(\text{multiplicity of } \psi \text{ in } \mathrm{Res}^H_{H_\sigma} \theta \right). \end{aligned}$$

(4) Setting $\xi := \iota_F$ and $\theta := \iota_H$ (the trivial representations), one deduces that $(F \wr H, F \times H)$ is a Gelfand pair (cf. [30, Section 13.3]) if and only if (F^X, F) is a Gelfand pair. For more on these Gelfand pairs, see [15, 104].

An interesting research project is the investigation of the representation theory of iterated wreath products of finite groups, in particular by considering their actions on finite trees. Preliminary results in this directions are in our paper [23] as well as in the paper [102] by Orellana, Orrison, and Rockmore. A general introduction to groups acting on finite trees may be found in the research monograph [10] by Bass, Otero-Espinar, Rockmore, and Tresser.

Fast Fourier Transform algorithms for wreath products (with potential applications to data analysis, molecular spectroscopy, and matched filter design in signal processing) are constructed and studied by Rockmore in [110].

When the rooted trees are infinite, typically spherically homogeneous, the study of actions of (infinite) groups by Grigorchuk and his school led to important results based on the notion of wreath product and the representation theory of (finite) groups. We refer, for instance, to the papers by Bartholdi and Grigorchuk [7, 8], as well as to the research expository paper [9] written in collaboration with Nekrashevych. The study of the actions on each level of an infinite rooted tree yield interesting examples of (finite) Gelfand pairs as in the papers [40, 41] by D'Angeli and Donno.

2.11 Multiplicity-Free Normal Subgroups

Let G be a finite group. We say that $H \leq G$ is a *multiplicity-free subgroup* if for all $\theta \in \widehat{G}$ its restriction $\mathrm{Res}^G_H\theta$ decomposes without multiplicity. Alternatively, by Frobenius reciprocity, this is equivalent to say that for all $\sigma \in \widehat{H}$, the induction $\mathrm{Ind}^G_H\sigma$ decomposes without multiplicity. See [24, Section 9.8] and [27, Section 2.1].

Form Proposition 2.32 it follows immediately that a normal subgroup N of G is multiplicity-free if and only if $H_G(\sigma) := I_G(\sigma)/N$ (cf. Definition 2.1) is Abelian and σ has an extension to $I_G(\sigma)$, for all $\sigma \in \widehat{N}$.

Example 2.63 Let $G = B \rtimes A$ with A and B Abelian. Then, B is a normal multiplicity-free subgroup of G. Indeed, for all $\xi \in \widehat{B}$ we have that $H_G(\xi) \leq A$, so that it is Abelian, and, by Theorem 2.54, ξ has an extension to its inertia group $I_G(\xi)$.

This example includes all the groups in [30, Chapter 12].

Chapter 3
Abelian Extensions

This chapter is based on [111, 112] by R. L. Roth. Previous papers with some results on this subject include [69, 71] by N. Iwahori – H. Matsumoto and G. J. Janusz, respectively. See also Sect. 2.4, where the more general case when $G/I_G(\sigma)$ is Abelian was studied.

In the following, G is a finite group and $N \leq G$ is a normal group with $A := G/N$ Abelian.

3.1 The Dual Action

Definition 3.1 We define a *dual action* of $\widehat{A}$ on $\widehat{G}$ by setting

$$(\varepsilon, \theta) \longmapsto \theta \otimes \overset{\vartriangle}{\varepsilon} \tag{3.1}$$

for all $\varepsilon \in \widehat{A}$ and $\theta \in \widehat{G}$, where $\overset{\vartriangle}{\varepsilon}$ is the inflation of ε to G.

Recall that for $\sigma \in \widehat{N}$ we defined $A_G(\sigma) := I_G(\sigma)/N$ (cf. Sect. 2.4).

Theorem 3.2 *Let $\theta \in \widehat{G}$. Then θ is a fixed point for the action (3.1) if and only if there exists $\sigma \in \widehat{N}$ such that $\widehat{G}(\sigma) = \{\theta\}$. Moreover, if this is the case, then*

$$\mathrm{Ind}_N^G \sigma \sim \ell\theta, \tag{3.2}$$

where ℓ is the inertia index of θ with respect to N, and

$$[I_G(\sigma) : N] = \ell^2. \tag{3.3}$$

T. Ceccherini-Silberstein et al., *Representation Theory of Finite Group Extensions*, Springer Monographs in Mathematics,
https://doi.org/10.1007/978-3-031-13873-7_3

Proof Assume first that $\widehat{G}(\sigma) = \{\theta\}$ so that, by (2.15), also (3.2) holds true. From (1.36) and $\mathrm{Res}^G_N \overset{\Delta}{\varepsilon} \equiv \iota_N$ we deduce that, for all $\varepsilon \in \widehat{A}$,

$$\ell\left(\theta \otimes \overset{\Delta}{\varepsilon}\right) \sim (\ell\theta) \otimes \overset{\Delta}{\varepsilon} \sim \mathrm{Ind}^G_N \sigma \otimes \overset{\Delta}{\varepsilon} \sim \mathrm{Ind}^G_N\left[\sigma \otimes \mathrm{Res}^G_N \overset{\Delta}{\varepsilon}\right] \sim \mathrm{Ind}^G_N \sigma \sim \ell\theta,$$

so that $\theta \otimes \overset{\Delta}{\varepsilon} \sim \theta$. This shows that θ is a fixed point for the action (3.1).

Conversely, suppose that $\theta \otimes \overset{\Delta}{\varepsilon} \sim \theta$ for all $\varepsilon \in \widehat{A}$. By the injectivity of Clifford's correspondence (Theorem 2.12) there exists $\eta \in \widehat{I_G}(\sigma)$ such that $\theta \sim \mathrm{Ind}^G_{I_G(\sigma)} \eta$. Again from (1.36) it follows that

$$\mathrm{Ind}^G_{I_G(\sigma)} \eta \sim \theta \sim \theta \otimes \overset{\Delta}{\varepsilon} \sim \mathrm{Ind}^G_{I_G(\sigma)} \eta \otimes \overset{\Delta}{\varepsilon} \sim \mathrm{Ind}^G_{I_G(\sigma)}\left(\eta \otimes \mathrm{Res}^G_{I_G(\sigma)} \overset{\Delta}{\varepsilon}\right).$$

Since $\left\{\mathrm{Res}^G_{I_G(\sigma)} \overset{\Delta}{\varepsilon} : \varepsilon \in \widehat{A}\right\} = \left\{\overset{\Delta}{\omega} : \omega \in \widehat{A}_G(\sigma)\right\}$ (now $\overset{\Delta}{\omega}$ is the inflation to $I_G(\sigma)$), it follows that

$$\mathrm{Ind}^G_{I_G(\sigma)}\left(\eta \otimes \overset{\Delta}{\omega}\right) \sim \mathrm{Ind}^G_{I_G(\sigma)} \eta,$$

for all $\omega \in \widehat{A}_G(\sigma)$. By Clifford's correspondence, we necessarily have $\eta \otimes \overset{\Delta}{\omega} \sim \eta$ for all $\omega \in \widehat{A}_G(\sigma)$, so that, by Theorem 2.30.(1), $\widehat{I_G}(\sigma) = \{\eta\}$. We deduce that $\widehat{G}(\sigma) = \{\theta\}$.

Finally, from Theorem 2.5.(1), (3.2), and Theorem 2.5.(3) we get

$$[I_G(\sigma) : N] \bigoplus_{r \in \mathcal{R}} {}^r\sigma \sim \mathrm{Res}^G_N \mathrm{Ind}^G_N \sigma \sim \ell\, \mathrm{Res}^G_N \theta \sim \ell^2 \bigoplus_{r \in \mathcal{R}} {}^r\sigma,$$

and this yields (3.3). ■

We now examine Gallagher's theorem (Theorem 2.18) in the context of Abelian extensions.

Proposition 3.3 *For a representation $\theta \in \widehat{G}$ the following conditions are equivalent.*

(a) $\mathrm{Res}^G_N \theta$ *is irreducible.*
(b) *The action of $\widehat{A}$ restricted to the orbit of θ is regular, that is, the representations $\theta \otimes \overset{\Delta}{\varepsilon}$, $\varepsilon \in \widehat{A}$, are pairwise inequivalent.*

Moreover, if this is the case and $\sigma \sim \mathrm{Res}^G_N\theta$*, then* $I_G(\sigma) = G$*,* $\widehat{G}(\sigma) = \{\theta \otimes \overset{\vartriangle}{\varepsilon} : \varepsilon \in \widehat{A}\}$*,* $\mathrm{Ind}^G_N\sigma \sim \bigoplus_{\varepsilon\in\widehat{A}} \theta \otimes \overset{\vartriangle}{\varepsilon}$*, and, for* $\tau \in \widehat{G}$*,*

$$\mathrm{Res}^G_N\tau \sim \sigma \iff \tau \sim \theta \otimes \overset{\vartriangle}{\varepsilon} \textit{ for some } \varepsilon \in \widehat{A}. \tag{3.4}$$

Proof The implication (b) ⇒ (a) is a particular case of Proposition 2.22. The reverse implication is an immediate consequence of Theorem 2.18. The consequences in the last statement are just particular cases of Gallagher's theorem and Corollary 2.20.■

Let now $\theta \in \widehat{G}$ and denote by

$$\Omega_\theta := \{\varepsilon \in \widehat{A} : \theta \otimes \overset{\vartriangle}{\varepsilon} \sim \theta\} \tag{3.5}$$

its stabilizer. We then define a subgroup $J_N(\theta)$ of G by setting

$$J_N(\theta) := \bigcap_{\varepsilon\in\Omega_\theta} \ker \overset{\vartriangle}{\varepsilon} \equiv \{g \in G : \varepsilon(gN) = 1 \text{ for all } \varepsilon \in \Omega_\theta\}. \tag{3.6}$$

Setting

$$C(\theta) := J_N(\theta)/N \equiv \bigcap_{\varepsilon\in\Omega_\theta} \ker \varepsilon, \tag{3.7}$$

then, from the third isomorphism theorem and from (2.42), we deduce that

$$\widehat{G/J_N(\theta)} \cong \widehat{A/C(\theta)} \cong \Omega_\theta. \tag{3.8}$$

In other words, for $\varepsilon \in \widehat{A}$ we have

$$\theta \otimes \overset{\vartriangle}{\varepsilon} \sim \theta \iff \overset{\vartriangle}{\varepsilon} = \overset{\vartriangle}{\xi} \text{ for some } \xi \in \widehat{G/J_N(\theta)}. \tag{3.9}$$

Definition 3.4 The subgroup $J_N(\theta)$ is the *dual inertia group* of θ with respect to N.

Clearly, $N \trianglelefteq J_N(\theta) \trianglelefteq G$. In what follows, we use the notation in (2.4).

Proposition 3.5 *Let* $\theta \in \widehat{G}$*. Then there exist* $\psi \in \widehat{J_N(\theta)}$ *and a positive integer* e_θ *such that*

$$\mathrm{Ind}^G_{J_N(\theta)}\psi \sim e_\theta\theta.$$

Moreover,

$$\mathrm{Res}^G_{J_N(\theta)} \theta \sim e_\theta \bigoplus_{r\in\mathcal{R}} {}^r\psi, \tag{3.10}$$

where $\mathcal{R} = \mathcal{R}(\psi)$*, so that also*

$$\mathrm{Ind}^G_{J_N(\theta)} {}^r\psi \sim e_\theta\, \theta. \tag{3.11}$$

Proof Keeping in mind (3.9), we apply Theorem 3.2 with σ (resp. N) replaced by ψ (resp. by $J_N(\theta)$) and get the first statement. Formula (3.10) follows from Theorem 2.5.(3). Finally, we deduce (3.11) from Frobenius' reciprocity. ■

Let E_θ be a complete system of representatives of Ω_θ in $\widehat{A}$, that is, $\widehat{A} = \sqcup_{\varepsilon\in E_\theta} \varepsilon\Omega_\theta$. We assume that E_θ contains the trivial character $\varepsilon \equiv 1$. Then

$$\left\{\theta \otimes \overset{\triangle}{\varepsilon} \colon \varepsilon \in E_\theta\right\} \text{ is the } \widehat{A}\text{-orbit of } \theta. \tag{3.12}$$

For $\varepsilon \in \widehat{A}$, we set

$$\varepsilon^\sharp := \mathrm{Res}^A_{J_N(\theta)/N}\, \varepsilon. \tag{3.13}$$

Proposition 3.6 *Let* $\theta \in \widehat{G}$*. Then the map*

$$\begin{array}{ccc} \widehat{A} & \longrightarrow & \widehat{J_N(\theta)/N} \\ \varepsilon & \longmapsto & \varepsilon^\sharp \end{array} \tag{3.14}$$

is a surjective homomorphism with kernel Ω_θ*. In particular,*

$$\widehat{J_N(\theta)/N} = \{\varepsilon^\sharp : \varepsilon \in E_\theta\}. \tag{3.15}$$

Proof The fact that (3.14) is a surjective homomorphism follows from Remark 2.24. Moreover, (3.6) ensures that its kernel contains Ω_θ. On the other hand, from (3.8) it follows that $|\Omega_\theta| = |G/J_N(\theta)|$ and therefore

$$|\widehat{A}/\Omega_\theta| = \frac{|G/N|}{|G/J_N(\theta)|} = \frac{|J_N(\theta)|}{|N|},$$

so that Ω_θ equals the whole of the kernel of (3.14). ■

Recall (cf. Definition 2.33) that given an action of a group G, one says that a G-orbit Ω is regular if the stabilizer $\mathrm{Stab}_G(x)$ is trivial for some (equivalently, for all) $x \in \Omega$.

Theorem 3.7 (Roth) *Let $\theta \in \widehat{G}$ and let E_θ be a complete system of representatives of the cosets of Ω_θ (cf. (3.12)) in $\widehat{A}$. Let also (cf. Proposition 3.5) $\psi \in \widehat{J_N(\theta)}$ such that $\mathrm{Ind}_{J_N(\theta)}^G \psi \sim e_\theta \theta$, where $J_N(\theta)$ is the dual inertia group of θ (cf. (3.6)), and let $\mathscr{R} = \mathscr{R}(\psi)$ be, as usual, a complete set of representatives of the left cosets of $I_G(\psi)$ in G. Finally, set $\sigma := \mathrm{Res}_N^{J_N(\theta)} \psi$ and, for all $\varepsilon \in E_\theta$, denote by $\overset{\Delta}{\varepsilon^\sharp}$ the inflation of $\varepsilon^\sharp$ to $J_N(\theta)$ (cf. (3.13)).*

(1) *For all $r \in \mathscr{R}$ and $\varepsilon \in E_\theta$ we have*

$$I_G\left({}^r\psi \otimes \overset{\Delta}{\varepsilon^\sharp}\right) = I_G(\psi). \tag{3.16}$$

In particular, for every $\varepsilon \in E_\theta$ the set $\left\{{}^r\psi \otimes \overset{\Delta}{\varepsilon^\sharp} : r \in \mathscr{R}\right\}$ is a regular $G/I_G(\psi)$-orbit.

(2) *We have*

$$\mathrm{Res}_{J_N(\theta)}^G \left(\theta \otimes \overset{\Delta}{\varepsilon}\right) \sim e_\theta \bigoplus_{r \in \mathscr{R}} \left({}^r\psi \otimes \overset{\Delta}{\varepsilon^\sharp}\right)$$

and

$$\mathrm{Ind}_{J_N(\theta)}^G \left({}^r\psi \otimes \overset{\Delta}{\varepsilon^\sharp}\right) \sim e_\theta (\theta \otimes \overset{\Delta}{\varepsilon}),$$

for all $\varepsilon \in E_\theta$ and $r \in \mathscr{R}$.

(3) *For all $r \in \mathscr{R}$ the set $\left\{{}^r\psi \otimes \overset{\Delta}{\varepsilon^\sharp} : \varepsilon \in E_\theta\right\}$ is a regular $\widehat{J_N(\theta)/N}$-orbit.*

(4) $\left\{{}^r\psi \otimes \overset{\Delta}{\varepsilon^\sharp} : r \in \mathscr{R}, \varepsilon \in E_\theta\right\}$ *is a set of pairwise inequivalent $J_N(\theta)$-representations.*

(5) *σ is irreducible. Moreover,*

$$\mathrm{Res}_N^{J_N(\theta)} \left({}^r\psi \otimes \overset{\Delta}{\varepsilon^\sharp}\right) \sim {}^r\sigma$$

for all $\varepsilon \in E_\theta$ and $r \in \mathscr{R}$.

(6) *$I_G(\sigma) = I_G(\psi)$; in particular, we can choose $\mathscr{R}(\sigma) = \mathscr{R} = \mathscr{R}(\psi)$, and the ${}^r\sigma$'s, $r \in \mathscr{R}$, are pairwise inequivalent.*

(7) $\mathrm{Res}_N^G \theta \sim e_\theta \bigoplus_{r \in \mathscr{R}} {}^r\sigma$, *so that $e_\theta = \ell_\theta$, the inertia index of θ with respect to N.*

(8) $[I_G(\sigma) : J_N(\theta)] = \ell_\theta^2$.
(9) *For all* $r \in \mathscr{R}$

$$\mathrm{Ind}_N^{J_N(\theta)}\, {}^r\sigma \sim \bigoplus_{\varepsilon \in E_\theta} \left({}^r\psi \otimes \overset{\Delta}{\varepsilon^\sharp} \right)$$

and

$$\mathrm{Ind}_N^G\, {}^r\sigma \sim \ell_\theta \bigoplus_{\varepsilon \in E_\theta} \left(\theta \otimes \overset{\Delta}{\varepsilon} \right),$$

so that $\widehat{G}({}^r\sigma) = \left\{ \theta \otimes \overset{\Delta}{\varepsilon} : \varepsilon \in E_\theta \right\}$.

Proof

(1) Let $g \in G$ and $r \in \mathscr{R}$. We have ${}^g({}^r\psi) \sim {}^r\psi$ if and only if ${}^{r^{-1}gr}\psi \sim \psi$, that is, if and only if $r^{-1}gr \in I_G(\psi)$. By the normality of $I_G(\psi)$ in G ($N \leq I_G(\psi)$ and G/N is Abelian), this is equivalent to $g \in I_G(\psi)$. We deduce that

$$I_G({}^r\psi) = I_G(\psi). \tag{3.17}$$

Moreover, for $g \in G$ and $h \in J_N(\theta)$ we have

$$\begin{aligned} {}^g(\overset{\Delta}{\varepsilon^\sharp})(h) &= \overset{\Delta}{\varepsilon^\sharp}(g^{-1}hg) \\ &= \varepsilon^\sharp(g^{-1}hgN) \\ &= \varepsilon(g^{-1}N)\varepsilon(hN)\varepsilon(gN) \\ &= \varepsilon(hN) = \overset{\Delta}{\varepsilon^\sharp}(h), \end{aligned}$$

that is, ${}^g(\overset{\Delta}{\varepsilon^\sharp}) = \overset{\Delta}{\varepsilon^\sharp}$, an therefore ${}^g(\psi \otimes \overset{\Delta}{\varepsilon^\sharp}) \sim {}^g\psi \otimes \overset{\Delta}{\varepsilon^\sharp}$. We deduce that $g \in I_G(\psi)$ if and only if ${}^g({}^r\psi \otimes \overset{\Delta}{\varepsilon^\sharp}) \sim {}^r\psi \otimes \overset{\Delta}{\varepsilon^\sharp}$, and (3.16) follows.

(2) The first equivalence is an immediate consequence of (3.10) and the obvious equality $\mathrm{Res}^G_{J_N(\theta)} \overset{\Delta}{\varepsilon} = \overset{\Delta}{\varepsilon^\sharp}$. The second one follows from Frobenius reciprocity and a comparison of dimensions (cf. (3.11)).

(3) From the second equivalence in (2) we deduce that, for each $r \in \mathscr{R}$, the ${}^r\psi \otimes \overset{\Delta}{\varepsilon^\sharp}$'s, with $\varepsilon \in E_\theta$, are pairwise inequivalent (cf. (3.12)). Therefore, from (3.15) it follows that they form a regular orbit of $\widehat{J_N(\theta)/N}$.

(4) From (1) we deduce that for each $\varepsilon \in E_\theta$ the ${}^r\psi \otimes \overset{\vartriangle}{\varepsilon^\sharp}$'s, with $r \in \mathscr{R}$, are pairwise inequivalent. We conclude by using (3).

(5) From (3) above, (3.15), and Proposition 3.3 (with G replaced by $J_N(\theta)$) we deduce that σ is irreducible. From $\mathrm{Res}_N^{J_N(\theta)} \overset{\vartriangle}{\varepsilon^\sharp} = 1$ and the definition of σ we deduce that

$$\mathrm{Res}_N^{J_N(\theta)}\left({}^r\psi \otimes \overset{\vartriangle}{\varepsilon^\sharp}\right) \sim \mathrm{Res}_N^{J_N(\theta)}\, {}^r\psi \sim {}^r\left(\mathrm{Res}_N^{J_N(\theta)}\psi\right) \sim {}^r\sigma,$$

for all $\varepsilon \in E_\theta$ and $r \in \mathscr{R}$.

(6) Let $g \in G$. If ${}^g\psi \sim \psi$ then ${}^g\sigma \sim \sigma$ and therefore $I_G(\psi) \le I_G(\sigma)$. Conversely, suppose that $g \in I_G(\sigma)$. Then ${}^g\psi$ is an extension of σ to $J_N(\theta)$ and therefore, by the last statement in Proposition 3.3 (with G replaced by $J_N(\theta)$) we have ${}^g\psi \sim \psi \otimes \overset{\vartriangle}{\varepsilon^\sharp}$, for some $\varepsilon \in E_\theta$. On the other hand, ${}^g\psi \sim {}^r\psi$ for some $r \in \mathscr{R}$. By (4), the equivalence ${}^r\psi \sim \psi \otimes \overset{\vartriangle}{\varepsilon^\sharp}$ implies that $r = 1_G$ and $\varepsilon = 1$. We deduce that ${}^g\psi \sim \psi$, so that $g \in I_G(\psi)$. This shows that $I_G(\sigma) = I_G(\psi)$. The remaining part of this statement is clear.

(7) This is an immediate consequence of (2) and (5) (cf. also Theorem 2.5.(3)).

(8) This follows from (3.3), with σ replaced by ψ and N replaced by $J_N(\psi)$, (6)–(7), (3.11), and (3.9).

(9) The first equivalence follows from (5) and the last statement in Proposition 3.3. Then the second equivalence follows from transitivity of induction, the second equivalence in (2), and the equality $e_\theta = \ell_\theta$ in (7).

■

Corollary 3.8 $\ell_\theta = 1$ *if and only if* $I_G(\sigma) = J_N(\theta)$.

Corollary 3.9 *The correspondence* (2.19) *now has the form (for* $\sigma \in \widehat{N}$ *and* $\theta \in \widehat{G}(\sigma)$*)*

$$\{{}^r\sigma : r \in \mathscr{R}(\sigma)\} \longmapsto \left\{\theta \otimes \overset{\vartriangle}{\varepsilon} : \varepsilon \in E_\theta\right\}.$$

In particular, the number of orbits of A *on* $\widehat{N}$ *is equal to the number of orbits of* $\widehat{A}$ *on* $\widehat{G}$.

Corollary 3.10 *Let* σ, ψ *and* θ *be as in Proposition* 3.5 *and Theorem* 3.7, *and let* $\varphi \in \widehat{J_N(\theta)}$.

(1) *Let* $r \in \mathscr{R}$. *Then,* $\mathrm{Res}_N^{J_N(\theta)}\varphi \sim {}^r\sigma$ *if and only if* $\varphi \sim {}^r\psi \otimes \overset{\vartriangle}{\varepsilon^\sharp}$ *for some* $\varepsilon \in E_\theta$.

(2) *Let* $\varepsilon \in E_\theta$. *Then,* $\mathrm{Ind}_{J_N(\theta)}^G \varphi \sim \ell_\theta(\theta \otimes \overset{\Delta}{\varepsilon})$ *if and only if* $\varphi \sim {}^r\psi \otimes \overset{\Delta}{\varepsilon^\sharp}$ *for some* $r \in \mathscr{R}$.

(3) *Let* $r \in \mathscr{R}$ *and* $\varepsilon \in E_\theta$. *Then* ${}^r\psi \otimes \overset{\Delta}{\varepsilon^\sharp}$ *is the unique irreducible representation* φ *of* $J_N(\theta)$ *such that*

$$\mathrm{Ind}_{J_N(\theta)}^G \varphi \sim \ell_\theta(\theta \otimes \overset{\Delta}{\varepsilon}) \quad and \quad \mathrm{Res}_N^{J_N(\theta)} \varphi \sim {}^r\sigma.$$

Proof We just sketch the proof of (1). Suppose that $\mathrm{Res}_N^{J_N(\theta)} \varphi \sim {}^r\sigma$. By Frobenius reciprocity and the first equivalence in Theorem 3.7.(9), we have $\varphi \sim {}^r\psi \otimes \overset{\Delta}{\varepsilon^\sharp}$ for some $\varepsilon \in E_\theta$. The converse follows from Theorem 3.7.(5). ■

3.2 The Conjugation Action

Proposition 3.11 *Assume all the hypotheses and notation in Theorem 3.7. Then there exists* $\eta \in \widehat{I_G}(\sigma)$ *such that* $\mathrm{Ind}_{I_G(\sigma)}^G \eta \sim \theta$ *and*

$$\mathrm{Ind}_{J_N(\theta)}^{I_G(\sigma)} {}^r\psi \sim \ell_\theta {}^r\eta, \tag{3.18}$$

$$\mathrm{Res}_{J_N(\theta)}^{I_G(\sigma)} {}^r\eta \sim \ell_\theta {}^r\psi \tag{3.19}$$

for all $r \in \mathscr{R}$. *In particular,* $d_\eta = \ell_\theta d_\psi$.

Proof Since $\sigma \sim \mathrm{Res}_N^{J_N(\theta)} \psi$, by Frobenius' reciprocity and transitivity of induction we deduce that $\mathrm{Ind}_{J_N(\theta)}^{I_G(\sigma)} \psi$ is contained in $\mathrm{Ind}_N^{I_G(\sigma)} \sigma$. Therefore, $\mathrm{Ind}_{J_N(\theta)}^{I_G(\sigma)} \psi$ contains only representations in $\widehat{I_G}(\sigma)$. On the other hand, from Proposition 3.5 and Theorem 3.7.(7) we deduce that $\mathrm{Ind}_{J_N(\theta)}^G \psi \sim \ell_\theta \theta$, so that (2.26) in Theorem 2.12 (Clifford correspondence), with Nreplaced by $J_N(\theta)$ recalling that $I_G(\sigma) = I_G(\psi) = I_G({}^r\psi)$ (cf. Theorem 3.7.(6) and (3.17)), ensure the existence of $\eta \in \widehat{I_G}(\sigma)$ satisfying (3.19). Finally, we deduce (3.18) from (3.19), by using Theorem 3.7.(8) to count dimensions. ■

For $\varepsilon \in \widehat{A}$ we set $\varepsilon^\flat := \mathrm{Res}_{I_G(\sigma)/N}^A \varepsilon$ (cf. (3.13)).

Theorem 3.12 (Roth) *Assume all the hypotheses and notation in Theorem 3.7 and Proposition 3.11. Also, for all* $\varepsilon \in E_\theta$ *denote by* $\overset{\Delta}{\varepsilon^\flat}$ *the inflation of* $\varepsilon^\flat$ *to* $I_G(\sigma)$. *For all* $r \in \mathscr{R}$ *and* $\varepsilon \in E_\theta$ *we have:*

(1)

$$\operatorname{Res}^{I_G(\sigma)}_{J_N(\theta)}\left({}^{r}\eta\otimes\overset{\triangle}{\varepsilon^{\flat}}\right)\sim\ell_\theta\left({}^{r}\psi\otimes\overset{\triangle}{\varepsilon^{\sharp}}\right)$$

and

$$\operatorname{Ind}^{I_G(\sigma)}_{J_N(\theta)}\left({}^{r}\psi\otimes\overset{\triangle}{\varepsilon^{\sharp}}\right)\sim\ell_\theta\left({}^{r}\eta\otimes\overset{\triangle}{\varepsilon^{\flat}}\right).$$

In particular, $\ell_{{}^{r}\eta\otimes\overset{\triangle}{\varepsilon^{\flat}}}=\ell_\theta$.

(2)

$$\operatorname{Res}^{I_G(\sigma)}_{N}\left({}^{r}\eta\otimes\overset{\triangle}{\varepsilon^{\flat}}\right)\sim\ell_\theta\,{}^{r}\sigma$$

and

$$\operatorname{Ind}^{I_G(\sigma)}_{N}\,{}^{r}\sigma\sim\ell_\theta\bigoplus_{\varepsilon\in E_\theta}\left({}^{r}\eta\otimes\overset{\triangle}{\varepsilon^{\flat}}\right).$$

In particular, for $r\in\mathscr{R}$, $\widehat{I_G}\,({}^{r}\sigma)=\left\{{}^{r}\eta\otimes\overset{\triangle}{\varepsilon^{\flat}}:\varepsilon\in E_\theta\right\}$, *and this is an* $\widehat{I_G(\sigma)/N}$-*orbit.*

(3)

$$I_G\left({}^{r}\eta\otimes\overset{\triangle}{\varepsilon^{\flat}}\right)=I_G(\sigma).$$

In particular, for $\varepsilon\in E_\theta$ *the set* $\left\{{}^{r}\eta\otimes\overset{\triangle}{\varepsilon^{\flat}}:r\in\mathscr{R}\right\}$ *is a regular* $G/I_G(\sigma)$*-orbit.*

(4)

$$\operatorname{Res}^{G}_{I_G(\sigma)}\left(\theta\otimes\overset{\triangle}{\varepsilon}\right)\sim\bigoplus_{r\in\mathscr{R}}\left({}^{r}\eta\otimes\overset{\triangle}{\varepsilon^{\flat}}\right)$$

and

$$\operatorname{Ind}^{G}_{I_G(\sigma)}\left({}^{r}\eta\otimes\overset{\triangle}{\varepsilon^{\flat}}\right)\sim\theta\otimes\overset{\triangle}{\varepsilon}.$$

(5) $J_N\left({}^r\eta \otimes \overset{\vartriangle}{\varepsilon^{\flat}}\right) = J_N(\theta)$, *where the first group is defined by replacing G with $I_G(\sigma)$.*

In particular, $\left\{{}^r\eta \otimes \overset{\vartriangle}{\varepsilon^{\flat}} : r \in \mathscr{R}, \varepsilon \in E_\theta\right\}$ *is a set of pairwise inequivalent $I_G(\sigma)$-representations.*

Proof

(1) The first equivalence is just (cf. (3.19)):

$$\mathrm{Res}^{I_G(\sigma)}_{J_N(\theta)}\left({}^r\eta \otimes \overset{\vartriangle}{\varepsilon^{\flat}}\right) \sim \left(\mathrm{Res}^{I_G(\sigma)}_{J_N(\theta)}\, {}^r\eta\right) \otimes \overset{\vartriangle}{\varepsilon^{\sharp}} \sim \ell_\theta\left({}^r\psi \otimes \overset{\vartriangle}{\varepsilon^{\sharp}}\right).$$

The second equivalence follows from Frobenius reciprocity and a comparison of dimensions (cf. (3.18)).

(2) These follows from (1), Theorem 3.7.(5), the first equivalence in Theorem 3.7.(9), and transitivity of induction (and restriction). The last statement is just (3.12) (see also Corollary 3.9) with G replaced by $I_G(\sigma)$.

(3) Argue as in the proof of Theorem 3.7.(1), by also taking into account Corollary 2.29.

(4) The first equivalence follows from (2.27) and $\mathrm{Res}^G_{I_G(\sigma)} \overset{\vartriangle}{\varepsilon} = \overset{\vartriangle}{\varepsilon^{\flat}}$. The second one follows from Frobenius reciprocity and a comparison of dimensions (recall that $\mathrm{Ind}^G_{I_G(\sigma)}\eta \sim \theta$).

(5) From the second equivalence in (1) and Theorem 3.2 (with G replaced by $I_G(\sigma)$) it follows that $\widehat{I_G(\sigma)/J_N(\theta)}$ fixes ${}^r\eta \otimes \overset{\vartriangle}{\varepsilon^{\flat}}$. Therefore, setting

$$\Omega = \Omega_{{}^r\eta \otimes \overset{\vartriangle}{\varepsilon^{\flat}}} := \left\{\omega \in \widehat{I_G(\sigma)/N} : \left({}^r\eta \otimes \overset{\vartriangle}{\varepsilon^{\flat}}\right) \otimes \overset{\vartriangle}{\omega} \sim {}^r\eta \otimes \overset{\vartriangle}{\varepsilon^{\flat}}\right\},$$

(cf. (3.5)), we have that $\widehat{I_G(\sigma)/J_N(\theta)}$ is contained in Ω (by inflation). Hence (cf. (3.6))

$$J_N\left({}^r\eta \otimes \overset{\vartriangle}{\varepsilon^{\flat}}\right) = \bigcap_{\omega \in \Omega} \ker \overset{\vartriangle}{\omega} \le \bigcap_{\xi \in \widehat{I_G(\sigma)/J_N(\theta)}} \ker \overset{\vartriangle}{\xi} \equiv J_N(\theta). \tag{3.20}$$

By Theorem 3.7.(8) we have

$$[I_G(\sigma) : J_N(\theta)] = \ell_\theta^2 = \left[I_G(\sigma) : J_N\left({}^r\eta \otimes \overset{\vartriangle}{\varepsilon^{\flat}}\right)\right]$$

(the second equality may be obtained by replacing G with $I_G(\sigma)$), so that in (3.20) an equality holds.

Finally, for the last statement use the first equivalence in (1) and Theorem 3.7.(4).

■

Corollary 3.13 *Let ψ and η be as in Theorems 3.7 and 3.12. Then ${}^r\eta \otimes \overset{\Delta}{\varepsilon^\flat}$ is the unique irreducible representation of $I_G(\sigma)$ satisfying the first equivalence in Theorem 3.12.(1).*

Proof Argue as in the proof of Corollary 3.10. ■

3.3 The Intermediary Representations

By Theorem 2.40 and Remark 2.41, in the present context there exist intermediary subgroups (cf. Definition 2.39). A direct proof of the existence of such intermediary subgroups is in [109, Theorem 4], where D. Rockmore developed a Fourier Analysis for Abelian group extensions.

We now examine their properties.

Theorem 3.14 (Roth) *Suppose that $A := G/N$ is Abelian. Let $\sigma \in \widehat{N}$ and $\theta \in \widehat{G}$, and suppose that M is an intermediary subgroup for σ and θ, with intermediary representation τ, that is,*

$$\mathrm{Ind}_M^G \tau \sim \theta \quad \textit{and} \quad \mathrm{Res}_N^M \tau \sim \sigma.$$

(1) *We have $J_N(\theta) \leq M \leq I_G(\sigma)$.*

(2) *Let η be the unique representation in $\widehat{I_G(\sigma)}$ such that $\mathrm{Ind}_{I_G(\sigma)}^G \eta \sim \theta$ (cf. Theorem 2.12) and let ψ be the unique irreducible $J_N(\theta)$-representation such that (cf. Corollary 3.13).*

$$\mathrm{Res}_{J_N(\theta)}^{I_G(\sigma)} \eta \sim \ell_\theta \psi \quad \textit{and} \quad \mathrm{Ind}_{J_N(\theta)}^{I_G(\sigma)} \psi \sim \ell_\theta \eta.$$

Then, M is an intermediary subgroup also for ψ and η, with intermediary representation τ. That is,

$$\mathrm{Res}_{J_N(\theta)}^M \tau \sim \psi \quad \textit{and} \quad \mathrm{Ind}_M^{I_G(\sigma)} \tau \sim \eta. \tag{3.21}$$

Conversely, if $\tau \in \widehat{M}$ satisfies the equivalences in (3.21), then it is an intermediary representation for σ and θ.

(3) *We have $[I_G(\sigma) : M] = [M : J_N(\theta)] = \ell_\theta$.*

(4) *There are ℓ_θ intermediary representations for σ and θ, and they form a regular orbit for both $I_G(\sigma)/M$ (by conjugation) and $\widehat{M/J_N(\theta)}$ (by tensor product). Moreover,*

$$I_G(\sigma)/M \;\cong\; \widehat{M/J_N(\theta)}.$$

(5) *Let $\varepsilon \in E_\theta$ and $r \in \mathscr{R}$. Then M is an intermediary subgroup for both*

$${}^r\psi \otimes \overset{\vartriangle}{\varepsilon^\sharp} \quad \textit{and} \quad {}^r\eta \otimes \overset{\vartriangle}{\varepsilon^\flat},$$

and

$${}^r\sigma \quad \textit{and} \quad \theta \otimes \overset{\vartriangle}{\varepsilon}.$$

In both cases, the intermediary representations are those in the $\widehat{M/J_N(\theta)}$-orbit (equivalently, in the $I_G(\sigma)/M$-orbit) of ${}^r\tau \otimes \overset{\vartriangle}{\varepsilon^\star}$, where $\varepsilon^\star := \mathrm{Res}^{G/N}_{M/N}\,\varepsilon$.

Proof

(1) The inclusion $M \le I_G(\sigma)$ follows from Proposition 2.4 with L replaced by M. Moreover, from $\mathrm{Ind}_M^G \tau \sim \theta$ we deduce that $\widehat{G}(\tau) = \{\theta\}$ and therefore, by Theorem 3.2 with N replaced by M, θ is a fixed point for the action of $\widehat{G/M}$. In other words, if $\omega \in \widehat{G/M}$, then $\theta \otimes \overset{\vartriangle}{\omega} \sim \theta$, that is, $\widetilde{\omega} \in \Omega_\theta$ (here, $\overset{\vartriangle}{\omega}$ is the inflation to G, $\widetilde{\omega}$ is the inflation to A, and Ω_θ is as defined in (3.5)). Thus, $\{\overset{\vartriangle}{\omega} \colon \omega \in \widehat{G/M}\} \subseteq \Omega_\theta$. By means of (2.46) with $C = M/N$ and $A = G/N$, and by (3.6), we thus get

$$M/N = \bigcap_{\omega \in \widehat{G/M}} \ker(\overset{\vartriangle}{\omega}) \;\ge\; \bigcap_{\varepsilon \in \Omega_\theta} \ker \varepsilon \;=\; J_N(\theta)/N.$$

This shows that $M \ge J_N(\theta)$.

(2) We use the inclusions proved in (1). Now, $\mathrm{Res}^M_{J_N(\theta)}\tau$ is an extension of σ to $J_N(\theta)$ and therefore, by Corollary 3.10.(1), $\mathrm{Res}^M_{J_N(\theta)}\tau \;\sim\; \psi \otimes \overset{\vartriangle}{\varepsilon^\sharp}$ for some $\varepsilon \in E_\theta$. By Frobenius' reciprocity, $\mathrm{Res}^G_M \theta$ contains τ and therefore $\mathrm{Res}^G_{J_N(\theta)}\theta$ contains $\mathrm{Res}^M_{J_N(\theta)}\tau \sim \psi \otimes \overset{\vartriangle}{\varepsilon^\sharp}$. Thus, Theorem 3.7.(2) forces $\varepsilon \equiv 1$, and the first equivalence in (3.21) follows. Moreover, by transitivity of induction,

$$\theta \sim \mathrm{Ind}_M^G \tau \sim \mathrm{Ind}^G_{I_G(\sigma)}\left[\mathrm{Ind}_M^{I_G(\sigma)}\tau\right]$$

and from bijectivity of Clifford correspondence (cf. Theorem 2.12) we deduce the second equivalence in (3.21).

The last statement follows from the equivalences $\mathrm{Res}_N^{J_N(\theta)}\psi \sim \sigma$ (cf. Theorem 3.7.(5)) and $\mathrm{Ind}_{I_G(\sigma)}^G \eta \sim \theta$ (cf. Theorem 3.12).

(3) From (3.21) we get

$$d_\tau = d_\psi \ \text{ and } \ [I_G(\sigma) : M]\, d_\tau = d_\eta.$$

Moreover, $d_\eta/d_\psi = \ell_\theta$ (cf. Proposition 3.11) and $[I_G(\sigma) : J_N(\theta)] = \ell_\theta^2$ (cf. Theorem 3.7.(8)). We deduce that

$$[M : J_N(\theta)] = \frac{[I_G(\sigma) : J_N(\theta)]}{[I_G(\sigma) : M]} = \frac{\ell_\theta^2}{d_\eta/d_\tau} = \ell_\theta.$$

(4) By invoking the results in (1) and (2), we may limit ourselves to the case where $G = I_G(\sigma)$ and $\theta = \eta$, and $N = J_N(\theta)$ and $\sigma = \psi$ (cf. (3.21)).

From the irreducibility of $\mathrm{Ind}_M^{I_G(\sigma)} \tau$ ($\sim \eta$) and by virtue of Proposition 2.27, it follows that the representations ${}^k\tau$, with $k \in I_G(\sigma)/M$, are pairwise inequivalent and the corresponding action of $I_G(\sigma)/M$ is regular. Moreover, Proposition 2.27 ensures also that for $\tau_1 \in \widehat{M}$ one has $\mathrm{Ind}_M^{I_G(\sigma)} \tau_1 \sim \eta$ if and only if τ_1 is in the $I_G(\sigma)/M$-orbit of τ.

From $\mathrm{Res}^M_{J_N(\theta)} \tau \sim \psi$ and Proposition 3.3 it follows that the representations $\tau \otimes \overset{\triangle}{\xi}$ with $\xi \in \widehat{M/J_N(\theta)}$ also are pairwise inequivalent, and the action $\widehat{M/J_N(\theta)}$ on this set is regular. Moreover, (3.4) ensures that for $\tau_1 \in \widehat{M}$ one has $\mathrm{Res}^M_{J_N(\theta)} \tau_1 \sim \psi$ if and only if τ_1 is in the $\widehat{M/J_N(\theta)}$-orbit of τ.

We deduce, in particular, that there are $|I_G(\sigma)/M| = |M/J_N(\theta)| = \ell_\theta$ (cf. (3)) intermediate representations $\tau_1 \in \widehat{M}$ for ψ and η.

Let now $\xi \in \widehat{M/J_N(\theta)}$ and denote by $\widetilde{\xi}$ an extension of ξ to $\widehat{I_G(\sigma)/J_N(\theta)}$ (recall that G/N is Abelian, so that $I_G(\sigma)/J_N(\theta)$ is also Abelian and apply Remark 2.24), by $\overset{\triangle}{\xi}$ the inflation to M, and by $\overset{\triangle}{\widetilde{\xi}}$ the inflation to $I_G(\sigma)$. From (1.36) it follows that

$$\mathrm{Ind}_M^{I_G(\sigma)} \left(\tau \otimes \overset{\triangle}{\xi} \right) \sim \eta \otimes \overset{\triangle}{\widetilde{\xi}} \sim \eta,$$

where, the last equivalence follows from (3.9) with G replaced by $I_G(\sigma)$. Therefore, there exists $\varphi(\xi) \in I_G(\sigma)/M$ such that $\tau \otimes \overset{\triangle}{\xi} \sim {}^{\varphi(\xi)}\tau$. This establishes a bijective correspondence

$$\widehat{M/J_N(\theta)} \ni \xi \longmapsto \varphi(\xi) \in I_G(\sigma)/M. \tag{3.22}$$

Indeed, injectivity follows from the fact, established above, that the representations $\tau \otimes \overset{\triangle}{\xi}$ with $\xi \in \widehat{M/J_N(\theta)}$ are pairwise inequivalent, while, surjectivity follows from the equality $|M/J_N(\theta)| = |I_G(\sigma)/M|$. We now show that (3.22) is a group homomorphism. Let $\gamma, \xi \in \widehat{M/J_N(\theta)}$. Then ${}^k\overset{\triangle}{\xi} = \overset{\triangle}{\xi}$ for all $k \in I_G(\sigma)/M$ and therefore

$$
\begin{aligned}
{}^{\varphi(\gamma\xi)}\tau &\sim \tau \otimes \overset{\triangle}{(\gamma\xi)} \sim \left(\tau \otimes \overset{\triangle}{\gamma}\right) \otimes \overset{\triangle}{\xi} \sim \left({}^{\varphi(\gamma)}\tau\right) \otimes \overset{\triangle}{\xi} \\
&\sim {}^{\varphi(\gamma)}\left(\tau \otimes \overset{\triangle}{\xi}\right) \sim {}^{\varphi(\gamma)}\left({}^{\varphi(\xi)}\tau\right) \sim {}^{\varphi(\gamma)\varphi(\xi)}\tau.
\end{aligned}
$$

This proves that (3.22) is a group isomorphism.

(5) From the first equivalence in (3.21) we deduce that

$$
\mathrm{Res}^M_{J_N(\theta)}\left({}^r\tau \otimes \overset{\triangle}{\varepsilon^{\bigstar}}\right) \sim {}^r\psi \otimes \overset{\triangle}{\varepsilon^{\sharp}}
$$

while, from the second one, we get

$$
\begin{aligned}
\mathrm{Ind}^{I_G(\sigma)}_M\left({}^r\tau \otimes \overset{\triangle}{\varepsilon^{\bigstar}}\right) &\sim \mathrm{Ind}^{I_G(\sigma)}_M\left({}^r\tau \otimes \mathrm{Res}^{I_G(\sigma)}_M \overset{\triangle}{\varepsilon^{\flat}}\right) \\
\text{(by Theorem 1.36)} \quad &\sim \left(\mathrm{Ind}^{I_G(\sigma)}_M {}^r\tau\right) \otimes \overset{\triangle}{\varepsilon^{\flat}} \\
\text{(by (2.6))} \quad &\sim {}^r\left(\mathrm{Ind}^{I_G(\sigma)}_M \tau\right) \otimes \overset{\triangle}{\varepsilon^{\flat}} \\
&\sim {}^r\eta \otimes \overset{\triangle}{\varepsilon^{\flat}}.
\end{aligned}
$$

This shows that M is an intermediary group (with intermediary representation ${}^r\tau \otimes \overset{\triangle}{\varepsilon^{\bigstar}}$) for ${}^r\eta \otimes \overset{\triangle}{\varepsilon^{\flat}}$ and ${}^r\psi \otimes \overset{\triangle}{\varepsilon^{\sharp}}$. Then, by restricting to N (resp. inducing to G) and applying Theorem 3.7.(5) (resp. the second equivalence in Theorem 3.12.(4)) we deduce that M is an intermediary group (with the same intermediary representation ${}^r\tau \otimes \overset{\triangle}{\varepsilon^{\bigstar}}$) for ${}^r\sigma$ and $\theta \otimes \overset{\triangle}{\varepsilon}$.

The last statement, follows from the results in (2) and (4).

■

Corollary 3.15 $I_G(\sigma)/M \cong M/J_N(\theta)$.

Corollary 3.16 *For* $\xi \in \widehat{M/J_N(\sigma)}$ *denote by* $\overset{\Delta}{\xi}$ *its inflation to* M. *Then, for all* $r \in \mathscr{R}$ *and* $\varepsilon \in E_\theta$,

$$\begin{aligned}
\mathrm{Res}^{I_G(\sigma)}_{M}\left({}^{r}\eta \otimes \overset{\Delta}{\varepsilon}{}^{\flat}\right) &\sim \bigoplus_{\xi \in \widehat{M/J_N(\sigma)}} \left[\left({}^{r}\tau \otimes \overset{\Delta}{\varepsilon}{}^{\bigstar}\right) \otimes \overset{\Delta}{\xi}\right] \\
&\sim \bigoplus_{s \in I_G(\sigma)/M} {}^{s}\left({}^{r}\tau \otimes \overset{\Delta}{\varepsilon}{}^{\bigstar}\right) \\
&\sim \mathrm{Ind}^{M}_{J_N(\theta)}\left({}^{r}\psi \otimes \overset{\Delta}{\varepsilon}{}^{\sharp}\right).
\end{aligned}$$

Proof These equivalences follow from Theorem 3.14.(5), Frobenius reciprocity, and counting dimensions. ■

3.4 Diagrammatic Summaries

In what follows we summarize some the results of the present section by means of four Bratteli type diagrams. We need to enumerate the elements of $\mathscr{R}$: $r_1 = 1_G, r_2, \ldots, r_k$, those of M_θ: $\varepsilon_1 \equiv 1, \varepsilon_2, \ldots, \varepsilon_h$ and those of $\widehat{M/J_N(\theta)}$: $\xi_1 \equiv 1_G, \xi_2, \ldots, \xi_m$. We use the indices $i = 1, 2, \ldots, h$ and $j = 1, 2, \ldots, k$.

The first diagram is on restrictions:

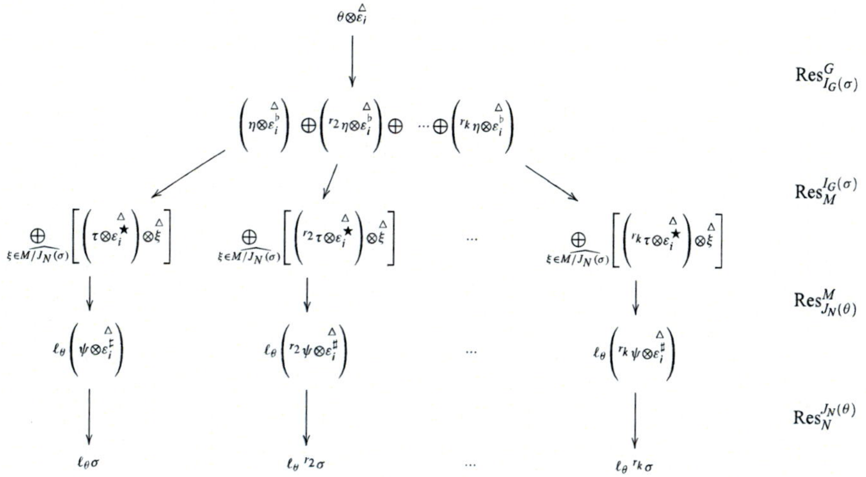

Then we describe more closely the bottom of the preceding diagram:

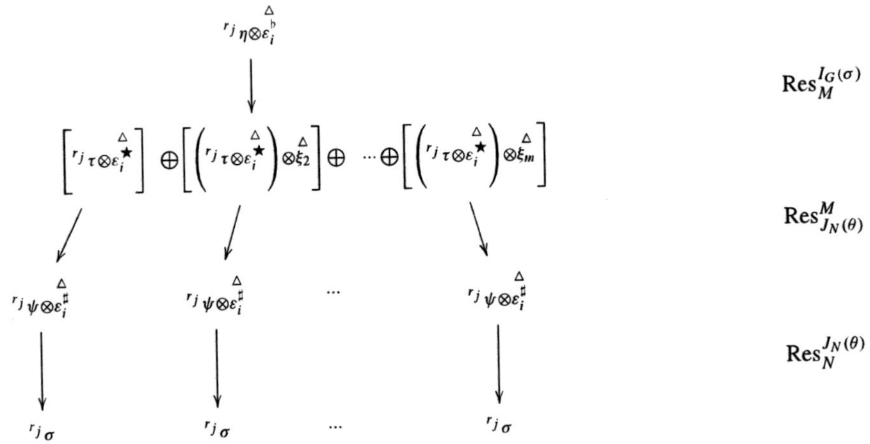

The third diagram is on induction:

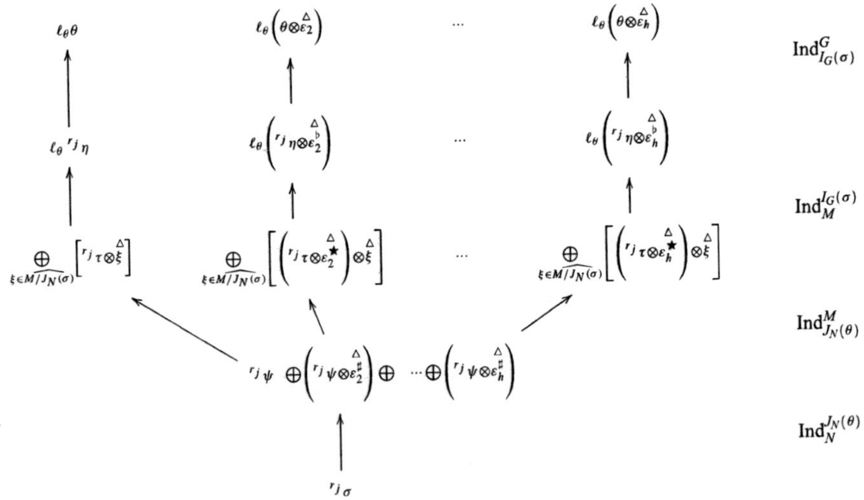

Finally, we examine more closely the top of the third diagram:

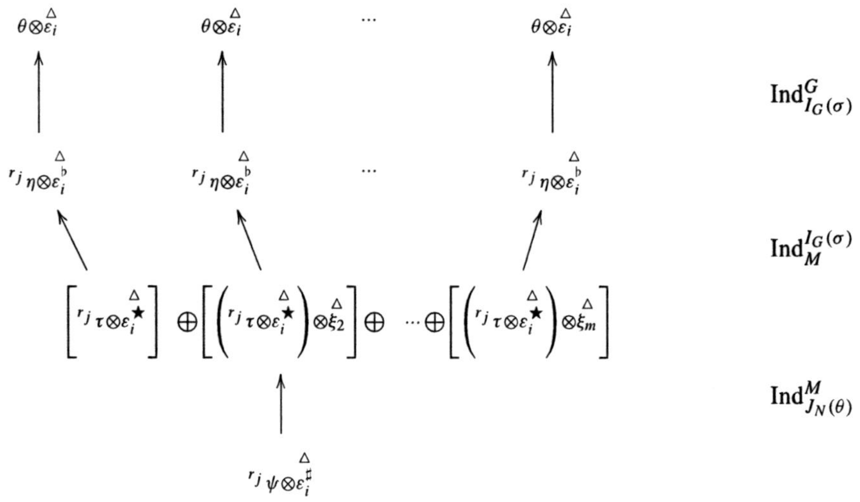

Chapter 4
The Little Group Method for Abelian Extensions

In this chapter we study the representation theory of the class of groups G with a normal subgroup N such that the following conditions are satisfied:

$$\begin{aligned} &(1)\ A := G/N \text{ is Abelian;} \\ &(2)\ \text{every } \sigma \in \widehat{N} \text{ has an extension } \widetilde{\sigma} \text{ to } I_G(\sigma). \end{aligned} \tag{4.1}$$

In other words, we consider the little group method (cf. Theorem 2.26.(2)) for Abelian extensions. Note that, under the assumptions in (4.1), N is a multiplicity-free subgroup of G (cf. Sect. 2.11 and Proposition 2.32). Now we give two basic examples where the conditions in (4.1) are satisfied.

Example 4.1 Suppose that G/N is cyclic. Then (1) is trivially satisfied, while (2) follows from the fact that also $I_G(\sigma)/N$ is cyclic, so that we may apply Proposition 2.23 with G replaced by $I_G(\sigma)$.

Example 4.2 Suppose that $G = B \rtimes A$ with A, B both Abelian (in other words, G is metabelian). Then $G/B \cong A$ is Abelian and, by Theorem 2.54, every $\xi \in \widehat{B}$ has an extension to its inertia group. This example includes all the groups in [30, Chapter 12]. See also Sect. 2.9 and Proposition 4.10 below.

4.1 General Theory

We recall the following elementary fact: if $B \leq A$ are Abelian groups, then every $\xi \in \widehat{B}$ has an extension to A; see Remark 2.24. Also, recall that when $A := G/N$ is Abelian, then (3.1) defines the dual action of $\widehat{A}$ (viewed as a group) on $\widehat{G}$.

T. Ceccherini-Silberstein et al., *Representation Theory of Finite Group Extensions*, Springer Monographs in Mathematics,
https://doi.org/10.1007/978-3-031-13873-7_4

By Corollary 3.8, the results in the following theorem and its corollary are particular cases of those in Chap. 3. Nevertheless, we prefer to present these results with independent statements and alternative proofs. In particular, we want to emphasize the use of the little group method and multiplicity freeness of N.

Theorem 4.3 *Let G be a group with a normal subgroup N such that $A := G/N$ is Abelian. Let $\sigma \in \widehat{N}$ and suppose that it has an extension $\widetilde{\sigma}$ to $I_G(\sigma)$. Then the following hold:*

(1) *Suppose that $N \lneqq I_G(\sigma) \lneqq G$ and set $A_G(\sigma) := I_G(\sigma)/N$ and $\theta := \mathrm{Ind}_{I_G(\sigma)}^G \widetilde{\sigma}$. For each $\omega \in \widehat{A_G(\sigma)}$ fix an extension $\widetilde{\omega}$ to A. Denote by $\overset{\vartriangle}{\omega}$ the inflation of ω to $I_G(\sigma)$ and by $\overset{\vartriangle}{\widetilde{\omega}}$ the inflation of $\widetilde{\omega}$ to G. Then:*

   (a) $\widehat{I_G}(\sigma) = \left\{\widetilde{\sigma} \otimes \overset{\vartriangle}{\omega} \colon \omega \in \widehat{A_G(\sigma)}\right\}$ *and* $\mathrm{Ind}_N^{I_G(\sigma)} \sigma \sim \bigoplus_{\omega \in \widehat{A_G(\sigma)}} \left(\widetilde{\sigma} \otimes \overset{\vartriangle}{\omega}\right)$.

   (b) $\widehat{G}(\sigma) = \left\{\mathrm{Ind}_{I_G(\sigma)}^G \left(\widetilde{\sigma} \otimes \overset{\vartriangle}{\omega}\right) : \omega \in \widehat{A_G(\sigma)}\right\} = \left\{\theta \otimes \overset{\vartriangle}{\widetilde{\omega}} \colon \omega \in \widehat{A_G(\sigma)}\right\}$, *so that*

$$\mathrm{Ind}_N^G \sigma \sim \bigoplus_{\omega \in \widehat{A_G(\sigma)}} \left(\theta \otimes \overset{\vartriangle}{\widetilde{\omega}}\right).$$

   In particular, $\ell_{\theta \otimes \overset{\vartriangle}{\widetilde{\omega}}} = 1$ for all $\omega \in \widehat{A_G(\sigma)}$.

   (c) $\mathrm{Res}_N^{I_G(\sigma)} \left(\widetilde{\sigma} \otimes \overset{\vartriangle}{\omega}\right) \sim \sigma$ *for all $\omega \in \widehat{A_G(\sigma)}$, that is, each $\widetilde{\sigma} \otimes \overset{\vartriangle}{\omega}$ is an extension of σ to $I_G(\sigma)$.*

   (d) $\mathrm{Res}_N^G \left(\theta \otimes \overset{\vartriangle}{\widetilde{\omega}}\right) \sim \bigoplus_{c \in G/I_G(\sigma)} {}^c\sigma$.

   (e) $\mathrm{Res}_{I_G(\sigma)}^G \left(\theta \otimes \overset{\vartriangle}{\widetilde{\omega}}\right) \sim \bigoplus_{c \in G/I_G(\sigma)} \left({}^c\widetilde{\sigma} \otimes \overset{\vartriangle}{\omega}\right)$.

   (f) *For $\varepsilon \in \widehat{A}$ we have $\theta \otimes \overset{\vartriangle}{\varepsilon} \sim \theta$ if and only if $\mathrm{Res}_{A_G(\sigma)}^A \varepsilon$ is the trivial character. That is, the stabilizer of θ with respect to the action (3.1) is $\widehat{G/I_G(\sigma)}$, which is a subgroup of $\widehat{A}$ by inflation.*

(2) *If $I_G(\sigma) = N$ then, setting $\theta := \mathrm{Ind}_N^G \sigma$:*

   (a) $\widehat{G}(\sigma) = \{\theta\}$;

   (b) $\theta \otimes \overset{\vartriangle}{\omega} \sim \theta$ *for all $\omega \in \widehat{A}$;*

   (c) $\mathrm{Res}_N^G \theta \sim \bigoplus_{a \in A} {}^a\sigma$, *and the ${}^a\sigma$, $a \in A$, are pairwise inequivalent.*

(3) *If $I_G(\sigma) = G$ then:*

   (a) $\mathrm{Ind}_N^G \sigma \sim \bigoplus_{\omega \in \widehat{A}} \left(\widetilde{\sigma} \otimes \overset{\vartriangle}{\omega}\right)$ *and* $\widehat{G}(\sigma) = \left\{\widetilde{\sigma} \otimes \overset{\vartriangle}{\omega} \colon \omega \in \widehat{A}\right\}$;

(b) $\mathrm{Res}^G_N\left(\widetilde{\sigma}\otimes\overset{\Delta}{\omega}\right)\sim\sigma$ *for each* $\omega\in\widehat{A}$*; in particular each* $\widetilde{\sigma}\otimes\overset{\Delta}{\omega}$ *is an extension of* σ *to* G.

Proof

(1) The first formula in (a) follows from (2.40) after taking $G=I_G(\sigma)$. We then deduce the second formula from the first one combined with Proposition 2.32 and (2.15). The first equality in (b) is nothing but (2.40). For the second equality, notice that for $\omega\in\widehat{A_G(\sigma)}$, since $\mathrm{Res}^A_{A_G(\sigma)}\widetilde{\omega}=\omega$, we have that $\mathrm{Res}^G_{I_G(\sigma)}\overset{\Delta}{\widetilde{\omega}}=\overset{\Delta}{\omega}$, so that, keeping in mind that $\theta:=\mathrm{Ind}^G_{I_G(\sigma)}\widetilde{\sigma}$ and using (1.36),

$$\mathrm{Ind}^G_{I_G(\sigma)}\left(\widetilde{\sigma}\otimes\overset{\Delta}{\omega}\right)\sim\mathrm{Ind}^G_{I_G(\sigma)}\left(\widetilde{\sigma}\otimes\mathrm{Res}^G_{I_G(\sigma)}\overset{\Delta}{\widetilde{\omega}}\right)\sim\theta\otimes\overset{\Delta}{\widetilde{\omega}}\,. \tag{4.2}$$

We then deduce the decomposition for $\mathrm{Ind}^G_N\sigma$ by using Proposition 2.32 and (2.15) again. (c) is trivial, while (d) follows from Theorem 2.5.(3) (recall that $\ell=1$, by Proposition 2.32). To prove (e), observe that from (4.2) it follows that

$$\mathrm{Res}^G_{I_G(\sigma)}\left(\theta\otimes\overset{\Delta}{\widetilde{\omega}}\right)\sim\mathrm{Res}^G_{I_G(\sigma)}\mathrm{Ind}^G_{I_G(\sigma)}\left(\widetilde{\sigma}\otimes\overset{\Delta}{\omega}\right)\sim\bigoplus_{c\in G/I_G(\sigma)}\left({}^c\widetilde{\sigma}\otimes\overset{\Delta}{\omega}\right),$$

where the last equivalence follows from Theorem 2.5.(1), with N replaced by $I_G(\sigma)$, Corollary 2.29, and Proposition 2.17. Finally, (f) follows from (a) and (b), by taking into account that $\widetilde{\omega}$ is an arbitrary character of A such that $\mathrm{Res}^A_{A_G(\sigma)}\widetilde{\omega}=\omega$.

(2) This is just a particular case of Proposition 2.28; it may be deduced also by taking $I_G(\sigma)=N$ in (1).

(3) This is just a particular case of Proposition 2.18; see also Corollary 2.35 and its proof. Clearly, it may be deduced also by taking $I_G(\sigma)=G$ in (1).

■

Corollary 4.4 *Let* G *and* N *be as before and suppose that the conditions in* (4.1) *are verified. Then the correspondence* (2.19) *becomes:*

$$\{{}^c\sigma : c\in G/I_G(\sigma)\}\mapsto\widehat{G}(\sigma)\equiv\left\{\theta\otimes\overset{\Delta}{\widetilde{\omega}}:\ \omega\in\widehat{A_G(\sigma)}\right\} \tag{4.3}$$

and it establishes a one–to–one correspondence between the A*-orbits (actually,* $G/I_G(\sigma)$*-orbits) on* $\widehat{N}$ *and the* $\widehat{A}$*-orbits (actually,* $\widehat{A_G(\sigma)}$*-orbits) on* $\widehat{G}$.

In particular, the correspondence becomes:

$$\{{}^a\sigma : a\in A\}\mapsto\widehat{G}(\sigma)\equiv\{\mathrm{Ind}^G_N\sigma\}\qquad \textit{when } I_G(\sigma)=N,$$

and

$$\{\sigma\} \mapsto \widehat{G}(\sigma) \equiv \left\{\theta \otimes \overset{\triangle}{\omega}\colon \omega \in \widehat{A}\right\} \qquad \text{when } I_G(\sigma) = G.$$

Finally, set $n := |G/N|$*. Then, to an A-orbit* $\mathscr{O}$ *in* $\widehat{N}$ *of cardinality* $k_{\mathscr{O}}$ *there corresponds an* $\widehat{A}$*-orbit* Ω *in* $\widehat{G}$ *of cardinality* $k_\Omega = n/k_{\mathscr{O}}$*. Moreover, if* $d_{\mathscr{O}}$ *is the dimension of the representations in* $\mathscr{O}$*, then the representations in* Ω *have dimension* $d_\Omega = k_{\mathscr{O}} d_{\mathscr{O}}$*. In particular:*

- *each single–element orbit* $\{\theta\}$ *in* $\widehat{G}$ *corresponds to an* $|A|$*–elements orbit in* $\widehat{N}$*, where each representation has dimension equal to* $d_\theta/|A|$*; we then say that we have a* "many–to–one" correspondence*;*
- *each single–element orbit* $\{\sigma\}$ *in* $\widehat{N}$ *corresponds to an* $|A|$*–elements orbit in* $\widehat{G}$*, where each representation has dimension equal to* d_σ*; we then say that we have a* " one–to–many" correspondence.

Proof This is an immediate consequence of Theorem 4.3. ■

Remark 4.5 One may deduce Theorem 4.3 from the theory developed in Chap. 3.

Remark 4.6 In the setting of Corollary 4.4, given $\sigma \in \widehat{N}$ let $\mathscr{O}(\sigma)$ denote the A-orbit of σ in $\widehat{N}$ and let $\Omega(\sigma)$ denote the corresponding $\widehat{A}$-orbit in $\widehat{G}$.

Then, simply writing k_σ for $k_{\mathscr{O}(\sigma)}$, given $\theta \in \Omega(\sigma)$, we have

$$|\mathscr{R}(\sigma)| = k_\sigma, \quad |\widehat{G}(\sigma)| = n/k_\sigma, \quad \text{and} \quad d_\theta = k_\sigma d_\sigma.$$

Then we can use (2.22) to check (1.20):

$$\sum_{\theta\in\widehat{G}} d_\theta^2 = \sum_{\sigma\in\widehat{N}} \frac{1}{k_\sigma} \cdot \frac{n}{k_\sigma} \cdot (k_\sigma d_\sigma)^2 = n \sum_{\sigma\in\widehat{N}} d_\sigma^2 = n \cdot |N| = |G|.$$

Remark 4.7 The relation between the real/complex/quaternionic type of two representations $\sigma \in \widehat{N}$ and $\theta \in \widehat{G}(\sigma)$ associated by the correspondence in Corollary 4.4 may be computed by means of the results in Sect. 2.7, taking also into account Remark 2.53 when G/N is cyclic. Indeed, when $I_G(\sigma) = N$ we may apply Theorem 2.50. When $I_G(\sigma) = G$ we may apply Theorem 2.52 for a fixed extension θ of σ and then we can obtain all the other elements in $\widehat{G}(\sigma)$ as tensor products by the characters of G/N. Finally, in the intermediate case $N \lneqq I_G(\sigma) \lneqq G$ we may apply Theorem 2.52 and then Theorem 2.50. In Theorem 4.22, Theorem 4.23, and Proposition 5.30, as well as in Chap. 5, we will see applications of the procedure described above. See also Sect. 5.5.

4.2 Normal Subgroups with the Prime Condition

In this section we examine a particular case of the conditions in (4.1) and we study its consequences. Our idea is inspired by Section III.11 in the monograph by Simon [125] which is devoted to the case of an index two subgroup. However, his treatment is more elementary: Simon does not use Clifford theory and he defines induced representations in this particular setting before developing the general theory in a later section. On the contrary, we use all the power of Clifford theory to get the relations between the representation theory of N and that of G.

Definition 4.8 We say that $N \lhd G$ satisfies the *prime condition* (or that it is a *prime couple*) if $N \neq G$, the quotient group $A := G/N$ is Abelian, and for each $\sigma \in \widehat{N}$ one (clearly, and only one) of the following conditions occurs:

(i) $I_G(\sigma) = N$;
(ii) $I_G(\sigma) = G$ and $\ell_\theta = 1$ for all $\theta \in \widehat{G}(\sigma)$; equivalently, σ has an extension $\widetilde{\sigma}$ to G.

Clearly, the conditions in Definition 4.8 are stronger than those in (4.1): essentially, we have added the requirement that $I_G(\sigma) = N$ or G. In particular, only cases (2) and (3) in Theorem 4.3 occur. These conditions also appear in Isaac's going down theorem (Theorem 2.37) with G replaced by M.

We now describe two remarkable classes of prime couples. A version of the following proposition (with the relative consequences in Theorem 4.3) is attributed to Burnside by Berkovich and Zhmud [16, p. 200]. Note also that it is a simple consequence of Isaac's going down theorem (Theorem 2.37), but we prefer to give a direct proof.

Proposition 4.9 *Suppose that* $[G : N]$ *is a prime number. Then* $N \lhd G$ *is a prime couple.*

Proof Set $p := [G : N]$, so that $G/N \cong C_p$, the cyclic group of order p. Since C_p is a simple group, the stabilizers of the action of C_p on $\widehat{N}$ may only be $\{1\}$ or the whole of C_p. Therefore, for each $\sigma \in \widehat{N}$, the inertia group $I_G(\sigma)$ is equal to either N or G. Suppose that $I_G(\sigma) = G$. Then, from (2.47) we deduce that, for each $\theta \in \widehat{G}(\sigma)$, the square ℓ_θ^2 of its inertia index divides p, so that necessarily $\ell_\theta = 1$. ∎

Sections 4.3 and 4.4 below will be devoted to a more detailed analysis of this case.

Proposition 4.10 *Suppose that* $B \lhd G$ *satisfies that* $A := G/B$ *is Abelian and* $B \subseteq Z(G)$*. Then* $B \lhd G$ *is a prime couple.*

Proof Let $\xi \in \widehat{B}$. It is straightforward to check that, since B is central, $I_G(\xi) = G$ and ξ admits an extension to the whole of G. Since $I_G(\xi)/B = A$ is Abelian, Proposition 2.32 ensures that $\ell_\theta = 1$ for all $\theta \in \widehat{G}(\xi)$. ∎

The group G in the above proposition is a central extension by an Abelian group (cf. Chap. 6) and therefore it is 2-step nilpotent (see Example 6.4).

We introduce some notation. Suppose that $N \lhd G$ is a prime couple. We set

$$\widehat{N}_I := \{\sigma \in \widehat{N} : I_G(\sigma) = N\} \quad \text{and} \quad \widehat{N}_{II} := \{\sigma \in \widehat{N} : I_G(\sigma) = G\}. \tag{4.4}$$

Moreover, accordingly with Theorem 4.3, for each $\sigma \in \widehat{N}_{II}$ we fix an extension $\widetilde{\sigma}$ to G. In order to prove a beautiful theorem of Mackey, we first give a technical result.

Lemma 4.11 *Suppose that $B \lhd G$ is a prime couple with B Abelian and set $A := G/B$. Then:*

- *for each $\theta \in \widehat{G}$ either $\dim\theta = 1$ or $\dim\theta = |A|$;*
- *if $\theta_1, \theta_2 \in \widehat{G}$ and $\dim\theta_1 = \dim\theta_2 = |A|$ then there exist $\xi_1, \xi_2 \in \widehat{B}_I$ such that: $\theta_i \sim \mathrm{Ind}_B^G \xi_i$, for $i = 1, 2$, and the decomposition of their tensor product into irreducible G-representations is:*

$$\theta_1 \otimes \theta_2 \sim \left[\bigoplus_{\substack{c\in A:\\ \xi_1\otimes {}^c\xi_2 \in \widehat{B}_I}} \mathrm{Ind}_B^G\left(\xi_1 \otimes {}^c\xi_2\right) \right] \bigoplus \left[\bigoplus_{\substack{c\in A:\\ \xi_1\otimes {}^c\xi_2 \in \widehat{B}_{II}}} \left[\bigoplus_{\omega\in\widehat{A}} \left(\widetilde{\xi_1 \otimes {}^c\xi_2}\right) \otimes \overset{\triangle}{\omega} \right] \right]. \tag{4.5}$$

Moreover, the second summand in the right hand side of (4.5) is multiplicity-free and has no common subrepresentations with the first summand.

Proof From Corollary 4.4 it follows immediately that each $\theta \in \widehat{G}$ has dimension equal to either 1 or $[G : B] = |A|$ (see also [30, Exercise 10.3.16]). Suppose that $\dim\theta_1 = \dim\theta_2 = |A|$, so that we are in the case of Theorem 4.3.(2). Then there exist $\xi_1, \xi_2 \in \widehat{B}$ such that, for $i = 1, 2$, one has: $\theta_i \sim \mathrm{Ind}_B^G \xi_i$, $\mathrm{Res}_B^G \theta_i \sim \bigoplus_{a\in A} {}^a\xi_i$, where $\{{}^a\xi_i : a \in A\}$ are pairwise inequivalent, and

$$\theta_i \otimes \overset{\triangle}{\omega} \sim \theta_i \qquad \text{for all } \omega \in \widehat{A}. \tag{4.6}$$

Therefore

$$\begin{aligned}
\mathrm{Res}_B^G(\theta_1 \otimes \theta_2) &\sim \left(\bigoplus_{a\in A} {}^a\xi_1\right) \bigotimes \left(\bigoplus_{b\in A} {}^b\xi_2\right) \\
(c = a^{-1}b) \quad &\sim \bigoplus_{c\in A} \left[\bigoplus_{a\in A} {}^a(\xi_1 \otimes {}^c\xi_2)\right] \\
&\sim \left[\bigoplus_{\substack{c\in A:\\ \xi_1\otimes {}^c\xi_2 \in \widehat{B}_I}} \left[\bigoplus_{a\in A} {}^a(\xi_1 \otimes {}^c\xi_2)\right] \right] \bigoplus \left[\bigoplus_{\substack{c\in A:\\ \xi_1\otimes {}^c\xi_2 \in \widehat{B}_{II}}} \left[\bigoplus_{a\in A} {}^a(\xi_1 \otimes {}^c\xi_2)\right] \right].
\end{aligned} \tag{4.7}$$

In order to compute the decomposition of $\theta_1 \otimes \theta_2$ into irreducible G-representations we use Theorem 4.3 and Corollary 4.4 which enable us to recover an irreducible G-representation from the decomposition of its restriction to B. Note also that the terms in the last expression in (4.7) are assembled in such a way that, for each $c \in A$, the representations ${}^a(\xi_1 \otimes {}^c\xi_2)$ appearing in the left and the right summands belong to distinct orbits of A on $\widehat{B}$. In particular,

- if $\xi_1 \otimes {}^c\xi_2 \in \widehat{B}_I$, then $\theta := \mathrm{Ind}_B^G(\xi_1 \otimes {}^c\xi_2)$ is irreducible and contained in $\theta_1 \otimes \theta_2$ because it is the unique irreducible G-representation such that:

$$\mathrm{Res}_B^G \theta \sim \bigoplus_{a \in A} {}^a(\xi_1 \otimes {}^c\xi_2)\,; \tag{4.8}$$

- if $\xi_1 \otimes {}^c\xi_2 \in \widehat{B}_{II}$, then $\xi_1 \otimes {}^c\xi_2$ has an extension $\widetilde{\xi_1 \otimes {}^c\xi_2}$ to G. Moreover, $\theta_1 \otimes \theta_2$ must contain one element from $\widehat{G}(\xi_1 \otimes {}^c\xi_2)$, namely, $\left(\widetilde{\xi_1 \otimes {}^c\xi_2}\right) \otimes \overset{\triangle}{\omega_0}$, for some $\omega_0 \in \widehat{A}$. But $(\theta_1 \otimes \theta_2) \otimes \overset{\triangle}{\omega} \sim \theta_1 \otimes (\theta_2 \otimes \overset{\triangle}{\omega}) \sim \theta_1 \otimes \theta_2$ by (4.6), so that $\theta_1 \otimes \theta_2$ must contain all the representations $\left(\widetilde{\xi_1 \otimes {}^c\xi_2}\right) \otimes \overset{\triangle}{\omega}$, with $\omega \in \widehat{A}$; the sum of these representations corresponds to the summand

$$\bigoplus_{a \in A} {}^a\left(\xi_1 \otimes {}^c\xi_2\right) \sim |A|(\xi_1 \otimes {}^c\xi_2). \tag{4.9}$$

The fact that the summands in (4.5) have no common subrepresentations is obvious. Moreover, if $c_1, c_2 \in A$, $c_1 \neq c_2$, and $\xi_1 \otimes {}^{c_j}\xi_2 \in \widehat{B}_{II}$, $j = 1, 2$, then these characters are fixed by A so that ${}^a(\xi_1 \otimes {}^{c_1}\xi_2) \neq \xi_1 \otimes {}^{c_2}\xi_2$, for all $a \in A$. This implies that the second summand in (4.5) is multiplicity-free; cf. (4.9). ■

On the other hand, the first summand in (4.5) may decompose with multiplicity. This happens if there exist $c_1, c_2, a \in A$, $c_1 \neq c_2$ such that ${}^a(\xi_1 \otimes {}^{c_1}\xi_2) = \xi_1 \otimes {}^{c_2}\xi_2$, cf. (4.8). In the example below we show that this is indeed possible.

Example 4.12 Consider the cyclic group of order two $C_2 = \{0, 1\}$ and set $B := C_2 \times C_2 \times C_2$. Let $\phi \in \mathrm{Aut}(B)$ defined by setting

$$\phi(\alpha, \beta, \gamma) := (\beta, \gamma, \alpha)$$

for all $(\alpha, \beta, \gamma) \in B$. Then ϕ has order three and we consider the semidirect product $G := B \rtimes_\phi C_3$; see Proposition 1.7 and Theorem 2.54 (in fact, B is the wreath product $C_2 \wr C_3$; see Sect. 2.10 and [28]). We parameterize the characters of B by setting:

$$\xi_{\alpha,\beta,\gamma}(\alpha', \beta', \gamma') := (-1)^{\alpha\alpha' + \beta\beta' + \gamma\gamma'},$$

for all $\alpha, \alpha', \beta, \beta', \gamma, \gamma' \in C_2$. Then the action of $C_3 = \{1, \phi, \phi^2\}$ on $\widehat{B}$ is given by

$$\begin{aligned} {}^{1}\xi_{\alpha,\beta,\gamma} &= \xi_{\alpha,\beta,\gamma} \\ {}^{\phi}\xi_{\alpha,\beta,\gamma} &= \xi_{\gamma,\alpha,\beta} \\ {}^{\phi^2}\xi_{\alpha,\beta,\gamma} &= \xi_{\beta,\gamma,\alpha}. \end{aligned}$$

The orbits of this action are:

- $\{\xi_{0,0,0}\}$ and $\{\xi_{1,1,1}\}$: each of these corresponds to three inequivalent one-dimensional representations of G and
- $\{\xi_{1,0,0}, \xi_{0,1,0}, \xi_{0,0,1}\}$ and $\{\xi_{1,1,0}, \xi_{0,1,1}, \xi_{1,0,1}\}$: we denote by θ_1 and θ_2 the corresponding irreducible three-dimensional G-representations.

Taking into account that $\xi_{\alpha,\beta,\gamma} \otimes \xi_{\alpha',\beta',\gamma'} = \xi_{\alpha+\alpha',\beta+\beta',\gamma+\gamma'}$, in the present case (4.7) becomes:

$$\begin{aligned} \mathrm{Res}^G_B(\theta_1 \otimes \theta_2) &\sim \oplus^2_{\epsilon,\delta=0} {}^{\phi^\epsilon}\left(\xi_{1,0,0} \otimes {}^{\phi^\delta}\xi_{1,1,0}\right) \\ &\sim \left(\oplus^2_{\epsilon=0} {}^{\phi^\epsilon}\xi_{0,1,0}\right) \oplus \left(\oplus^2_{\epsilon=0} {}^{\phi^\epsilon}\xi_{1,1,1}\right) \oplus \left(\oplus^2_{\epsilon=0} {}^{\phi^\epsilon}\xi_{0,0,1}\right) \\ &\sim 2\left(\oplus^2_{\epsilon=0} {}^{\phi^\epsilon}\xi_{1,0,0}\right) \oplus 3\,\xi_{1,1,1}. \end{aligned}$$

It follows that $\theta_1 \otimes \theta_2$ contains θ_1 (with multiplicity 2) and each one-dimensional representation associated with $\xi_{1,1,1}$ (with multiplicity 1). Similarly, it is easy to check that $\theta_1 \otimes \theta_1 \sim \theta_2 \otimes \theta_2$ because both of these products contain θ_2 (with multiplicity 2) and each representation associated with $\xi_{0,0,0}$ (with multiplicity 1).

By using Lemma 4.11, we are now in a position to deduce Mackey's theorem [93] we alluded to above.

Theorem 4.13 (Mackey) *Suppose that G contains a (normal) Abelian subgroup B of index two. Then, for any $\theta_1, \theta_2 \in \widehat{G}$, the tensor product $\theta_1 \otimes \theta_2$ decomposes multiplicity-freely.*

Proof Let $\theta_1, \theta_2 \in \widehat{G}$. Note that θ_1 and θ_2 have dimension ≤ 2. If θ_1 (or θ_2) has dimension one, then the tensor product $\theta_1 \otimes \theta_2$ is clearly irreducible. Suppose that $\dim\theta_1 = \dim\theta_2 = 2$. Then Lemma 4.11 ensures that there exist $\xi_1, \xi_2 \in \widehat{B}_I$ such that $\theta_i \sim \mathrm{Ind}^G_B\xi_i$ for $i = 1, 2$. Let $h \in G \setminus B$, so that $G = B \sqcup hB$. Note that now the right hand side of (4.5) is the direct sum of two representations (parameterized by $\xi_1 \otimes \xi_2$ and $\xi_1 \otimes {}^h\xi_2$, respectively). If one of these belongs to the second summand therein (that is, one of $\xi_1 \otimes \xi_2$ and $\xi_1 \otimes {}^h\xi_2$ is in $\widehat{B}_{II}$) then we are done (recall that the second summand in (4.5) is always multiplicity-free). On the other hand, if both $\xi_1 \otimes \xi_2$ and $\xi_1 \otimes {}^h\xi_2$ are in $\widehat{B}_I$, then

$${}^h(\xi_1 \otimes \xi_2) \equiv {}^h\xi_1 \otimes {}^h\xi_2 \neq (\xi_1 \otimes {}^h\xi_2) \neq \xi_1 \otimes \xi_2$$

and Clifford correspondence ensures that $\mathrm{Ind}_B^G(\xi_1 \otimes \xi_2) \not\sim \mathrm{Ind}_B^G(\xi_1 \otimes {}^h\xi_2)$. We deduce that (4.5) is multiplicity-free also in this outstanding case. ∎

An exposition of Mackey's paper [93] is in [29].

Proposition 4.14 *Suppose that $B \lhd G$ is a prime couple and $B \leq Z(G)$ (cf. Chap. 6). Then for any $\theta_1, \theta_2 \in \widehat{G}$, the tensor product $\theta_1 \otimes \theta_2$ decomposes multiplicity-free in the following way:*

$$\theta_1 \otimes \theta_2 \sim \bigoplus_{c \in G/B} \left[\bigoplus_{\omega \in \widehat{G/B}} \left(\widetilde{\xi_1 \otimes {}^c\xi_2} \right) \otimes \overset{\Delta}{\omega} \right].$$

Proof As B is contained in the center of G, one has $I_G(\sigma) = G$ for all $\sigma \in \widehat{B}$ and therefore $\widehat{B}_I = \varnothing$. In other words, the right hand side of (4.5) consists only of the second summand which, as we know, is always multiplicity-free. ∎

4.3 Normal Subgroups of Prime Index

In this section, we assume the hypotheses in Proposition 4.9. Moreover, we use the notation introduced in Remark 2.53: we choose $h \in G$ such that Nh generates $G/N \cong C_p$, so that $G = \sqcup_{j=0}^{p-1} Nh^j$ and the corresponding section is $Nh^j \mapsto h^j$, $j = 0, 1, \ldots, p-1$.

We first recall two well known and elementary facts. The first is that an index two subgroup is normal; this is not true for prime indices greater than two: for instance, in the symmetric group S_3, the subgroup generated by the transposition $(1, 2)$ has index three and is not a normal subgroup. The second fact is that if N is normal of index p prime in G, then G/N is isomorphic to the cyclic group of order p (this is the unique (up to isomorphisms) group of order p).

We now show that a relationship similar to those in Theorem 4.3 is also valid for the conjugacy classes of N and G. Indeed, consider the C_p-action on the conjugacy classes of N: for a conjugacy class $\mathscr{C}$ of N and $0 \leq j \leq p-1$ set ${}^{h^j}\mathscr{C} := h^{-j}\mathscr{C}h^j$ (cf. Sect. 2.6). Moreover, $\bigcup_{j=0}^{p-1} {}^{h^j}\mathscr{C}$ is a conjugacy class of G.

The stabilizer of $\mathscr{C}$ is equal to either $\{1_G\}$ or C_p. In the first case, $\sqcup_{j=0}^{p-1} {}^{h^j}\mathscr{C}$ is a disjoint union and we say that $\mathscr{C}$ is a *split* class of N; we also say that we have a *p-tuple of split classes*. In the second case, $\mathscr{C} = {}^{h^j}\mathscr{C}$ for all $j = 0, 1, \ldots, p-1$, so that $\mathscr{C}$ is a conjugacy class of G, and we say that $\mathscr{C}$ is *nonsplit*.

We now refine Corollary 2.48 in the present setting and show an interesting duality.

Proposition 4.15

(1) *The number of split classes in* N *is equal to* $|\widehat{N}_I|$.
(2) *The number of nonsplit classes in* N *equals both* $|\widehat{N}_{II}|$ *and* $\frac{c}{p-1}$, *where* c *is the number of conjugacy classes of* G *not contained in* N.

Proof Denote by a (resp. b) the number of nonsplit classes (resp. p-tuples of split classes) and set $a' := |\widehat{N}_{II}|$ and $b' := |\widehat{N}_I|/p$. Also denote by $\mathscr{A}$ the vector space of N-central functions $f \in L(N)$ such that ${}^{h^j}f = f$, for all $j = 0, 1, \ldots, p-1$, where ${}^{h^j}f(n) := f(h^{-j}nh^j)$ for all $n \in N$. In other words, setting ${}^{g}f(n) := f(g^{-1}ng)$, for all $f \in L(N)$, $g \in G$, and $n \in N$, then $\mathscr{A} = \{f \in L(N) : {}^{g}f = f$ for all $g \in G\}$.

Then

$$a + pb = |\widehat{N}| = a' + pb'.$$

Moreover,

$$a + b = \dim \mathscr{A} = a' + b'$$

(for the second equality, note that a basis for $\mathscr{A}$ is constituted by the functions χ^σ, with $\sigma \in \widehat{N}_{II}$, together with the functions of the form $\Sigma_{j=0}^{p} \chi^{{}^{h^j}\sigma}$, with $\sigma \in \widehat{N}_I$). It follows that $a = a'$ and $b = b'$.

Finally, $a+b+c = |\widehat{G}| = pa'+b' = pa+b$ (where the second equality follows from Corollary 4.4), so that $a = c/(p-1)$. ■

We now give an explicit matrix realization of $\mathrm{Ind}_N^G \sigma$.

Lemma 4.16 *In the setting of Theorem 4.3.(2), let* $A(n)$ *with* $n \in N$ *be a matrix realization of* σ. *Then,* ${}^{h^j}A(n) := A(h^{-j}nh^j)$ *is a matrix realization of* ${}^{h^j}\sigma$. *Moreover,* $B(g)$ *with* $g \in G$, *defined by*

$$B(nh^j) := \begin{array}{c} \\ 0 \\ 1 \\ 2 \\ \\ j-1 \\ j \\ j+1 \\ \\ p-1 \end{array} \begin{array}{ccccccccc} 0 & 1 & & p-j-1 & p-j & p-j+1 & p-j+2 & & p-1 \\ \hline 0 & 0 & \cdots & 0 & A(nh^p) & 0 & 0 & \cdots & 0 \\ 0 & 0 & \cdots & 0 & 0 & {}^{h}A(nh^p) & 0 & \cdots & 0 \\ 0 & 0 & \cdots & 0 & 0 & 0 & {}^{h^2}A(nh^p) & & 0 \\ \vdots & \vdots & & \vdots & \vdots & \vdots & & \ddots & \\ 0 & 0 & \cdots & 0 & 0 & 0 & 0 & & {}^{h^{j-1}}A(nh^p) \\ {}^{h^j}A(n) & 0 & \cdots & 0 & 0 & 0 & 0 & \cdots & 0 \\ 0 & {}^{h^{j+1}}A(n) & & 0 & 0 & 0 & 0 & \cdots & 0 \\ \vdots & & \ddots & & \vdots & \vdots & \vdots & & \vdots \\ 0 & 0 & & {}^{h^{p-1}}A(n) & 0 & 0 & 0 & \cdots & 0 \end{array}$$

for all $n \in N$ *and* $j = 0, 1, 2, \ldots, p-1$*, is a matrix realization of* $\mathrm{Ind}_N^G \sigma$.

Proof This follows immediately from (1.31). Just note that for $0 \le i, k \le p-1$, the (i,k)-block is given by

$$A(h^{-i} n h^{j+k}) = \begin{cases} 0 & \text{if } i \not\equiv j+k \bmod p \\ {}^{h^i}A(n) & \text{if } i \equiv j+k \bmod p \text{ and } 0 \le k \le p-j-1 \\ {}^{h^i}A(nh^p) & \text{if } i \equiv j+k \bmod p \text{ and } p-j \le k \le p-1. \end{cases}$$

■

We now give the formulas that relate the characters of $\sigma \in \widehat{N}$ and $\theta \in \widehat{G}(\sigma)$.

Proposition 4.17

(1) *Suppose that* $\sigma \in \widehat{N}_I$ *and set* $\theta := \mathrm{Ind}_N^G \sigma$*. Then*

$$\chi^\theta(nh^j) = \begin{cases} 0 & \text{if } 1 \le j \le p-1, n \in N, \\ p\chi^\sigma(n) & \text{if } j = 0 \text{ and } n \text{ belongs to a nonsplit class,} \\ \sum_{j=0}^{p-1} \chi^{{}^{h^j}\sigma}(n) & \text{if } j = 0 \text{ and } n \text{ belongs to a split class.} \end{cases}$$

Moreover, for each choice of $0 \le i < j \le p-1$ *there exist a split class on which* $\chi^{{}^{h^i}\sigma}$ *and* $\chi^{{}^{h^j}\sigma}$ *disagree.*

(2) *Suppose that* $\sigma \in \widehat{N}_{II}$, $\theta \in \widehat{G}(\sigma)$*, and* $\omega \in \widehat{C_p}$*. Then*

$$\chi^{\theta \otimes \overset{\triangle}{\omega}}(nh^j) = \begin{cases} \chi^\theta(nh^j)\omega(h)^j & \text{if } 1 \le j \le p-1, n \in N, \\ \chi^\sigma(n) & \text{if } j = 0 \text{ and } n \in N. \end{cases}$$

Proof

(1) This corresponds to the situation described in Theorem 4.3.(2). It follows from Proposition 2.9.(4), taking into account that if n belongs to a nonsplit class then $\chi^{h^j\sigma}(n) = \chi^\sigma(h^{-j} n h^j) = \chi^\sigma(n)$ for $1 \le j \le p-1$. See also Lemma 4.16. Finally, for each choice of $0 \le i < j \le p-1$ we have $\chi^{{}^{h^i}\sigma} \ne \chi^{{}^{h^j}\sigma}$.

(2) This corresponds to the situation described in Theorem 4.3.(3). Just note that, for all $n \in N$,

$$\chi^{\theta \otimes \overset{\triangle}{\omega}}(nh^j) = \chi^\theta(nh^j)\, \overset{\triangle}{\omega}\, (nh^j) = \chi^\theta(nh^j)\omega(h^j)$$

and that $\chi^\theta(n) = \chi^\sigma(n)$.

■

4.4 The Case of Index Two Subgroups

It is convenient to state the results of Sects. 4.2 and 4.3 in the particular case of subgroups of index two. In this setting, the theory is more elementary and transparent. A particular but beautiful result that we have already presented is Mackey's theorem (Theorem 4.13).

Let then N be a subgroup of G of index two. Taking $h \in G \setminus N$ we have the coset decomposition

$$G = N \sqcup hN.$$

Moreover, we can define the *alternating representation* of G (with respect to N) as the one–dimensional representation $(\varepsilon, \mathbb{C})$ defined by

$$\varepsilon(g) := \begin{cases} 1 & \text{if } g \in N \\ -1 & \text{otherwise.} \end{cases}$$

We define two actions of $C_2 = \{0, 1\}$ on $\widehat{N}$ and $\widehat{G}$ as follows. The element 1 acts on $\widehat{N}$ by $\widehat{N} \ni \sigma \mapsto {}^h\sigma \in \widehat{N}$ where, as usual, ${}^h\sigma(n) = \sigma(h^{-1}nh)$ for all $n \in N$. Also, 1 acts on $\widehat{G}$ by $\widehat{G} \ni \theta \mapsto \theta \otimes \varepsilon \in \widehat{G}$. Clearly, both $\widehat{N}$ and $\widehat{G}$ are partitioned into their C_2-orbits. Moreover, every such an orbit consists of one or two representations.

We now reformulate Theorem 4.3 and Corollary 4.4 in the present setting.

Theorem 4.18 *Let N be a subgroup of index two in G and let $h \in G \setminus N$. Let $\sigma \in \widehat{N}$.*

(1) *If $I_G(\sigma) = N$, then*

$$\theta := \mathrm{Ind}_N^G \sigma \in \widehat{G}(\sigma), \ \theta \otimes \varepsilon \sim \theta, \ \textit{and } \mathrm{Res}_N^G \theta \sim \sigma \oplus {}^h\sigma \textit{ with } \sigma \not\sim {}^h\sigma.$$

(2) *If $I_G(\sigma) = G$, then for $\theta \in \widehat{G}(\sigma)$ we have $\mathrm{Ind}_N^G(\sigma) \sim \theta \oplus (\theta \otimes \varepsilon)$ with $\theta \not\sim \theta \otimes \varepsilon$ and $\mathrm{Res}_N^G \theta \sim \mathrm{Res}_N^G(\theta \otimes \varepsilon) \sim \sigma$.*

(3) *Setting*

$$\{\sigma, {}^h\sigma\} \mapsto \widehat{G}(\sigma) \equiv \{\mathrm{Ind}_N^G \sigma\} \qquad \textit{when } I_G(\sigma) = N$$

and

$$\{\sigma\} \mapsto \widehat{G}(\sigma) \equiv \{\theta, \theta \otimes \varepsilon\} \qquad \textit{when } I_G(\sigma) = G,$$

we get a one–to–one correspondence between the C_2-orbits on $\widehat{N}$ and on $\widehat{G}$. In particular, to each single–element orbit on $\widehat{N}$ (resp. on $\widehat{G}$) there corresponds a two–elements orbit on $\widehat{G}$ (resp. on $\widehat{N}$).

We now reformulate the analogue of Proposition 4.15. Let $\mathscr{C}$ be an N-conjugacy class of N. In the present setting, if $\mathscr{C} \neq {}^h\mathscr{C}$ (and therefore $\mathscr{C} \cap {}^h\mathscr{C} = \varnothing$) we say that $\mathscr{C}$ is a *split* class of N. One also says that $\{\mathscr{C}, {}^h\mathscr{C}\}$ is a *pair of split classes* or that $\mathscr{C} \cup {}^h\mathscr{C}$ *splits* into two N-classes. If $\mathscr{C} = {}^h\mathscr{C}$ (and therefore $\mathscr{C}$ is a G-conjugacy class) we say that $\mathscr{C}$ is *nonsplit*.

Proposition 4.19

(1) *The number of split classes in N is equal to* (*cf.* (4.4))

$$|\widehat{N}_I| \equiv |\{\sigma \in \widehat{N} : \sigma \not\sim {}^h\sigma\}|.$$

(2) *The number of nonsplit classes in N is equal to both the number of conjugacy classes of G but not in N and to*

$$|\widehat{N}_{II}| \equiv |\{\sigma \in \widehat{N} : \sigma \sim {}^h\sigma\}|.$$

Then the analogue of Lemma 4.16 states as follows.

Lemma 4.20 *In the setting of Theorem 4.18.(1), let $A(n)$ with $n \in N$ be a matrix realization of σ. Then ${}^hA(n) := A(h^{-1}nh)$ is a matrix realization of ${}^h\sigma(h)$. Moreover, $B(g)$ with $g \in G$, defined by*

$$B(n) := \begin{pmatrix} A(n) & 0 \\ 0 & {}^hA(n) \end{pmatrix} \quad \text{and} \quad B(nh) := \begin{pmatrix} 0 & A(nh^2) \\ {}^hA(n) & 0 \end{pmatrix}$$

for all $n \in N$, is a matrix realization of $\mathrm{Ind}_N^G \sigma$.

Proposition 4.17 takes the following form.

Proposition 4.21

(1) *Let $\sigma \in \widehat{N}_I$ and set $\theta := \mathrm{Ind}_N^G \sigma$. Then $\chi^\theta(g) = 0$ if $g \notin N$ and $\chi^\theta(n) = \chi^\sigma(n) + \chi^{h_\sigma}(n)$, for all $n \in N$. Moreover, if $\mathscr{C}$ is a nonsplit class of N and $n \in \mathscr{C}$, then $\chi^\sigma(n) = \chi^{h_\sigma}(n) = \frac{1}{2}\chi^\theta(n)$. If $\mathscr{C} \cup {}^h\mathscr{C}$ is a pair of split classes and $n_1 \in \mathscr{C}, n_2 \in {}^h\mathscr{C}$, then $\chi^\sigma(n_1) = \chi^{h_\sigma}(n_2)$ and $\chi^\sigma(n_2) = \chi^{h_\sigma}(n_1)$.*
Finally, there exists a split class on which χ^σ and χ^{h_σ} disagree (because $\chi^\sigma \neq \chi^{h_\sigma}$).

(2) *Let $\sigma \in \widehat{N}_{II}$ and let $\theta \in \widehat{G}(\sigma)$. Then we have $\chi^{\theta\otimes\varepsilon}(nh) = -\chi^\theta(nh)$ and $\chi^\sigma(n) = \chi^\theta(n) = \chi^{\theta\otimes\varepsilon}(n)$, for all $n \in N$.*

We end this section with a reformulation of the results of Sect. 2.7 in the present setting. See also Remark 4.7. In what follows, $A(n)$ and $B(g)$, $n \in N$ and $g \in G$, will be matrix realizations of σ and θ, respectively, as in Lemma 4.20.

The following is the index-two version of Theorem 2.50. Recall that given $\sigma \in \widehat{N}$, we denote by σ' the adjoint of σ.

Theorem 4.22 *Let $\sigma \in \widehat{N}$ such that $I_G(\sigma) = N$ and set $\theta := \mathrm{Ind}_N^G \sigma$.*

(1) *If σ is real (resp. quaternionic) then θ is real (resp. quaternionic).*
(2) *If σ is complex and $\sigma' \not\sim {}^h\sigma$ then θ is complex.*
(3) *If σ is complex and $\sigma' \sim {}^h\sigma$ then θ is selfadjoint. Moreover. if $\overline{A(n)} = U^h A(n) U^*$, with U unitary, then $A(h^2) = +\overline{U}U / -\overline{U}U$ and correspondingly θ is real/quaternionic.*

The above criterion is particularly useful when $\overline{U}U = I$; for instance, this is the case when σ is one–dimensional.

We now state Theorem 2.52 for the index-two case. In what follows, $A(n)$ and $C(g)$, $n \in N$ and $g \in G$, will be matrix realizations of σ and θ, respectively, as in Theorem 2.52. Just note that now if $V := C(h)$, then $A(h^2) = V^2$, and $C(nh) = A(n)V$ for all $n \in N$. Also, if σ is selfadjoint let W be a unitary matrix such that $\overline{A(n)} = WA(n)W^*$ for all $n \in N$ (cf. (1.42)).

Theorem 4.23 *Let $\sigma \in \widehat{N}$ such that $I_G(\sigma) = G$ and let $\theta \in \widehat{G}(\sigma)$.*

(1) *If σ is complex then θ and $\theta \otimes \varepsilon$ are complex as well.*
(2) *If σ is selfadjoint then $W^*\overline{V}W = \pm V$. Moreover,*
    (2.i) *if σ is real, then θ is real/complex according to $W^*\overline{V}W = +V / -V$;*
    (2.ii) *if σ is quaternionic, then θ is quaternionic/complex according to $W^*\overline{V}W = +V / -V$.*

Chapter 5
Examples and Applications

The purpose of this chapter is to present some examples and applications of the little group method for Abelian extensions. This includes the representation theory of the *alternating groups* $\mathfrak{A}_n$, by using an approach based on Clifford theory (Sects. 5.1–5.4), Isaac's going down theorem (Sect. 5.5), an analysis of p^2-extensions (Sect. 5.6), and the representation theory of finite metacyclic groups, including the dihedral groups and the generalized quaternion groups (Sects. 5.7 and 5.8).

While the representation theory of the symmetric groups $\mathfrak{S}_n$ is most popular and widely studied, the representation theory of $\mathfrak{A}_n$ is often neglected and only a few monographs provide an exhaustive and fully detailed treatment of it. Among these exceptional monographs we mention the one by Fulton and Harris [51] where some important results are, however, left as (clearly outlined) exercises, and the one by James and Kerber [70] which is implicitly based on Clifford theory and refers to the book by Curtis and Reiner [39] for a description of this theory and the proofs of its main results.

In the following, we first collect some background material (without proofs, but with appropriate references): we recall the basic results of the representation theory of the symmetric groups $\mathfrak{S}_n$ including the Murnaghan–Nakayama rule. We describe the conjugacy classes and the irreducible representations of $\mathfrak{A}_n$; in particular, we present a detailed proof of the Frobenius formula (Theorem 5.7) for the corresponding characters. Finally, we discuss ambivalence of the groups $\mathfrak{A}_n$, providing a complete analysis.

T. Ceccherini-Silberstein et al., *Representation Theory of Finite Group Extensions*, Springer Monographs in Mathematics,
https://doi.org/10.1007/978-3-031-13873-7_5

5.1 Representation Theory and Conjugacy Classes of the Symmetric Groups $\mathfrak{S}_n$

We review the main facts on the representation theory and the conjugacy classes of $\mathfrak{S}_n$.

When $G = \mathfrak{S}_n$ both the conjugacy classes and the irreducible representations of $\mathfrak{S}_n$ are parameterized by the set of *partitions* of n, that is, by the set of tuples $\lambda = (\lambda_1, \lambda_2, \dots, \lambda_\ell)$, where $\lambda_1 \geq \lambda_2 \geq \cdots \geq \lambda_\ell$ are positive integers satisfying $\lambda_1 + \lambda_2 + \cdots + \lambda_\ell = n$.

If λ is a partition of n, we then write $\lambda \vdash n$ and denote by S^λ the associated irreducible representation of $\mathfrak{S}_n$. The correspondence $\lambda \mapsto S^\lambda$ is standard (see [70] and [27]).

For instance, $S^{(n)}$ is the trivial representation, while $S^{(1^n)}$, $(1^n) = (1, 1, \dots, 1)$, is the *alternating representation* (or *sign* representation).

Pictorially, one associates with a partition $\lambda \vdash n$ a diagram, called the *Young frame of shape* λ consisting of n boxes with ℓ left-justified rows, the ith row containing exactly λ_i boxes.

Given a partition $\lambda = (\lambda_1, \lambda_2, \dots, \lambda_\ell) \vdash n$, the associated *conjugate partition* $\lambda' = (\lambda'_1, \lambda'_2, \dots, \lambda'_{\ell'}) \vdash n$ is defined by setting $\ell' = \lambda_1$ and $\lambda'_j = |\{i \in \{1, 2, \dots, \ell\} : \lambda_i \geq j\}|$ for all $j = 1, 2, \dots, \ell'$. It follows from the definitions that the Young frames associated with conjugate partitions λ and λ' are one the transposed (that is, obtained by exchanging rows and columns) of the other. For instance, one has $(n)' = (1^n)$.

If $\lambda = \lambda'$ one says that λ is *self-conjugate* and the associated Young frame is *symmetric*.

We have (cf. [24, Theorem 10.4.1] or [27, Lemma 3.6.10])

$$S^\lambda \otimes S^{(1^n)} \sim S^{\lambda'}. \tag{5.1}$$

As a consequence, $S^\lambda \otimes S^{(1^n)} \sim S^\lambda$ if and only if λ is self-conjugate.

Let now $\lambda \vdash n$ and $\mu = (\mu_1, \mu_2, \dots, \mu_k) \vdash n$. A *rim hook tableau* of *shape* λ and *content* μ is a Young frame T of shape λ filled with the numbers $1, 2, \dots, k$ (with possible repetitions) in such a way that

- μ_j is the number of j's in T, $j = 1, 2, \dots, k$;
- the numbers are weakly increasing along the rows and the columns;
- for $j = 1, 2, \dots, k$ the diagram T_j consisting of the boxes occupied by j is connected (given two boxes b and b' in T_j there exists a sequence $b_0, b_1, \dots, b_k$, $k \geq 0$, of boxes of T_j such that $b_0 = b$ and $b_k = b'$, and b_i and b_{i+1} share a common edge, for $i = 0, 1, \dots, k-1$) and intersects the diagonal of T (that is, the set of boxes of T whose coordinates (i.e. row and column numbers) coincide) at most once.

Fig. 5.1 The two rim hook tableaux of shape $(5, 4, 3)$ and content $(1, 2, 3, 6)$

$$T = \begin{array}{|c|c|c|c|c|}\hline 1 & 3 & 3 & 4 & 4 \\ \hline 2 & 3 & 4 & 4 \\ \cline{1-4} 2 & 4 & 4 \\ \cline{1-3}\end{array} \qquad T' = \begin{array}{|c|c|c|c|c|}\hline 1 & 2 & 2 & 4 & 4 \\ \hline 3 & 3 & 4 & 4 \\ \cline{1-4} 3 & 4 & 4 \\ \cline{1-3}\end{array}$$

Denoting by $\langle T_j \rangle$ the number of rows of T_j decreased by one, the *height* $\langle T \rangle$ of T is the integer number defined by

$$\langle T \rangle := \sum_{j=1}^{k} \langle T_j \rangle.$$

The following is the celebrated Murnaghan–Nakayama rule (cf. [27, Sect. 3.5.5]).

Theorem 5.1 *Let $\lambda \vdash n$ and let $\pi \in \mathfrak{S}_n$ have cyclic structure μ. Then, denoting by χ^λ the corresponding character, we have*

$$\chi^\lambda(\pi) = \sum_T (-1)^{\langle T \rangle} \tag{5.2}$$

where the sum runs over all rim hook tableaux T of shape λ and content μ.

Example 5.2 Take $\lambda = (5, 4, 3) \vdash 12$ and suppose that $\pi \in \mathfrak{S}_{12}$ has cycle structure $\mu = (1, 2, 3, 6) \vdash 12$. There are only two rim hook tableaux of shape λ and content μ, namely those in Fig. 5.1.

Since $\langle T \rangle = 4$ and $\langle T' \rangle = 3$, we have $\chi^\lambda(\pi) = (-1)^4 + (-1)^3 = 1 - 1 = 0$.

For $\mu \vdash n$ we denote by $\mathscr{C}_\mu$ the conjugacy class of $\mathfrak{S}_n$ consisting of all permutations whose cyclic structure is μ.

Recall that a conjugacy class $\mathscr{C}$ in a group G is said to be *ambivalent* if $g \in \mathscr{C}$ infers $g^{-1} \in \mathscr{C}$, equivalently, if every element in $\mathscr{C}$ is conjugate to its own inverse. The group G itself is said to be *ambivalent* if every conjugacy class is ambivalent, equivalently, if every group element is conjugate to its inverse. It is an easy exercise to check that $\mathfrak{S}_n$ is ambivalent (cf. [24, Corollary 9.8.22]).

It is well known that all representations of $\mathfrak{S}_n$ are real (see, for instance [27, Corollary 3.4.3]).

5.2 Conjugacy Classes of $\mathfrak{A}_n$

For $\mu = (\mu_1, \mu_2, \ldots, \mu_m) \vdash n$ we have that $\mathscr{C}_\mu \subseteq \mathfrak{A}_n$ if and only if $n - m$ is even. Indeed, any cycle $(a_1, a_2, \ldots, a_k)$ of length k may be written as the product of $k-1$ transpositions:

$$(a_1, a_2, \ldots, a_k) = (a_1, a_2)(a_1, a_3) \cdots (a_1, a_k)$$

and therefore any $\pi \in \mathscr{C}_\mu$ may be written as the product of $(\mu_1 - 1) + (\mu_2 - 1) + \cdots + (\mu_n - 1) = n - m$ transpositions.

Lemma 5.3 *Let $\mathscr{C}_\mu$ be the conjugacy class of $\mathfrak{S}_n$ associated with the partition $\mu = (\mu_1, \mu_2, \ldots, \mu_m) \vdash n$. Suppose that $n - m$ is even. Then $\mathscr{C}_\mu$ splits into two $\mathfrak{A}_n$ classes if and only if $\mu_1 > \mu_2 > \cdots > \mu_m$ and $\mu_1, \mu_2, \ldots, \mu_m$ are odd numbers.*

Proof Take $\pi \in \mathscr{C}_\mu$ and let $\mathscr{C}$ be the $\mathfrak{A}_n$-conjugacy class of π. Then $\mathscr{C}_\mu$ does not split in $\mathfrak{A}_n$ (that is $\mathscr{C} = \mathscr{C}_\mu$) if and only if for any $\sigma \in \mathfrak{S}_n$ we have $\sigma^{-1}\pi\sigma \in \mathscr{C}$. This is equivalent to the following condition: there exists $\theta \in \mathfrak{S}_n \setminus \mathfrak{A}_n$ such that: $\theta^{-1}\pi\theta = \pi$. Indeed, if $\sigma^{-1}\pi\sigma \in \mathscr{C}$ with $\sigma \in \mathfrak{S}_n \setminus \mathfrak{A}_n$, then there exists $\rho \in \mathfrak{A}_n$ such that $\sigma^{-1}\pi\sigma = \rho^{-1}\pi\rho$ and we can take $\theta := \sigma\rho^{-1}$.

Let $1 \leq i \leq m - 1$. Suppose that $\mu_i = \mu_{i+1}$ is odd. Then, denoting by $(a_1, a_2, \ldots, a_{\mu_i})$ and $(b_1, b_2, \ldots, b_{\mu_{i+1}})$ the corresponding cycles in π, take $\theta := (a_1, b_1)(a_2, b_2) \cdots (a_{\mu_i}, b_{\mu_{i+1}})$. On the other hand, if μ_i is even then the corresponding cycle $(a_1, a_2, \ldots, a_{\mu_i})$ is odd, take $\theta := (a_1, a_2, \ldots, a_{\mu_i})$. We then deduce the necessary condition.

Conversely, suppose that $\mu_1, \mu_2, \ldots, \mu_m$ are distinct odd numbers. Then, $\theta^{-1}\pi\theta = \pi$ if and only if every cycle of θ is also a cycle of π or is trivial. Since these are only even permutations, there exist no $\theta \in \mathfrak{S}_n \setminus \mathfrak{A}_n$ satisfying $\theta^{-1}\pi\theta = \pi$, and the sufficient condition follows as well. ■

We denote by Split_n the set of all partitions $\mu = (\mu_1, \mu_2, \ldots, \mu_m) \vdash n$ such that $\mu_1, \mu_2, \ldots, \mu_n$ are odd and distinct.

5.3 The Irreducible Representations of $\mathfrak{A}_n$

For any 2-set $\{\lambda, \lambda'\}$ where $\lambda \vdash n$ such that $\lambda \not\sim \lambda'$, choose a representative, and denote by $\mathscr{N}_n$ the set of these representatives. On the other hand, we denote by SI_n the set of all self-conjugate partitions of n. From Theorem 4.18, Lemma 5.3, and Proposition 4.15 we then deduce the following:

Theorem 5.4 *For $\lambda \in \mathscr{N}_n$ set $\widetilde{S}^\lambda := \mathrm{Res}^{\mathfrak{S}_n}_{\mathfrak{A}_n} S^\lambda$, and for $\lambda \in \mathrm{SI}_n$ denote by $\mathrm{Res}^{\mathfrak{S}_n}_{\mathfrak{A}_n} S^\lambda = S^\lambda_- \oplus S^\lambda_+$ the corresponding $\mathfrak{A}_n$-irreducible decomposition. Then*

$$\{\widetilde{S}^\lambda : \lambda \in \mathscr{N}_n\} \cup \{S^\lambda_-, S^\lambda_+ : \lambda \in \mathrm{SI}_n\}$$

is a complete set of irreducible pairwise inequivalent representations of $\mathfrak{A}_n$. Moreover, $|\mathrm{SI}_n| = |\mathrm{Split}_n|$.

The following lemma provides a natural and useful bijection between SI_n and Split_n (thus recovering the last assertion of the previous theorem) (see Fig. 5.2).

Fig. 5.2 With $\lambda = (6, 5, 5, 4, 3, 1) \in \mathrm{SI}_{24}$ one associates $\mu = (11, 7, 5, 1) \in \mathrm{Split}_{24}$

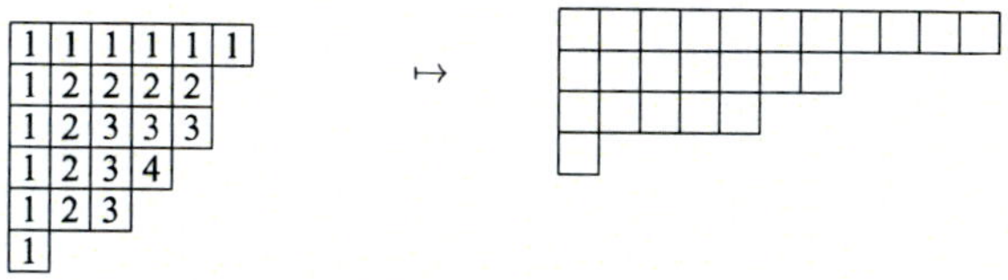

Lemma 5.5 *The map*

$$\begin{aligned} \mathrm{SI}_n &\to \quad \mathrm{Split}_n \\ \lambda &\mapsto \mu = \mu(\lambda) \end{aligned} \tag{5.3}$$

that associates with $\lambda = (\lambda_1, \lambda_2, \dots, \lambda_\ell) \in \mathrm{SI}_n$ *the partition* $\mu = (\mu_1, \mu_2, \dots, \mu_k)$ *defined by setting* $k := \max\{i \in \{1, 2, \dots, \ell\} : 2\lambda_i - 2i + 1 > 0\}$ *and* $\mu_i := 2\lambda_i - 2i + 1$ *for* $i = 1, 2, \dots, k$*, is a bijection.*

Proof There is a nice pictorial description of this correspondence, from which the lemma easily follows. Denote by r_i the set of boxes in the ith row of the diagram of λ, $i = 1, 2, \dots, \lambda_1$, and by c_j the set of boxes in the jth column, $j = 1, 2, \dots, \lambda_1$. Then we have:

$$\begin{aligned} 2\lambda_1 - 1 &= |c_1 \cup r_1| \\ 2\lambda_2 - 3 &= |(c_2 \cup r_2) \setminus (c_1 \cup r_1)| \\ 2\lambda_3 - 5 &= |(c_3 \cup r_3) \setminus ((c_2 \cup r_2) \cup (c_1 \cup r_1))\,| \\ \cdots &= \cdots \end{aligned}$$

Note that k can be interpreted as the maximum $i \in \{1, 2, \dots, \lambda_1\}$ such that $r_i \cap c_i \neq \emptyset$. We then write 1 inside the boxes of $c_1 \cup r_1$, 2 inside those of $(c_2 \cup r_2) \setminus (c_1 \cup r_1)$, 3 inside those of $|(c_3 \cup r_3) \setminus [(c_2 \cup r_2) \cup (c_1 \cup r_1)]$, and so on. This way, the number of boxes labelled with i is exactly $2\lambda_i - 2i + 1$. In particular, there is exactly one box labelled by k and there are no boxes labelled by $k + 1, k + 2, \dots$ ∎

Remark 5.6 There is an alternative description of the bijection presented in the previous lemma. Given the Young frame corresponding to a partition $\lambda \vdash n$, we denote by $H^\lambda_{i,j}$ the set consisting of the box b of coordinates (i, j) together with all boxes that are either to the right (but in the same row) or lower down (but in the same column) with respect to box b. We call $H^\lambda_{i,j}$ the*(i,j)-hook* associated with λ and we denote by $h^\lambda_{i,j}$ its *lenght*, i.e. the number of boxes in it. Then the correspondence in (5.3) can be rewritten as:

$$(\lambda_1, \lambda_2, \dots, \lambda_\ell) \mapsto (h^\lambda_{1,1}, h^\lambda_{2,2}, \dots, h^\lambda_{k,k}).$$

For $\lambda \vdash n$ we now denote by χ^λ the character of the irreducible representation S^λ of $\mathfrak{S}_n$, and, when $\lambda \in \mathrm{SI}_n$, we denote by χ^λ_+ and χ^λ_- the characters of S^λ_+ and S^λ_-, respectively. By Proposition 4.21 we know that the values of the characters of $\mathfrak{A}_n$ are completely determined by those of $\mathfrak{S}_n$ when the corresponding partition either is in $\mathcal{N}_n$ or is in SI_n but the conjugacy class at which the character is evaluated is non-split. In order to determine the values of $\chi_\pm^\lambda$ on pairs of split conjugacy classes we need the following deep result of Frobenius. For its formulation, since any character χ takes a constant value on each conjugacy class $\mathscr{C}$, we denote this value by $\chi(\mathscr{C})$.

Theorem 5.7 (Frobenius) *Let $\lambda \in \mathrm{SI}_n$ and denote by χ^λ (resp. χ_+^λ and χ_-^λ) the corresponding characters of $\mathfrak{S}_n$ (resp. $\mathfrak{A}_n$). Let also $\mu = (\mu_1, \mu_2, \dots, \mu_k) \in \mathrm{Split}_n$ and denote by $\mathscr{C}_\mu$ (resp. $\mathscr{C}_\mu^+$ and $\mathscr{C}_\mu^-$) the corresponding conjugacy classes in $\mathfrak{S}_n$ (resp. $\mathfrak{A}_n$). Then*

(i) *if μ does not correspond to λ under the map (5.3) then*

$$\chi_+^\lambda(\mathscr{C}_\mu^+) = \chi_+^\lambda(\mathscr{C}_\mu^-) = \chi_-^\lambda(\mathscr{C}_\mu^+) = \chi_-^\lambda(\mathscr{C}_\mu^-) = \frac{1}{2}\chi^\lambda(\mathscr{C}_\mu);$$

(ii) *if μ corresponds to λ then*

$$\chi_+^\lambda(\mathscr{C}_\mu^+) = \chi_-^\lambda(\mathscr{C}_\mu^-) = \frac{1}{2}\left[(-1)^m + \sqrt{(-1)^m \mu_1\mu_2\cdots\mu_k}\right] \tag{5.4}$$

and

$$\chi_+^\lambda(\mathscr{C}_\mu^-) = \chi_-^\lambda(\mathscr{C}_\mu^+) = \frac{1}{2}\left[(-1)^m - \sqrt{(-1)^m \mu_1\mu_2\cdots\mu_k}\right] \tag{5.5}$$

where $m = m(\mu) = \sum_{j=1}^k (\mu_j - 1)/2 \equiv (n-k)/2$.

Proof In the following, we denote by $\nu = (\nu_1, \nu_2, \dots, \nu_h) \in \mathrm{Split}_n$ a general partition of n whose associated conjugacy class is split, and by $\mu = \mu(\lambda)$ the partition corresponding to λ. We shall first prove (ii) under the assumption of (i), and then we prove (i). We divide the proof into several steps.

Step 0. Under our assumptions, from the Murnaghan–Nakayama rule (Theorem 5.1) one immediately deduces the following special case:

$$\chi^\lambda(\mathscr{C}_\mu) = (-1)^{(n-k)/2}. \tag{5.6}$$

Step 1. From Proposition 4.21, we know that $\chi_+^\lambda(\mathscr{C}_\nu^+) = \chi_-^\lambda(\mathscr{C}_\nu^-)$ (resp. $\chi_+^\lambda(\mathscr{C}_\nu^-) = \chi_-^\lambda(\mathscr{C}_\nu^+)$) and that, denoting by x_ν (resp. y_ν) this value, one has $x_\nu + y_\nu = \chi^\lambda(\mathscr{C}_\nu)$.

Step 2. Let $\pi = (a_1, a_2, \ldots, a_{\nu_1})(b_1, b_2, \ldots, b_{\nu_2}) \cdots (c_1, c_2, \ldots, c_{\nu_k})$ denote the cyclic decomposition of $\pi \in \mathscr{C}_\nu$ and set

$$\xi := \big((a_2, a_{\nu_1})(a_3, a_{\nu_1 - 1}) \cdots (a_{(\nu_1+1)/2}, a_{(\nu_1+3)/2})\big) \cdots \big((b_2, b_{\nu_2})(b_3, b_{\nu_2-1}) \cdots (b_{(\nu_2+1)/2}, b_{(\nu_2+3)/2})\big) \cdots \big((c_2, c_{\nu_k})(c_3, c_{\nu_k-1}) \cdots (c_{(\nu_k+1)/2}, c_{(\nu_k+3)/2})\big).$$

Then $\xi^{-1}\pi\xi = \pi^{-1}$. Moreover, $\xi \in \mathfrak{A}_n$ if and only if $m = m(\nu)$ is even. Therefore, π is $\mathfrak{A}_n$-conjugate to π^{-1} if and only if m is even (cf. the end of the proof of Lemma 5.3).

Step 3. Suppose m is even. Fix $\pi \in \mathscr{C}_\nu^+$; then $\pi^{-1} \in \mathscr{C}_\nu^+$ and therefore $x_\nu = \chi_+^\lambda(\pi) = \overline{\chi_+^\lambda(\pi^{-1})} = \overline{x_\nu}$. Similarly, by taking $\pi \in \mathscr{C}_\nu^-$, one deduces $y_\nu = \overline{y_\nu}$. It follows that $x_\nu, y_\nu \in \mathbb{R}$.

Suppose now that m is odd. Fix $\pi \in \mathscr{C}_\nu^+$; then $\pi^{-1} \in \mathscr{C}_\nu^-$ and therefore $x_\nu = \chi_+^\lambda(\pi) = \overline{\chi_+^\lambda(\pi^{-1})} = \overline{y_\nu}$.

Step 4. Assume, for the moment that $x_\nu = y_\nu$ if (and only if) $\nu \neq \mu$. Let us set $\theta := \chi_+^\lambda - \chi_-^\lambda$. Then $\langle \theta, \theta \rangle = 2$ (recall that characters corresponding to inequivalent irreducible representations are orthogonal). We thus have

$$\begin{aligned}
\langle \theta, \theta \rangle &= \frac{1}{|\mathfrak{A}_n|} \sum_{g \in \mathfrak{A}_n} \theta(g)\overline{\theta(g)} \\
&= \frac{2}{n!} \left(\sum_{g \in \mathscr{C}_\mu} \theta(g)\overline{\theta(g)} + \sum_{\substack{\nu \vdash n \\ \nu \neq \mu}} \sum_{g \in \mathscr{C}_\nu} \theta(g)\overline{\theta(g)} \right) \\
\text{(by our assumption)} &= \frac{2}{n!} \sum_{g \in \mathscr{C}_\mu} \theta(g)\overline{\theta(g)} \\
&= \frac{2}{n!} \left(\sum_{g \in \mathscr{C}_\mu^+} \theta(g)\overline{\theta(g)} + \sum_{g \in \mathscr{C}_\mu^-} \theta(g)\overline{\theta(g)} \right) \\
&= \frac{4}{n!} \sum_{g \in \mathscr{C}_\mu^+} |x_\mu - y_\mu|^2 \\
&= \frac{4|\mathscr{C}_\mu^+|}{n!} |x_\mu - y_\mu|^2.
\end{aligned}$$

Since $|\mathscr{C}_\mu^+| = \frac{n!}{2\mu_1\mu_2\cdots\mu_k}$, we deduce that $|x_\mu - y_\mu|^2 = \mu_1\mu_2\cdots\mu_k$. Collecting together our information so far, we get the systems:

$$\begin{cases} x_\mu + y_\mu = 1 & \text{by virtue of (5.6)} \\ x_\mu, y_\mu \in \mathbb{R} & \text{by Step. 3} \\ |x_\mu - y_\mu|^2 = \mu_1\mu_2\cdots\mu_k \end{cases}$$

if m is even, and

$$\begin{cases} x_\mu + y_\mu = -1 & \text{by virtue of (5.6)} \\ x_\mu = \overline{y_\mu} & \text{by Step. 3} \\ |x_\mu - y_\mu|^2 = \mu_1\mu_2\cdots\mu_k \end{cases}$$

if m is odd. Then (ii) follows. Note that we have made an arbitrary choice for the split classes $\mathscr{C}_\mu^+$ and $\mathscr{C}_\mu^-$ so that, in fact, the values of x_μ and y_μ are interchangeable.

Step 5. We now show that if $\epsilon \in \mathrm{Split}_n$ and $x_\nu = y_\nu$ for all $\nu \in \mathrm{Split}_n$ such that $\nu \neq \epsilon$, then necessarily $\epsilon = \mu$. By contradiction, if $\epsilon \neq \mu$ we would have $x_\mu = y_\mu$. Since $x_\mu + y_\mu = (-1)^m$ (by virtue of (5.6)), we would have $x_\mu = (-1)^m/2$. But this is impossible since characters take only algebraic integer values (this holds for any finite group G: indeed if $\sigma \in \widehat{G}$ and $g \in G$, one has $\chi^\sigma(g)$ is the sum of the eigenvalues of $\sigma(g)$ which are $|G|$-roots of the unity).

Step 6. Given two partitions $\eta = (\eta_1, \eta_2, \ldots, \eta_\ell), \gamma = (\gamma_1, \gamma_2, \ldots, \gamma_j) \vdash n$ we write $\eta \trianglelefteq \gamma$ if $j \leq \ell$ and $\sum_{p=1}^i \eta_p \leq \sum_{p=1}^i \gamma_p$ for all $i = 1, 2, \ldots, j$. It is well known (see, for instance, [27, Sect. 3.6.1]) that $\trianglelefteq$ is a partial order on the set of all partitions of n. Now, given $\eta \in \mathrm{SI}_n$, from the Murnaghan–Nakayama rule (Theorem 5.1) it follows that if $\chi^\eta(C_\nu) \neq 0$, one necessarily has $\nu \trianglelefteq (h_{1,1}^\eta, h_{2,2}^\eta, \ldots, h_{k,k}^\eta)$. Indeed, a rim hook tableau of shape η and content ν should have all the 1's in the hook $H_{1,1}^\eta$, all the 1's and the 2's in $H_{1,1}^\eta \sqcup H_{2,2}^\eta$, and so on.

Step 7 With θ as in Step 4, we set $\phi := \theta\overline{\theta}$. Observe that ϕ is a so-called *generalized character* of $\mathfrak{A}_n$, i.e., a $\mathbb{Z}$-linear combination of characters of irreducible representations (this follows from the fact that the conjugate of a character is the character of the conjugate representation and the product of two characters is the character of the tensor product of the corresponding representations). By Step 1, $\phi(\mathscr{C}_\nu^+) = \phi(\mathscr{C}_\nu^-)$ if $\nu \in \mathrm{Split}_n$ and ϕ vanishes on all other $\mathfrak{A}_n$-conjugacy classes. We now define $\psi\colon \mathfrak{S}_n \to [0, +\infty)$ by setting

$$\psi(g) := \begin{cases} \phi(g) & \text{if } g \in \mathfrak{A}_n \\ 0 & \text{if } g \in \mathfrak{S}_n \setminus \mathfrak{A}_n \end{cases} \tag{5.7}$$

for all $g \in \mathfrak{S}_n$.

Let $\eta \in \mathrm{SI}_n$. Then, by Step 4 we have that $\langle \phi, \chi_+^\eta \rangle = \langle \phi, \chi_-^\eta \rangle$, moreover, since $\mathrm{Ind}_{\mathfrak{A}_n}^{\mathfrak{S}_n} S_+^\eta = \mathrm{Ind}_{\mathfrak{A}_n}^{\mathfrak{S}_n} S_-^\eta = S^\eta$ we deduce that

$$\langle \psi, \chi^\eta \rangle \in \mathbb{Z}. \tag{5.8}$$

The class function ψ vanishes on all nonsplit conjugacy classes, but it is not identically zero (since $\chi_+^\lambda \neq \chi_-^\lambda$). Therefore, we can select a partition $\epsilon = (\epsilon_1, \epsilon_2, \ldots, \epsilon_k) \in \mathrm{Split}_n$ such that $\psi(\mathscr{C}_\epsilon) \neq 0$ and it is minimal with respect to the partial order $\trianglelefteq$ (cf. Step 6). Denote by $\eta = (\eta_1, \eta_2, \ldots, \eta_\ell)$ the selfconjugate partition corresponding to ϵ. Then (arguing as in Step 4)

$$\begin{aligned}
\langle \psi, \chi^\eta \rangle &= \frac{1}{n!} \sum_{\mu \vdash n} |\mathscr{C}_\mu| \psi(\mathscr{C}_\mu) \chi^\eta(\mathscr{C}_\mu) \\
&= \frac{1}{n!} \sum_{\mu \in \mathrm{Split}_n} |\mathscr{C}_\mu| \psi(C_\mu) \chi^\eta(C_\mu) \\
\text{(by minimality of } \varepsilon \text{ and by Step 6)} \quad &= \frac{1}{n!} |\mathscr{C}_\epsilon| \psi(\mathscr{C}_\epsilon) \chi^\eta(\mathscr{C}_\epsilon) \\
\text{(by the Murnaghan -- Nakayama rule)} \quad &= \frac{1}{n!} |\mathscr{C}_\epsilon| \psi(\mathscr{C}_\epsilon) (-1)^{n-k}.
\end{aligned}$$

From (5.8) we then deduce that

$$\frac{1}{n!} |\mathscr{C}_\epsilon| \psi(\mathscr{C}_\epsilon) \in \{1, 2, \ldots\}. \tag{5.9}$$

On the other hand, reasoning again as in Step 4, we have

$$\begin{aligned}
1 &= \frac{1}{2} \langle \theta, \theta \rangle \\
&= \frac{1}{n!} \sum_{\substack{\nu \vdash n \\ \mathscr{C}_\nu \subseteq \mathfrak{A}_n}} |\mathscr{C}_\nu| \phi(\mathscr{C}_\nu) \\
\text{(by (5.7))} \quad &= \frac{1}{n!} \sum_{\nu \vdash n} |\mathscr{C}_\nu| \psi(\mathscr{C}_\nu).
\end{aligned} \tag{5.10}$$

Finally, since ψ takes only nonnegative values (ψ can be expressed as the squared absolute value of a function), from (5.9) and (5.10), we deduce that

$$\frac{1}{n!} |C_\mu| \psi(C_\mu) = \delta_{\mu,\epsilon}.$$

This shows that the condition in Step 4 (taking into account Step 5) is not restrictive, and therefore the proof is complete. ■

Corollary 5.8 *Let* $\lambda \vdash n$.

(i) *If* $\lambda \in \mathcal{N}_n$, *then the corresponding irreducible representation* $\tilde{S}^\lambda$ *of* $\mathfrak{A}_n$ *is real;*
(ii) *if* $\lambda \in \mathrm{SI}_n$ *and* $m = m(\mu(\lambda))$ *is even, then* $S^\lambda_\pm$ *are real;*
(iii) $f\,\lambda \in \mathrm{SI}_n$ *and* $m = m(\mu(\lambda))$ *is odd, then* $S^\lambda_\pm$ *are complex.*

Proof We know that the representations of $\mathfrak{S}_n$ are real (see [27, Corollary 3.4.3]). In [33, Theorem 3.3] it is shown that every irreducible representation of $\mathfrak{A}_n$ is either real or complex, and then Frobenius theorem (Theorem 5.7) exactly determines when these two possibilities occur. ■

Example 5.9 The restriction of the $\mathfrak{S}_4$ representation $S^{2,2}$ to $\mathfrak{A}_4$ splits into the two irreducible representations $S^{2,2}_+$ and $S^{2,2}_-$. These are both complex since $\lambda = (2, 2)$ corresponds to $\mu = (3, 1)$ and therefore $m = \frac{1}{2}(2 + 0) = 1$ is odd.

5.4 Ambivalence of the Groups $\mathfrak{A}_n$

Theorem 5.10 *Let* $\mu = (\mu_1, \mu_2, \ldots, \mu_k) \in \mathrm{Split}_n$. *As above, we denote by* $\mathscr{C}^+_\mu$, $\mathscr{C}^-_\mu$ *the corresponding split classes in* $\mathfrak{A}_n$ *and set* $m := \frac{1}{2}\sum_{j=1}^k(\mu_j - 1) \equiv \frac{1}{2}(n-k)$. *If* m *is even, then* $\mathscr{C}^+_\mu$ *and* $\mathscr{C}^-_\mu$ *are both ambivalent. If* m *is odd then* $\mathscr{C}^+_\mu = \{\pi^{-1} : \pi \in \mathscr{C}^-_\mu\}$; *in particular,* $\mathscr{C}^+_\mu$ *and* $\mathscr{C}^-_\mu$ *are both non-ambivalent.*

Proof Let $\pi \in \mathscr{C}^-_\mu$ and $\xi \in \mathfrak{S}_n$ as in Step 2 of the proof of Theorem 5.7. Then $\xi \in \mathfrak{A}_n$ if and only if m is even. Therefore, if m is even then $\mathscr{C}^-_\mu$ (and similarly $\mathscr{C}^+_\mu$) is ambivalent. On the other hand, if m is odd, $\mathscr{C}^-_\mu$ cannot be ambivalent: if $\sigma\pi\sigma^{-1} = \pi^{-1}$ with $\sigma \in \mathfrak{A}_n$, then $(\xi\sigma)^{-1}\pi(\xi\sigma) = \pi$ and $\xi\sigma \in \mathfrak{S}_n \setminus \mathfrak{A}_n$; but this is impossible since $\mathscr{C}_\mu$ is split (cf. the beginning of the proof of Lemma 5.3). Finally, since $\pi^{-1} \in \mathscr{C}^+_\mu$, the last two assertions for $\mathscr{C}^+_\mu$ follow as well. ■

Corollary 5.11 *The group* $\mathfrak{A}_n$ *is ambivalent if and only if* $n = 1, 2, 5, 6, 10, 14$.

Proof For $n \in \{1, 2, 5, 6, 10, 14\}$, all conjugacy classes are non-split (and therefore ambivalent) except for the following ones:

- $(5) \vdash 5$ $(m = 2)$
- $(5, 1) \vdash 6$ $(m = 2)$
- $(9, 1), (7, 3) \vdash 10$ $(m = 4)$
- $(13, 1), (11, 3), (9, 5) \vdash 14$ $(m = 6)$.

Since m is even, from Theorem 5.10 it follows that these split classes are also ambivalent.

Conversely, for the other values of n, we consider the following split classes:

- $(4k-1, 1) \vdash n = 4k \ (m = 2k-1), k \geq 1$
- $(4k-3, 3, 1) \vdash n = 4k+1 \ (m = 2k-1), k \geq 2$
- $(4(k-1)-3, 5, 3, 1) \vdash n = 4k+2 \ (m = 2k-1), k \geq 4$
- $(4k+3) \vdash n = 4k+3 \ (m = 2k+1), k \geq 0.$

Since now m is odd, we deduce, again from Theorem 5.10, that these split classes are non-ambivalent. ■

5.5 An Application to Isaacs' Going Down Theorem

In this section we examine more closely case (2) in Isaacs' going down theorem (Theorem 2.37). More precisely, throughout the section we assume that:

$$A := G/N \text{ is Abelian, } \sigma \in \widehat{N} \text{ with } I_G(\sigma) = G, \text{ and } \theta \in \widehat{G} \text{ with } \widehat{G}(\sigma) = \{\theta\}. \tag{5.11}$$

See also "Case $I_G(\sigma) = G$ and $|\widehat{G}(\sigma)| = 1$" in Sect. 5.6, where $A := G/N$ is of order p^2.
Clearly, we are no more in the hypotheses of Theorem 4.3, that is, we do not assume that σ has an extension to G. But, at the end of the section, in Remark 5.14, we shall need the machinery from Sect. 4.4. Anyway, from the assumptions (5.11) it follows that:

$$\mathrm{Ind}_N^G \sigma \sim \underbrace{\theta \oplus \theta \oplus \cdots \oplus \theta}_{\ell \text{ times}} \tag{5.12}$$

and

$$\mathrm{Res}_N^G \theta \sim \underbrace{\sigma \oplus \sigma \oplus \cdots \oplus \sigma}_{\ell \text{ times}}, \tag{5.13}$$

where ℓ is the inertia index of θ, and

$$\ell^2 = [G : N] \equiv |A|. \tag{5.14}$$

These facts are immediate consequences of Theorem 2.5, Eq. (2.15), and Remark 2.16. In particular, from $\mathrm{Res}_N^G \mathrm{Ind}_N^G \sigma \sim \ell\, \mathrm{Res}_N^G \theta \sim \ell^2 \sigma$ we deduce (5.14).

Consider the factorization $A = A_0 \times A_1$ where

$$A_0 := \{a \in A : a \text{ is of order } 2^k, k \geq 1\},$$

$$A_1 := \{a \in A : a \text{ is of order } p^k, k \geq 1, \text{ with } p \text{ prime } \geq 3\}.$$

In other words, A_0 is the Sylow 2-subgroup of A while A_1 is the direct product of all the Sylow p-subgroups with $p \geq 3$; see [30, Theorem 1.3.7]. Let also M be an intermediate normal subgroup, that is, $N \leq M \trianglelefteq G$, such that

$$M/N \cong A_0. \tag{5.15}$$

Note that M exists by the correspondence theorem.

Warning: the subgroup M in the present situation does *not* correspond to the subgroup M in Theorem 2.37. Indeed, by our assumptions, it is the whole group G that corresponds to the subgroup M in Isaacs' Theorem.

We now prove a general result on the real/complex/quaternionic type of θ.

Theorem 5.12 *Suppose that $|A|$ is odd, so that A_0 is trivial (and, in the terminology of Sect. 6, A is 2-divisible). Then both σ and θ are complex.*

Proof From Theorem 2.30.(1), with $I_G(\sigma)$ replaced by G, and (5.14) it follows that

$$\theta \otimes \overset{\Delta}{\omega} \sim \theta \quad \text{for all} \quad \omega \in \widehat{A}. \tag{5.16}$$

Suppose that $\pi : G \to A$ is the projection and that $t : A \to G$ is a section (cf. (1.54)). Then (5.16) implies that

$$\omega(a)\chi^{\theta}(t(a)n) = \chi^{\theta}(t(a)n),$$

for all $\omega \in \widehat{A}$, $a \in A$, and $n \in N$. This in turn implies that

$$\chi^{\theta}(t(a)n) = 0 \quad \text{if } a \neq 1_A. \tag{5.17}$$

Now, if $g \in G$ satisfies that $g^2 \in N$, then $\pi(g^2) = \pi(g)^2 = 1_A$. Since A_0 is trivial, we necessarily have $\pi(g) = 1_A$, that is, $g \in N$. In other words,

$$g^2 \in N \iff g \in N. \tag{5.18}$$

We have:

$$\begin{aligned}\frac{1}{|G|}\sum_{g \in G}\chi^{\theta}(g^2) &= \frac{1}{|G|}\sum_{n \in N}\chi^{\theta}(n^2) && \text{(by (5.17) and (5.18))}\\ &= \frac{1}{\ell|N|}\sum_{n \in N}\chi^{\sigma}(n^2) && \text{(by (5.13) and (5.14)).}\end{aligned}$$

We now apply Frobenius-Schur's Theorem, namely formula (1.39): only the case where both σ and θ are complex is admissible (yielding vanishing of both corresponding averages), otherwise, we would have $\pm 1 = \pm\frac{1}{\ell}$, yielding a contradiction, since $\ell \neq 1$. ■

In the following theorem, we use the following elementary fact: two G-conjugate representations are of the same real/complex/quaternionic type.

Theorem 5.13 *Suppose now that both A_0 and A_1 are nontrivial and that M is as in (5.15). Suppose also that*

$$\mathrm{Ind}_N^M \sigma \sim \bigoplus_{j=1}^{t} \ell_j \psi_j, \tag{5.19}$$

with $\psi_1, \psi_2, \ldots, \psi_t \in \widehat{M}$ pairwise nonequivalent. Then

- *$\{\psi_1, \psi_2, \ldots, \psi_t\}$ is a G-orbit on $\widehat{M}$;*
- *θ is real/complex/quaternionic accordingly to the ψ_j's are real /complex/quaternionic;*
- *if the ψ_j's are real or quaternionic, then $I_G(\psi_j) = M$ and $\theta \sim \mathrm{Ind}_M^G \psi_j$ for all $j = 1, 2, \ldots, t$.*

Proof From (5.19) it follows that $\widehat{M}(\sigma) = \{\psi_1, \psi_2, \ldots, \psi_t\}$ and ℓ_j is the inertia index of ψ_j with respect to N. Then

$$\begin{aligned}
\ell\theta &\sim \mathrm{Ind}_N^G \sigma && \text{(by (5.12))}\\
&\sim \mathrm{Ind}_M^G \mathrm{Ind}_N^M \sigma && \text{(by (1.33))}\\
&\sim \bigoplus_{j=1}^{t} \ell_j \mathrm{Ind}_M^G \psi_j && \text{(by (5.19))}
\end{aligned}$$

implies that $\widehat{G}(\psi_j) = \{\theta\}$ for $j = 1, 2, \ldots, t$. Moreover, from (5.19), Frobenius reciprocity, (5.13), and by the correspondence (2.19), it follows that $\{\psi_1, \ldots, \psi_t\}$ is a G-orbit on $\widehat{M}$ and, by Theorem 2.5.(3),

$$\mathrm{Res}_M^G \theta \sim \widetilde{\ell} \bigoplus_{j=1}^{t} \psi_j, \tag{5.20}$$

where $\widetilde{\ell}$ denotes the inertia index of θ with respect to M. By (2.15),

$$\mathrm{Ind}_M^G \psi_j \sim \widetilde{\ell}\theta, \quad \text{for } j = 1, 2, \ldots, t. \tag{5.21}$$

Then

$$\mathrm{Res}_M^G \mathrm{Ind}_M^G \psi_j \sim \widetilde{\ell}^2 \bigoplus_{j=1}^t \psi_j$$

so that, by Theorem 2.5.(1),

$$\widetilde{\ell}^2 = \left[I_G(\psi_j) : M\right] \tag{5.22}$$

and

$$t = \left[G : I_G(\psi_j)\right], \tag{5.23}$$

for all $j = 1, 2, \ldots, t$. By the observation preceding the statement of the theorem, the quantity

$$F := \frac{1}{|M|}\sum_{m\in M}\chi^{\psi_j}(m^2) \equiv \begin{cases} 1 & \text{if the } \psi_j' s \text{ are real,} \\ 0 & \text{if the } \psi_j' s \text{ are complex,} \\ -1 & \text{if the } \psi_j' s \text{ are quaternionic} \end{cases} \tag{5.24}$$

is independent of $j = 1, 2, \ldots, t$. We can now invoke (5.17) because $G/M \cong A_1$ is 2-divisible. Therefore, the Frobenius-Schur theorem yields

$$\begin{aligned}
\frac{1}{|G|}\sum_{g\in G}\chi^\theta(g^2) &= \frac{1}{|G|}\sum_{m\in M}\chi^\theta(m^2) && \text{(by (5.17))} \\
&= \frac{1}{|G|}\sum_{m\in M}\widetilde{\ell}\sum_{j=1}^t \chi^{\psi_j}(m^2) && \text{(by (5.20))} \\
&= \frac{\widetilde{\ell}}{[G:M]}\sum_{j=1}^t \frac{1}{|M|}\sum_{m\in M}\chi^{\psi_j}(m^2) \\
&= F/\widetilde{\ell} && \text{(by (5.22), (5.23) and (5.24)).}
\end{aligned}$$

We deduce that, on the one hand, the ψ_j's are complex if and only if $F = 0$ and this in turn is equivalent to θ being complex.

On the other hand, the ψ_j's are real/quaternionic if and only if $F = \pm 1$ and this in turn is equivalent to $\widetilde{\ell} = 1$ and θ being real/quaternionic. Moreover, if this is the case, by (5.22) we have $I_G(\psi_j) = M$ and, by (5.21), $\mathrm{Ind}_M^G \psi_j \sim \theta$. ■

Remark 5.14 When A_1 is trivial, Theorem 5.13 does not cover the relationship between the real/complex/quaternionic type of σ and that of the ψ_j's. However, we may construct a tower of groups $B_0 = \{1_A\} \leq B_1 \leq B_2 \leq \cdots \leq B_h = A_0$

such that $B_k/B_{k-1} \cong C_2$ for $k = 1, 2, \ldots, h$, thus yielding, by the correspondence theorem, a tower of subgroups

$$M_0 = N \leq M_1 \leq M_2 \leq \cdots \leq M_h = M$$

such that $M_k/M_{k-1} \cong B_k/B_{k-1} \cong C_2$ for all $k = 1, 2, \ldots, h$. Then, by means of a repeated application of Theorems 4.22 and 4.23 we may investigate the relationship between the types of σ and of the ψ_j's.

5.6 Another Application: Analysis of p^2-Extensions

In this section we apply the results in Sects. 2.2, 2.3, and 4.3 to the following case: $N \lhd G$ and $[G : N] = p^2$, where p is a prime. It is well known that, under these assumptions, G/N is Abelian: indeed, either $G/N \cong C_{p^2}$ or $G/N \cong C_p \times C_p$; see, for instance, [59, Corollary to Theorem 2.11.2] and [30, Sect. 1.3].

We now show that Clifford theory leads to the following four possible cases for $\sigma \in \widehat{N}$.

Case $I_G(\sigma) = N$

We have:

- $\theta := \mathrm{Ind}_N^G \sigma$ is irreducible and $d_\theta = p^2 d_\sigma$;
- $\widehat{G}(\sigma) = \{\theta\}$ and $\theta \otimes \overset{\vartriangle}{\omega} \sim \theta$ for all $\omega \in \widehat{G/N}$;
- $\mathrm{Res}_N^G \theta \sim \bigoplus_{a \in G/N} {}^a\sigma$ and the irreducible representations ${}^a\sigma$, $a \in G/N$, are pairwise nonequivalent.

This is just a particular case of Theorem 4.3.(2).

Case $I_G(\sigma) = M$ with $N \lneqq M \lneqq G$

Now, M is normal in G and $G/M \cong M/N \cong C_p$. Fix $\psi \in \widehat{M}(\sigma)$ and $\theta \in \widehat{G}(\psi)$, so that (by transitivity of induction) $\theta \in \widehat{G}(\sigma)$. Fix a generator c of G/M and a generator ω of $\widehat{M/N}$; recall that $\overset{\vartriangle}{\omega} \in \widehat{M}$ denotes its inflation to M. Then choose a set of representatives for the $\widehat{G/M}$-cosets in $\widehat{G/N}$ (since $G/M \cong (G/N)/(M/N)$, we may view $\widehat{G/M}$ as a subgroup of $\widehat{G/N}$, via the inflation: this is indeed the subgroup consisting of all $\psi \in \widehat{G/N}$ such that $\mathrm{Res}^{G/N}_{M/N} \psi \equiv 1$). More precisely, choose $\varepsilon \in \widehat{G/N}$ such that $\mathrm{Res}^{G/N}_{M/N} \varepsilon \sim \omega$ (ε is an extension of ω to G/N; cf. Remark 2.24) and the representations are $\varepsilon^j : j = 0, 1, \ldots, p-1$; let $\overset{\vartriangle}{\varepsilon} \in \widehat{G}$ denote its inflation to G. From Proposition 4.9 applied to the couple $N \lhd M$ it follows that ψ is an extension of σ to M, and therefore we may apply Theorem 4.3.(1). We may also apply Theorem 4.3.(3) to the couple $N \lhd M$ and Theorem 4.3.(2) to the couple $M \lhd G$. Finally, note that in the case $G/N \cong C_{p^2}$ the subgroup $M \leq G$ with $[G : M] = p$ is unique and therefore $I_G(\sigma) = M$ for all $\sigma \in \widehat{N}$ such that $N \lneqq I_G(\sigma) \lneqq G$. We therefore have:

- $\{{}^{c^k}\sigma : 0 \leq k \leq p-1\}$ is the G-conjugacy class of σ;
- $\mathrm{Ind}_N^M {}^{c^k}\sigma \sim \bigoplus_{j=0}^{p-1} \left({}^{c^k}\psi \otimes \overset{\triangle}{\omega}{}^j\right)$ and $\widehat{M}({}^{c^k}\sigma) = \left\{{}^{c^k}\psi \otimes \overset{\triangle}{\omega}{}^j : 0 \leq j \leq p-1\right\}$ for all $0 \leq k \leq p-1$;
- $\mathrm{Res}_N^M \left({}^{c^k}\psi \otimes \overset{\triangle}{\omega}{}^j\right) \sim {}^{c^k}\sigma$, for $0 \leq j, k \leq p-1$; in particular, each ${}^{c^k}\psi \otimes \overset{\triangle}{\omega}{}^j$ is an extension of ${}^{c^k}\sigma$ to M;
- $\mathrm{Res}_M^G \left(\theta \otimes \overset{\triangle}{\varepsilon}{}^j\right) \sim \bigoplus_{k=0}^{p-1} \left[\left({}^{c^k}\psi\right) \otimes \overset{\triangle}{\omega}{}^j\right]$ for all $0 \leq j \leq p-1$;
- $\mathrm{Ind}_M^G \left({}^{c^k}\psi \otimes \overset{\triangle}{\omega}{}^j\right) \sim \theta \otimes \overset{\triangle}{\varepsilon}{}^j$, for all $0 \leq j, k \leq p-1$;
- $\mathrm{Ind}_N^G \left({}^{c^k}\sigma\right) \sim \bigoplus_{j=0}^{p-1} \left(\theta \otimes \overset{\triangle}{\varepsilon}{}^j\right)$ for all $0 \leq k \leq p-1$;
- $\widehat{G}(\sigma) = \left\{\theta \otimes \overset{\triangle}{\varepsilon}{}^j : j = 0, 1, \ldots, p-1\right\}$ and $\ell_{\theta \otimes \overset{\triangle}{\varepsilon}{}^j} = 1$, for all $0 \leq j \leq p-1$;
- $\dim \left({}^{c^k}\psi \otimes \overset{\triangle}{\omega}{}^j\right) = d_\sigma$ and $\dim \left(\theta \otimes \overset{\triangle}{\varepsilon}{}^j\right) = p\, d_\sigma$, for all $0 \leq j, k \leq p-1$;
- For all $0 \leq j \leq p-1$, the stabilizer of $\theta \otimes \overset{\triangle}{\varepsilon}{}^j$ with respect to the action of $\widehat{G/N}$ on $\widehat{G}$ is given by $\widehat{G/M}$, which is a subgroup of $\widehat{G/N}$ (via inflation).
- M is an intermediary subgroup for ${}^{c^k}\sigma$ and $\theta \otimes \overset{\triangle}{\varepsilon}{}^j$, $j, k = 0, 1, \ldots, p-1$ (cf. Definition 2.38); the intermediary representation is ${}^{c^k}\psi \otimes \overset{\triangle}{\omega}{}^j$.

The following Bratteli diagram summarizes the restriction rules. By reversing the arrows, we get a description of the induction rules.

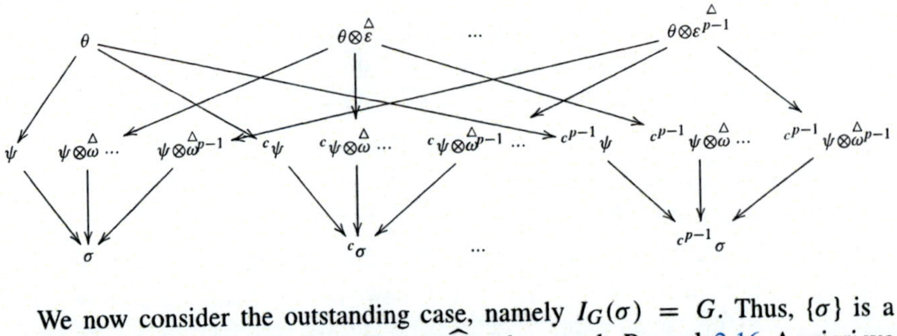

We now consider the outstanding case, namely $I_G(\sigma) = G$. Thus, $\{\sigma\}$ is a conjugacy class for the G-action of G on $\widehat{N}$ and we apply Remark 2.16. A priori we have three subcases, depending on the cardinality of $\widehat{G}(\sigma)$. However, we first show that we cannot have $|\widehat{G}(\sigma)| = p$. Indeed, by (2.47) with $I_G(\sigma)$ replaced by G, this condition leads to $\ell_\theta^2 = \frac{|G/N|}{|\widehat{G}(\sigma)|} = p$, which is impossible.

Also note that in the case $G/N \cong C_{p^2}$, by virtue of Proposition 2.23, σ has an extension to G and therefore, by Theorem 4.3.(3), necessarily $|\widehat{G}(\sigma)| = p^2$.

Case $I_G(\sigma) = G$ and $|\widehat{G}(\sigma)| = p^2$

The action of $\widehat{G/N}$ on $\widehat{G}(\sigma)$ is regular and therefore, by Corollary 2.34, σ has an extension $\widetilde{\sigma}$ to G. Therefore we may apply Theorem 4.3.(3), which describes precisely the relations between $\{\sigma\}$ and $\widehat{G}(\sigma)$.

Case $I_G(\sigma) = G$ and $|\widehat{G}(\sigma)| = 1$

We are not in one of the situations analyzed in Theorem 4.3 but in one described by Isaacs' going down theorem, namely that in Theorem 2.37.(2) and in (5.11) Sect. 5.5. Indeed:

- $\widehat{G}(\sigma) = \{\theta\}$ and $\theta \otimes \overset{\triangle}{\omega} \sim \theta$ for all $\omega \in \widehat{G/N}$;
- $\mathrm{Res}^G_N\theta \sim \ell\sigma$, where $\ell = \ell_\theta$ (cf. Remark 2.16);
- $\mathrm{Ind}^G_N\sigma \sim \ell\theta$ (by (2.15));
- $\ell = p$ (by (2.47)) and therefore $d_\theta = p\, d_\sigma$.

Moreover, taking a subgroup $N \lneqq M \lneqq G$, then σ has an extension ψ to M, because $M/N \cong C_p$ and $I_M(\sigma) = M$. Then ψ does not have an extension to G (otherwise, σ would) and therefore $\theta \sim \mathrm{Ind}^G_M\psi$, that is, M is an intermediary subgroup for σ and θ.

Example 5.15 Examples where the first three cases occur, may be provided by using the methods in 2.61, both for $H = C_{p^2}$ and for $H = C_p \times C_p$. For the fourth case, which may occur only for $H = C_p \times C_p$, wreath products do not suffice and in order to describe a relative example, we first consider the following abstract situation.

Let G be a finite group, let Z denote its center, and suppose that $G/Z \cong C_p \times C_p$ for some prime p and that G contains an Abelian normal subgroup $B \cong Z \times C_p$. Since G/Z is Abelian, we deduce that Z is contained in the derived subgroup G'. Therefore a one-dimensional representation ψ of G is trivial on Z (cf. (1.24)) so that a nontrivial character $\xi \in \widehat{Z}$ cannot have an extension to the whole of G. On the other hand, $I_G(\xi) = G$ (the action of G on its center is trivial; cf. (2.16)), ξ has an extension $\widetilde{\xi}$ to B (which represents the subgroup M in the fourth case) and $I_G(\widetilde{\xi}) = B$: it is impossible that $I_G(\widetilde{\xi}) = G$ because it would imply that $\widetilde{\xi}$ has an extension to the whole of G. In particular, $\theta := \mathrm{Ind}^G_B\widetilde{\xi}$ is a G-irreducible representation, nd necessarily we are in the fourth case and $\widetilde{G} = \{\theta\}$.

A concrete example is given by the Heisenberg group (cf. Sect. 6.9) and may be obtained either by setting $n = p$ in Example 6.10 or setting $h = 1$ in Sects. 6.9, 6.11, and 7.4.

5.7 Representation Theory of Finite Metacyclic Groups

In this section, we study a remarkable class of couples $N \lhd G$ satisfying the conditions (4.1), namely that of finite metacyclic groups. In comparison with [38], our analysis not only stresses the relevance of Clifford theory (which is not directly used in [38]), but also applies to the entire class of metacyclic groups (and not only to metacyclic groups satisfying the (CR) condition, see below, as in [38]). For the theory of cyclic groups and related arithmetic questions, we refer to [30, Chap. 1]. For more properties of metacyclic groups and their classification see [58, 124].

Definition 5.16 A finite group G is *metacyclic* if it contains a normal cyclic subgroup C such that G/C is also cyclic.

In the following we recall some basic properties of metacyclic groups.

Suppose that G is metacyclic. Then we can find $c \in G$ which generates a normal cyclic subgroup C such that G/C is cyclic. Let also $a \in G$ such that aC is a generator of G/C. Also let γ denote the automorphism of C given by $\gamma(c) = a^{-1}ca$. We define the following five integers:

- m is the order of C;
- s is the order of G/C;
- r is the smallest positive integer such that $\gamma(c) \equiv a^{-1}ca = c^r$;
- u is the order of γ;
- $d := \gcd(r-1, m)$;
- t is the smallest nonnegative integer such that $a^s = c^t$ (note that $a^s \in C$ because $G/C \equiv \langle aC \rangle$ is cyclic of order s).

The integers m, s, r, u, t satisfy the following arithmetic conditions. Recall that $\gcd(x, y)$ denotes the greatest common divisor of the integers x and y, while $x \mid y$ means that x divides y.

Lemma 5.17

(1)

$$r^k - 1 \begin{cases} \not\equiv 0 \pmod{m} & \textit{for } k = 0, 1, \ldots, u-1, \\ \equiv 0 \pmod{m} & \textit{for } k = u. \end{cases}$$

(2) $\gcd(m, r) = 1$, *that is, m and r are relatively prime;*
(3) $m \mid t(r-1)$*;*
(4) $u \mid s$ *and* $m \mid (r^s - 1)$.

Proof

(1) For $0 \le k \le u-1$ we recursively have $\gamma^k(c) = a^{-k}ca^k = c^{r^k}$. Thus, $\gamma^k(c) = c$ if and only if $r^k \equiv 1 \pmod{m}$. This is indeed the case if $k = u$, the order of γ.
(2) Since γ is a group homomorphism, $\gamma(c) = c^r$ must be a generator of C. As a consequence, $\gcd(m, r) = 1$.

(3) We have

$$c^{rt} = (c^r)^t = (a^{-1}ca)^t = a^{-1}c^t a = a^{-1}a^s a = a^s = c^t$$

and therefore $c^{(r-1)t} = 1$. Since m is the order of c, this shows that $m \mid t(r-1)$.

(4) $\gamma^s(c) = a^{-s}ca^s = c^{-t}cc^t = c$. Thus, s must be divisible by u, the order of γ. By (1), $m \mid r^u - 1$, so that from $u \mid s$, we deduce that $m \mid (r^s - 1)$.

■

We now determine $G' = [G, G]$, the derived (or commutator) subgroup of G.

Proposition 5.18 *The derived subgroup G' of G is the cyclic group generated by c^d and coincides with $\langle c^{r-1}\rangle$. Moreover, its order is equal to m/d.*

Proof We first show that c^d and c^{r-1} generate the same cyclic group of order $\frac{m}{d}$. Indeed, for $k \in \mathbb{Z}$ we have, on the one hand

$$(c^d)^k = 1 \iff c^{dk} = 1 \iff m \mid dk \iff \frac{m}{d} \mid k$$

and, on the other hand,

$$\left(c^{r-1}\right)^k = 1 \iff c^{(r-1)k} = 1 \iff m \mid (r-1)k \iff \frac{m}{d} \mid \frac{r-1}{d}k \iff \frac{m}{d} \mid k,$$

where the nontrivial part of the last equivalence follow from $\gcd(\frac{r-1}{d}, \frac{m}{d}) = 1$. This shows that the order of c^d and c^{r-1} is $\frac{m}{d}$. Since in a cyclic group C there exists a unique (cyclic) subgroup of a given order t (dividing $|C|$) we deduce that the (cyclic) subgroups $\langle c^d\rangle$ and $\langle c^{r-1}\rangle$ coincide. Alternatively, this equality can be deduced from the previous argument combined with the Bézout identity: there exist α and β in $\mathbb{Z}$ such that $\alpha(r-1) + \beta m = d$, so that $c^d = c^{\alpha(r-1)}$.

Then note that

$$c^{r-1} = c^{-1}c^r = c^{-1}a^{-1}ca = [c, a] \in G'.$$

Moreover, any $g \in G$ may be written in the form $g = c^i a^j$, for unique $0 \leq i \leq m-1$ and $0 \leq j \leq s-1$, and we have

$$\begin{aligned}
[c^i a^j, c^h a^k] &= a^{-j}c^{-i}a^{-k}c^{-h}c^i a^j c^h a^k \\
&= c^{-ir^j}a^{-(j+k)}c^{-h+i}a^j c^h a^k \\
&= c^{-ir^j}c^{(-h+i)r^{j+k}}a^{-k}c^h a^k \\
&= c^{-ir^j}c^{(-h+i)r^{j+k}}c^{hr^k} \\
&= c^{-ir^j+(-h+i)r^{j+k}+hr^k}
\end{aligned}$$

$$= c^{-ir^j - hr^j r^k + ir^j r^k + hr^k}$$

$$= c^{ir^j(r^k-1) - hr^k(r^j-1)}$$

$$= c^{(r-1)w},$$

where $w := ir^j(r^{k-1} + \cdots + r + 1) - hr^k(r^{j-1} + \cdots + r + 1)$. This shows that every commutator in G is a power of c^{r-1}. We deduce that $G' = \langle c^{r-1} \rangle$. ■

Corollary 5.19 *G has exactly sd one-dimensional representations.*

Proof The map $\chi \mapsto \chi'$, where

$$\chi'(gG') := \chi(g)$$

for all $g \in G$, yields a bijective correspondence between the one-dimensional representations of (any group) G and the characters of the abelianization G/G' (cf. [24, Proposition 9.2.5]). The statement then follows from the equality $|G/G'| = ms/(m/d) = sd$. ■

Set $\zeta := \exp\left(\frac{2\pi i}{m}\right)$. Then the characters of $C = < c >$ are $\xi_0, \xi_1, \ldots, \xi_{m-1}$, where

$$\xi_j(c^h) = \zeta^{hj}, \tag{5.25}$$

for all $j, h = 0, 1, \ldots, m-1$.

Proposition 5.20

(1) *For $h = 0, 1, \ldots, s-1$ we have:*

$$^{a^h}\xi_j = \xi_{r^h j}. \tag{5.26}$$

(2) *The inertia group $I_G(\xi_j)$ of ξ_j is given by*

$$I_G(\xi_j) = \sqcup_{h \in \mathscr{C}_j} Ca^h \tag{5.27}$$

where $\mathscr{C}_j := \{h : r^h j \equiv j \mod m\}$.

(3) *The number of fixed points of the G-action on $\widehat{C}$ is equal to d. More precisely, these fixed points are the characters $\xi_{hm/d}$, for $h = 0, 1, \ldots, d-1$.*

Proof For all $c \in C$, $h = 0, 1, \ldots, s-1$, and $j = 0, 1, \ldots, m-1$ we have

$$^{a^h}\xi_j(c) = \xi_j(a^{-h}ca^h) = \xi_j(c^{r^h}) = \zeta^{r^h j} = \xi_{r^h j}(c),$$

which proves (5.26). Then (5.27) is an immediate consequence. Moreover from (5.26) we deduce that $\xi_j \in \widehat{C}$ is fixed by G if and only if $rj \equiv j \pmod{m}$, that is, if and only if $j(r-1) \equiv 0 \pmod{m}$. Recalling that $1 = \gcd\left(\frac{r-1}{d}, \frac{m}{d}\right)$, this is equivalent to $\frac{m}{d} \mid j$. Since $0 \le j \le m-1$, we deduce that these fixed points are the ξ_j with $j = h\frac{m}{d}$ and $h = 0, 1, \ldots, d-1$. ∎

We now apply the little group method (cf. Theorems 2.26.(2), 2.54, and 4.3.(1b); also recall Proposition 2.23) which immediately yields the following.

Proposition 5.21 *Let Ξ be a system of representatives for the G-orbits on $\widehat{C}$ (cf. (2.5)) and set $A_G(\xi) := I_G(\xi)/C$. Then*

$$\widehat{G} = \left\{ \mathrm{Ind}_{I_G(\xi)}^{G} \left(\widetilde{\xi} \otimes \overset{\vartriangle}{\omega} \right) : \xi \in \Xi, \omega \in \widehat{A_G(\xi)} \right\}. \tag{5.28}$$

Moreover, $\dim \mathrm{Ind}_{I_G(\xi)}^{G} \left(\widetilde{\xi} \otimes \overset{\vartriangle}{\omega} \right) = |G/I_G(\xi)|$ *and* $|\widehat{G}(\xi)| = |I_G(\xi)/C|$. ∎

Taking into account that the G-orbit of $\xi \in \Xi$ contains $|G/I_G(\xi)|$ elements and recalling the correspondence (4.3), we now check (1.20) in the present situation:

$$\begin{aligned}
\sum_{\xi \in \Xi} |G/I_G(\xi)|^2 \cdot |I_G(\xi)/C| &= \sum_{j=0}^{m-1} \frac{|G|^2}{|I_G(\xi_j)| \cdot |C|} \cdot \frac{1}{|G/I_G(\xi_j)|} \\
&= \sum_{j=0}^{m-1} \frac{|G|}{|C|} \\
&= |G|.
\end{aligned}$$

Definition 5.22 A metacyclic group G satisfies the *Curtis-Reiner* (briefly (CR)) *condition* if, for $j = 0, 1, \ldots, m-1$, the inertia group $I_G(\xi_j)$ equals either C or G, where the characters $\xi_j \in \widehat{C}$ are as in (5.25).

In the following we present some characterizations of the (CR) condition.

Proposition 5.23 *A metacyclic group satisfies the (CR) condition if and only if for all $h = 1, 2, \ldots, s-1$ and $j = 0, 1, \ldots, m-1$ the following holds:*

$$r^h j \equiv j \pmod{m} \Rightarrow rj \equiv j \pmod{m}. \tag{5.29}$$

Proof Let G be a metacyclic group and let $0 \le j \le m-1$. Suppose first that condition (5.29) is satisfied. If $I_G(\xi_j) \ne C$, then by (5.27) there exists $0 \le h \le s-1$ such that $r^h j \equiv j \bmod m$. By (5.29) we in fact have $rj \equiv j \bmod m$, that is, with the notation in Proposition 5.20.(2), $\mathscr{C}_j = \{0, 1, \ldots, s-1\}$, and (5.27) gives $I_G(\xi_j) = \sqcup_{h \in \mathscr{C}_j} Ca^h = G$. Thus, G satisfies the Curtis-Reiner condition.

Conversely, suppose that G satisfies the (CR) condition. If $I_G(\xi_j) = C$, then (5.27) gives us $\mathscr{C}_j = \{0\}$, and (5.29) trivially holds true. If $I_G(\xi_j) = G$, then (5.27) gives us $\mathscr{C}_j = \{0, 1, \ldots, s-1\}$, that is, $r^h j \equiv j \pmod m$ for all $h = 0, 1, 2, \ldots, s-1$ and, again, (5.29) trivially holds true. ■

Proposition 5.24 *A metacyclic group G satisfies the (CR) condition if and only if $C \lhd G$ is a prime couple.*

Proof It follows immediately from the definition that if $C \lhd G$ is a prime couple then G satisfies the (CR) condition. To prove the converse implication, we only need to show that if G satisfies the (CR) condition, then each character $\xi_j \in \widehat{C}$ such that $I_G(\xi_j) = G$ has an extension to G. But this follows from Proposition 2.23 since G/C is cyclic. An alternative proof is the following: from Corollary 2.21.(2) and Proposition 5.18 we deduce that there are $[C : G'] = \frac{m}{m/d} = d$ characters $\xi \in \widehat{C}$ with $I_G(\xi) = G$ which have an extension to G. Proposition 5.20.(3) then ensures that there are exactly d characters $\xi \in \widehat{C}$ with $I_G(\xi) = G$. ■

Remark 5.25 Let G be a group with a normal cyclic subgroup C such that $|G/C|$ is prime. Then G is metacyclic and satisfies the (CR) condition. Indeed, for $\xi_i \in \widehat{C}$ one has $C \leq I_G(\xi_i) \leq G$ and the correspondence between subgroups of G/C and subgroups of G containing C ensures that either $I_G(\xi_i) = C$ or $I_G(\xi_i) = G$.

Alternatively, one may deduce this from Propositions 4.9 and 5.24.

Theorem 5.26 *Let G be a metacyclic group. Let Ξ' be a set of representatives for the G-orbits on $\widehat{C}$ consisting of s elements and set $\mathscr{B} := \{\mathrm{Ind}_C^G \xi : \xi \in \Xi'\}$. Let also $\mathscr{A}$ denote the set of all one dimensional representations of G.*

Then $\widehat{G} = \mathscr{A} \cup \mathscr{B}$ if and only if G satisfies the (CR) condition.

Proof First observe that $\mathscr{B} \subseteq \widehat{G}$. Indeed, if $\xi \in \Xi'$, so that the G-orbit of ξ consists of s elements (and therefore it is the largest possible), then necessarily $I_G(\xi_i) = C$ and this is equivalent to $\mathrm{Ind}_C^G \xi_i$ being irreducible (cf. the first statement in Proposition 2.27). Moreover, the representations in $\mathscr{B}$ are pairwise inequivalent (cf. the second statement in Proposition 2.27).

We present two proofs of the theorem.

First proof: Suppose that G satisfies the (CR) condition. Let Ξ be a system of representatives for the G-orbits on $\widehat{C}$ such that $\Xi \supseteq \Xi'$. If $\xi_i \in \Xi'$, the quotient $A_G(\xi_i) := I_G(\xi_i)/C$ is trivial and, in the notation as in (5.28), we have $\mathrm{Ind}_{I_G(\xi_i)}^G(\widetilde{\xi} \otimes \overset{\Delta}{\omega}) = \mathrm{Ind}_C^G \xi_i \in \mathscr{B}$. If $\xi_i \in \Xi \setminus \Xi'$, by the (CR) condition we necessarily have $I_G(\xi_i) = G$ and $A_G(\xi_i) = I_G(\xi_i)/C = G/C$. Again in the notation as in (5.28) we have $\mathrm{Ind}_{I_G(\xi_i)}^G(\widetilde{\xi} \otimes \overset{\Delta}{\omega}) = \xi_i \otimes \overset{\Delta}{\omega} \in \mathscr{A}$ for all $\omega \in \widehat{A_G(\xi_i)}$. Then the equality $\widehat{G} = \mathscr{A} \cup \mathscr{B}$ follows from (5.28). Vice versa, suppose that $\widehat{G} = \mathscr{A} \cup \mathscr{B}$, so that, in particular, every irreducible representation of G is either one-dimensional (if in $\mathscr{A}$) or s-dimensional (if in $\mathscr{B}$). Let $\xi_i \in \widehat{C}$. Pick $\xi \in \Xi$ in the G-orbit of ξ_i. By Proposition 5.21, we have

$\dim\mathrm{Ind}_{I_G(\xi)}^G \left(\widetilde{\xi}\otimes \overset{\Delta}{\omega}\right) = |G/I_G(\xi)|$, for all $\omega \in \widehat{A_G(\xi_i)}$. This forces $I_G(\xi) = G$ (in the one-dimensional case) or $I_G(\xi) = C$ (in the s-dimensional case). We deduce that also $I_G(\xi_i) = G$ or $I_G(\xi_i) = C$, and the (CR) condition follows.

Second proof. Assume that the (CR) condition is verified. Proposition 5.20.(3) ensures as that there are exactly d fixed points for the action of G on $\widehat{C}$. Since by the (CR) condition, the remaining points in $\widehat{C}$ all have a G-orbit of size s, we deduce that

$$|\mathscr{B}| = |\Xi'| = |G\text{-orbits on } \widehat{C} \text{ of size } s| = \frac{|\widehat{C}| - d}{s} = \frac{m-d}{s}.$$

Since $d_\rho = s$ for all $\rho \in \mathscr{B}$ and $d_\sigma = 1$ for all $\sigma \in \mathscr{A}$, we have

$$\sum_{\rho\in\mathscr{B}} d_\rho^2 = (m-d)s \quad \text{and} \quad \sum_{\sigma\in\mathscr{A}} d_\sigma^2 = |\mathscr{A}| = ds,$$

where the last equality follows from Corollary 5.19. We deduce that

$$\sum_{\sigma\in\mathscr{A}} d_\sigma^2 + \sum_{\rho\in\mathscr{B}} d_\rho^2 = ms = |G|$$

and (1.20) ensures that $\widehat{G} = \mathscr{A} \cup \mathscr{B}$.

Conversely, suppose that $\widehat{G} = \mathscr{A} \cup \mathscr{B}$. Since $|\mathscr{A}| = ds$ and $d_\rho = s$ for all $\rho \in \mathscr{B}$, by invoking again (1.20) we deduce that $|\mathscr{B}| = (ms - sd)/s^2 = (m-d)/s$. Therefore, the number of $\xi_j \in \widehat{C}$ with $I_G(\xi_j) = C$ (i.e., the G-orbit of ξ_j has size s) equals $s \cdot (m-d)/s = m - d$. By Proposition 5.20.(3) the G-action on $\widehat{C}$ has exactly d fixed points. As a consequence, the size of any G-orbit must be equal to either 1 or s, equivalently $I_G(\xi)$ equals either G or C for all $\xi \in \widehat{C}$. Thus, the (CR) condition is satisfied. ∎

Corollary 5.27 *Let G be a metacyclic group and suppose that it satisfies the (CR) condition. Then $\widehat{G}$ is made up of sd one-dimensional representations and $(m-d)/s$ representations of dimension s.*

Let G be a metacyclic group with cyclic normal subgroup $C = \langle c\rangle$ such that G/C is cyclic. Let $a \in G$ such that aC generates G/C so that $\{1, a, \ldots, a^{s-1}\}$ is a set of representatives for the cosets of C in G. Then, for $0 \le h, k \le s-1$, the element $a^{-h}ca^k = c^{r^h}a^{k-h}$ belongs to C if and only if $h = k$. Also, $a^{-h}aa^k = a^{k-h+1} \in C$ if and only if $k - h + 1 \in \{0, s\}$. Since for $\xi_j \in \widehat{C}$ as in (5.25) one has $\xi_j(a^s) = \xi_j(c^t) = \zeta^{jt}$ (recall that t is the smallest nonnegative integer such that $a^s = c^t$), the following yields a matrix expression U_j for $\rho_j := \mathrm{Ind}_C^G \xi_j$ (cf. (1.31))

$$U_j(c) = \begin{pmatrix} \zeta^j & 0 & \cdots & 0 \\ 0 & \zeta^{rj} & \cdots & 0 \\ \vdots & \vdots & \ddots & \vdots \\ 0 & 0 & \cdots & \zeta^{r^{s-1}j} \end{pmatrix} \quad \text{and} \quad U_j(a) = \begin{pmatrix} 0 & 0 & \cdots & \zeta^{jt} \\ 1 & & \cdots & 0 \\ \vdots & \vdots & \ddots & \vdots \\ 0 & \cdots & 1 & 0 \end{pmatrix}. \tag{5.30}$$

5.8 Examples: Dihedral and Generalized Quaternion Groups

We now present two applications of the theory developed in Sect. 5.7.

Example 5.28 (The Dihedral Group) Let $G := D_n = \langle a, c : c^n = a^2 = 1, aca = c^{n-1} \rangle$ be the *dihedral* group of degree n, i.e., the group of isometries of a regular polygon with n vertices. It is a metacyclic group (with $C = \langle c \rangle$ and $G/C = \{C, aC\} \cong C_2$) that satisfies the (CR) condition: indeed $s = |G/C| = 2$ is prime (cf. Remark 5.25).

Clearly, $G = \{a^h c^k : h = 0, 1 \text{ and } k = 0, 1, \ldots, n-1\}$ with multiplication law

$$a^{h_1}c^{k_1} \cdot a^{h_2}c^{k_2} := \begin{cases} a^{h_1}c^{k_1+k_2} & \text{if } h_2 = 0 \\ a^{h_1+h_2}c^{k_2-k_1} & \text{otherwise,} \end{cases}$$

for all $0 \leq h_1, h_2 \leq 1$ and $0 \leq k_1, k_2 \leq n-1$ (with the exponents of a and b in the right hand side taken modulo 2 and n, respectively). It is also easy to show that D_n is a semidirect product: $D_n = \langle c \rangle \rtimes \langle a \rangle \cong C_n \rtimes C_2$.

We now determine all the irreducible representations of $G = D_n$. We only consider the case when n is even and leave the odd case as an exercise. In the notation as in Sect. 5.7, we have $m = |C| = n$, $r = n-1$, $t = 0$, and $d = \gcd(r-1, m) = \gcd(n-2, n) = 2$. Therefore (cf. Proposition 5.18), $G' = \langle c^2 \rangle$ and $G/G' = \langle cG', aG' \rangle \cong C_2 \times C_2$. Thus, there are $4 = |G/G'|$ (pairwise inequivalent) one-dimensional representations of G, namely (we identify these with the corresponding characters), the inflations $\widetilde{\chi_{ij}}$, $0 \leq i, j \leq 1$ of the characters of $C_2 \times C_2$, that is,

$$\begin{aligned} \widetilde{\chi_{00}}(a^h c^k) &= 1 \\ \widetilde{\chi_{10}}(a^h c^k) &= (-1)^h \\ \widetilde{\chi_{01}}(a^h c^k) &= (-1)^k \\ \widetilde{\chi_{11}}(a^h c^k) &= (-1)^{h+k}, \end{aligned} \tag{5.31}$$

for all $h = 0, 1$ and $k = 0, 1, \ldots, n-1$.

Also, for $j = 1, 2, \ldots, (n-2)/2$ formula (5.30) yields (pairwise inequivalent) two-dimensional representations ρ_j given in matrix form by

$$U_j(c^k) = \begin{pmatrix} \zeta^{jk} & 0 \\ 0 & \zeta^{-jk} \end{pmatrix} \quad \text{and} \quad U_j(ac^k) = \begin{pmatrix} 0 & \zeta^{-jk} \\ \zeta^{jk} & 0 \end{pmatrix}$$

for $k = 0, 1, \ldots, n-1$, where $\zeta := \exp(2\pi i/n)$.

Example 5.29 (The Generalized Quaternion Group) The group $G := Q_n = \langle c, a : a^2 = c^n, a^{-1}ca = c^{-1} \rangle$ is called the *generalized quaternion group*. From the defining relations we immediately deduce that $a^2 = a^{-1}a^2a = a^{-1}c^na = c^{-n}$ and therefore $a^4 = 1$ and $c^{2n} = 1$. Moreover, from $a = a^2a^{-1} = c^na^{-1}$ and $a^{-1}ca = c^{-1}$ it follows that

$$aca^{-1} = c^{-1}. \tag{5.32}$$

Indeed,

$$a \cdot ca^{-1} = c^na^{-1} \cdot ca^{-1} = c^n(a^{-1}ca)a^{-2} = c^{n-1}c^{-n} = c^{-1}.$$

Any element $g \in Q_n$ may be uniquely written in the form $g = c^ka^h$ with $k \in \{0, 1, \ldots, 2n-1\}$ and $h \in \{0, 1\}$; in particular, Q_n has exactly $4n$ elements. An elementary proof of these facts may be obtained by the following concrete realization of Q_n

$$c := \begin{pmatrix} \omega & 0 \\ 0 & \overline{\omega} \end{pmatrix} \quad \text{and} \quad a := \begin{pmatrix} 0 & -1 \\ 1 & 0 \end{pmatrix},$$

where $\omega := \exp(\pi i/n)$ is a primitive $2n$-th root of 1. Then, it is easy to check that $a^2 = c^n$, $a^{-1}ca = c^{-1}$, and that the elements c^ka^h with $k = 0, 1, \ldots, 2n-1$ and $h = 0, 1$ are all distinct and form a group (clearly isomorphic to Q_n).

For $n = 2$ this is the ordinary *quaternion* group. For $n = 1$ it is the *Klein* group $C_2 \times C_2$.

Q_n is metacyclic (here $C = \langle c \rangle \cong C_{2n}$ and $G/C = \langle aC \rangle \cong C_2$) and satisfies the (CR) condition: indeed $s = |G/C| = 2$ is prime (cf. Remark 5.25). Note that, differently from the dihedral group D_n from the previous example, Q_n is *not* a semidirect product of the form $C_n \rtimes C_2$. Indeed, there is no element of order two in Q_n/C_n: from (5.32) we deduce that, for all $k = 0, 1, \ldots, 2n-1$,

$$(c^ka)^2 = c^kac^ka = c^kac^ka^{-1}a^2 = c^kc^{-k}a^2 = a^2,$$

showing that the order of c^ka is 4.

Now, $r = 2n - 1$, $m = |C| = 2n$, $s = |G/C| = 2$, $t = n$, and $d = \gcd(r - 1, m) = \gcd(2n - 2, 2n) = 2$. Therefore (cf. Proposition 5.18), $G' = \langle c^2 \rangle$ and

$$G/G' \cong \begin{cases} C_2 \times C_2 & \text{if } n \text{ is even} \\ C_4 & \text{if } n \text{ is odd,} \end{cases}$$

where $C_2 \times C_2 = \langle cG' \rangle \times \langle aG' \rangle$ when n is even, and $C_4 = \langle aG' \rangle$ when n is odd. Thus, there are $4 = |G/G'|$ (pairwise inequivalent) one-dimensional representations of G, namely (we identify these with the corresponding characters), the inflations of the characters of $C_2 \times C_2$ as in (5.31) if n is even, and the inflations $\widetilde{\chi}_j$, $j = 0, 1, 2, 3$ of the characters of C_4, that is,

$$\widetilde{\chi}_j(a^h c^k) = i^{jh} \tag{5.33}$$

for all $j, h = 0, 1, 2, 3$, and $k = 0, 1, \ldots, n - 1$, if n is odd.

The other irreducible representations of G are 2-dimensional and their number equals $(m - d)/s = n - 1$. Let $\xi_j \in \widehat{C}$ be given by $\xi_j(c^h) := \omega^{hj}$ for all $j, h = 0, 1, \ldots, 2n - 1$ (recall that $\omega := \exp(\pi i/n)$), then (5.26) becomes ${}^a\xi_j = \xi_{2n-j}$, so that the orbits of G on $\widehat{C}$ are $\{\xi_0\}$ and $\{\xi_n\}$ (1-element orbits, i.e. fixed points), and $\{\xi_1, \xi_{2n-1}\}, \{\xi_2, \xi_{2n-2}\}, \ldots, \{\xi_{n-1}, \xi_{n+1}\}$ (2-elements orbits).

Therefore, if we set $\rho_j := \mathrm{Ind}_C^G \xi_j$ for $j = 1, 2, \ldots, n-1$, then $\rho_1, \rho_2, \ldots, \rho_{n-1}$ are exactly the 2-dimensional irreducible representations of G. From (5.30) we immediately get the following matrix realizations U_j of ρ_j for $j = 1, 2, \ldots, n - 1$:

$$U_j(c) = \begin{pmatrix} \omega^j & 0 \\ 0 & \omega^{-j} \end{pmatrix} \quad \text{and} \quad U_j(a) = \begin{pmatrix} 0 & (-1)^j \\ 1 & 0 \end{pmatrix}.$$

We can apply Theorem 4.22 in this situation obtaining the following.

Proposition 5.30 *The irreducible representation $\rho_j \in \widehat{G}$ is real/quaternionic if and only if j is even/odd.*

Proof This follows from Theorem 4.22 by taking $G = Q_n$, $N = C$, $\sigma = A = \xi_j$, $\theta = \rho_j$, and $h = a$. Indeed, one has ${}^a\xi_j = \xi_{2n-j} \equiv \overline{\xi_j}$, the complex conjugate of ξ_j. This corresponds to case (3) of the theorem and, since $\xi_j(a^2) = \xi_j(c^n) = \omega^{nj} = (-1)^j$, we deduce that ρ_j is real/quaternionic if and only if j is even/odd. ■

Chapter 6
Central Extensions and the Orbit Method

In this chapter we treat central extensions of groups, a particular case of the construction in Sect. 1.5, and we give a complete cohomological characterization of these. Following Mihailovs [98] and Kokhas [81] with a 2-step nilpotent group with 2-divisible center we associate a 2-step nilpotent Lie ring. This is a key construction for the definition and application of the orbit method.

6.1 Central Extensions

Let H be a finite group and let $\mathscr{A}$ be an Abelian, possibly infinite, group (written in multiplicative notation): for instance, $\mathscr{A} = \mathbb{T} = \mathbb{R}/\mathbb{Z}$. Actually, also the finiteness hypothesis on H may be dropped for most of the initial results of this section.

Definition 6.1 A 2-*cocycle*, briefly a *cocycle* (or *factor set*) for the pair $(H, \mathscr{A})$, is a map $\tau\colon H \times H \to \mathscr{A}$ such that

$$\tau(h_1h_2, h_3)\tau(h_1, h_2) = \tau(h_1, h_2h_3)\tau(h_2, h_3) \quad \text{(cocycle identity)} \tag{6.1}$$

and

$$\tau(1_H, h) = \tau(h, 1_H) = 1_{\mathscr{A}} \qquad \text{(normalization)} \tag{6.2}$$

for all $h, h_1, h_2, h_3 \in H$. We denote by $\mathscr{C}^2(H, \mathscr{A})$ the set of all cocycles relative to the pair $(H, \mathscr{A})$. When $\mathscr{A} = \mathbb{T}$ we say that $\tau \in \mathscr{C}^2(H, \mathbb{T})$ is a *unitary* cocycle.

Note that, up to a slight change of notation, (6.1) is just (1.59) in the case where $h^\sharp$ is the trivial automorphism of N for all $h \in H$.

T. Ceccherini-Silberstein et al., *Representation Theory of Finite Group Extensions*, Springer Monographs in Mathematics,
https://doi.org/10.1007/978-3-031-13873-7_6

Proposition 6.2

(1) $\mathscr{C}^2(H, \mathscr{A})$ *is an Abelian group under multiplication.*
(2) *Let* $\tau \in \mathscr{C}^2(H, \mathscr{A})$. *Then*

$$\tau(h, h^{-1}) = \tau(h^{-1}, h)$$

for all $h \in H$.
(3) *Let* $\rho\colon H \to \mathscr{A}$ *be a map such that* $\rho(1_H) = 1_{\mathscr{A}}$. *Then the map* $\tau_\rho\colon H \times H \to \mathscr{A}$ *defined by setting*

$$\tau_\rho(h_1, h_2) := \rho(h_1 h_2)\,[\rho(h_1)\rho(h_2)]^{-1}, \tag{6.3}$$

for all $h_1, h_2 \in H$, *is a cocycle.*
(4) *If* H *is Abelian, then the cocycle* τ_ρ *is* symmetric*:*

$$\tau_\rho(h_1, h_2) = \tau_\rho(h_2, h_1) \tag{6.4}$$

for all $h_1, h_2 \in H$.

Proof

(1) The product $\tau_1\tau_2$ of two cocycles $\tau_1, \tau_2 \in \mathscr{C}^2(H, \mathscr{A})$ is defined by setting

$$(\tau_1\tau_2)(h_1, h_2) = \tau_1(h_1, h_2)\tau_2(h_1, h_2),$$

for all $h_1, h_2 \in H$, and conditions (6.1) and (6.2) are trivially satisfied by $\tau_1\tau_2$.
(2) Set $h_1 = h_3 = h$ and $h_2 = h^{-1}$ in (6.1) and use (6.2).
(3) Let $h, h_1, h_2, h_3 \in H$. Then

$$\begin{aligned}
\tau_\rho(h_1h_2, h_3)\tau_\rho(h_1, h_2) &= \rho(h_1h_2h_3)\,[\rho(h_1h_2)\rho(h_3)]^{-1}\,\rho(h_1h_2)\,[\rho(h_1)\rho(h_2)]^{-1}\\
&= \rho(h_1h_2h_3)\,[\rho(h_1)\rho(h_2)\rho(h_3)]^{-1}\\
&= \rho(h_1h_2h_3)\,[\rho(h_1)\rho(h_2h_3)]^{-1}\,\rho(h_2h_3)\,[\rho(h_2)\rho(h_3)]^{-1}\\
&= \tau_\rho(h_1, h_2h_3)\tau_\rho(h_2, h_3).
\end{aligned}$$

Moreover,

$$\tau_\rho(h, 1_H) = \rho(h)\,[\rho(h)\rho(1_H)]^{-1} = 1_{\mathscr{A}},$$

and analogously, one shows that $\tau_\rho(1_H, h) = 1_{\mathscr{A}}$.
(4) Finally, since H and $\mathscr{A}$ are both Abelian, for $h_1, h_2 \in H$ we have

$$\tau_\rho(h_1, h_2) = \rho(h_1h_2)\,[\rho(h_1)\rho(h_2)]^{-1} = \rho(h_2h_1)\,[\rho(h_2)\rho(h_1)]^{-1} = \tau_\rho(h_2, h_1)$$

and (6.4) follows. ∎

The 2-cocycles of the form τ_ρ as in (6.3) are called *2-coboundaries*. We denote by $\mathscr{B}^2(H, \mathscr{A})$ the set of all 2-coboundaries. Clearly, $\mathscr{B}^2(H, \mathscr{A})$ is a subgroup of $\mathscr{C}^2(H, \mathscr{A})$: indeed, $\tau_{\rho_1}\tau_{\rho_2} = \tau_{\rho_1\rho_2}$ and $(\tau_{\rho_3})^{-1} = \tau_{\rho_3^{-1}}$ for all $\rho_j\colon H \to \mathscr{A}$ with $\rho_j(1_H) = 1_{\mathscr{A}}$, $j = 1, 2, 3$, where $\rho_3^{-1}(h) = (\rho_3(h))^{-1}$ for all $h \in H$.

Definition 6.3 The quotient group $\mathscr{H}^2(H, \mathscr{A}) := \mathscr{C}^2(H, \mathscr{A})/\mathscr{B}^2(H, \mathscr{A})$ is called the *second cohomology group of H with values in $\mathscr{A}$*. The elements of $\mathscr{H}^2(H, \mathscr{A})$ are called *cocycle classes*, and two cocycles belonging to the same class are said to be *cohomologous*.

Let N and H be two groups. An extension $(G; \iota, \pi)$ of H by N (cf. Definition 1.8) is said to be *central* provided that $\iota(N)$ is contained in the center $Z(G)$ of G, and therefore N is Abelian. It is straightforward to check that an extension which is equivalent to a central extension is itself central. Indeed, if $(G_1; \iota_1, \pi_1)$ and $(G_2; \iota_2, \pi_2)$ are two equivalent extensions of H by N, the first one being central, and $\eta\colon G_1 \to G_2$ realizes the required isomorphism, then

$$\iota_2(N) = \eta(\iota_1(N)) \subseteq \eta(Z(G_1)) = Z(G_2),$$

showing that the second extension is central as well.

Example 6.4 A group G is called *2-step nilpotent* provided that

$$G \gneqq [G, G] \gneqq [G, [G, G]] = \{1_G\}, \tag{6.5}$$

where, for two subgroups $H, K \leq G$, we denote by $[H, K] \subseteq G$ the subgroup generated by all commutators $[h, k] = h^{-1}k^{-1}hk$ with $h \in H$ and $k \in K$. Equivalently, a group G is 2-step nilpotent if it is a *nontrivial central extension by an Abelian group*, i.e., we have:

$$\begin{cases} 1 \longrightarrow Z(G) \stackrel{\iota}{\longrightarrow} G \stackrel{\pi}{\longrightarrow} G/Z(G) \longrightarrow 1 \\ \text{with } \{1_G\} \neq Z(G) \neq G \text{ and } G/Z(G) \text{ Abelian.} \end{cases} \tag{6.6}$$

Indeed,

$$[G, [G, G]] = \{1_G\} \iff [G, G] \leq Z(G) \iff G/Z(G) \text{ is Abelian}$$

so that (6.5) is equivalent to $\{1_G\} \lneqq Z(G) \lneqq G$ and $[G, G] \leq Z(G)$, that is,

$$\{1_G\} \lneqq Z(G) \lneqq G \text{ and } G/Z(G) \text{ is Abelian.} \tag{6.7}$$

The above discussion actually shows that if G is 2-step nilpotent, then, for every subgroup A such that $[G, G] \leq A \leq Z(G)$ (so that A is Abelian and normal in G) we have a short exact sequence

$$1 \longrightarrow A \longrightarrow G \longrightarrow G/A \longrightarrow 1,$$

where G/A is Abelian, that is, G is a central extension by an Abelian group. In particular, if $A = [G, G]$ we have

$$1 \longrightarrow [G, G] \longrightarrow G \longrightarrow G/[G, G] \longrightarrow 1$$

(note that $G/[G, G]$ is the Abelianization of G) while, for $A = Z(G)$ we have (6.6).

Proposition 6.5 *Let H be a finite group, let $\mathscr{A}$ an Abelian group, and let $\tau \in \mathscr{C}^2(H, \mathscr{A})$. Define a multiplication $*$ on the set $\mathscr{A} \times H$ by setting:*

$$(a_1, h_1) * (a_2, h_2) := \left(a_1 a_2 \tau(h_1, h_2)^{-1}, h_1 h_2\right), \tag{6.8}$$

for all $a_1, a_2 \in \mathscr{A}$ and $h_1, h_2 \in H$. Then, with respect to (6.8), $\mathscr{A} \times H$ is a group, that we denote by $G = G(\mathscr{A}, H, \tau)$. We have $1_G = (1_{\mathscr{A}}, 1_H)$ and

$$(a, h)^{-1} = (a^{-1}\tau(h, h^{-1}), h^{-1}) \tag{6.9}$$

for all $(a, h) \in G$.

Moreover, denoting by $\iota \colon \mathscr{A} \to G$ and $\pi : G \to H$ the maps defined by

$$\iota(a) := (a, 1_H) \text{ and } \pi(a, h) := h \tag{6.10}$$

for all $a \in \mathscr{A}$ and $h \in H$, we have that ι is an injective homomorphism and π is a surjective homomorphism with $\iota(\mathscr{A}) = \ker(\pi)$, and $(G; \iota, \pi)$ is a central extension of H by $\mathscr{A}$.

Proof Let $a_j, a \in \mathscr{A}$ and $h_j, h \in H$, $j = 1, 2, 3$. We have

$$\begin{aligned}
[(a_1, h_1) * (a_2, h_2)] * (a_3, h_3) &= \left(a_1 a_2 \tau(h_1, h_2)^{-1}, h_1 h_2\right) * (a_3, h_3) \\
&= \left(a_1 a_2 \tau(h_1, h_2)^{-1} \cdot a_3 \tau(h_1 h_2, h_3)^{-1}, h_1 h_2 h_3\right) \\
\text{(by (6.1))} \qquad &= \left(a_1 \tau(h_1, h_2 h_3)^{-1} \cdot a_2 a_3 \tau(h_2, h_3)^{-1}, h_1 h_2 h_3\right) \\
&= (a_1, h_1) * \left(a_2 a_3 \tau(h_2, h_3)^{-1}, h_2 h_3\right) \\
&= (a_1, h_1) * [(a_2, h_2) * (a_3, h_3)],
\end{aligned}$$

and the associativity property is proved. Moreover, from (6.2) it follows immediately that

$$(a, h) * (1_{\mathscr{A}}, 1_H) = (a, h) = (1_{\mathscr{A}}, 1_H) * (a, h),$$

showing that $1_G = (1_{\mathscr{A}}, 1_H)$. Finally,

$$(a, h) * (a^{-1}\tau(h, h^{-1}), h^{-1}) = \left(a \cdot a^{-1}\tau(h, h^{-1}) \cdot \tau(h, h^{-1})^{-1}, h \cdot h^{-1}\right) = (1_{\mathscr{A}}, 1_H),$$

and similarly $(a^{-1}\tau(h, h^{-1}), h^{-1}) * (a, h) = (1_{\mathscr{A}}, 1_H)$ (cf. Proposition 6.2.(2)), so that each element has an inverse and (6.9) holds. This shows that $G(\mathscr{A}, H, \tau)$ is a group.

The properties of the maps ι and π are immediate as well as the equality $\ker(\pi) = \iota(\mathscr{A})$. Moreover, from (6.2) it follows that

$$(a_1, 1_H) * (a_2, h_2) = (a_1a_2, h_2) = (a_2a_1, h_2) = (a_2, h_2) * (a_1, 1_H),$$

for all $a_1, a_2 \in \mathscr{A}$ and $h \in H$, showing that $\iota(\mathscr{A}) \leq Z(G)$. This shows that $(G; \iota, \pi)$ is indeed a central extension. ∎

We are now in a position to state and prove the basic result on central extensions. In what follows, we write the left hand side of (6.8) in the form $(a_1, h_1) *_\tau (a_2, h_2)$ when we are dealing with products defined by means of different cocycles. Note also that part of the proof of Theorem 6.6 below is contained, in a more general setting, in the last part of Sect. 1.5 (with a slightly different notation), but we prefer to present here a complete independent treatment.

Theorem 6.6 *Let H be a finite group and let $\mathscr{A}$ be an Abelian group.*

(1) *Let $(G'; \iota', \pi')$ be a central extension of H by $\mathscr{A}$. Then there exists $\tau \in \mathscr{C}^2(H, \mathscr{A})$ such that if $G = G(\mathscr{A}, H, \tau)$ and ι and π are as in (6.10), then $(G; \iota, \pi)$ and $(G'; \iota', \pi')$ are equivalent as group extensions of H by $\mathscr{A}$.*
(2) *Let $\tau_1, \tau_2 \in \mathscr{C}^2(H, \mathscr{A})$. Then setting $G_j = G(\mathscr{A}, H, \tau_j)$, $j = 1, 2$, the corresponding group extensions $(G_1; \iota_1, \pi_1)$ and $(G_2; \iota_2, \pi_2)$ of H by $\mathscr{A}$ are equivalent if and only if τ_1 and τ_2 are cohomologous.*

Proof

(1) Let $t\colon H \to G'$ be a section (so that $\pi'(t(h)) = h$ for all $h \in H$) satisfying that $t(1_H) = 1_{G'}$. Then, for all $h_1, h_2 \in H$,

$$\begin{aligned}\pi'\left[t(h_1h_2)t(h_2)^{-1}t(h_1)^{-1}\right] &= \pi'[t(h_1h_2)] \cdot \pi'[t(h_2)]^{-1} \cdot \pi'[t(h_1)]^{-1}\\ &= (h_1h_2) \cdot h_2^{-1} \cdot h_1^{-1}\\ &= 1_H,\end{aligned}$$

which implies that $t(h_1h_2)t(h_2)^{-1}t(h_1)^{-1} \in \ker(\pi') = \iota'(\mathscr{A})$. Therefore

$$t(h_1h_2)t(h_2)^{-1}t(h_1)^{-1} \cdot t(h) = t(h) \cdot t(h_1h_2)t(h_2)^{-1}t(h_1)^{-1}, \qquad (6.11)$$

for all $h \in H$, because $\iota'(\mathscr{A}) \leq Z(G')$. We then define $\tau\colon H \times H \to \mathscr{A}$ by setting

$$\tau(h_1, h_2) := \iota'^{-1}\left(t(h_1h_2)t(h_2)^{-1}t(h_1)^{-1}\right) \tag{6.12}$$

for all $h_1, h_2 \in H$, and prove that τ is a cocycle. The proof is similar to that of Proposition 6.2.(3), but now, as the values of t belong to a noncommutative group, we must use (6.11).

$$\begin{aligned}
\tau(h_1h_2, h_3) \cdot \tau(h_1, h_2) &= \iota'^{-1}\left(t(h_1h_2h_3)t(h_3)^{-1}t(h_1h_2)^{-1}\right)\\
&\quad \cdot \iota'^{-1}\left(t(h_1h_2)t(h_2)^{-1}t(h_1)^{-1}\right)\\
&= \iota'^{-1}\left(t(h_1h_2h_3)t(h_3)^{-1}t(h_1h_2)^{-1}\right.\\
&\quad \left.\cdot\, t(h_1h_2)t(h_2)^{-1}t(h_1)^{-1}\right)\\
&= \iota'^{-1}\left(t(h_1h_2h_3)t(h_3)^{-1}t(h_2)^{-1}t(h_1)^{-1}\right)\\
&= \iota'^{-1}\left(t(h_1h_2h_3)t(h_2h_3)^{-1}\right.\\
&\quad \left.\cdot t(h_2h_3)t(h_3)^{-1}t(h_2)^{-1} \cdot t(h_1)^{-1}\right)\\
\text{(by (6.11))} \qquad &= \iota'^{-1}\left(t(h_1h_2h_3)t(h_2h_3)^{-1}t(h_1)^{-1}\right.\\
&\quad \left.\cdot t(h_2h_3)t(h_3)^{-1}t(h_2)^{-1}\right)\\
&= \iota'^{-1}\left(t(h_1h_2h_3)t(h_2h_3)^{-1}t(h_1)^{-1}\right)\\
&\quad \cdot \iota'^{-1}\left(t(h_2h_3)t(h_3)^{-1}t(h_2)^{-1}\right)\\
&= \tau(h_1, h_2h_3) \cdot \tau(h_2, h_3),
\end{aligned}$$

for all $h_1, h_2, h_3 \in H$. Moreover, for all $h \in H$,

$$\tau(h, 1_H) = \iota'^{-1}\left(t(h)t(1_H)^{-1}t(h)^{-1}\right) = \iota'^{-1}\left(t(h)1_{G'}t(h)^{-1}\right) = \iota'^{-1}(1_{G'}) = 1_{\mathscr{A}}$$

and, similarly, $\tau(1_H, h) = 1_{\mathscr{A}}$, proving (6.2). This shows that τ is a cocycle.

Let us show that the map $\eta\colon G \to G'$ given by

$$\eta(a, h) := \iota'(a)t(h)$$

for all $a \in \mathscr{A}$ and $h \in H$, yields an equivalence between the group extensions $(G; \iota, \pi)$ and $(G'; \iota', \pi')$. Indeed,

$$\begin{aligned}
&\eta\,[(a_1, h_1) * (a_2, h_2)] \\
&= \eta\left(a_1 a_2 \tau(h_1, h_2)^{-1}, h_1 h_2\right) && \text{(by (6.8))} \\
&= \iota'(a_1 a_2 \tau(h_1, h_2)^{-1}) \cdot t(h_1 h_2) \\
&= \iota'(a_1)\iota'(a_2) t(h_1) t(h_2) t(h_1 h_2)^{-1} \cdot t(h_1 h_2) && \text{(by (6.12))} \\
&= \iota'(a_1) t(h_1) \cdot \iota'(a_2) t(h_2) && (\iota'(a_2) \in Z(G')) \\
&= \eta(a_1, h_1) \cdot \eta(a_2, h_2),
\end{aligned}$$

for all $a_1, a_2 \in \mathscr{A}$ and $h_1, h_2 \in H$, showing that η is a group homomorphism. Moreover, since every element $g \in G'$ can be expressed in the form $g = \iota'(a)t(h)$ for unique $a \in \mathscr{A}$ and $h \in H$, we deduce that η is a bijection, and therefore a group isomorphism. Finally, if $a \in \mathscr{A}$ and $h \in H$, we have

$$\eta\,(\iota(a)) = \eta(a, 1_H) = \iota'(a) t(1_H) = \iota'(a),$$

showing that $\iota' = \eta\iota$, and

$$\pi'\,(\eta(a, h)) = \pi'\left(\iota'(a) t(h)\right) = \pi'\left(\iota'(a)\right) \cdot \pi'\,(t(h)) = 1_H \cdot h = h = \pi(a, h)$$

showing that $\pi = \pi'\eta$. Thus, η satisfies (1.53) and yields the desired equivalence.

(2) Let now $\tau_1, \tau_2 \in \mathscr{C}^2(H, \mathscr{A})$. Suppose first that τ_1 and τ_2 are cohomologous. Then there exists $\rho\colon H \to \mathscr{A}$ such that $\rho(1_H) = 1_{\mathscr{A}}$ and

$$\tau_2(h_1, h_2) = \tau_1(h_1, h_2)\rho(h_1)\rho(h_2)\rho(h_1 h_2)^{-1}, \tag{6.13}$$

for all $h_1, h_2 \in H$. We claim that the map $\eta\colon G(\mathscr{A}, H, \tau_1) \to G(\mathscr{A}, H, \tau_2)$ given by

$$\eta(a, h) := (a\rho(h), h)$$

for all $a \in \mathscr{A}$ and $h \in H$, yields an equivalence of group extensions. Indeed,

$$\begin{aligned}
\eta\left((a_1, h_1) *_{\tau_1} (a_2, h_2)\right) &= \eta\left(a_1 a_2 \tau_1(h_1, h_2)^{-1}, h_1 h_2\right) \\
&= (a_1 a_2 \tau_1(h_1, h_2)^{-1} \rho(h_1 h_2), h_1 h_2) \\
\text{(by (6.13))} \qquad &= (a_1 a_2 \rho(h_1)\rho(h_2)\tau_2(h_1, h_2)^{-1}, h_1 h_2) \\
&= (a_1 \rho(h_1) a_2 \rho(h_2) \tau_2(h_1, h_2)^{-1}, h_1 h_2) \\
&= \eta(a_1, h_1) *_{\tau_2} \eta(a_2, h_2),
\end{aligned}$$

for all $a_1, a_2 \in \mathscr{A}$ and $h_1, h_2 \in H$. Moreover, we have $\eta(a_1, h_1) = (a_2, h_2)$ if and only if $h_1 = h_2$ and $a_1 = a_2\rho(h_2)^{-1}$, so that η is a bijection. This shows that η is a group isomorphism. Finally, we show that η satisfies the conditions in (1.53). For all $a \in \mathscr{A}$ and $h \in H$ we have

$$\eta(\iota_1(a)) = \eta(a, 1_H) = (a\rho(1_H), 1_H) = (a1_{\mathscr{A}}, 1_H) = (a, 1_H) = \iota_2(a), \tag{6.14}$$

showing that $\iota_2 = \eta\iota_1$, and

$$\pi_2(\eta(a, h)) = \pi_2(a\rho(h), h) = h = \pi_1(a, h), \tag{6.15}$$

showing that $\pi_1 = \pi_2\eta$.

Conversely, suppose that there exists a map $\eta\colon G(\mathscr{A}, H, \tau_1) \to G(\mathscr{A}, H, \tau_2)$ yielding an equivalence of group extensions. Then, by (6.15), for all $h \in H$ we have $\pi_2\,(\eta(1_{\mathscr{A}}, h)) = \pi_1(1_{\mathscr{A}}, h) = h$, so that there exists $\rho(h) \in \mathscr{A}$ such that

$$\eta(1_{\mathscr{A}}, h) = (\rho(h), h). \tag{6.16}$$

We have $(1_{\mathscr{A}}, 1_H) = 1_{G_2} = \eta(1_{G_1}) = \eta(1_{\mathscr{A}}, 1_H) = (\rho(1_H), 1_H)$, where the last equality follows from (6.16), showing that $\rho(1_H) = 1_{\mathscr{A}}$. Moreover, arguing as in (6.14), we get $\eta(a, 1_H) = (a, 1_H)$ so that

$$\begin{aligned}\eta(a, h) &= \eta\big((a, 1_H) *_{\tau_1} (1_{\mathscr{A}}, h)\big) = \eta(a, 1_H) *_{\tau_2} \eta(1_{\mathscr{A}}, h)\\ &= (a, 1_H) *_{\tau_2} (\rho(h), h) = (a\rho(h)\tau_2(1_H, h), 1_H h) = (a\rho(h), h),\end{aligned}$$

for all $a \in \mathscr{A}$ and $h \in H$. Therefore, for arbitrary $h_1, h_2 \in H$,

$$\begin{aligned}\big(\tau_1(h_1, h_2)^{-1}\rho(h_1h_2), h_1h_2\big) &= \eta\big(\tau_1(h_1, h_2)^{-1}, h_1h_2\big)\\ &= \eta\big((1_{\mathscr{A}}, h_1) *_{\tau_1} (1_{\mathscr{A}}, h_2)\big)\\ &= \eta(1_{\mathscr{A}}, h_1) *_{\tau_2} \eta(1_{\mathscr{A}}, h_2)\\ &= (\rho(h_1), h_1) *_{\tau_2} (\rho(h_2), h_2)\\ &= (\rho(h_1)\rho(h_2)\tau_2(h_1, h_2)^{-1}, h_1h_2).\end{aligned}$$

This forces (6.13) and shows that τ_1 and τ_2 are cohomologous. ■

Remark 6.7 In the proof of Theorem 6.6.(1), we fixed a section $t\colon H \to G'$ satisfying $t(1_H) = 1_{G'}$ and defined a cocycle τ by means of (6.12). We now show that if $t'\colon H \to G'$ is another section satisfying $t'(1_H) = 1_{G'}$, then the corresponding cocycle τ' is cohomologous to τ. For $h \in H$ we have $\pi'(t(h)^{-1}t'(h)) = \pi(t(h))^{-1}\pi(t'(h)) = h^{-1}h = 1_H$ so that there exists $\rho(h) \in \mathscr{A}$

such that $t'(h) = t(h)\iota'(\rho(h))$. Note that, by our assumptions on the sections t and t', we have $\rho(1_H) = 1_{\mathscr{A}}$. Then (cf. (6.12)),

$$\begin{aligned}
\tau'(h_1, h_2) &= \iota'^{-1}\left(t'(h_1h_2)t'(h_2)^{-1}t'(h_1)^{-1}\right) \\
&= \iota'^{-1}\left(t(h_1h_2)\iota'(\rho(h_1h_2)) \cdot (t(h_2)\iota'(\rho(h_2)))^{-1} \cdot (t(h_1)\iota'(\rho(h_1)))^{-1}\right) \\
&= \iota'^{-1}\left(t(h_1h_2)t(h_2)^{-1}t(h_1)^{-1}\right) \cdot \rho(h_1h_2)\rho(h_2)^{-1}\rho(h_1)^{-1} \\
&= \tau(h_1, h_2)\tau_\rho(h_1, h_2)
\end{aligned}$$

for all $h_1, h_2 \in H$, so that the two 2-cocycles τ and τ' are cohomologous. In other words, a change of section does not change the cohomology class of the corresponding 2-cocycle.

Suppose that G is a central extension of H by $\mathscr{A}$. In the present situation, if G is a split extension (cf. Exercise 1.14) then actually G is isomorphic to the direct product of $\mathscr{A}$ and H. Indeed, a semidirect product $\mathscr{A} \rtimes H$, where $\mathscr{A}$ is a central subgroup, is a genuine direct product; see Proposition 1.7. Therefore, we consider the extension

$$1 \longrightarrow \mathscr{A} \stackrel{\iota}{\longrightarrow} \mathscr{A} \times H \stackrel{\pi}{\longrightarrow} H \longrightarrow 1, \tag{6.17}$$

where $\iota(a) = (a, 1_H)$ and $\pi(a, h) = h$, for all $a \in \mathscr{A}, h \in H$.

Proposition 6.8 *The extension $G(\mathscr{A}, H, \tau)$ in Proposition 6.5 is equivalent to (6.17) if and only if τ is a coboundary.*

Proof Clearly, if in (6.17) we choose the section $t\colon H \to \mathscr{A} \times H$ defined by setting $t(h) := (1_{\mathscr{A}}, h)$, for all $h \in H$, then $t(h_1h_2) = t(h_1)t(h_2)$ and the corresponding cocycle is identically equal to $1_{\mathscr{A}}$. Then, by Theorem 6.6, $G(\mathscr{A}, H, \tau)$ is equivalent to (6.17) if and only if τ is a coboundary. ∎

We now present three remarkable examples of 2-cocycles with a description of the corresponding group extensions. The first one is Abelian, while the second and third ones are step-2 nilpotent.

Example 6.9 Fix integers $n, m > 1$ and let $C_n = \langle a \rangle$, $C_m = \langle b \rangle$, and $C_{nm} = \langle c \rangle$ cyclic groups of order n, m, nm, respectively, with the indicated generators. We then have the exact sequence

$$1 \longrightarrow C_m \stackrel{\iota}{\longrightarrow} C_{nm} \stackrel{\pi}{\longrightarrow} C_n \longrightarrow 1,$$

where $\iota(b^h) := c^{nh}$ and $\pi(c^k) := a^k$, for all $h, k \in \mathbb{Z}$. In other words, C_{nm} is an extension of C_n by C_m. Define a section $t\colon C_n \to C_{nm}$ by setting

$$t(a^k) := c^k$$

for al $k = 0, 1, \dots, n-1$. Then the 2-cocycle associated to this extension is given by (6.12):

$$\tau(a^j, a^k) = \iota^{-1}\left(t(a^{j+k})t(a^j)^{-1}t(a^k)^{-1}\right) = \begin{cases} 1 & \text{if } j+k \le n-1 \\ b^{-1} & \text{if } j+k \ge n, \end{cases}$$

for all $0 \le j, k \le n-1$, because $j+k \ge n$ implies that $t(a^{j+k}) = t(a^{j+k-n}) = c^{j+k-n}$ and $\iota^{-1}(c^{-n}) = b^{-1}$. We now recover C_{nm} by means of the group law (6.8). For $0 \le i, h \le m-1$ and $0 \le j, k \le n-1$, it yields:

$$(b^i, a^j) * (b^h, a^k) = \begin{cases} (b^{i+h}, a^{j+k}) & \text{if } j+k \le n-1 \\ (b^{i+h+1}, a^{j+k-n}) & \text{if } j+k \ge n, \end{cases}$$

noting that if $i+h \ge m$ then $b^{i+h} = b^{i+h-m}$. This is indeed C_{nm}: we may uniquely represent an integer $0 \le r \le nm-1$ in the form $r = in + j$, with $0 \le i \le m-1$ $0 \le j \le n-1$, and if also $s = hn + k$ then the product of c^r and c^s is given by

$$c^r c^s = c^{in+j} c^{hn+k} = \begin{cases} c^{(i+h)n+j+k} & \text{if } j+k \le n-1 \\ c^{(i+h+1)n+(j+k-n)} & \text{if } j+k \ge n. \end{cases}$$

Again, if $i+h \ge m$ then $c^{(i+h)n} = c^{(i+h-m)n}$ and $(b^i, a^j) \to c^{in+j}$ is an isomorphism. These calculations are particular cases of those in [90, Sect. 7.3.2] for the second cohomology group of a finite cyclic group.

Example 6.10 Consider the group $A := C_n \times C_n \equiv \langle a \rangle \times \langle b \rangle$, where a, b are the generators of the two C_n's.

We define a unitary 2-cocycle $\tau \in \mathscr{C}^2(A, \mathbb{T})$ by setting

$$\tau(a^h b^t, a^m b^s) := \varepsilon^{-hs}, \tag{6.18}$$

for all $h, t, m, s \in \mathbb{Z}$, where $\varepsilon := \exp\left(\frac{2\pi i}{n}\right)$. This expression is a particular case of the "canonical form" for unitary 2-cocycles on finite Abelian groups; see Corollary 10.22 and Proposition 10.23. Clearly, it is well defined because $a^n = b^n = 1_A$ and $\varepsilon^n = 1$. It is a 2-cocycle:

$$\begin{aligned} \tau(a^h b^t \cdot a^k b^j, a^m b^s)\tau(a^h b^t, a^k b^j) &= \varepsilon^{-(h+k)s-hj} \\ &= \varepsilon^{-h(j+s)s-ks} = \tau(a^h b^t, a^k b^j \cdot a^m b^s)\tau(a^k b^j, a^m b^s) \end{aligned}$$

and

$$\tau(1_A, a^m b^s) = 1 = \tau(a^h b^t, 1_A),$$

for all $h, t, m, s \in \mathbb{Z}$. Note that, actually, $\tau \in \mathscr{C}^2(A, C_n)$ so that the corresponding extension is given by the group $G_n := G(C_n, A, \tau) \equiv \{(\varepsilon^j, a^h b^t) : 0 \le j, h, t \le n-1\}$ with the multiplication rule

$$(\varepsilon^j, a^h b^t) * (\varepsilon^k, a^m b^s) = (\varepsilon^{j+k+hs}, a^{h+m} b^{t+s}),$$

for all $h, t, m, s, j, k \in \mathbb{Z}$. The inverse and the power formulas are given by

$$(\varepsilon^j, a^h b^t)^{-1} = (\varepsilon^{-j+ht}, a^{-h} b^{-t})$$

and

$$(\varepsilon^j, a^h b^t)^{*k} = \left(\varepsilon^{kj+k(k-1)ht/2}, a^{kh} b^{kt}\right),$$

for all $h, t, m, s, j, k \in \mathbb{Z}$, respectively. The latter may be proved by induction. In particular, if n is odd, then the maximum period of an element $g \in G_n$ is n while, if n is even, then $n(n-1)/2 \equiv n(n-2)/2 + n/2$ is not divisible by n and the maximum period of and element $g \in G_n$ is $2n$.

Note that τ is not a coboundary: indeed,

$$\tau(a^h b^t, a^m b^s) = \tau(a^m b^s, a^h b^t) \iff \varepsilon^{hs} = \varepsilon^{mt} \iff hs \equiv mt \mod n,$$

while a coboundary τ_ρ satisfies $\tau_\rho(h,k) = \tau_\rho(k,h)$ for all h,k; cf. Proposition 6.2.(4). Moreover, the same argument shows that G_n is not commutative. By Proposition 6.8, G_n is not the semidirect product of C_n by $C_n \times C_n$. It is easy to check that the center of G_n is $Z = \{(\varepsilon^j, 1_A) : 0 \le j \le n-1\}$ and that $B := \{(\varepsilon^j, a^h) : 0 \le j, h \le n-1\}$ is a normal subgroup isomorphic to $C_n \times C_n$ containing Z. In particular, for $n = p$ prime we get an example of the groups described, in an abstract manner, in Example 5.15. Moreover, G_n is isomorphic to the Heisenberg group on $\mathbb{Z}/n\mathbb{Z}$ (now the cyclic group is written additively) described in [30, Sect. 12.3]); see also Sect. 6.9.

Example 6.11 Let A be as in Example 6.10 and let τ be the cocycle (6.18). Set also

$$\tau_1(a^h b^t, a^m b^s) := \begin{cases} 1 & \text{if } h+m \le n-1 \\ \varepsilon^{-1} & \text{if } h+m \ge n, \end{cases}$$

for $0 \le h, m, t, s \le n-1$. That is, τ_1 is the 2-cocycle in Example 6.9 on the first factor C_n of A and it is trivial on the second factor C_n. Finally, we set $\eta := \tau\tau_1$. Then, η is not cohomologous to τ because τ_1 is not a coboundary so that the group extension $H_n := G(C_n, C_n \times C_n, \eta)$ is not equivalent to G_n in Example 6.10. Moreover, the groups G_n and H_n are not isomorphic if $n > 2$: indeed, H_n contains a subgroup $C_{n^2} \equiv \{(\varepsilon^j, a^h) : 0 \le j, h \le n-1\}$ (cf. Example 6.9), while every element in G_n satisfies $g^{2n} = 1$. Clearly $H_n \equiv \{(\varepsilon^j, a^h b^t) : 0 \leq j, h, t \leq n-1\}$ with the multiplication rule

$$(\varepsilon^j, a^h b^t) * (\varepsilon^k, a^m b^s) = \begin{cases} (\varepsilon^{j+k+hs}, a^{h+m} b^{t+s}) & \text{if } h+m \leq n-1 \\ (\varepsilon^{j+k+hs+1}, a^{h+m-n} b^{t+s}) & \text{if } h+m \geq n \end{cases}$$

and the subgroup C_{n^2} is normal:

$$(1, b^{-1}) * (\varepsilon^j, a^h) * (1, b) = (\varepsilon^{j+h}, a^h). \tag{6.19}$$

Setting $c := (1, a, 1)$ and denoting $(1, 1, b)$ simply by b, it is easy to check that $C_{n^2} = \langle c \rangle$ (cf. again Example 6.9), and (6.19) implies that

$$b^{-1} c b = c^{1+n}. \tag{6.20}$$

It follows that $H_n \cong C_n \ltimes C_{n^2}$, where $C_n = \langle b \rangle$ and the left action of C_n on C_{n^2} is given by (6.20). It is left as an exercise to check that H_n is isomorphic to a subgroup of the affine group $\mathrm{Aff}(\mathbb{Z}/n^2\mathbb{Z})$ (cf. [30, Sect. 12.2]), namely to

$$\left\{ \left(\begin{smallmatrix} a & b \\ 0 & 1 \end{smallmatrix}\right) : a, b \in \mathbb{Z}/n^2\mathbb{Z}, a \equiv 1 \bmod n \right\}.$$

Remark 6.12 If p is an odd prime, there are exactly 2 noncommutative groups of order p^3: they are given by G_p (cf. Example 6.10) and H_p (cf. Example 6.11). See the monograph [57] by Hall and the expository notes [36, 88].

Remark 6.13 Recall (cf. Example 1.9) that two group extensions giving rise to the same group may not be equivalent, that is, equivalence is a finer invariant than isomorphism; see also [90, Chap. 7].

6.2 2-Divisible Abelian Groups, Equalized Cocycles, and Schur Multipliers

Let H be a finite group and let $\mathscr{A}$ be an Abelian group (written in multiplicative notation).

Definition 6.14 A cocycle $\tau \in \mathscr{C}^2(H, \mathscr{A})$ is called *equalized* (cf. [98]) if

$$\tau(h, h^{-1}) = 1_{\mathscr{A}}$$

for all $h \in H$.

Moreover, $\mathscr{A}$ is called 2-*divisible* if the map $a \mapsto a^2$ is an isomorphism. If this is the case, we denote by $a \mapsto \sqrt{a}$ the inverse map, so that the square root of any element $a \in \mathscr{A}$ is well and uniquely defined (in $\mathscr{A}$). Note that, in general (since $\mathscr{A}$ is Abelian), the map $a \mapsto a^2$ is a group homomorphism whose kernel is made up of exactly the elements of order 2. Suppose now that $\mathscr{A}$ is *finite*. It is a consequence of the theorems of Lagrange and Cauchy that $\mathscr{A}$ has no elements of order 2—that is, $a \mapsto a^2$ is an isomorphism, equivalently, $\mathscr{A}$ is 2-divisible—if and only if $\mathscr{A}$ is of *odd* order.

The following is our reformulation of [98, Lemma 2.5].

Proposition 6.15 *Let H be a finite group and let $\mathscr{A}$ be an Abelian group.*

(1) *If $\mathscr{A}$ is 2-divisible then any $\tau \in \mathscr{C}^2(H, \mathscr{A})$ is cohomologous to an equalized cocycle.*
(2) *Every unitary cocycle is cohomologous to an equalized cocycle.*

Proof Let $\tau \in \mathscr{C}^2(H, \mathscr{A})$. Suppose that for every $h \in H$ we can find $a_h \in \mathscr{A}$ such that $a_h^2 = \tau(h, h^{-1})$. Since $\tau(h, h^{-1}) = \tau(h^{-1}, h)$ by Proposition 6.2.(2), and $\tau(1_H, 1_H) = 1_{\mathscr{A}}$, we may suppose that $a_{h^{-1}} = a_h$ for all $h \in H$ and that $a_{1_H} = 1_{\mathscr{A}}$. We then set $\rho(h) = a_h$ for all $h \in H$. Let us show that the 2-cocycle $\tau' = \tau\tau_\rho$ (which, by definition, is cohomologous to τ) is equalized. Given $h \in H$ we have

$$\begin{aligned}\tau'(h, h^{-1}) = \tau(h, h^{-1})\tau_\rho(h, h^{-1}) &= \tau(h, h^{-1})\rho(1_H)[\rho(h)\rho(h^{-1})]^{-1}\\ &= \tau(h, h^{-1})1_{\mathscr{A}}(\tau(h, h^{-1}))^{-1} = 1_{\mathscr{A}}.\end{aligned}$$

This shows that τ' is equalized.

(1) Since $\mathscr{A}$ is 2-divisible, in the above argument, set $a_h := \sqrt{\tau(h, h^{-1})}$ for all $h \in H$.
(2) Recall that now $\mathscr{A} = \mathbb{T}$. For $h \in H$, let $\theta_h \in [0, 2\pi)$ such that $\tau(h, h^{-1}) = \exp(i\theta_h)$. Then, in the above argument, set $a_h := \exp(i\theta_h/2)$. ∎

Example 6.16 We apply the procedure in Proposition 6.15 to the cocycle in Example 6.10; see also Proposition 10.23. Actually, it is more convenient to assume that n is odd and then apply the procedure for 2-divisible groups. Indeed, if n is odd we may set $\varepsilon^{1/2} := \varepsilon^{(n+1)/2}$; this is the unique square root of $\varepsilon \in \mathbb{T}$ which is in C_n. This way, since $\tau(a^h b^t, a^{-h} b^{-t}) = \varepsilon^{ht}$, we set

$$\rho(a^h b^t) := \varepsilon^{ht/2} \equiv \left(\varepsilon^{(n+1)/2}\right)^{ht}$$

for all $h, t \in \mathbb{Z}$, so that

$$\tau_\rho(a^h b^t, a^m b^s) = \varepsilon^{(h+m)(t+s)/2} \cdot \varepsilon^{-(ht+ms)/2} = \varepsilon^{(hs+mt)/2},$$

and the equalized cocycle τ' cohomologous to τ is given by:

$$\begin{aligned}\tau'(a^h b^t, a^m b^s) &= \tau(a^h b^t, a^m b^s)\tau_\rho(a^h b^t, a^m b^s)\\ &= \varepsilon^{-hs+(hs+mt)/2}\\ &= \varepsilon^{(mt-hs)/2}\\ &\equiv \left(\varepsilon^{(n+1)/2}\right)^{mt-hs},\end{aligned}$$

for all $h, t, m, s \in \mathbb{Z}$.

On the other hand, if n is even we take $\omega := \exp\left(\frac{\pi i}{n}\right)$ as a square root of ε and we set $\rho(a^h b^t) := \omega^{ht}$ for $0 \le h, t \le n-1$. But the problem is that $h \mapsto \omega^h$ is $2n$-periodic, while $h \mapsto a^h$ is n-periodic. Then to express $\rho(a^h b^t)$ for arbitrary values of h, t we set

$$[t]_n := j$$

if $t \equiv j \bmod n$ and $0 \le j \le n-1$, so that

$$\rho(a^h b^t) = \omega^{[h]_n [t]_n}$$

for all $h, t \in \mathbb{Z}$. Then the equalized cocycle is given by

$$\begin{aligned}\tau'(a^h b^t, a^m b^s) &= \tau(a^h b^t, a^m b^s)\rho(a^{h+m} b^{t+s})\rho(a^h, b^t)^{-1}\rho(a^m, b^s)^{-1}\\ &= \omega^{-2hs+[h+m]_n[t+s]_n-[h]_n[t]_n-[m]_n[s]_n}.\end{aligned}$$

Now τ' has its values in $C_{2n} = \langle\omega\rangle$ (this agrees with Theorem 6.19, where the values are in C_{n^2} which contains C_{2n} because n is even) and we do not have the neat expressions of the n odd case.

In the following remark, we show that if $\mathscr{A}$ is not 2-divisible, the existence of an equalized 2-cocycle in $\mathscr{C}^2(H, \mathscr{A})$ (as in the theorem above) is not guaranteed.

Remark 6.17 Let $\tau \in \mathscr{C}^2(H, \mathscr{A})$ and suppose that it is equalized. Then the expression (6.9) for the inverse of the element $(a, h) \in G(\mathscr{A}, H, \tau)$ simplifies as follows:

$$(a, h)^{-1} = (a^{-1}, h^{-1}). \tag{6.21}$$

Let now $G = D_4 = \langle r, s : r^4 = s^2 = srsr = 1\rangle$ be the dihedral group of order 8 (cf. Example 5.28). Let also $\mathscr{A} = C_2 = \{1, t\}$ and $H = C_2 \times C_2 = \{(1,1), (t_1, 1), (1, t_2), (t_1, t_2)\}$ and define the injective homomorphism $\iota\colon \mathscr{A} \to G$ and the surjective homomorphism $\pi : G \to H$ by setting $\iota(t) := r^2$ and $\pi(r) := (t_1, 1)$, $\pi(s) := (1, t_2)$, respectively. We clearly have $\iota(\mathscr{A}) = \{1, r^2\} = \ker(\pi) =$

$Z(G)$. Now, there exist no equalized 2-cocycle $\tau \in \mathscr{C}^2(H, \mathscr{A})$ such that in $G \cong G(\mathscr{A}, H, \tau)$. Otherwise, since in both $\mathscr{A}$ and H all nontrivial elements have order two, the same would hold in G by (6.21), a contradiction since r has order 4.

The following elementary identities for the inverse of an equalized cocycle are very useful.

Lemma 6.18 *Let $\tau \in \mathscr{C}^2(H, \mathscr{A})$ be an equalized cocycle. Then for all $h, k \in H$ we have:*

$$\tau(h,k)^{-1} = \tau(h^{-1}, hk) \tag{6.22}$$

and

$$\tau(h,k)^{-1} = \tau(k^{-1}, h^{-1}). \tag{6.23}$$

Proof Setting $h_1 = h^{-1}$, $h_2 = h$, and $h_3 = k$ in (6.1), we get

$$\tau(h^{-1}, hk)\tau(h,k) = \tau(1_H, k)\tau(h^{-1}, h) = 1_{\mathscr{A}}$$

and this proves (6.22). Similarly,

$$\begin{aligned}
\tau(h,k)\tau(k^{-1},h^{-1}) &= \tau(h,k)\tau(hk, k^{-1}h^{-1})\tau(k^{-1},h^{-1}) \\
(h_1 = h, h_2 = k, h_3 = k^{-1}h^{-1} \text{ in } (6.1)) \quad &= \tau(h,h^{-1})\tau(k, k^{-1}h^{-1})\tau(k^{-1},h^{-1}) \\
(h_1 = k, h_2 = k^{-1}, h_3 = h^{-1} \text{ in } (6.1)) \quad &= \tau(1,h^{-1})\tau(k,k^{-1}) \\
&= 1_{\mathscr{A}},
\end{aligned}$$

yields (6.23). ■

Theorem 6.19 *Let H be a finite group of order n. Then every unitary cocycle $\tau \in \mathscr{C}^2(H, \mathbb{T})$ is cohomologous to a unitary cocycle $\tau' \in \mathscr{C}^2(H, C_n)$. If τ is equalized then τ' may be chosen to be equalized as well.*

The importance of the above result is that any unitary cocycle $\tau \in \mathscr{C}^2(H, \mathbb{T})$ is in fact a 2-cocycle for (H, C_n), where $C_n \leq \mathbb{T}$ is the (finite!) subgroup of the nth roots of unity, so that, in particular, $\tau'(h,k)^n = 1$ for all $h, k \in H$. Note also that, in fact, if $m \in \mathbb{N}$ is any multiple of $n = |H|$, then the above theorem yields that every unitary cocycle $\tau \in \mathscr{C}^2(H, \mathbb{T})$ is cohomologous to a unitary cocycle $\tau' \in \mathscr{C}^2(H, C_m)$.

Proof of Theorem 6.19 Let $\tau \in \mathscr{C}^2(H, \mathbb{T})$. From the fundamental identity $\tau(hk, h')\tau(h,k) = \tau(h, kh')\tau(k, h')$ and recalling that $|H| = n$, we get

$$\left[\prod_{h\in H} \tau(hk, h')\right]\left[\prod_{h\in H} \tau(h,k)\right] = \left[\prod_{h\in H} \tau(h, kh')\right]\tau(k,h')^n \tag{6.24}$$

for all $h, k, h' \in H$. Define $\rho\colon H \to \mathbb{T}$ by setting

$$\rho(k) := \prod_{h \in H} \tau(h, k)$$

for all $k \in H$. Note that $\rho(1_H) = 1$, since $\tau(h, 1_H) = 1 \in \mathbb{T}$ for all $h \in H$. Then the identity (6.24) becomes:

$$\rho(h')\rho(k) = \rho(kh')\tau(k, h')^n, \tag{6.25}$$

for all $h', k \in H$, which implies that τ^n is a coboundary.

By Proposition 6.15.(2) we may suppose that τ is equalized. We define $\rho'\colon H \to \mathbb{T}$ as follows. First observe that $\rho(h)\rho(h^{-1}) = \tau(h^{-1}, h)^n = 1$ so that

$$\rho(h^{-1}) = \rho(h)^{-1} \tag{6.26}$$

for all $h \in H$. Moreover, for every $h \in H$ there exists a unique $\theta = \theta(h) \in (-\pi, \pi]$ such that $\rho(h) = e^{i\theta}$. Let $h \in H$. If $\rho(h) \neq -1$ (equivalently, $\rho(h^{-1}) \neq -1$), that is, $\theta(h) \neq \pi$ (equivalently, $\theta(h^{-1}) \neq \pi$), it follows from (6.26) that $\theta(h^{-1}) = -\theta(h)$ and we set $\rho'(h) := e^{-i\theta(h)/n}$. Note that

$$\rho'(h)^n = \rho(h)^{-1} \tag{6.27}$$

(thus, $\rho'(h)$ is an n-th root of $\rho(h)^{-1}$) and

$$\rho'(h^{-1}) = \rho'(h)^{-1}. \tag{6.28}$$

On the other hand, if $h \neq h^{-1}$ and $\rho(h) = \rho(h^{-1}) = -1$ (the case $\theta(h) = \theta(h^{-1}) = \pi$) we can choose one of h, h^{-1}, say h, and set $\rho'(h) := e^{-i\pi/n}$ and then $\rho'(h^{-1}) = e^{i\pi/n}$. Finally, if $h^2 = 1_H$, then necessarily $\rho(h) = 1$. Indeed, we now have $\tau(h, h) = \tau(h, h^{-1}) = 1$ and the cocycle identity yields

$$\tau(kh, h)\tau(k, h) = \tau(k, 1_H)\tau(h, h) = 1$$

for all $k \in H$, so that if $\mathcal{T}$ is a system of representatives for the left cosets of the subgroup $\{1_H, h\} \subseteq H$, so that $H = \sqcup_{t \in \mathcal{T}} t\{1_H, h\}$, then

$$\rho(h) = \prod_{k \in H} \tau(k, h) = \prod_{t \in \mathcal{T}} (\tau(th, h)\tau(t, h)) = 1.$$

In this last case we then set $\rho'(h) := 1$.

Setting

$$\tau'(h, k) = \tau(h, k)\rho'(h)\rho'(k)\rho'(hk)^{-1}$$

yields a cocycle $\tau' \in \mathscr{C}^2(H, \mathbb{T})$ which is cohomologous to τ and

$$\begin{aligned}\tau'(h,k)^n &= \tau(h,k)^n \rho'(h)^n \rho'(k)^n \rho'(hk)^{-n} \\ \text{(by (6.25) and (6.27))} \qquad &= \rho(h)\rho(k)\rho(hk)^{-1} \cdot \rho(h)^{-1}\rho(k)^{-1}\rho(hk) \\ &= 1,\end{aligned}$$

for all $h, k \in H$. Since τ is equalized, (6.28) infers that

$$\tau'(h, h^{-1}) = \tau(h, h^{-1})\rho'(h)\rho'(h^{-1}) = 1,$$

for all $h \in H$, so that τ' is equalized as well. ∎

Definition 6.20 The *Schur multiplier* of a finite group G is the Abelian group $\mathscr{M}(G) := \mathscr{H}^2(G, \mathbb{T})$.

From the remark preceding the proof of the above theorem we immediately obtain the following.

Corollary 6.21 *The Schur multiplier of a finite group is always finite; in particular, if $|G| = n$ then $\mathscr{M}(G) = \mathscr{H}^2(G, C_n)$, where $C_n \leq \mathbb{T}$ is the group of n-th roots of unity.*

Remark 6.22 Actually, in Theorem 6.19 it may happen that $(\tau')^m = 1$ for an integer $1 \leq m < n$ (clearly $m|n$): for instance, if τ is cohomologous to the trivial cocycle, then $m = 1$.

From now on, we suppose that we have an exact sequence $1 \longrightarrow \mathscr{A} \stackrel{\iota}{\longrightarrow} G \stackrel{\pi}{\longrightarrow} H \to 1$, where $\mathscr{A}$ and H are Abelian groups and $\iota(\mathscr{A}) = Z(G)$, so that, cf. Example 6.4, G is a 2-step nilpotent group.

Definition 6.23 A cocycle $\tau \in \mathscr{C}^2(H, \mathscr{A})$ is *nondegenerate* if for every $h_1 \in H \setminus \{1_H\}$ there exists $h_2 \in H$ such that

$$\tau(h_1, h_2) \neq \tau(h_2, h_1). \tag{6.29}$$

On the other hand, τ is *symmetric* if

$$\tau(h_1, h_2) = \tau(h_2, h_1)$$

for all $h_1, h_2 \in H$.

Let $\tau \in \mathscr{C}^2(H, \mathscr{A})$. We observe that if τ is nondegenerate, an element $g = (a, h_1) \in G(\mathscr{A}, H, \tau)$ with $h_1 \neq 1_H$ cannot be central (i.e., $g \notin Z(G)$). Indeed if $h_2 \in H$ satisfies (6.29) we have

$$(a, h_1) *_\tau (1_{\mathscr{A}}, h_2) = (a\tau(h_1, h_2)^{-1}, h_1h_2) \neq (a\tau(h_2, h_1)^{-1}, h_2h_1) = (1_{\mathscr{A}}, h_2) *_\tau (a, h_1).$$

Moreover, since any coboundary in $\mathscr{C}^2(H, \mathscr{A})$ with H Abelian is symmetric (cf. Proposition 6.2.(4)), every cocycle τ' which is cohomologous to τ is nondegenerate if and only if τ is nondegenerate.

6.3 Lie Rings

Definition 6.24 (Lie Ring) A *Lie ring* is an Abelian group $\mathfrak{g}$ (written additively) equipped with a *bi-additive nilpotent* map, called the *Lie bracket*, $\mathfrak{g} \times \mathfrak{g} \to \mathfrak{g}$, denoted $(g_1, g_2) \mapsto [g_1, g_2]$, satisfying the Jacobi identity. This means that

(1) $[g_1 + g_2, g] = [g_1, g] + [g_2, g]$ and $[g, g_1 + g_2] = [g, g_1] + [g, g_2]$ for all $g, g_1, g_2 \in \mathfrak{g}$;
(2) $[g, g] = 0$ for all $g \in \mathfrak{g}$ (nilpotence);
(3) $[g_1, [g_2, g_3]] + [g_2, [g_3, g_1]] + [g_3, [g_1, g_2]] = 0$ for all $g_1, g_2, g_3 \in \mathfrak{g}$.

The element $[g_1, g_2]$ is called the *commutator* of $g_1, g_2 \in \mathfrak{g}$. Note that from (2) we have $[g_1 + g_2, g_1 + g_2] = 0$ so that $[g_2, g_1] = -[g_1, g_2]$ for all $g_1, g_2 \in \mathfrak{g}$, that is, the Lie bracket is *antisymmetric*. Conversely, if $\mathfrak{g}$ is 2-divisible, every antisymmetric bi-additive map is nilpotent: $[g, g] = -[g, g] \Rightarrow 2[g, g] = 0$, so that, by 2-divisibility, $[g, g] = 0$, for all $g \in G$. A subset $\mathfrak{h} \subseteq \mathfrak{g}$ which is a subgroup and is closed with respect to the Lie bracket operation is called a *Lie subring* and we write $\mathfrak{h} \leq \mathfrak{g}$. Given a subset $B \subseteq \mathfrak{g}$, the smallest Lie subring (equivalently, the intersection of all Lie subrings of $\mathfrak{g}$) containing B is called the Lie subring *generated* by B. Given two Lie subrings $\mathfrak{h}, \mathfrak{k} \leq \mathfrak{g}$ one denotes by $[\mathfrak{h}, \mathfrak{k}]$ the Lie subring generated by all elements $[h, k]$ with $h \in \mathfrak{h}$ and $k \in \mathfrak{k}$. The *center* of a Lie ring $\mathfrak{g}$ is the Lie subring $Z(\mathfrak{g}) := \{g \in \mathfrak{g} : [g, h] = 0 \text{ for all } h \in \mathfrak{g}\}$. When $Z(\mathfrak{g}) = \mathfrak{g}$ one says that $\mathfrak{g}$ is *Abelian*. Given two Lie rings $\mathfrak{g}$ and $\mathfrak{h}$, a map (resp. bijective map) $f : \mathfrak{g} \to \mathfrak{h}$ is called a *Lie ring homomorphism* (resp. *Lie ring isomorphism*) provided it yields a group homomorphism, i.e., $f(g_1+g_2) = f(g_1)+f(g_2)$ for all $g_1, g_2 \in \mathfrak{g}$ and preserves the Lie bracket (i.e., $f([g_1, g_2]) = [f(g_1), f(g_2)]$ for all $g_1, g_2 \in \mathfrak{g}$ (resp., in addition it is bijective). Given a Lie ring $\mathfrak{g}$, we denote by $\mathrm{Aut}(\mathfrak{g})$ the group of all Lie ring isomorphisms $f : \mathfrak{g} \to \mathfrak{g}$, called the *automorphisms* of $\mathfrak{g}$.

Let $\mathfrak{g}$ be a Lie ring. We say that $\mathfrak{g}$ is 2-*step nilpotent* if it is not Abelian but there exist Abelian Lie rings $\mathfrak{a}$ and $\mathfrak{h}$ and a Lie ring short exact sequence

$$0 \longrightarrow \mathfrak{a} \xrightarrow{\iota} \mathfrak{g} \xrightarrow{\pi} \mathfrak{h} \longrightarrow 0$$

such that $\iota(\mathfrak{a}) = Z(\mathfrak{g})$ $(= \ker(\pi))$. This is equivalent (cf. Example 6.4) to

$$\mathfrak{g} \gneqq [\mathfrak{g}, \mathfrak{g}] \gneqq [\mathfrak{g}, [\mathfrak{g}, \mathfrak{g}]] = \{0\}. \tag{6.30}$$

Suppose that $\mathfrak{g}$ is 2-step nilpotent. Then, simply viewing $\mathfrak{a}$, $\mathfrak{g}$, and $\mathfrak{h}$ as Abelian groups, we can find a cocycle $\tau_{\mathrm{sym}} \in \mathscr{C}^2(\mathfrak{h}, \mathfrak{a})$ such that the central extension

$(\mathfrak{g};\iota,\pi)$ is equivalent to $G(\mathfrak{a},\mathfrak{h},\tau_{\text{sym}})$. Since $\mathfrak{g}$ is Abelian, we have that τ_{sym} is necessarily symmetric, i.e. (cf. Definition 6.23),

$$\tau_{\text{sym}}(h_1,h_2)=\tau_{\text{sym}}(h_2,h_1)$$

for all $h_1,h_2\in\mathfrak{h}$, whence the notation. Indeed, Formula (6.8) written in additive notation yields, for all $a_1,a_2\in\mathfrak{a}$ and $h_1,h_2\in\mathfrak{h}$,

$$(a_1+a_2-\tau_{\text{sym}}(h_1,h_2),h_1+h_2)=(a_1,h_1)+_{\tau_{\text{sym}}}(a_2,h_2)=(a_2,h_2)+_{\tau_{\text{sym}}}(a_1,h_1)$$
$$=(a_2+a_1-\tau_{\text{sym}}(h_2,h_1),h_2+h_1)=(a_1+a_2-\tau_{\text{sym}}(h_2,h_1),h_1+h_2).$$

Let $t\colon\mathfrak{h}\to\mathfrak{g}$ be a section of π such that $t(0_{\mathfrak{h}})=0_{\mathfrak{g}}$ (in the present additive notation we thus have $\mathfrak{g}=\iota(\mathfrak{a})+t(\mathfrak{h}))$. Let $h_1,h_2\in\mathfrak{h}$. We have

$$\pi\left([t(h_1),t(h_2)]\right)=[\pi(t(h_1)),\pi(t(h_2))]=[h_1,h_2]=0$$

(recall that $\mathfrak{h}$ is Abelian, as a Lie ring), so that $[t(h_1),t(h_2)]\in\iota(\mathfrak{a})$. We thus set

$$\tau_{\text{b,n}}(h_1,h_2):=\iota^{-1}([t(h_1),t(h_2)])\in\mathfrak{a}. \tag{6.31}$$

Lemma 6.25 *The map* $\tau_{\text{b,n}}\colon\mathfrak{h}\times\mathfrak{h}\to\mathfrak{a}$ *defined in* (6.31) *is bi-additive nilpotent (i.e.,* $\tau_{\text{b,n}}(h,h)=0_{\mathfrak{a}}$ *for all* $h\in\mathfrak{h}$*), and therefore also antisymmetric. Moreover,* $\tau_{\text{b,n}}\in\mathscr{C}^2(\mathfrak{h},\mathfrak{a})$.

Proof Let $h_1,h_2,h_3,h\in\mathfrak{h}$. We have

$$\begin{aligned}
\tau_{\text{b,n}}(h_1+h_2,h_3)&=\iota^{-1}([t(h_1+h_2),t(h_3)])\\
\text{(for a unique } a(h_1,h_2)\in\mathfrak{a})\quad&=\iota^{-1}([t(h_1)+t(h_2)+\iota(a(h_1,h_2)),t(h_3)])\\
&=\iota^{-1}([t(h_1),t(h_3)]+[t(h_2),t(h_3)]+[\iota(a(h_1,h_1)),t(h_3)])\\
\text{(since } \iota(\mathfrak{a})=Z(\mathfrak{g}))\quad&=\iota^{-1}([t(h_1),t(h_3)]+[t(h_2),t(h_3)])\\
&=\iota^{-1}([t(h_1),t(h_3)])+\iota^{-1}[t(h_2),t(h_3)])\\
&=\tau_{\text{b,n}}(h_1,h_3)+\tau_{\text{b,n}}(h_2,h_3),
\end{aligned}$$

that is,

$$\tau_{\text{b,n}}(h_1+h_2,h_3)=\tau_{\text{b,n}}(h_1,h_3)+\tau_{\text{b,n}}(h_2,h_3) \tag{6.32}$$

and, analogously $\tau_{\mathfrak{b},\mathrm{n}}(h_1, h_2 + h_3) = \tau_{\mathfrak{b},\mathrm{n}}(h_1, h_2) + \tau_{\mathfrak{b},\mathrm{n}}(h_2, h_3)$. This proves that $\tau_{\mathfrak{b},\mathrm{n}}$ is bi-additive. As a consequence, we have

$$\begin{aligned}\tau_{\mathfrak{b},\mathrm{n}}(h_1 + h_2, h_3) + \tau_{\mathfrak{b},\mathrm{n}}(h_1, h_2) &= \tau_{\mathfrak{b},\mathrm{n}}(h_1, h_3) + \tau_{\mathfrak{b},\mathrm{n}}(h_2, h_3) + \tau_{\mathfrak{b},\mathrm{n}}(h_1, h_2)\\ &= \tau_{\mathfrak{b},\mathrm{n}}(h_1, h_2 + h_3) + \tau_{\mathfrak{b},\mathrm{n}}(h_2, h_3),\end{aligned}$$

that is, $\tau_{\mathfrak{b},\mathrm{n}}$ satisfies the additive version of the cocycle identity (6.1). Moreover,

$$\tau_{\mathfrak{b},\mathrm{n}}(0_{\mathfrak{h}}, h) = \iota^{-1}([t(0_{\mathfrak{h}}), t(h)]) = \iota^{-1}([0_{\mathfrak{g}}, t(h)]) = \iota^{-1}(0_{\mathfrak{g}}) = 0_{\mathfrak{a}}$$

and, analogously, $\tau_{\mathfrak{b},\mathrm{n}}(h, 0_{\mathfrak{h}}) = 0_{\mathfrak{a}}$, that is, $\tau_{\mathfrak{b},\mathrm{n}}$ satisfies the additive version of the cocycle identity (6.2). This shows that $\tau_{\mathfrak{b},\mathrm{n}} \in \mathscr{C}^2(\mathfrak{h}, \mathfrak{a})$. Finally,

$$\tau_{\mathfrak{b},\mathrm{n}}(h, h) = \iota^{-1}([t(h), t(h)]) = \iota^{-1}(0_{\mathfrak{g}}) = 0_{\mathfrak{a}}, \tag{6.33}$$

and this shows that $\tau_{\mathfrak{b},\mathrm{n}}$ is nilpotent (and therefore antisymmetric). ∎

Theorem 6.26 *Let $\overline{\mathfrak{g}} := \mathfrak{a} \times \mathfrak{h}$ denote the set theoretical Cartesian product of $\mathfrak{a}$ and $\mathfrak{h}$. Given two elements (a_1, h_1) and (a_2, h_2) in $\overline{\mathfrak{g}}$ we define their sum by*

$$(a_1, h_1) +_{\tau_{\mathrm{sym}}} (a_2, h_2) := (a_1 + a_2 - \tau_{\mathrm{sym}}(h_1, h_2), h_1 + h_2) \tag{6.34}$$

and their Lie bracket by

$$[(a_1, h_1), (a_2, h_2)]_{\tau_{\mathfrak{b},\mathrm{n}}} := \left(\tau_{\mathfrak{b},\mathrm{n}}(h_1, h_2), 0_{\mathfrak{h}}\right). \tag{6.35}$$

With these operations, $\overline{\mathfrak{g}}$ is a 2-step nilpotent Lie ring, that we denote by $\mathfrak{g}(\mathfrak{a}, \mathfrak{h}; \tau_{\mathrm{sym}}, \tau_{\mathfrak{b},\mathrm{n}})$.

Moreover, the maps $\overline{\iota}\colon \mathfrak{a} \to \overline{\mathfrak{g}}$ and $\overline{\pi}\colon \overline{\mathfrak{g}} \to \mathfrak{h}$ defined by

$$\overline{\iota}(a) := (a, 0_{\mathfrak{h}}) \text{ and } \overline{\pi}(a, h) := h$$

for all $a \in \mathfrak{a}$ and $h \in \mathfrak{h}$, are Lie ring homomorphisms.

Finally, the map $\eta\colon \overline{\mathfrak{g}} \to \mathfrak{g}$ defined by

$$\eta(a, h) := \iota(a) + t(h) \tag{6.36}$$

for all $a \in \mathfrak{a}$ and $h \in \mathfrak{h}$, yields an equivalence of Lie ring extensions from $\mathfrak{g}(\mathfrak{a}, \mathfrak{h}; \tau_{\mathrm{sym}}, \tau_{\mathfrak{b},\mathrm{n}}) \equiv (\overline{\mathfrak{g}}; \overline{\iota}, \overline{\pi})$ onto $(\mathfrak{g}; \iota, \pi)$.

Proof The definition of the sum operation given by (6.34) just says that $\overline{\mathfrak{g}}$, as an (Abelian) group, is nothing but $G(\mathfrak{a}, \mathfrak{h}, \tau_{\mathrm{sym}})$. Let us check that (6.35) defines a bi-additive nilpotent map satisfying the Jacobi identity. Let $a_1, a_2, a_3, a \in \mathfrak{a}$ and $h_1, h_2, h_3, h \in \mathfrak{h}$. We have

$$\begin{aligned}
[(a_1, h_1) +_{\tau_{\mathrm{sym}}} (a_2, h_2), (a, h)]_{\tau_{\mathfrak{b},\mathfrak{n}}} &= [(a_1 + a_2 - \tau_{\mathrm{sym}}(h_1, h_2), h_1 + h_2), (a, h)]_{\tau_{\mathfrak{b},\mathfrak{n}}} \\
&= \big(\tau_{\mathfrak{b},\mathfrak{n}}(h_1 + h_2, h), 0_{\mathfrak{h}}\big) \\
\text{(since } \tau_{\mathfrak{b},\mathfrak{n}} \text{ is bi-additive)} &= \big(\tau_{\mathfrak{b},\mathfrak{n}}(h_1, h) + \tau_{\mathfrak{b},\mathfrak{n}}(h_2, h), 0_{\mathfrak{h}}\big) \\
&= \big(\tau_{\mathfrak{b},\mathfrak{n}}(h_1, h), 0_{\mathfrak{h}}\big) +_{\tau_{\mathrm{sym}}} \big(\tau_{\mathfrak{b},\mathfrak{n}}(h_2, h), 0_{\mathfrak{h}}\big) \\
&= [(a_1, h_1), (a, h)]_{\tau_{\mathfrak{b},\mathfrak{n}}} +_{\tau_{\mathrm{sym}}} [(a_2, h_2), (a, h)]_{\tau_{\mathfrak{b},\mathfrak{n}}}
\end{aligned}$$

and, analogously,

$$[(a, h), (a_1, h_1) +_{\tau_{\mathrm{sym}}} (a_2, h_2)]_{\tau_{\mathfrak{b},\mathfrak{n}}} = [(a, h), (a_1, h_1)]_{\tau_{\mathfrak{b},\mathfrak{n}}} +_{\tau_{\mathrm{sym}}} [(a, h), (a_2, h_2)]_{\tau_{\mathfrak{b},\mathfrak{n}}}.$$

This shows that the Lie bracket is bi-additive.

Recalling that $\tau_{\mathfrak{b},\mathfrak{n}}$ is nilpotent, we have

$$[(a, h), (a, h)]_{\tau_{\mathfrak{b},\mathfrak{n}}} = \big(\tau_{\mathfrak{b},\mathfrak{n}}(h, h), 0_{\mathfrak{h}}\big) = (0_{\mathfrak{a}}, 0_{\mathfrak{h}}).$$

This shows that the Lie bracket $[\ ,\ ]_{\tau_{\mathfrak{b},\mathfrak{n}}}$ is nilpotent.

Finally, we have

$$\begin{aligned}
&[(a_1, h_1), [(a_2, h_2), (a_3, h_3)]_{\tau_{\mathfrak{b},\mathfrak{n}}}]_{\tau_{\mathfrak{b},\mathfrak{n}}} \\
&\quad +_{\tau_{\mathrm{sym}}} [(a_2, h_2), [(a_3, h_3), (a_1, h_1)]_{\tau_{\mathfrak{b},\mathfrak{n}}}]_{\tau_{\mathfrak{b},\mathfrak{n}}} +_{\tau_{\mathrm{sym}}} [(a_3, h_3), [(a_1, h_1), (a_2, h_2)]_{\tau_{\mathfrak{b},\mathfrak{n}}}]_{\tau_{\mathfrak{b},\mathfrak{n}}} \\
&\quad = [(a_1, h_1), (\tau_{\mathfrak{b},\mathfrak{n}}(h_2, h_3), 0_{\mathfrak{h}})]_{\tau_{\mathfrak{b},\mathfrak{n}}} +_{\tau_{\mathrm{sym}}} [(a_2, h_2), (\tau_{\mathfrak{b},\mathfrak{n}}(h_3, h_1), 0_{\mathfrak{h}})]_{\tau_{\mathfrak{b},\mathfrak{n}}} \\
&\qquad\qquad +_{\tau_{\mathrm{sym}}} [(a_3, h_3), (\tau_{\mathfrak{b},\mathfrak{n}}(h_1, h_2), 0_{\mathfrak{h}})]_{\tau_{\mathfrak{b},\mathfrak{n}}} \\
&\quad = (\tau_{\mathfrak{b},\mathfrak{n}}(h_1, 0_{\mathfrak{h}}), 0_{\mathfrak{h}}) +_{\tau_{\mathrm{sym}}} (\tau_{\mathfrak{b},\mathfrak{n}}(h_2, 0_{\mathfrak{h}}), 0_{\mathfrak{h}}) +_{\tau_{\mathrm{sym}}} (\tau_{\mathfrak{b},\mathfrak{n}}(h_3, 0_{\mathfrak{h}}), 0_{\mathfrak{h}}) \\
&\quad = (0_{\mathfrak{a}}, 0_{\mathfrak{h}})
\end{aligned}$$

and the Jacobi identity follows as well. This shows that $\overline{\mathfrak{g}}$ is a Lie ring. The fact that it is 2-step nilpotent follows from our assumptions on $\mathfrak{g}$ and the remaining of the proof.

In order to show that $\overline{\iota}\colon \mathfrak{a} \to \overline{\mathfrak{g}}$ and $\overline{\pi}\colon \overline{\mathfrak{g}} \to \mathfrak{h}$ are Lie rings homomorphisms, we first observe that these are Abelian group homomorphisms (cf. Proposition 6.5). Moreover, keeping in mind that the Lie rings $\mathfrak{a}$ and $\mathfrak{h}$ are Abelian, we have, for all $a_1, a_2 \in \mathfrak{a}$ and $h_1, h_2 \in \mathfrak{h}$

$$\overline{\iota}([a_1, a_2]) = \overline{\iota}(0_{\mathfrak{a}}) = 0_{\overline{\mathfrak{g}}} = (\tau_{\mathfrak{b},\mathfrak{n}}(0_{\mathfrak{h}}, 0_{\mathfrak{h}}), 0_{\mathfrak{h}}) = [(a_1, 0_{\mathfrak{h}}), (a_2, 0_{\mathfrak{h}})]_{\tau_{\mathfrak{b},\mathfrak{n}}} = [\overline{\iota}(a_1), \overline{\iota}(a_2)]_{\tau_{\mathfrak{b},\mathfrak{n}}}$$

and

$$\overline{\pi}\left([(a_1,h_1),(a_2,h_2)]_{\tau_{\mathfrak{b},\mathfrak{n}}}\right) = \overline{\pi}(\tau_{\mathfrak{b},\mathfrak{n}}(h_1,h_2),0_{\mathfrak{h}})) = 0_{\mathfrak{h}} = [h_1,h_2] = [\overline{\pi}(a_1,h_1),\overline{\pi}(a_2,h_2)].$$

Finally, in order to show that η yields the alluded equivalence of Lie ring extensions, we observe that conditions (1.53), namely $\iota = \eta\overline{\iota}$ and $\pi = \overline{\pi}\eta$, are satisfied (we already know it at the group level). Thus, as η is a bijection and a group homomorphism, we only need to check that η preserves the Lie bracket. We have

$$\begin{aligned}
\eta\left([(a_1,h_1),(a_2,h_2)]_{\tau_{\mathfrak{b},\mathfrak{n}}}\right) &= \eta\left(\tau_{\mathfrak{b},\mathfrak{n}}(h_1,h_2),0_{\mathfrak{h}}\right)\\
&= \eta\overline{\iota}(\tau_{\mathfrak{b},\mathfrak{n}}(h_1,h_2))\\
&= \iota(\tau_{\mathfrak{b},\mathfrak{n}}(h_1,h_2))\\
(\text{by } (6.31)) \quad &= [t(h_1),t(h_2)]\\
(\text{since } \iota(\mathfrak{a}) = Z(\mathfrak{g})) \quad &= [\iota(a_1)+t(h_1),\iota(a_2)+t(h_2)]\\
(\text{by } (6.36)) \quad &= [\eta(a_1,h_1),\eta(a_2,h_2)].
\end{aligned}$$

This completes the proof of the theorem. ■

6.4 The Cocycle Decomposition

Let H be a finite Abelian group and let A be an Abelian group, both written in additive notation. We denote by $\mathscr{C}^2_{\mathrm{sym}}(H,A)$ (resp. $\mathscr{C}^2_{\mathrm{b,n}}(H,A)$) the subgroup of $\mathscr{C}^2(H,A)$ consisting of all symmetric cocycles (resp. bi-additive nilpotent maps $H \times H \to A$; cf. (6.32) and (6.33)).

From now on, we suppose that A is 2-divisible (note that in the present setting we use additive notation so that $\sqrt{a}$ is now $a/2$ for all $a \in A$). We then have an (Abelian) group isomorphism

$$\Phi\colon \mathscr{C}^2(H,A) \to \mathscr{C}^2_{\mathrm{sym}}(H,A) \bigoplus \mathscr{C}^2_{\mathrm{b,n}}(H,A)$$

given by $\Phi(\tau) := (\tau_{\mathrm{sym}}, \tau_{\mathrm{b,n}})$, where

$$\begin{aligned}
\tau_{\mathrm{sym}}(h_1,h_2) &:= \frac{\tau(h_1,h_2)+\tau(h_2,h_1)}{2}\\
\tau_{\mathrm{b,n}}(h_1,h_2) &:= \tau(h_2,h_1)-\tau(h_1,h_2)
\end{aligned} \tag{6.37}$$

for all $\tau \in \mathscr{C}^2(H, A)$, and $h_1, h_2 \in H$. Let us check that $\tau_{\mathrm{b,n}}$, which clearly satisfies the nilpotence condition $\tau_{\mathrm{b,n}}(h, h) = 0_A$ for all $h \in H$, is indeed bi-additive, whence the notation). Let $h_1, h_2, h \in H$. Then

$$\begin{aligned}
\tau_{\mathrm{b,n}}(h_1 + h_2, h) &= \tau(h, h_1 + h_2) - \tau(h_1 + h_2, h) \\
(\text{by } (6.1)) &= (\tau(h + h_1, h_2) + \tau(h, h_1) - \tau(h_1, h_2)) \\
&\quad - (\tau(h_1, h_2 + h) + \tau(h_2, h) - \tau(h_1, h_2)) \\
&= \tau(h_1 + h, h_2) + \tau(h, h_1) - \tau(h_1, h + h_2) - \tau(h_2, h) \\
&= \tau(h_1 + h, h_2) + \tau(h_1, h) - \tau(h_1, h) + \tau(h, h_1) \\
&\quad - \tau(h_1, h + h_2) - \tau(h, h_2) + \tau(h, h_2) - \tau(h_2, h) \\
(\text{by } (6.1)) &= \tau(h_1, h + h_2) + \tau(h, h_2) - \tau(h_1, h) + \tau(h, h_1) \\
&\quad - \tau(h_1, h + h_2) - \tau(h, h_2) + \tau(h, h_2) - \tau(h_2, h) \\
&= (\tau(h, h_1) - \tau(h_1, h)) + (\tau(h, h_2) - \tau(h_2, h)) \\
&= \tau_{\mathrm{b,n}}(h_1, h) + \tau_{\mathrm{b,n}}(h_2, h).
\end{aligned}$$

This shows that $\tau_{\mathrm{b,n}}(h_1 + h_2, h) = \tau_{\mathrm{b,n}}(h_1, h) + \tau_{\mathrm{b,n}}(h_2, h)$ and similar calculations show that $\tau_{\mathrm{b,n}}(h, h_1 + h_2) = \tau_{\mathrm{b,n}}(h, h_1) + \tau_{\mathrm{b,n}}(h, h_2)$. We deduce that $\tau_{\mathrm{b,n}}$ is bi-additive.

Note that $\Phi^{-1} \colon \mathscr{C}^2_{\mathrm{sym}}(H, A) \bigoplus \mathscr{C}^2_{\mathrm{b,n}}(H, A) \to \mathscr{C}^2(H, A)$ is given by

$$\Phi^{-1}(\tau_{\mathrm{sym}}, \tau_{\mathrm{b,n}}) = \tau_{\mathrm{sym}} - \frac{\tau_{\mathrm{b,n}}}{2} \tag{6.38}$$

for all $(\tau_{\mathrm{sym}}, \tau_{\mathrm{b,n}}) \in \mathscr{C}^2_{\mathrm{sym}}(H, A) \bigoplus \mathscr{C}^2_{\mathrm{b,n}}(H, A)$.

Since $\mathscr{B}^2(H, A) \subseteq \mathscr{C}^2_{\mathrm{sym}}(H, A)$ we have

$$\mathscr{H}^2(H, A) = \mathscr{H}^2_{\mathrm{sym}}(H, A) \bigoplus \mathscr{H}^2_{\mathrm{b,n}}(H, A),$$

where $\mathscr{H}^2_{\mathrm{sym}}(H, A) := \mathscr{C}^2_{\mathrm{sym}}(H, A)/\mathscr{B}^2(H, A)$ and $\mathscr{H}^2_{\mathrm{b,n}}(H, A) := \mathscr{C}^2_{\mathrm{b,n}}(H, A)$. In other words, τ and σ in $\mathscr{C}^2(H, A)$ are cohomologous if and only if τ_{sym} is cohomologous to σ_{sym} and $\tau_{\mathrm{b,n}} = \sigma_{\mathrm{b,n}}$.

6.5 The Malcev Correspondence

The results from the previous section yield a correspondence between 2-step nilpotent groups with 2-divisible center and 2-step nilpotent Lie rings with 2-divisible center, that we now describe. Given a 2-step nilpotent group G with

2-divisible center, we have an exact sequence $1 \to A \xrightarrow{\iota} G \xrightarrow{\pi} H \to 1$ with $A = Z(G)$ (2-divisible by hypotheses) and H Abelian, and we associate a unique cocycle class $[\tau] \in \mathscr{H}^2(H, A)$ such that $(G; \iota, \pi) \cong G(A, H, \tau)$. We denote by

$$\eta\colon G(A, H, \tau) \to G = (G; \iota, \pi) \tag{6.39}$$

a group isomorphism (this depends on the section $t\colon H \to G$ for π).

Let $\Phi(\tau) = (\tau_{\mathrm{sym}}, \tau_{\mathrm{b,n}})$. Then, denoting by $\mathfrak{a}$ (resp. $\mathfrak{h}$) the Abelian Lie ring with underlying group structure A (resp. H) (and therefore with trivial Lie brackets), the associated 2-step nilpotent Lie ring is $\mathfrak{g} = \mathfrak{g}(\mathfrak{a}, \mathfrak{h}, \tau_{\mathrm{sym}}, \tau_{\mathrm{b,n}})$ (whose center $\mathfrak{a}$ is 2-divisible as an Abelian group).

Vice versa, given a 2-step nilpotent Lie ring $\mathfrak{g}$ with 2-divisible center, we have an exact sequence $0 \to \mathfrak{a} \xrightarrow{\iota} \mathfrak{g} \xrightarrow{\pi} \mathfrak{h} \to 0$ with $\mathfrak{a} = Z(\mathfrak{g})$ 2-divisible (as an Abelian group) and $\mathfrak{h}$ Abelian Lie rings. Then, denoting by A (resp. H) the underlying Abelian group of $\mathfrak{a}$ (resp. $\mathfrak{h}$) we can find unique elements $[\tau_{\mathrm{sym}}] \in \mathscr{H}^2_{\mathrm{sym}}(H, A)$ and $\tau_{\mathrm{b,n}} \in \mathscr{H}^2_{\mathrm{b,n}}(H, A)$ such that $\mathfrak{g} \cong \mathfrak{g}(\mathfrak{a}, \mathfrak{h}, \tau_{\mathrm{sym}}, \tau_{\mathrm{b,n}})$. We denote by

$$\overline{\eta}\colon \mathfrak{g}(\mathfrak{a}, \mathfrak{h}, \tau_{\mathrm{sym}}, \tau_{\mathrm{b,n}}) \to \mathfrak{g} \tag{6.40}$$

a (Lie ring) isomorphism (this depends on the section $t\colon H \to G$ for π).

Then the element $\tau = \Phi^{-1}(\tau_{\mathrm{sym}}, \tau_{\mathrm{b,n}}) \in \mathscr{C}^2(H, A)$ yields a 2-step nilpotent group $G = G(A, H, \tau)$ (whose center $Z(G) = A$ is 2-divisible).

Note that, the above correspondence yields, set theoretically, a bijection $\phi\colon G \to \mathfrak{g}$ given by

$$\phi := \overline{\eta} \circ i \circ \eta^{-1}, \tag{6.41}$$

where η and $\overline{\eta}$ are given by (6.39) and (6.40), respectively, and $i : A \times H \to \mathfrak{a} \times \mathfrak{h}$ is the identity map between the underlying sets (in fact it is an isomorphism between Abelian groups). We call ϕ the *logarithmic map*. The following constitutes a finite 2-step nilpotent analogue of the well known *Baker-Campbell-Hausdorff formula* (see [44] for a proof of the classical case and [1] for the history of the formula).

Theorem 6.27 *Let G be a finite 2-step nilpotent group with 2-divisible center. Let $\mathfrak{g}$ denote the associated 2-step nilpotent Lie ring with 2-divisible center and let $\phi\colon G \to \mathfrak{g}$ denote the logarithmic map as in (6.41). Then for $g, g_1, g_2 \in G$ we have:*

(1) $\phi(g^{-1}) = -\phi(g)$,
(2) $\phi(g_1 g_2) = \phi(g_1) + \phi(g_2) + \frac{1}{2}[\phi(g_1), \phi(g_2)]$,
(3) $\phi(g_1 g_2 g_1^{-1}) = \phi(g_2) + [\phi(g_1), \phi(g_2)]$,
(4) $\phi(g_1 g_2 g_1^{-1} g_2^{-1}) = [\phi(g_1), \phi(g_2)]$.

In particular, if g_1 and g_2 commute, then $\phi(g_1 g_2) = \phi(g_1) + \phi(g_2)$.

Proof Let $\tau \in \mathscr{C}^2(H, A)$ be an equalized cocycle (cf. Proposition 6.15) such that $G \cong_\eta G(A, H, \tau)$. Let $g \in G$ and let $a \in A$ and $h \in H$ such that $g = \eta(a, h)$. Then we have

$$\begin{aligned}\phi(g^{-1}) &= \overline{\eta}(i(\eta^{-1}(g^{-1}))) = \overline{\eta}(i((a, h)^{-1}))\\ \text{(by (6.21))} &= \overline{\eta}(-a, -h)\\ &= \overline{\eta}(-(a, h)) = -\overline{\eta}(a, h) = -\overline{\eta}(i(\eta^{-1}(g)))\\ &= -\phi(g)\end{aligned}$$

an this proves (1).

Let now $g_i \in G$ and let $a_i \in A$ and $h_i \in H$ such that $g_i = \eta(a_i, h_i)$ for $i = 1, 2$. Then

$$\begin{aligned}\phi(g_1 g_2) &= \overline{\eta}(i\left(\eta^{-1}(g_1 g_2)\right))\\ &= \overline{\eta}\,(i((a_1, h_1) *_\tau (a_2, h_2)))\\ \text{(by (6.8))} &= \overline{\eta}(i(a_1 + a_2 - \tau(h_1, h_2), h_1 + h_2))\\ \text{(by (6.38))} &= \overline{\eta}\left(i(a_1 + a_2 - \tau_{\mathrm{sym}}(h_1, h_2) + \tau_{\mathrm{b,n}}(h_1, h_2)/2, h_1 + h_2)\right)\\ &= \overline{\eta}\left(i(a_1 + a_2 - \tau_{\mathrm{sym}}(h_1, h_2), h_1 + h_2)\right) + \overline{\eta}\left(i(\tau_{\mathrm{b,n}}(h_1, h_2)/2, 0_H)\right)\\ \text{(by (6.34) and (6.35))} &= \overline{\eta}\left(i(a_1, h_1) +_{\tau_{\mathrm{sym}}} i(a_2, h_2)\right) + \frac{1}{2}\overline{\eta}\left([i(a_1, h_1), i(a_2, h_2)]_{\tau_{\mathrm{b,n}}}\right)\\ &= \overline{\eta}\,(i(a_1, h_1)) + \overline{\eta}\,(i(a_2, h_2)) + \frac{1}{2}[\overline{\eta}(i(a_1, h_1)), \overline{\eta}(i(a_2, h_2))]\\ &= \phi(g_1) + \phi(g_2) + \frac{1}{2}[\phi(g_1), \phi(g_2)].\end{aligned}$$

Then (2) follows. Formulas (3) and (4) are easily deduced from the previous ones:

$$\begin{aligned}\phi(g_1 g_2 g_1^{-1}) &= \phi(g_1 g_2) - \phi(g_1) + \frac{1}{2}[\phi(g_1 g_2), -\phi(g_1)]\\ &= \phi(g_1) + \phi(g_2) + \frac{1}{2}[\phi(g_1), \phi(g_2)] - \phi(g_1)\\ &\qquad + \frac{1}{2}\left[\phi(g_1) + \phi(g_2) + \frac{1}{2}[\phi(g_1), \phi(g_2)], -\phi(g_1)\right]\\ \text{(by (6.30))} &= \phi(g_2) + \frac{1}{2}[\phi(g_1), \phi(g_2)] - \frac{1}{2}[\phi(g_2), \phi(g_1)]\\ &= \phi(g_2) + [\phi(g_1), \phi(g_2)]\end{aligned}$$

and

$$\phi(g_1g_2g_1^{-1}g_2^{-1}) = \phi(g_1g_2g_1^{-1}) - \phi(g_2) + \frac{1}{2}[\phi(g_1g_2g_1^{-1}), \phi(g_2)]$$

$$= \phi(g_2) + [\phi(g_1), \phi(g_2)] - \phi(g_2) + \frac{1}{2}[\phi(g_2) + [\phi(g_1), \phi(g_2)], \phi(g_2)]$$

$$(\text{by } (6.30)) = [\phi(g_1), \phi(g_2)].$$

■

More on the relations between Lie rings and nilpotent groups may be found in the important paper [87] by M. Lazard and in Chap. VIII, Sect. 9, of the classical monograph [66] by B. Huppert and N. Blackburn. The specific case of 2-step nilpotent groups in which every element has a square root is treated in [66, Lemma 9.16].

6.6 The Orbit Method

In this section we present the so-called orbit method for determining all the irreducible representations of a 2-step nilpotent group with 2-divisible center (we refer to Kirillov's survey [78] for a treatment of the classical case of nilpotent Lie groups).

Let G be a 2-step nilpotent group with 2-divisible center. Let $\mathfrak{g} := \phi(G)$ denote the corresponding Lie ring.

Lemma 6.28 *The map $g \mapsto \sigma(g)$, where*

$$\sigma(g)(\overline{g}) := \phi(g\phi^{-1}(\overline{g})g^{-1}) = \overline{g} + [\phi(g), \overline{g}] \tag{6.42}$$

for all $g \in G$ and $\overline{g} \in \mathfrak{g}$, defines a group homomorphism $\sigma : G \to \mathrm{Aut}(\mathfrak{g})$.

Proof Let $g_1, g_2 \in G$ and $\overline{g} \in \mathfrak{g}$. Then, applying two times the definition (6.42), we get

$$\sigma(g_1g_2)(\overline{g}) = \phi(g_1g_2\phi^{-1}(\overline{g})g_2^{-1}g_1^{-1})$$

$$= \phi(g_1\phi^{-1}(\sigma(g_2)(\overline{g}))g_1^{-1})$$

$$= \sigma(g_1)(\sigma(g_2)(\overline{g})),$$

which gives $\sigma(g_1g_2) = \sigma(g_1) \circ \sigma(g_2)$. In particular, since $\sigma(1_G) = \mathrm{id}_{\mathfrak{g}}$, we deduce that $\sigma(g^{-1}) = \sigma(g)^{-1}$ for all $g \in G$, showing that $\sigma(g)$ is bijective.

For $g \in G$ and $\overline{g}_1, \overline{g}_2 \in \mathfrak{g}$ we have, using Theorem 6.27.(3),

$$\begin{aligned}\sigma(g)(\overline{g}_1 + \overline{g}_2) &= \phi(g\phi^{-1}(\overline{g}_1 + \overline{g}_2)g^{-1}) \\ &= (\overline{g}_1 + \overline{g}_2) + [\phi(g), \overline{g}_1 + \overline{g}_2] \\ &= (\overline{g}_1 + \overline{g}_2) + [\phi(g), \overline{g}_1] + [\phi(g), \overline{g}_2] \\ &= \overline{g}_1 + [\phi(g), \overline{g}_1] + \overline{g}_2 + [\phi(g), \overline{g}_2] \\ &= \phi(g\phi^{-1}(\overline{g}_1)g^{-1}) + \phi(g\phi^{-1}(\overline{g}_2)g^{-1}) \\ &= \sigma(g)(\overline{g}_1) + \sigma(g)(\overline{g}_2)\end{aligned}$$

and, using Theorem 6.27.(4),

$$\begin{aligned}\sigma(g)([\overline{g}_1, \overline{g}_2]) &= \phi(g\phi^{-1}([\overline{g}_1, \overline{g}_2])g^{-1}) \\ &= \phi(g\phi^{-1}(\overline{g}_1)\phi^{-1}(\overline{g}_2)\phi^{-1}(\overline{g}_1)^{-1}\phi^{-1}(\overline{g}_2)^{-1}g^{-1}) \\ &= \phi(g\phi^{-1}(\overline{g}_1)g^{-1} \cdot g\phi^{-1}(\overline{g}_2)g^{-1} \cdot (g\phi^{-1}(\overline{g}_1)g^{-1})^{-1} \cdot (g\phi^{-1}(\overline{g}_2)g^{-1})^{-1}) \\ &= [\phi(g\phi^{-1}(\overline{g}_1)g^{-1}), \phi(g\phi^{-1}(\overline{g}_2)g^{-1})] \\ &= [\sigma(g)(\overline{g}_1), \sigma(g)(\overline{g}_2)].\end{aligned}$$

This shows that $\sigma(g) \in \mathrm{Aut}(\mathfrak{g})$ for all $g \in G$. ■

We denote by $\mathscr{O}(\mathfrak{g})$ the set of all $\sigma(G)$-orbits of $\mathfrak{g}$. Note that, by (6.42), given $g \in G$, the $\sigma(G)$-orbit of the element $\overline{g} := \phi(g) \in \mathfrak{g}$ coincides with $\phi(\mathscr{C}_g)$, where $\mathscr{C}_g := \{hgh^{-1} : h \in G\} \subseteq G$ is the conjugacy-class of g, so that

$$\mathscr{O}(\mathfrak{g}) = \{\phi(\mathscr{C}_g) : g \in G\}. \tag{6.43}$$

Regarding $\mathfrak{g}$ as an Abelian group, we denote by $\widehat{\mathfrak{g}}$ the dual group consisting of all characters $\chi : \mathfrak{g} \to \mathbb{T}$. Given an automorphism $\alpha \in \mathrm{Aut}(\mathfrak{g})$ and $\chi \in \widehat{\mathfrak{g}}$, we denote by ${}^{\alpha}\chi \in \widehat{\mathfrak{g}}$ the character defined by ${}^{\alpha}\chi(\overline{g}) := \chi(\alpha^{-1}(\overline{g}))$. Note that if $\alpha_1, \alpha_2 \in \mathrm{Aut}(\mathfrak{g})$, then

$${}^{\alpha_1\alpha_2}\chi = {}^{\alpha_1}({}^{\alpha_2}\chi). \tag{6.44}$$

Given $g \in G$ and $\chi \in \widehat{\mathfrak{g}}$ we set $\sigma^*(g)(\chi) := {}^{\sigma(g)}\chi$, that is,

$$[\sigma^*(g)(\chi)](\overline{g}) = \chi(\sigma(g^{-1})(\overline{g})) = \chi(\overline{g} - [\phi(g), \overline{g}]) \tag{6.45}$$

for all $\overline{g} \in \mathfrak{g}$. From Lemma 6.28 and (6.44) we deduce that

$$\sigma^*(g_1g_2) = \sigma^*(g_1)\sigma^*(g_2)$$

for all $g_1, g_2 \in G$. Thus, σ^* yields an action of G on $\widehat{\mathfrak{g}}$ and we denote by $\mathscr{O}(\widehat{\mathfrak{g}})$ the corresponding set of orbits. Finally, we denote by $\pi_\sigma : G \to \mathrm{GL}(L(\mathfrak{g}))$ and $\pi_{\sigma^*} : G \to \mathrm{GL}(L(\widehat{\mathfrak{g}}))$ the permutation representations corresponding to the actions σ and σ^*, respectively. The latter is called the *co-adjoint* representation of G. Note that when regarding $\mathfrak{g}$ as an Abelian group one has $|\widehat{\mathfrak{g}}| = |\mathfrak{g}|$ so that $\dim \pi_{\sigma^*} = \dim \pi_\sigma$.

Consider the group $G \times G$ and its diagonal subgroup $\widetilde{G} := \{(g, g) : g \in G\} \cong G$. We define an action γ of $G \times G$ on G by setting $\gamma(g_1, g_2)g := g_1 g g_2^{-1}$ for all $g_1, g_2, g \in G$. The corresponding permutation representation $\pi_\gamma : G \to \mathrm{GL}(L(G))$ is given by

$$[\pi_\gamma(g_1, g_2)f](g) := f(\gamma(g_1, g_2)^{-1}g) = f(g_1^{-1} g g_2) \equiv [\lambda(g_1)\rho(g_2)f](g)$$

for all $g_1, g_2, g \in G$, where $\lambda, \rho : G \to \mathrm{GL}(L(G))$ are the left and right regular representations of G (recall that these commute!). We observe that a subspace $W \leq L(G)$ is π_γ-invariant if and only if it is bi-invariant, that is, it is both λ and ρ-invariant. Moreover, the orbits of the subgroup $\widetilde{G}$ on G are exactly the conjugacy classes $\mathscr{C}_g$, $g \in G$, of G.

It is a general fact that, given an action α of a group G on a set X with orbit set $\mathscr{O}_G(X)$, then the corresponding permutation representation $\pi_\alpha : G \to \mathrm{GL}(L(X))$ yields a decomposition of the isotypic component $L_0(X)$ of $L(X)$ corresponding to the trivial representation ι of G given by

$$L_0(X) = \bigoplus_{\Omega \in \mathscr{O}_G(X)} W_\Omega,$$

where $W_\Omega = \mathbb{C}\chi_\Omega$ is the one-dimensional G-invariant subspace spanned by the characteristic function $\mathbf{1}_\Omega$ of $\Omega \in \mathscr{O}_G(X)$. In particular, $|\mathscr{O}_G(X)| = \dim L_0(X) = \dim\mathrm{Hom}_G(\iota, \pi_\alpha)$.

The following result is well known. We include the easy proof for the convenience of the reader and the sake of completeness.

Lemma 6.29 $|\mathscr{O}(\widehat{\mathfrak{g}})| = |\mathscr{O}(\mathfrak{g})| = |\widehat{G}|$.

Proof Regarding $L(\widehat{\mathfrak{g}})$ as the dual vector space of $L(\mathfrak{g})$ by setting

$$\langle \phi, f \rangle := \sum_{\overline{g} \in \mathfrak{g}} \sum_{\chi \in \widehat{\mathfrak{g}}} \phi(\chi)\chi(\overline{g})f(\overline{g})$$

for all $\phi \in \widehat{\mathfrak{g}}$ and $f \in L(\mathfrak{g})$, then (1.40) yields

$$\begin{aligned}\langle[\pi'_\sigma(g)\phi], f\rangle &= \langle\phi, [\pi_\sigma(g^{-1})f]\rangle \\ &= \sum_{\overline{g}\in\mathfrak{g}}\sum_{\chi\in\widehat{\mathfrak{g}}}\phi(\chi)\chi(\overline{g})[\pi_\sigma(g^{-1})f](\overline{g}) \\ &= \sum_{\overline{g}\in\mathfrak{g}}\sum_{\chi\in\widehat{\mathfrak{g}}}\phi(\chi)\chi(\overline{g})f(\sigma(g)\overline{g}) \\ &= \sum_{\overline{g}\in\mathfrak{g}}\sum_{\chi\in\widehat{\mathfrak{g}}}\phi(\chi)\chi(\sigma(g^{-1})\overline{g})f(\overline{g}) \\ &= \sum_{\overline{g}\in\mathfrak{g}}\sum_{\chi\in\widehat{\mathfrak{g}}}\phi(\chi)[\sigma^*(g)\chi](\overline{g})f(\overline{g}) \\ &= \sum_{\overline{g}\in\mathfrak{g}}\sum_{\chi\in\widehat{\mathfrak{g}}}\phi(\sigma^*(g^{-1})\chi)\chi(\overline{g})f(\overline{g}) \\ &= \sum_{\overline{g}\in\mathfrak{g}}\sum_{\chi\in\widehat{\mathfrak{g}}}[\pi_{\sigma^*}(g)\phi](\chi)\chi(\overline{g})f(\overline{g}) \\ &= \langle[\pi_{\sigma^*}(g)\phi], f\rangle.\end{aligned}$$

This shows that $\pi'_\sigma = \pi_{\sigma^*}$. Since all permutation representations are self-adjoint, we deduce that $\pi_\sigma \sim \pi_{\sigma^*}$. It follows that

$$|\mathscr{O}(\widehat{\mathfrak{g}})| = \mathrm{dimHom}_G(\iota, \pi_{\sigma^*}) = \mathrm{dimHom}_G(\iota, \pi_\sigma) = |\mathscr{O}(\mathfrak{g})|$$

and the first equality follows. Finally, the second one follows from (6.43) and the well known fact that $|\widehat{G}|$ equals the number of conjugacy classes of G. ■

Definition 6.30 A group G is said to be 2-*rootable* provided the square map $g \mapsto g^2$ is a bijection. If this is the case, given $g \in G$ there exists a unique element in G, denoted $\sqrt{g}$, such that $(\sqrt{g})^2 = g$.

Note that for Abelian groups the notions of 2-divisibility and 2-rootability are equivalent.

Lemma 6.31 *Let G be a finite 2-step nilpotent group of odd order. Then G is 2-rootable.*

Proof Denote, as usual, by $\phi \colon G \to \mathfrak{g}$ the canonical bijection between the group G and its Lie ring $\mathfrak{g}$. Since the Abelian group $(\mathfrak{g}, +)$ is 2-divisible (cf. Sect. 6.2), given $g \in G$ we can find $\overline{g}_1 \in \mathfrak{g}$ such that $\phi(g) = 2\overline{g}_1$. Setting $g_1 := \phi^{-1}(\overline{g}_1) \in G$ and keeping in mind Theorem 6.27.(2), we have

$$\phi(g_1^2) = 2\phi(g_1) = 2\overline{g}_1 = \phi(g)$$

so that $g_1^2 = g$. ■

It follows from the proof of the previous lemma that indeed

$$\frac{\phi(g)}{2} = \phi(\sqrt{g}) \tag{6.46}$$

for all $g \in G$ (a finite 2-step nilpotent group of odd order).

We are now in a position to state and prove the most relevant result of this section. Let us first introduce some notation. Given $\chi \in \widehat{\mathfrak{g}}$ we set

$$X_\chi := \chi \circ \phi \in L(G) \tag{6.47}$$

where $\phi \colon G \to \mathfrak{g}$ is the bijection as in (6.41). Also, for every $\Omega \in \mathcal{O}(\widehat{\mathfrak{g}})$ we denote by

$$V_\Omega := \operatorname{span}\{X_\chi : \chi \in \Omega\} \le L(G) \tag{6.48}$$

the subspace generated by the X_χs with $\chi \in \Omega$.

Theorem 6.32 (The Orbit Method) *Let G be a 2-step nilpotent group of odd order and let $\mathfrak{g} := \phi(G)$ denote the corresponding Lie ring. The functions*

$$\frac{1}{\sqrt{|G|}} X_\chi, \tag{6.49}$$

$\chi \in \widehat{\mathfrak{g}}$, constitute an orthonormal basis of $L(G)$. For every $\Omega \in \mathcal{O}(\widehat{\mathfrak{g}})$, the subspace V_Ω is bi-invariant—with respect to the left regular representation $(\lambda, L(G))$ and to the right regular representation $(\rho, L(G))$—and, indeed,

$$L(G) = \bigoplus_{\Omega \in \mathcal{O}(\mathfrak{g})} V_\Omega \tag{6.50}$$

is the canonical decomposition of $L(G)$ into its isotypic components.

Proof Let $\chi_1, \chi_2 \in \widehat{\mathfrak{g}}$. Then

$$\begin{aligned}
\langle X_{\chi_1}, X_{\chi_2}\rangle_{L(G)} &= \sum_{g \in G} X_{\chi_1}(g)\overline{X_{\chi_2}(g)} \\
\text{(by (6.47))} \quad &= \sum_{g \in G} \chi_1(\phi(g))\overline{\chi_2(\phi(g))} \\
&= \sum_{\overline{g} \in \mathfrak{g}} \chi_1(\overline{g})\overline{\chi_2(\overline{g})} \\
&= \langle \chi_1, \chi_2\rangle_{L(\mathfrak{g})}.
\end{aligned}$$

Keeping in mind the orthogonality relations for characters (cf. [30, Proposition 2.3.5]) we deduce that the system (6.49) is orthonormal and that the decomposition (6.50) is orthogonal. We are only left to show that the V_Ωs are G-invariant and indeed constitute the isotypic components of $L(G)$. Let $g \in G$, $\Omega \in \mathscr{O}(\widehat{\mathfrak{g}})$, and $X_\chi \in V_\Omega$. Then

$$\begin{aligned}
[\lambda(g)X_\chi](h) &= X_\chi(g^{-1}h) \\
\text{(by (6.47))} \quad &= \chi(\phi(g^{-1}h)) \\
\text{(by Theorem 6.27.(1–2))} \quad &= \chi(-\phi(g) + \phi(h) - \frac{1}{2}[\phi(g), \phi(h)]) \\
\text{(by bilinearity of } [\cdot,\cdot] \text{ and (6.46))} \quad &= \chi(-\phi(g))\chi(\phi(h) - [\phi(\sqrt{g}), \phi(h)]) \\
\text{(by (6.45))} \quad &= \chi(-\phi(g))[\sigma^*(\sqrt{g})(\chi)](\phi(h)) \\
\text{(by (6.47))} \quad &= \chi(-\phi(g))[X_{\sigma^*(\sqrt{g})(\chi)}](h)
\end{aligned}$$

that is,

$$\lambda(g)X_\chi = \chi(-\phi(g))X_{\sigma^*(\sqrt{g})(\chi)}$$

which belongs to V_Ω as $\sigma^*(\sqrt{g})(\chi) \in \Omega$.

Similarly, one obtains

$$\rho(g)X_\chi = \chi(\phi(g))X_{\sigma^*(\sqrt{g})(\chi)} \in V_\Omega. \tag{6.51}$$

This shows bi-invariance of the subspaces V_Ω, $\Omega \in \mathscr{O}(\widehat{\mathfrak{g}})$. By virtue of the discussion preceding Lemma 6.29, we have that each V_Ω is π_γ-invariant for the action of $G \times G$. It follows from the fact that $(G \times G, \widetilde{G})$ is a Gelfand pair (cf. [24, Theorem 9.5.4]) and Lemma 6.29 that the V_Ωs are π_γ-irreducible and pairwise inequivalent (as π_γ representations). From [24, Theorem 9.5.2.(i)] it then follows that (6.50) is indeed the decomposition of $L(G)$ into its λ-isotypic components. ■

By virtue of the Peter-Weyl theorem (cf. [24, Theorem 3.7.11] and [24, Theorem 9.5.2.(ii)]) we have the decomposition $L(G) = \bigoplus_{\theta \in \widehat{G}} d_\theta W_\theta$, with respect to the right action ρ. Thus, with every $\Omega \in \mathscr{O}(\widehat{\mathfrak{g}})$ we associate a unique $\theta = \theta(\Omega) \in \widehat{G}$ such that $V_\Omega = d_\theta W_\theta$. In particular, since $\dim V_\Omega = |\Omega|$, we have $d_\theta = \sqrt{|\Omega|}$. In the following, we shall not make a distinction between an orbit Ω and the corresponding irreducible representation $\theta(\Omega)$ and we shall write W_Ω (resp. d_Ω) instead of W_θ (resp. d_θ). This way, we have $V_\Omega = d_\Omega W_\Omega$ and $d_\Omega = \sqrt{|\Omega|}$.

We complete the section with a useful character formula. First we need a lemma.

Lemma 6.33 *Let $\Omega \in \mathscr{O}(\widehat{\mathfrak{g}})$. Then, with respect to the action σ^* of G on $\widehat{\mathfrak{g}}$, for all $\chi_1, \chi_2 \in \Omega$ one has*

$$\mathrm{Stab}\chi_1 = \mathrm{Stab}\chi_2$$

and

$$\chi_1(\phi(g)) = \chi_2(\phi(g))$$

for all $g \in \mathrm{Stab}\chi_1 = \mathrm{Stab}\chi_2$. *In particular,* $\mathrm{Stab}\chi$ *is a normal subgroup of* G *for all* $\chi \in \widehat{\mathfrak{g}}$.

Proof Let $\chi_1, \chi_2 \in \Omega$ and $\overline{g} \in \mathfrak{g}$. Then for all $g \in \mathrm{Stab}\chi_1$ we have

$$\chi_1(\overline{g}) = [\sigma^*(g)\chi_1](\overline{g}) = \chi_1(\overline{g} - [\phi(g), \overline{g}]) = \chi_1(\overline{g}) \cdot \chi_1([\phi(g), \overline{g}])^{-1}$$

which gives

$$\chi_1([\phi(g), \overline{g}]) = 1. \tag{6.52}$$

As χ_1 and χ_2 belong to the same orbit, we can find $h \in G$ such that $\chi_2 = \sigma^*(h)\chi_1$. Then, for $g \in \mathrm{Stab}\chi_1$ and $\overline{g} \in \mathfrak{g}$, using (6.45) and (6.52) we get

$$\chi_2([\phi(g), \overline{g}]) = [\sigma^*(h)\chi_1]([\phi(g), \overline{g}]) = \chi_1([\phi(g), \overline{g}] - [\phi(h), [\phi(g), \overline{g}]]) = \chi_1([\phi(g), \overline{g}]) = 1.$$

As a consequence, for $g \in \mathrm{Stab}\chi_1$ and $\overline{g} \in \mathfrak{g}$, using again (6.45) we obtain

$$[\sigma^*(g)\chi_2](\overline{g}) = \chi_2(\overline{g} - [\phi(g), \overline{g}]) = \chi_2(\overline{g})\chi_2([\phi(g), \overline{g}])^{-1} = \chi_2(\overline{g}),$$

that is, $g \in \mathrm{Stab}\chi_2$. By symmetry we deduce that $\mathrm{Stab}\chi_1 = \mathrm{Stab}\chi_2$.

Finally, if $g \in \mathrm{Stab}\chi_1 = \mathrm{Stab}\chi_2$ and $h \in G$ satisfies that $\chi_2 = \sigma^*(h)\chi_1$, we get

$$\begin{aligned}\chi_2(\phi(g)) &= [\sigma^*(h)\chi_1](\phi(g)) = \chi_1(\phi(g) - [\phi(h), \phi(g)]) \\ &= \chi_1(\phi(g))\chi_1([\phi(g), \phi(h)]) = \chi_1(\phi(g)),\end{aligned}$$

where the last equality follows from (6.52).

The last statement follows from the general fact that stabilizers of elements in the same orbit are conjugate. ■

Given $\Omega \in \mathcal{O}(\widehat{\mathfrak{g}})$, we denote by χ^Ω the character of W_Ω.

Theorem 6.34 *Let* $\Omega \in \mathcal{O}(\widehat{\mathfrak{g}})$ *and let* $\chi \in \Omega$. *Then, for all* $g \in G$,

$$\chi^\Omega(g) = \begin{cases} \sqrt{|\Omega|}\, X_\chi(g) & \text{if } g \in \mathrm{Stab}\chi \\ 0 & \text{otherwise.} \end{cases}$$

Proof We have already observed that if $\theta = \theta(\Omega)$, then $d_\Omega\theta \sim \rho|_{d_\Omega W_\Omega} = \rho|_{V_\Omega}$, where $d_\Omega = \sqrt{|\Omega|}$. Also, it follows from the definitions that the character χ^{V_Ω} of V_Ω in (6.50) by means of the orthonormal basis (6.49) with $\chi \in \Omega$ is given by

$$\chi^{V_\Omega}(g) = \frac{1}{|G|} \sum_{\chi' \in \Omega} \langle \rho(g) X_{\chi'}, X_{\chi'} \rangle_{L(G)}$$

for all $g \in G$. Since $\chi^{V_\Omega} = d_\Omega \chi^\Omega$, we deduce that

$$\chi^\Omega(g) = \frac{1}{d_\Omega |G|} \sum_{\chi' \in \Omega} \langle \rho(g) X_{\chi'}, X_{\chi'} \rangle_{L(G)}. \tag{6.53}$$

Let $g \in G$. If $g \in \mathrm{Stab}\chi$, so that $g \in \mathrm{Stab}\chi'$ for all $\chi' \in \Omega$, taking into account (6.51), we have

$$\rho(g) X_{\chi'} = \chi'(\phi(g)) X_{\sigma^*(\sqrt{g})(\chi')} = \chi'(\phi(g)) X_{\chi'},$$

because the cyclic group generated by g is of odd order and therefore it contains $\sqrt{g}$ which therefore belongs to $\mathrm{Stab}\chi'$. From (6.53) and Lemma 6.33 we deduce that

$$\chi^\Omega(g) = \frac{1}{d_\Omega} \sum_{\chi' \in \Omega} \chi'(\phi(g)) = \frac{1}{d_\Omega} \sum_{\chi' \in \Omega} X_{\chi'}(g) = \frac{|\Omega|}{d_\Omega} X_\chi(g) = \sqrt{|\Omega|}\, X_\chi(g).$$

On the other hand, if $g \in G \setminus \mathrm{Stab}\chi$ then, keeping in mind that (6.49) is an orthonormal system,

$$\rho(g) X_\chi = \chi(\phi(g)) X_{\sigma^*(\sqrt{g})(\chi)} \perp X_\chi$$

so that (6.53) yields $\chi^\Omega(g) = 0$. ■

6.7 More on the Orbit Method: Induced Representations

In this section we further investigate the orbit method and show that the isotypic components arise as induced representations of suitable stabilizer subgroups. With the notation from the previous section, let $\Omega \in \mathcal{O}(\widehat{\mathfrak{g}})$, and denote by $S_\Omega = \mathrm{Stab}\chi \leq G$ (where $\chi \in \Omega$ which we then fix once and for all) the corresponding stabilizer (cf. Lemma 6.33). Thus, for $s \in G$ and $\overline{g} \in \mathfrak{g}$ we have, keeping in mind (6.45),

$$[\sigma^*(s)\chi](\overline{g}) = \chi(\overline{g} - [\phi(s), \overline{g}]) = \chi(\overline{g})\, \overline{\chi}([\phi(s), \overline{g}]),$$

so that

$$s \in S_\Omega \iff [\phi(s), \mathfrak{g}] \subseteq \ker \chi. \tag{6.54}$$

As a consequence of Theorem 6.27.(2) and (6.54) we then have, for $s_1, s_2 \in S_\Omega$,

$$\begin{aligned}\chi(\phi(s_1 s_2)) &= \chi\left(\phi(s_1) + \phi(s_2) + \frac{1}{2}[\phi(s_1), \phi(s_2)]\right) \\ &= \chi(\phi(s_1) + \phi(s_2)) = \chi(\phi(s_1))\,\chi(\phi(s_2))\end{aligned}$$

that is, the map $s \mapsto \chi(\phi(s))$ is a one-dimensional character of S_Ω. Note that, as a consequence of (6.54) and Theorem 6.27.(4) we have

$$\chi(\phi(gsg^{-1}s^{-1})) = 1 \quad \text{for all } g \in G \text{ and } s \in S_\Omega. \tag{6.55}$$

We claim that $[G, G] \subseteq S_\Omega$. Indeed, if $g_1, g_2 \in G$ and $\overline{g} \in \mathfrak{g}$, then, by (6.45), Theorem 6.27.(4), and (6.5)

$$[\sigma^*(g_1 g_2 g_1^{-1} g_2^{-1})\chi](\overline{g}) = \chi(\overline{g} - [\phi(g_1 g_2 g_1^{-1} g_2^{-1}), \overline{g}]) = \chi(\overline{g} - [[\phi(g_1), \phi(g_2)], \overline{g}]) = \chi(\overline{g})$$

so that $[g_1, g_2] = g_1 g_2 g_1^{-1} g_2^{-1} \in S_\Omega$.

Moreover, it follows from Lemma 6.33 that S_Ω is a normal subgroup of G. As a consequence of the claim and this last fact, $X := G/S_\Omega$ is an Abelian group.

Let

$$\begin{aligned} V &:= \{f \in L(G) : f(s_1 g s_2) = \overline{\chi}(\phi(s_1))\,\overline{\chi}(\phi(s_2)) f(g) \text{ for all } g \in G \text{ and } s_1, s_2 \in S_\Omega\} \\ V_\ell &:= \{f \in L(G) : f(gs) = \overline{\chi}(\phi(s)) f(g) \text{ for all } g \in G \text{ and } s \in S_\Omega\} \\ V_r &:= \{f \in L(G) : f(sg) = \overline{\chi}(\phi(s)) f(g) \text{ for all } g \in G \text{ and } s \in S_\Omega\}. \end{aligned} \tag{6.56}$$

We claim that $V = V_\ell = V_r$. Indeed, it is clear that $V = V_\ell \cap V_r$. Moreover, if $f \in V_\ell$ we have, keeping in mind (6.55) and that $[G, G] \subseteq S_\Omega$,

$$\begin{aligned} f(sg) &= f(gs(s^{-1}g^{-1}sg)) = \overline{\chi}(\phi(s(s^{-1}g^{-1}sg)) f(g) \\ &= \overline{\chi}(\phi(s))\,\overline{\chi}(\phi(s^{-1}g^{-1}sg))\, f(g) = \overline{\chi}(\phi(s))\, f(g) \end{aligned}$$

showing that $f \in V_r$. Analogously, one shows that $V_r \subseteq V_\ell$. Note that, in addition, for $f \in V$ one has

$$f(sg) = f(gs),$$

equivalently,

$$f(sgs^{-1}) = f(g),$$

for all $s \in S_\Omega$ and $g \in G$, that is, V consists of S_Ω-conjugacy invariant functions.

We now consider the representation $\lambda_\Omega := \mathrm{Ind}_{S_\Omega}^G(\chi \circ \phi)$. Recall that its representation space is V, and

$$[\lambda_\Omega(g)f](g') := f(g^{-1}g')$$

for all $f \in V$ and $g, g' \in G$.

We now introduce a vector basis for V (a not normalized special case of the *Curtis and Fossum basis* as in [30, Definition 13.2.5]). For each $x \in X := G/S_\Omega$ we fix an element $t = t(x) \in G$ such that $tS_\Omega = x$. We denote by $T := \{t(x) : x \in X\} \subseteq G$ the corresponding *complete transversal*. Given $g \in G$ and $t \in T$, there exist unique $t' \in T$ and $s' \in S_\Omega$ such that $gt = t's'$: we then write $t' := g \cdot t$ and $s' := s_{g,t}$. Note that

$$g \cdot t = t \iff g \in S_\Omega. \tag{6.57}$$

Indeed,

$$gt = t(t^{-1}gt) \in tS_\Omega \iff t^{-1}gt \in S_\Omega \iff g \in S_\Omega$$

and, if this is the case, $s_{g,t} = t^{-1}gt$. Note also that, for $g \in G$ and $t_1, t_2 \in T$,

$$g \cdot t_1 = t_2 \iff t_1 = g^{-1} \cdot t_2. \tag{6.58}$$

Indeed,

$$g \cdot t_1 = t_2 \iff gt_1 = t_2s_2 \text{ with } s_2 \in S_\Omega \iff t_1s_2^{-1} = g^{-1}t_2 \text{ with } s_2 \in S_\Omega \iff t_1 = g^{-1} \cdot t_2.$$

For every $t \in T$ we define $a_t \in V$ by setting

$$a_t(g) := \begin{cases} \overline{\chi}(\phi(s)) & \text{if } g = ts \text{ with } s \in S_\Omega \\ 0 & \text{otherwise.} \end{cases}$$

It is straightforward to check that $f = \sum_{t \in T} f(t)a_t$, for all $f \in V$, thus showing that $\{a_t : t \in T\}$ is a vector basis of V. Moreover, it follows from the fact that distinct cosets are disjoint and from the definition of the scalar product (1.29) that it is an orthonormal basis. In particular, each $f \in V$ is uniquely determined by its values on T. We now describe the action of λ_Ω on this basis.

Let $t, t' \in T$ and $g \in G$. We then have

$$[\lambda_\Omega(g)a_t](t') = a_t(g^{-1}t') = a_t((g^{-1}\cdot t')s_{g^{-1},t'}) = \overline{\chi}(\phi(s_{g^{-1},t'}))\, a_t(g^{-1}\cdot t')$$

so that, keeping in mind that the support of a_t is the coset tS_Ω, and using (6.58),

$$\lambda_\Omega(g)a_t = \overline{\chi}(\phi(s_{g^{-1},g\cdot t}))\, a_{g\cdot t}. \tag{6.59}$$

Let us compute the character χ^{λ_Ω} of λ_Ω. First of all observe that, as it immediately follows from (6.59) and (6.57),

$$\chi^{\lambda_\Omega}(g) = 0 \text{ if } g \in G \setminus S_\Omega. \tag{6.60}$$

On the other hand, for $g \in S_\Omega$ we can compute χ^{λ_Ω} by means of (6.59) (recall that $\{a_t : t \in T\}$ is an orthonormal basis), getting

$$\begin{aligned}
\chi^{\lambda_\Omega}(g) &= \sum_{t\in T}\overline{\chi}(\phi(s_{g^{-1},t}))\\
&= \sum_{t\in T}\overline{\chi}(\phi(t^{-1}g^{-1}t))\\
\text{(by Theorem 6.27.(3))}\quad &= \sum_{t\in T}\overline{\chi}\left(\phi(g^{-1}) + [\phi(t^{-1}), \phi(g^{-1})]\right)\\
\text{(by Theorem 6.27.(4) and (6.55))}\quad &= \sum_{t\in T}\overline{\chi}(\phi(g^{-1}))\\
\text{(by Theorem 6.27.(1))}\quad &= \sum_{t\in T}\chi(\phi(g))\\
&= |T|\,\chi(\phi(g))\\
&= |\Omega|\,\chi(\phi(g)).
\end{aligned} \tag{6.61}$$

Recalling that $\theta = \theta(\Omega)$ has multiplicity $d_\theta = \dim W_\theta = \sqrt{|\Omega|}$ in $L(G)$ and the character χ^Ω is given by Theorem 6.34, from (6.60) and (6.61) we deduce that

$$\rho|_{V_\Omega} \sim \lambda_\Omega = \mathrm{Ind}_{S_\Omega}^G(\chi\circ\phi).$$

We end this section by presenting a more enlightening description of S_Ω. First of all, from (6.34) we deduce that χ satisfies

$$\chi\left((a_1 + a_2 - \tau_{\mathrm{sym}}(h_1, h_2), h_1 + h_2)\right) = \chi(a_1, h_1)\chi(a_2, h_2),$$

for all $a_1, a_2 \in \mathfrak{a}$ and $h_1, h_2 \in \mathfrak{h}$. In particular, $\xi(a) := \chi(a, 0_{\mathfrak{h}})$, for $a \in \mathfrak{a}$, is a character of $\mathfrak{a}$. From (6.54), (6.35), and (6.37) we deduce that and element $s \in G$ with $\phi(s) = (a, h)$ belongs to S_Ω, if and only if

$$\chi([(a, h), (a_1, h_1)]) = \chi\left((\tau_{\mathfrak{b},\mathfrak{n}}(h, h_1), 0_{\mathfrak{a}})\right) = \xi(\tau(h_1, h) - \tau(h, h_1)) = 1,$$

that is,

$$\xi(\tau(h_1, h)) = \xi(\tau(h, h_1)) \text{ for all } h_1 \in \mathfrak{h}. \tag{6.62}$$

In other words, S_Ω is the extension of H_0, the subgroup of H consisting of all $h \in H$ satisfying (6.62), by A, obtained by using the cocycle τ. Moreover,

$$\chi \circ \phi|_{S_\Omega} \text{ is an extension of } \xi \circ (\phi|_A) \text{ from } A \text{ to } S_\Omega. \tag{6.63}$$

Finally, we remark that, in the notation of (6.48) and (6.56), we have $V_\ell = \overline{V_\Omega}$: just consider (6.51) with $g = s \in S_\Omega$. In other words, $\mathrm{Ind}_{S_\Omega}^G(\chi \circ \phi)$ is the adjoint isotypic component of V_Ω.

6.8 More on the Orbit Method: Restricting to a Subgroup

Suppose now that we are given a group G and a subgroup $G_0 \leq G$ which are both 2-step nilpotent with a 2-divisible center. We denote by $\mathfrak{g}$ and $\mathfrak{g}_0$ the corresponding 2-step nilpotent Lie rings. Suppose that $\mathfrak{g}_0$ is a Lie subring of $\mathfrak{g}$, then restricting a character $\chi \in \widehat{\mathfrak{g}}$ to $\mathfrak{g}_0$ yields a character $P(\chi) \in \widehat{\mathfrak{g}_0}$. We refer to $P: \widehat{\mathfrak{g}} \to \widehat{\mathfrak{g}_0}$ as to the *canonical projection*. In the following, given $\chi \in \widehat{\mathfrak{g}}$, we denote by $\Omega = \Omega(\chi) \subseteq \widehat{\mathfrak{g}}$ its G-orbit and by $S_\Omega \leq G$ its stabilizer. Analogously, we denote by $\Omega(P(\chi)) \subseteq \widehat{\mathfrak{g}_0}$ the G_0-orbit of $P(\chi)$ and by $S_{\Omega(P(\chi))} \leq G_0$ its stabilizer. Sometimes, to simplify notation, we shall not indicate the map $\phi\colon G \to \mathfrak{g}$. For instance, given a character $\chi \in \widehat{\mathfrak{g}}$ and $g \in G$ (resp. $S \subseteq G$), we shall write $\chi(g)$ instead of $\chi(\phi(g))$ (resp. $\chi|_S$ instead of $\chi|_{\phi(S)}$), thus regarding χ as an element in $L(G)$.

Lemma 6.35 *Let $\Omega \subseteq \widehat{\mathfrak{g}}$ be a G-orbit and suppose that $\chi_1, \chi_2 \in \Omega$.*

(1) *The canonical projections of χ_1 and χ_2 have the same G_0-stabilizers, in formulæ, $S_{\Omega(P(\chi_1))} = S_{\Omega(P(\chi_2))}$.*
(2) *The restriction of χ_1 and χ_2 to $S_\Omega \cap G_0$ coincide, in formulæ, $\chi_1|_{S_\Omega \cap G_0} = \chi_2|_{S_\Omega \cap G_0}$.*
(3) *$P(\Omega) := \{P(\chi) : \chi \in \Omega\} \subseteq \widehat{\mathfrak{g}_0}$ decomposes as a disjoint union of G_0-orbits all of the same cardinality.*
(4) *$|P^{-1}(\chi_0)| = |\Omega|/|P(\Omega)|$ for all $\chi_0 \in P(\Omega)$.*

Proof

(1) Since χ_1 and χ_2 belong to the same G-orbit, we can find $g \in G$ such that $\chi_2 = \sigma^*(g)(\chi_1)$, that is,

$$\chi_2(\overline{g}) = \chi_1(\overline{g} - [\phi(g), \overline{g}]) \equiv \chi_1(\overline{g})\,\overline{\chi}_1([\phi(g), \overline{g}]) \tag{6.64}$$

for all $\overline{g} \in \mathfrak{g}$. By 2-step nilpotence, we deduce that

$$\chi_2|_{[\mathfrak{g},\mathfrak{g}]} = \chi_1|_{[\mathfrak{g},\mathfrak{g}]}. \tag{6.65}$$

When restricting to $\mathfrak{g}_0$, we have, for $s \in G_0$,

$$\begin{aligned} s \in S_{\Omega(P(\chi_1))} &\iff [\phi(s), \mathfrak{g}_0] \subseteq \ker P(\chi_1) && \text{(by (6.54))}\\ &\iff [\phi(s), \mathfrak{g}_0] \subseteq \ker P(\chi_2) && \text{(by (6.65))}\\ &\iff s \in S_{\Omega(P(\chi_2))} && \text{(by (6.54)).} \end{aligned}$$

(2) This follows from the fact that χ_1 and χ_2 coincide on S_Ω (cf. Lemma 6.33).
(3) By using (1), we have,

$$|\Omega(P(\chi_1))| = |G_0|/|S_{\Omega(P(\chi_1))}| = |G_0|/|S_{\Omega(P(\chi_2))}| = |\Omega(P(\chi_2))|.$$

(4) From (6.64) we deduce that for $\chi \in \Omega$ and $g \in G$

$$P\left(\sigma^*(g)\chi\right) = P(\chi) \iff [\phi(g), \mathfrak{g}_0] \subseteq \ker \chi. \tag{6.66}$$

By (6.65), the subgroup $H \leq G$ consisting of all $g \in G$ satisfying (6.66) does not depend on χ and therefore $|P^{-1}(\chi_0)| = |G/H|$. Clearly, $S_\Omega \subseteq H$, $|\Omega| = |G|/|S_\Omega|$, and $|P(\Omega)| = |H|/|S_\Omega|$.

■

Let $\Omega \subseteq \widehat{\mathfrak{g}}$ be a G-orbit and denote by $V_\Omega \leq L(G)$ the corresponding isotypic component (cf. (6.48)). Recall that $V_\Omega = d_\Omega W_\Omega$ with $d_\Omega = \sqrt{|\Omega|}$ and $W_\Omega \in \widehat{G}$. Analogously, for $\chi_0 \in P(\Omega) \subseteq \widehat{\mathfrak{g}_0}$ we denote by $\Omega_0 = \Omega(\chi_0) \subseteq \widehat{\mathfrak{g}_0}$ the corresponding G_0-orbit and by $V_{\Omega_0} \leq L(G_0)$ the corresponding isotypic component. Again, we have $V_{\Omega_0} = d_{\Omega_0} W_{\Omega_0}$ with $d_{\Omega_0} = \sqrt{|\Omega_0|}$ and $W_{\Omega_0} \in \widehat{G_0}$. For $\chi \in \Omega$, we denote by $\chi^{\Omega(P(\chi))}$ the corresponding character. Note that by Lemma 6.35.(3), the multiplicity d_{Ω_0} does not depend on the particular $\chi_0 \in P(\Omega)$ and we thus denote it simply by d_0. We are in a position to establish the main result of this section.

Theorem 6.36 *With the above notation we have*

$$\operatorname{Res}^G_{G_0} W_\Omega = \frac{d_\Omega d_0}{|P(\Omega)|} \bigoplus_{\Omega_0 \subseteq P(\Omega)} W_{\Omega_0} \tag{6.67}$$

where the sum is over all the G_0-orbits Ω_0 contained in $P(\Omega)$.

Proof We denote by χ^Ω (resp. χ_0^Ω) the character of W_Ω (resp. $\operatorname{Res}^G_{G_0} W_\Omega$). We claim that

$$\chi_0^\Omega(g_0) = \frac{1}{d_\Omega d_0} \sum_{\chi \in \Omega} \chi^{\Omega(P(\chi))}(g_0) \tag{6.68}$$

for all $g_0 \in G_0$. On the one hand, by virtue of Theorem 6.34, for $g_0 \in G_0$ we have

$$\chi_0^\Omega(g_0) = \begin{cases} d_\Omega \chi'(g_0) & \text{if } g_0 \in S_\Omega \cap G_0 \\ 0 & \text{otherwise,} \end{cases} \tag{6.69}$$

where $\chi' \in \Omega$ is arbitrary (cf. Lemma 6.35.(2)). On the other hand, we first observe that, if $\chi_0 \in \Omega_0$ then there exists $\chi \in \Omega$ such that $\chi_0 = P(\chi)$. Thus, if $g_0 \in G_0$, we have

$$\begin{aligned} g_0 \in S_\Omega &\iff [\phi(g_0), \mathfrak{g}] \subseteq \ker \chi && \text{(by (6.54))} \\ &\implies [\phi(g_0), \mathfrak{g}_0] \subseteq \ker \chi_0 && (\text{since } \mathfrak{g}_0 \subseteq \mathfrak{g}, [\phi(g_0), \mathfrak{g}_0] \subseteq \mathfrak{g}_0, \text{ and } \chi_0 = P(\chi)) \\ &\iff g_0 \in S_{\Omega_0} && \text{(by (6.54)).} \end{aligned}$$

In other words,

$$S_\Omega \cap G_0 \subseteq S_{\Omega_0}.$$

We distinguish three cases.

- Suppose $g_0 \in S_\Omega \cap G_0$. Then, by the above, $g_0 \in S_{\Omega_0}$, and Theorem 6.34 (for the group G_0) gives

$$\chi^{\Omega(P(\chi'))}(g_0) = \sqrt{|\Omega(P(\chi'))|}\, P(\chi')(g_0) = \sqrt{|\Omega_0|}\, \chi'(g_0),$$

  where $\chi' \in \Omega$ is arbitrary (we used the fact that $S_{\Omega_0} = S_{\Omega(P(\chi'))}$ and $|\Omega_0| = |\Omega(P(\chi'))|$ by Lemma 6.35) so that the right hand side of (6.68) equals (recall that $d_0 = d_{\Omega_0} = \sqrt{|\Omega_0|}$ and $d_\Omega = \sqrt{|\Omega|}$)

$$\frac{1}{d_\Omega d_0} |\Omega| \cdot \sqrt{|\Omega_0|}\, \chi'(g_0) = d_\Omega\, \chi'(g_0)$$

in agreement with (6.69).

- Suppose $g_0 \in S_{\Omega_0} \setminus S_\Omega \cap G_0$. Then, keeping in mind that $g_0 \in G_0$ and fixing an arbitrary $\chi' \in \Omega$,

$$\begin{aligned}
\sum_{\chi\in\Omega} \chi^{\Omega(P(\chi))}(\phi(g_0)) &= \frac{1}{|S_\Omega|}\sum_{g\in G} \chi^{\Omega(P(\sigma^*(g)\chi'))}(\phi(g_0)) \\
&= \frac{\sqrt{|\Omega(P(\chi'))|}}{|S_\Omega|}\sum_{g\in G} P(\sigma^*(g)\chi')(g_0) \quad \text{(by Theorem 6.34)} \\
&= \frac{\sqrt{|\Omega_0|}}{|S_\Omega|}\sum_{g\in G} (\sigma^*(g)\chi')(\phi(g_0)) \quad \text{(by Lemma 6.35)} \\
&= \frac{\sqrt{|\Omega_0|}}{|S_\Omega|}\sum_{g\in G} \chi'(\phi(g_0) - [\phi(g), \phi(g_0)]) \quad \text{(by (6.45))} \\
&= \frac{\sqrt{|\Omega_0|}}{|S_\Omega|}\chi'(\phi(g_0))\sum_{g\in G} \chi'([\phi(g_0), \phi(g)]).
\end{aligned}$$

Now observe that the map $\chi'([\phi(g_0), \cdot\,])$ is a non-trivial character of $\mathfrak{g}$: indeed,

$$\chi'([\phi(g_0), \overline{g}+\overline{h}]) = \chi'([\phi(g_0), \overline{g}]+[\phi(g_0), \overline{h}]) = \chi'([\phi(g_0), \overline{g}])\, \chi'([\phi(g_0), \overline{h}])$$

for all $\overline{g}, \overline{h} \in \mathfrak{g}$, and, moreover, since $g_0 \notin S_\Omega$, by (6.54) there exists $\overline{g}_0 \in \mathfrak{g}_0$ such that $[\phi(g_0), \overline{g}_0] \notin \ker\chi'$. From the orthogonality relations for characters, we have $\sum_{g\in G}\chi'([\phi(g_0), \phi(g)]) = 0$. Again, this agrees with (6.69).

- Suppose $g_0 \in G_0 \setminus S_{\Omega_0}$. Then (6.68) is trivially satisfied since both the left hand side and each summand in the right hand side vanish.

The claim is proved. Now the right hand side of (6.68) is equal to:

$$\begin{aligned}
\frac{1}{d_\Omega d_0}\sum_{\chi\in\Omega}\chi^{\Omega(P(\chi))} &= \frac{1}{d_\Omega d_0}\frac{|\Omega|}{|P(\Omega)|}\sum_{\chi_0\in P(\Omega)}\chi^{\Omega_0} \quad \text{(by Lemma 6.35.(4))} \\
&= \frac{1}{d_\Omega d_0}\frac{|\Omega|}{|P(\Omega)|}|\Omega_0|\sum_{\Omega_0\subseteq P(\Omega)}\chi^{\Omega_0} \\
&= \frac{d_\Omega d_0}{|P(\Omega)|}\sum_{\Omega_0\subseteq P(\Omega)}\chi^{\Omega_0}.
\end{aligned}$$

This is the character of the right hand side of (6.67), and the proof is complete. ■

6.9 The Orbit Method for the Finite Heisenberg Group

Let $\mathbb{F}_q$ denote the finite field with q elements ($q = p^h$ with p an *odd* prime and $h \geq 1$ an integer). We denote by $G = H_q$ the *Heisenberg group* over $\mathbb{F}_q$ which we identify with the set of triples $(m, k, \ell) \in \mathbb{F}_q^3$ equipped with the multiplication

$$(m_1, k_1, \ell_1) \cdot (m_2, k_2, \ell_2) := (m_1 + m_2 + k_1\ell_2, k_1 + k_2, \ell_1 + \ell_2) \tag{6.70}$$

for all $m_1, m_2, k_1, k_2, \ell_1, \ell_2 \in \mathbb{F}_q$. It is a straightforward calculation to check that the multiplication (6.70) is associative, that the identity element is $1_G = (0, 0, 0)$ and that the inverse is given by

$$(m, k, \ell)^{-1} = (-m + k\ell, -k, -\ell)$$

for all $m, k, \ell \in \mathbb{F}_q$. The map $G \to \mathrm{GL}(3, \mathbb{F}_q)$ defined by

$$(m, k, \ell) \longrightarrow \begin{pmatrix} 1 & k & m \\ 0 & 1 & \ell \\ 0 & 0 & 1 \end{pmatrix} \tag{6.71}$$

for all $m, k, \ell \in \mathbb{F}_q$ is an injective group homomorphism. In the literature one often defines the Heisenberg group as the corresponding image subgroup in $\mathrm{GL}(3, \mathbb{F}_q)$.

The center $\mathscr{A} = Z(G)$ consists of the elements $(m, 0, 0)$ with $m \in \mathbb{F}_q$ and therefore is isomorphic to $(\mathbb{F}_q, +)$. Then $G/Z(G)$ can be identified with the Abelian group $H = (\mathbb{F}_q, +) \oplus (\mathbb{F}_q, +)$ and we have the exact sequence

$$1 \longrightarrow \mathbb{F}_q \stackrel{\iota}{\longrightarrow} G \stackrel{\pi}{\longrightarrow} \mathbb{F}_q \oplus \mathbb{F}_q \longrightarrow 0,$$

where

$$\iota(m) := (m, 0, 0)$$

and

$$\pi(m, k, \ell) := (k, \ell)$$

for all $m, k, \ell \in \mathbb{F}_q$. Consider the section $t\colon H \to G$ defined by

$$t(k, \ell) := (0, k, \ell)$$

for all $k, \ell \in \mathbb{F}_q$. According with (6.12) we define $\tau\colon H \times H \to \mathscr{A}$ by setting

$$\begin{aligned}\tau((k_1,\ell_1),(k_2,\ell_2)) &:= \iota^{-1}\left(t(k_1+k_2,\ell_1+\ell_2)t(k_2,\ell_2)^{-1}t(k_1,\ell_1)^{-1})\right)\\ &= \iota^{-1}\left((0,k_1+k_2,\ell_1+\ell_2)\cdot(0,k_2,\ell_2)^{-1}\cdot(0,k_1,\ell_1)^{-1}\right)\\ &= \iota^{-1}\left((0,k_1+k_2,\ell_1+\ell_2)\cdot(k_2\ell_2,-k_2,-\ell_2)\cdot(k_1\ell_1,-k_1,-\ell_1)\right)\\ &= \iota^{-1}(-k_1\ell_2,0,0)\\ &= -k_1\ell_2,\end{aligned}$$

for all $k_1, k_2, \ell_1\ell_2 \in \mathbb{F}_q$. Then $G(\mathscr{A}, H, \tau) \equiv G(\mathbb{F}_q, \mathbb{F}_q \oplus \mathbb{F}_q, \tau)$ is the group $(\mathbb{F}_q \oplus \mathbb{F}_q \oplus \mathbb{F}_q, *_\tau)$ with (cf. (6.8))

$$\begin{aligned}(m_1,k_1,\ell_1) *_\tau (m_2,k_2,\ell_2) &:= (m_1+m_2-\tau((k_1,\ell_1),(k_2,\ell_2)),k_1+k_2,\ell_1+\ell_2)\\ &= (m_1+m_2+k_1\ell_2,k_1+k_2,\ell_1+\ell_2),\end{aligned} \tag{6.72}$$

for all $k_1, k_2, \ell_1\ell_2 \in \mathbb{F}_q$. In other words, cf. (6.70), $G = G(\mathscr{A}, H, \tau)$.

Moreover (6.37) becomes

$$\begin{aligned}\tau_{\mathrm{sym}}((k_1,\ell_1),(k_2,\ell_2)) &= -\frac{k_1\ell_2+k_2\ell_1}{2}\\ \tau_{\mathrm{b,n}}((k_1,\ell_1),(k_2,\ell_2)) &= -k_2\ell_1+k_1\ell_2\end{aligned}$$

for all $k_1, k_2, \ell_1, \ell_2 \in \mathbb{F}_q$. See also Example 6.10.

Consider the map $q\colon H \to \mathscr{A}$ defined by

$$q(k,\ell) := \frac{k\ell}{2}$$

for all $k, \ell \in \mathbb{F}_q$ (recall, cf. Sect. 6.2, that $(\mathbb{F}_q, +)$ is 2-divisible, since $|\mathbb{F}_q| = q$ is odd). Then, for all $k_1, k_2, \ell_1, \ell_2 \in \mathbb{F}_q$, we have,

$$\begin{aligned}q(k_1,\ell_1)+q(k_2,\ell_2)-q(k_1+k_2,\ell_1+\ell_2) &= \frac{1}{2}(k_1\ell_1+k_2\ell_2-k_1\ell_1-k_1\ell_2-k_2\ell_1-k_2\ell_2)\\ &= -\frac{k_1\ell_2+k_2\ell_1}{2}\\ &= \tau_{\mathrm{sym}}((k_1,\ell_1),(k_2,\ell_2)),\end{aligned}$$

so that τ_{sym} is cohomologous to the trivial cocycle $\tau_0\colon H \times H \to \mathscr{A}$ defined by $\tau_0(h_1, h_2) = 0_{\mathscr{A}}$ for all $h_1, h_2 \in H$. As a consequence of Theorem 6.6 we have that $G(\mathscr{A}, H, \tau_{\mathrm{sym}}) \cong G(\mathscr{A}, H, \tau_0) \cong (\mathbb{F}_q, +) \oplus (\mathbb{F}_q, +) \oplus (\mathbb{F}_q, +)$. Then, denoting by $\mathfrak{a}$ (resp. $\mathfrak{h}$) the Abelian Lie ring with underling group structure $\mathscr{A}$ (resp.

H), the associated 2-step nilpotent Lie ring $\mathfrak{g}(\mathfrak{a}, \mathfrak{h}, \tau_0, \tau_{\mathrm{b,n}})$ is the Lie ring $\mathfrak{g} = (\mathbb{F}_q \oplus \mathbb{F}_q \oplus \mathbb{F}_q, +_{\tau_0}, [\ ,\ ]_{\tau_{\mathrm{b,n}}})$ with

$$(m_1, k_1, \ell_1) +_{\tau_0} (m_2, k_2, \ell_2) = (m_1 + m_2, k_1 + k_2, \ell_1 + \ell_2)$$

and (cf. (6.35))

$$[(m_1, k_1, \ell_1), (m_2, k_2, \ell_2)]_{\tau_{\mathrm{b,n}}} = (\tau_{\mathrm{b,n}}((k_1, \ell_1), (k_2, \ell_2)), 0, 0) = (k_1\ell_2 - k_2\ell_1, 0, 0)$$

for all $m_1, m_2, k_1, k_2, \ell_1, \ell_2 \in \mathbb{F}_q$.

We equip $\mathrm{M}(3, \mathbb{F}_q)$, the set of all 3×3 matrices with coefficients in $\mathbb{F}_q$, with a structure of a Lie ring $(\mathrm{M}(3, \mathbb{F}_q), +, [\ ,\ ])$, where $+$ is the usual matrix sum, and the bracket is defined by setting

$$[A, B] := AB - BA$$

for all $A, B \in \mathrm{M}(3, \mathbb{F}_q)$. Then the subset

$$\mathrm{SUT}(3, \mathbb{F}_q) = \left\{ \begin{pmatrix} 0 & k & m \\ 0 & 0 & \ell \\ 0 & 0 & 0 \end{pmatrix} : m, k, \ell \in \mathbb{F}_q \right\}$$

of strictly upper-triangular matrices, is a Lie subring of $\mathrm{M}(3, \mathbb{F}_q)$ and the map $\mathfrak{g} \to \mathrm{SUT}(3, \mathbb{F}_q)$ defined by

$$(m, k, \ell) \longrightarrow \begin{pmatrix} 0 & k & m \\ 0 & 0 & \ell \\ 0 & 0 & 0 \end{pmatrix} \tag{6.73}$$

for all $m, k, \ell \in \mathbb{F}_q$, is a Lie ring isomorphism. We leave the straightforward check of this fact to the reader.

In the following, in order to simplify notation, we drop the bijection $\phi \colon G \to \mathfrak{g} = (\mathbb{F}_q \oplus \mathbb{F}_q \oplus \mathbb{F}_q, +_{\tau_0}, [\ ,\ ]_{\tau_{\mathrm{b,n}}})$ (cf. (6.41)) and we write $\phi(m, k, \ell) = (m, k, \ell)$ for all $m, k, \ell \in \mathbb{F}_q$.

Then, for $g = (m, k, \ell) \in G$ and $\overline{g} = (\overline{m}, \overline{k}, \overline{\ell}) \in \mathfrak{g}$ we have (cf. Lemma 6.28)

$$\begin{aligned}
\sigma(g)(\overline{g}) &= (m, k, \ell) \cdot (\overline{m}, \overline{k}, \overline{\ell}) \cdot (m, k, \ell)^{-1} \\
&= (\overline{m} + k\overline{\ell} - \ell\overline{k}, \overline{k}, \overline{\ell}) \\
&= (\overline{m}, \overline{k}, \overline{\ell}) +_{\tau_0} (k\overline{\ell} - \ell\overline{k}, 0, 0) \\
&= (\overline{m}, \overline{k}, \overline{\ell}) +_{\tau_0} (\tau_{\mathrm{b,n}}((k, \ell), (\overline{k}, \overline{\ell})), 0, 0) \\
&= \overline{g} +_{\tau_0} [g, \overline{g}]_{\tau_{\mathrm{b,n}}},
\end{aligned}$$

in agreement with (6.42).

We now apply the orbit method to $G = H_q$.

First recall that the *trace* $\mathrm{Tr} = \mathrm{Tr}_{\mathbb{F}_q/\mathbb{F}_p}$ is a surjective $\mathbb{F}_p$-linear map from $\mathbb{F}_q$ (viewed as an extension of $\mathbb{F}_p$) onto $\mathbb{F}_p$ (cf. [30, Sect. 6.7]). Also, we denote by

$$\widehat{\mathfrak{g}} = \widehat{\mathbb{F}_q \oplus \mathbb{F}_q \oplus \mathbb{F}_q} \cong \widehat{\mathbb{F}_q} \oplus \widehat{\mathbb{F}_q} \oplus \widehat{\mathbb{F}_q} \cong (\mathbb{F}_q, +) \oplus (\mathbb{F}_q, +) \oplus (\mathbb{F}_q, +)$$

the dual of the Abelian group underlying $\mathfrak{g}$. More explicitly, for $\rho, \alpha, \beta \in \mathbb{F}_q$ we denote by $\chi_{\rho,\alpha,\beta} \in \widehat{\mathfrak{g}}$ the character defined by setting

$$\chi_{\rho,\alpha,\beta}(\overline{m}, \overline{k}, \overline{\ell}) = \exp\left((2\pi i/p)\mathrm{Tr}(\rho\overline{m} + \alpha\overline{k} + \beta\overline{\ell})\right) \tag{6.74}$$

for all $(\overline{m}, \overline{k}, \overline{\ell}) \in \mathfrak{g}$ (cf. [30, Sect. 7.1]). For $(m, k, \ell) \in G$ we have (cf. (6.45)),

$$\begin{aligned}
[\sigma^*(m, k, \ell)\chi_{\rho,\alpha,\beta}](\overline{m}, \overline{k}, \overline{\ell}) &= \chi_{\rho,\alpha,\beta}\left((\overline{m}, \overline{k}, \overline{\ell}) - [(m, k, \ell), (\overline{m}, \overline{k}, \overline{\ell})]\right) \\
&= \chi_{\rho,\alpha,\beta}\left((\overline{m}, \overline{k}, \overline{\ell}) - (k\overline{\ell} - \ell\overline{k}, 0, 0)\right) \\
&= \chi_{\rho,\alpha,\beta}\left(\overline{m} - k\overline{\ell} + \ell\overline{k}, \overline{k}, \overline{\ell}\right) \\
&= \exp\left((2\pi i/p)\mathrm{Tr}(\rho(\overline{m} - k\overline{\ell} + \ell\overline{k}) + \alpha\overline{k} + \beta\overline{\ell})\right) \\
&= \exp\left((2\pi i/p)\mathrm{Tr}(\rho\overline{m} + (\rho\ell + \alpha)\overline{k} + (-\rho k + \beta)\overline{\ell})\right) \\
&= \chi_{\rho,\rho\ell+\alpha,-\rho k+\beta}(\overline{m}, \overline{k}, \overline{\ell})
\end{aligned}$$

(where, as usual, we drop the ϕ's), showing that

$$\sigma^*(m, k, \ell)\chi_{\rho,\alpha,\beta} = \chi_{\rho,\rho\ell+\alpha,-\rho k+\beta}. \tag{6.75}$$

If $\rho \neq 0$, the orbit of $\chi_{\rho,\alpha,\beta}$ is given by

$$\Omega_\rho = \Omega(\chi_{\rho,\alpha,\beta}) = \{\chi_{\rho,\alpha',\beta'} : \alpha', \beta' \in \mathbb{F}_q\}$$

and the corresponding stabilizer is given by

$$S_\rho = S_{\Omega(\chi_{\rho,\alpha,\beta})} = \{(m, 0, 0) : m \in \mathbb{F}_q\} = \mathscr{A} = Z(G), \tag{6.76}$$

while, if $\rho = 0$, the character $\chi_{0,\alpha,\beta}$ is a fixed point and therefore

$$\Omega_{\alpha,\beta} = \Omega(\chi_{0,\alpha,\beta}) = \{\chi_{0,\alpha,\beta}\} \text{ and } S_{\alpha,\beta} = S_{\Omega(\chi_{0,\alpha,\beta})} = G. \tag{6.77}$$

Thus, if $\rho = 0$, the orbit method yields the representation space $V_{\alpha,\beta} = V_{\Omega_{\alpha,\beta}} = \mathbb{C}\chi_{0,\alpha,\beta} \leq L(G)$ and the corresponding one-dimensional representation is given by

$\lambda_{\alpha,\beta} = \lambda|_{V_{\alpha,\beta}}$, where λ is the regular representation of G. Moreover, recalling that $S_{\alpha,\beta} = G$, we have (cf. (6.61))

$$\lambda_{\alpha,\beta} = \chi_{0,\alpha,\beta} \circ \phi$$

($\chi_{0,\alpha,\beta} \circ \phi$ is a character of the whole group G).

On the other hand, if $\rho \neq 0$, then the orbit method yields (cf. (6.48))

$$\begin{aligned} V_\rho = V_{\Omega_\rho} &= \mathrm{span}\{\chi_{\rho,\alpha,\beta} \circ \phi : \alpha, \beta \in \mathbb{F}_q\} \\ &= \{f \in L(G) : f(m,k,\ell) = \exp((2\pi i/p)\mathrm{Tr}(\rho m)) f'(k,\ell) \text{ with } f' \in L(H)\} \\ &\equiv \mathbb{C}\exp((2\pi i/p)\mathrm{Tr}(\rho\,\cdot)) \otimes L(H). \end{aligned}$$

With the notation as in (6.56) with $\Omega = \Omega_\rho$, we now check that $V_{-\rho} = V_\ell$. Since

$$\dim V_{-\rho} = |\Omega_{-\rho}| = q^2 = [G : Z(G)] = [G : S_\rho] = \dim \mathrm{Ind}_{S_\rho}^G \chi_\rho = \dim V_\ell,$$

it suffices to check that $V_{-\rho} \subseteq V_\ell$. If $f \in V_{-\rho}$ we have $f(m,k,\ell) = \overline{\exp((2\pi i/p)\mathrm{Tr}(\rho m))} f'(k,\ell)$ with $f' \in L(H)$, so that, for $g = (m,k,\ell) \in G$ and $s = (m', 0, 0) \in S_\rho$,

$$\begin{aligned} f(gs) = f((m,k,\ell)\cdot(m',0,0)) &= f(m+m',k,\ell) \\ &= \overline{\exp((2\pi i/p)\mathrm{Tr}(\rho(m+m')))} f'(k,\ell) \\ &= \overline{\exp((2\pi i/p)\mathrm{Tr}(\rho m'))} \cdot \overline{\exp((2\pi i/p)\mathrm{Tr}(\rho m))} f'(k,\ell) \\ &= \overline{\exp((2\pi i/p)\mathrm{Tr}(\rho m'))} f(m,k,\ell) \\ &= \overline{\chi}_\rho(m',0,0) f(m,k,\ell) \\ &= \overline{\chi}_\rho(s) f(g). \end{aligned}$$

Theorem 6.37 *Let $G = H_q$ denote the Heisenberg group over the field $\mathbb{F}_q$ (with $q = p^h$, p odd prime). Then*

$$\widehat{G} = \{W_{\Omega_\rho}, W_{\Omega_{\alpha,\beta}} : \rho \in \mathbb{F}_q^*, \alpha, \beta \in \mathbb{F}_q\}.$$

The corresponding characters are

$$\chi^{\Omega_\rho} = q\,\chi_{\rho,0,0}\mathbf{1}_{Z(G)} \quad \text{and} \quad \chi^{\Omega_{\alpha,\beta}} = \chi_{0,\alpha,\beta}. \tag{6.78}$$

Proof The description of $\widehat{G}$ is an immediate consequence of the orbit method and the above discussion. Formula (6.78) follows from Theorem 6.34, taking into account (6.76) and (6.77), and the fact that $|\Omega_\rho| = q^2$. ■

For applications, see [22].

6.10 Restricting from H_{q^t} to H_q

In this section we present the decomposition into isotypic components of the restriction $\mathrm{Res}^{H_{q^t}}_{H_q} \mu$ to the Heisenberg group H_q over $\mathbb{F}_q$ of any representation μ of the Heisenberg group H_{q^t} over $\mathbb{F}_{q^t}$, where $t \geq 2$.

We denote by $\mathrm{Tr}_{\mathbb{F}_{q^t}/\mathbb{F}_q} \colon \mathbb{F}_{q^t} \to \mathbb{F}_q$ (resp. $\mathrm{Tr}_{\mathbb{F}_{q^t}/\mathbb{F}_p} \colon \mathbb{F}_{q^t} \to \mathbb{F}_p$, resp. $\mathrm{Tr} = \mathrm{Tr}_{\mathbb{F}_q/\mathbb{F}_p} \colon \mathbb{F}_q \to \mathbb{F}_p$) the *trace* of the extension $\mathbb{F}_{q^t}/\mathbb{F}_q$ (resp. $\mathbb{F}_{q^t}/\mathbb{F}_p$, resp. $\mathbb{F}_q/\mathbb{F}_p$), cf. [30, Sect. 6.7]. Then $\mathrm{Tr}_{\mathbb{F}_{q^t}/\mathbb{F}_q}$ (resp. $\mathrm{Tr}_{\mathbb{F}_{q^t}/\mathbb{F}_p}$, resp. Tr) is a surjective $\mathbb{F}_q$-linear (resp. $\mathbb{F}_p$-linear) map and the following *transitivity* property holds

$$\mathrm{Tr}_{\mathbb{F}_{q^t}/\mathbb{F}_p} = \mathrm{Tr} \circ \mathrm{Tr}_{\mathbb{F}_{q^t}/\mathbb{F}_q}. \tag{6.79}$$

We apply the methods from Sect. 6.8 with $G = H_{q^t}$ and $G_0 = H_q$ and, again, to simplify notation, we shall not indicate the map $\phi \colon G \to \mathfrak{g}$ (resp. $\phi_0 \colon G_0 \to \mathfrak{g}_0$).

Let $P \colon \widehat{\mathfrak{g}} \to \widehat{\mathfrak{g}_0}$ denote, as usual, the canonical projection (cf. Sect. 6.8). We have, for $\rho, \alpha, \beta \in \mathbb{F}_{q^t}$ and $(\overline{m}, \overline{k}, \overline{\ell}) \in \mathfrak{g}_0$

$$\begin{aligned}
P(\chi_{\rho,\alpha,\beta})(\overline{m}, \overline{k}, \overline{\ell}) &= \chi_{\rho,\alpha,\beta}(\overline{m}, \overline{k}, \overline{\ell}) \\
\text{(by (6.74))} \quad &= \exp\left((2\pi i/p)\mathrm{Tr}_{\mathbb{F}_{q^t}/\mathbb{F}_p}(\rho\overline{m} + \alpha\overline{k} + \beta\overline{\ell})\right) \\
\text{(by (6.79))} \quad &= \exp\left((2\pi i/p)[\mathrm{Tr} \circ \mathrm{Tr}_{\mathbb{F}_{q^t}/\mathbb{F}_q}](\rho\overline{m} + \alpha\overline{k} + \beta\overline{\ell})\right) \\
&= \exp\left((2\pi i/p)[\mathrm{Tr}(\mathrm{Tr}_{\mathbb{F}_{q^t}/\mathbb{F}_q}(\rho)\overline{m}) + \mathrm{Tr}(\mathrm{Tr}_{\mathbb{F}_{q^t}/\mathbb{F}_q}(\alpha)\overline{k}) + \mathrm{Tr}(\mathrm{Tr}_{\mathbb{F}_{q^t}/\mathbb{F}_q}(\beta)\overline{\ell})]\right) \\
&= \chi_{\mathrm{Tr}_{\mathbb{F}_{q^t}/\mathbb{F}_q}(\rho), \mathrm{Tr}_{\mathbb{F}_{q^t}/\mathbb{F}_q}(\alpha), \mathrm{Tr}_{\mathbb{F}_{q^t}/\mathbb{F}_q}(\beta)}(\overline{m}, \overline{k}, \overline{\ell}),
\end{aligned}$$

that is,

$$P(\chi_{\rho,\alpha,\beta}) = \chi_{\mathrm{Tr}_{\mathbb{F}_{q^t}/\mathbb{F}_q}(\rho), \mathrm{Tr}_{\mathbb{F}_{q^t}/\mathbb{F}_q}(\alpha), \mathrm{Tr}_{\mathbb{F}_{q^t}/\mathbb{F}_q}(\beta)}. \tag{6.80}$$

In order to simplify the notation, for $\gamma \in \mathbb{F}_{q^t}$ we write $\widetilde{\gamma}$ instead of $\mathrm{Tr}_{\mathbb{F}_{q^t}/\mathbb{F}_q}(\gamma)$. Moreover, for $\rho \in \mathbb{F}_{q^t}^*$ and $\alpha, \beta \in \mathbb{F}_{q^t}$ we denote by Ω_ρ^t and $\Omega_{\alpha,\beta}^t$ the corresponding orbit in $\widehat{\mathfrak{g}}$, omitting the superscript t if it is equal to 1. In the present setting (6.78) becomes

$$\chi^{\Omega_\rho^t} = q^t \chi_{\rho,0,0}\mathbf{1}_{Z(G)} \quad \text{and} \quad \chi^{\Omega_{\alpha,\beta}^t} = \chi_{0,\alpha,\beta}.$$

Observing that $Z(G_0) = Z(G) \cap G_0$ from (6.80) we deduce that

$$\chi^{\Omega_\rho^t}|_{G_0} = q^t \chi_{\widetilde{\rho},0,0}\mathbf{1}_{Z(G_0)} = \begin{cases} q^{t-1}(q\chi_{\widetilde{\rho},0,0}\mathbf{1}_{Z(G_0)}) = q^{t-1}\chi^{\Omega_{\widetilde{\rho}}} & \text{if } \widetilde{\rho} \neq 0 \\ q^t \mathbf{1}_{Z(G_0)} & \text{if } \widetilde{\rho} = 0 \end{cases} \tag{6.81}$$

and

$$\chi^{\Omega^t_{\alpha,\beta}}|_{G_0} = \chi^{\Omega_{\widetilde{\alpha},\widetilde{\beta}}}. \tag{6.82}$$

We are now in a position to state the main result of this section.

Theorem 6.38 *Let $\rho, \alpha, \beta \in \mathbb{F}_{q^t}$ with $\rho \neq 0$. Then the decomposition of the restriction to the subgroup G_0 of an irreducible G-representation (cf. Theorem 6.37) is given by*

$$\mathrm{Res}^G_{G_0} W_{\Omega^t_\rho} = \begin{cases} q^{t-1} W_{\Omega_{\widetilde{\rho}}} & \text{if } \widetilde{\rho} \neq 0 \\ q^{t-2} \bigoplus_{\alpha',\beta' \in \mathbb{F}_q} W_{\Omega_{\alpha',\beta'}} & \text{if } \widetilde{\rho} = 0 \end{cases} \tag{6.83}$$

and

$$\mathrm{Res}^G_{G_0} W_{\Omega^t_{\alpha,\beta}} = W_{\Omega_{\widetilde{\alpha},\widetilde{\beta}}}. \tag{6.84}$$

Proof Formulæ (6.83) for $\widetilde{\rho} \neq 0$ and (6.84) follow immediately from (6.81) and (6.82), respectively. We are only left with (6.83) for $\widetilde{\rho} = 0$ (cf. the second line of (6.81)). Given $\alpha', \beta' \in \mathbb{F}_q$ the multiplicity of $W_{\Omega_{\alpha',\beta'}}$ as a subrepresentation of $\mathrm{Res}^G_{G_0} W_{\Omega^t_\rho}$ is

$$\frac{1}{|G_0|}\langle \chi^{\Omega^t_\rho}|_{G_0}, \chi^{\Omega_{\alpha',\beta'}}\rangle_{L(G_0)} = \frac{1}{q^3} \sum_{g_0 \in Z(G_0)} q^t \overline{\chi^{\Omega_{\alpha',\beta'}}(g_0)} = \frac{1}{q^3} \sum_{g_0 \in Z(G_0)} q^t = \frac{q^{t+1}}{q^3} = q^{t-2},$$

while, for $\rho' \in \mathbb{F}_q^*$, the multiplicity of $W_{\Omega_{\rho'}}$ is

$$\frac{1}{|G_0|}\langle \chi^{\Omega^t_\rho}|_{G_0}, \chi^{\Omega_{\rho'}}\rangle_{L(G_0)} = \frac{1}{|G_0|}\langle \chi^{\Omega^t_\rho}|_{Z(G_0)}, \chi^{\Omega_{\rho'}}|_{Z(G_0)}\rangle_{L(Z(G_0))} = 0,$$

where the last equality follows from the orthogonality relations for characters of $Z(G_0)$. ■

Remark 6.39 Theorem 6.38 can be alternatively deduced from Theorem 6.36. Indeed, using again (6.80), it is easy to show that

$$P(\Omega^t_\rho) = \begin{cases} \Omega_{\widetilde{\rho}} & \text{if } \widetilde{\rho} \neq 0 \\ \bigsqcup_{\alpha',\beta' \in \mathbb{F}_q} \Omega_{\alpha',\beta'} & \text{if } \widetilde{\rho} = 0 \end{cases}$$

and

$$P(\Omega^t_{\alpha,\beta}) = \Omega_{\widetilde{\alpha},\widetilde{\beta}}.$$

Then Theorem 6.36 provides us with the corresponding multiplicities. We leave it to the reader to check for the remaining details.

6.11 The Little Group Method for the Heisenberg Group

In this section we apply the little group method to the Heisenberg group $G = H_q$. Now there is no restriction on the parity of the characteristic p of $\mathbb{F}_q$. We use the same notation as in Sect. 6.9. Consider the Abelian subgroups

$$B := \{(m, k, 0) : m, k \in \mathbb{F}_q\} \cong \mathbb{F}_q \oplus \mathbb{F}_q$$

and

$$H := \{(0, 0, \ell) : \ell \in \mathbb{F}_q\} \cong \mathbb{F}_q.$$

We claim that $G = B \rtimes H$. Indeed, it is clear that $G = BH$ and $B \cap H = \{(0, 0, 0)\} = \{1_G\}$. Moreover, recalling that $(0, 0, \ell)^{-1} = (0, 0, -\ell)$, we have

$$(0, 0, \ell)^{-1}(m, k, 0)(0, 0, \ell) = (m + k\ell, k, 0)$$

for all $m, k, \ell \in \mathbb{F}_q$, showing that B is a normal subgroup. We leave it to the reader to complete the proof of the claim, by determining the homomorphism $H \to \mathrm{Aut}(B)$ which implements the semi-direct product $B \rtimes H$.

We are now in a position to apply the methods from Sect. 2.8.

The dual of B is $\widehat{B} = \{\chi_{\rho,\alpha} : \rho, \alpha \in \mathbb{F}_q\}$, where the character $\chi_{\rho,\alpha} : B \to \mathbb{T}$ is defined by

$$\chi_{\rho,\alpha}(m, k, 0) := \exp\left((2\pi i/p)\mathrm{Tr}(\rho m + \alpha k)\right) \tag{6.85}$$

for all $m, k \in \mathbb{F}_q$. The action of H on $\widehat{B}$ (cf. (2.1)) is given by

$$\begin{aligned} {}^{(0,0,\ell)}\chi_{\rho,\alpha}(m, k, 0) &= \chi_{\rho,\alpha}(m + k\ell, k, 0) \\ &= \exp\left((2\pi i/p)\mathrm{Tr}(\rho((m + k\ell) + \alpha k)\right) \\ &= \exp\left((2\pi i/p)\mathrm{Tr}(\rho m + (\alpha + \rho\ell)k)\right) \\ &= \chi_{\rho,\alpha+\rho\ell}(m, k, 0) \end{aligned}$$

that is,

$${}^{(0,0,\ell)}\chi_{\rho,\alpha} = \chi_{\rho,\alpha+\rho\ell}. \tag{6.86}$$

If $\rho \neq 0$, the orbit of $\chi_{\rho,\alpha}$ is given by

$$\Omega_\rho = \Omega(\chi_{\rho,\alpha}) = \{\chi_{\rho,\alpha'} : \alpha' \in \mathbb{F}_q\},$$

and the corresponding inertia subgroup is

$$I_\rho = I_G(\chi_{\rho,\alpha}) = B,$$

while, if $\rho = 0$, the character $\chi_{0,\alpha}$ is a fixed point and therefore

$$\Omega_\alpha = \Omega(\chi_{0,\alpha}) = \{\chi_{0,\alpha}\} \quad \text{and} \quad I_\alpha = I_G(\chi_{0,\alpha}) = G.$$

Thus, if $\rho \neq 0$ we have that $H_\rho = I_\rho/B = B/B$ is a trivial group, while, if $\rho = 0$, $H_\alpha = I_\alpha/B = G/B \cong H$. We have $\widehat{H} = \{\chi_\beta : \beta \in \mathbb{F}_q\}$ where the character $\chi_\beta \colon H \to \mathbb{T}$ is defined by

$$\chi_\beta(0, 0, \ell) := \exp\left((2\pi i/p)\mathrm{Tr}(\beta\ell)\right) \tag{6.87}$$

for all $\ell \in \mathbb{F}_q$. Thus the inflation of χ_β si given by

$$\overset{\triangle}{\chi_\beta}(m, k, \ell) = \exp\left((2\pi i/p)\mathrm{Tr}(\beta\ell)\right).$$

We choose

$$\Sigma := \{\chi_{\rho,0} : \rho \in \mathbb{F}_q^*\} \sqcup \{\chi_{0,\alpha} : \alpha \in \mathbb{F}_q\} \subseteq \widehat{B}$$

(note that $|\Sigma| = (q-1) + q = 2q - 1$); cf. (2.5) and Theorem 2.54.

We thus obtain the following two families of irreducible representations:

$$\mathrm{Ind}_B^G \widetilde{\chi_{\rho,0}}, \quad \rho \in \mathbb{F}_q^* \tag{6.88}$$

and

$$\widetilde{\chi_{0,\alpha}}\, \overset{\triangle}{\chi_\beta}, \quad \alpha, \beta \in \mathbb{F}_q, \tag{6.89}$$

where the $q-1$ representations (6.88) of dimension q (resp. the q^2 one-dimensional representations (6.89)) correspond to the W_{Ω_ρ} (resp. $W_{\Omega_{\alpha,\beta}}$) in Theorem 6.37.

Remark 6.40 It would be interesting, and we leave it to the reader, to compute an explicit isomorphism between the representations W_{Ω_ρ} and $\mathrm{Ind}_B^G \widetilde{\chi}_{\rho,0}$.

Chapter 7
Representations of Finite Group Extensions via Projective Representations

In this chapter we give a formulation of the little group method (cf. Theorem 2.26) for arbitrary group extensions, that is, we do not assume existence of an extension of $\sigma \in \widehat{N}$ to its inertia group $I_G(\sigma)$. We also give a sufficient condition for the existence of such an extension. All these facts require a suitable cohomological machinery. It is due to Mackey (who actually proved more general results for topological groups), but some aspects of it were anticipated by Schur and others within the theory of projective representations; see [16, 65, 67]. Actually, we follow Mackey's approach as described in [48]; see also [128].

7.1 Mackey Obstruction

Now we assume *all* the notation in Sect. 1.5 and Chaps. 2 and 6, but we think of N as a (normal) subgroup of G and we omit the ι in (1.57) and the notation $\overline{n}$. In particular, we fix a normalized section $t\colon H := G/N \to G$, so that $t(h)N = h$ for all $h \in H$ and $t(1_H) = 1_G$, and, moreover, every element $g \in G$ can be (uniquely) written as $t(h)n$ for some $h \in H$ and $n \in N$.

Let $\sigma \in \widehat{N}$ and set $H_G(\sigma) := I_\sigma(G)/N$. Given $h \in H$, we defined in (2.2) ${}^h\sigma := {}^{t(h)}\sigma$ (recall that although ${}^h\sigma$ depends on the section t, its equivalence class is in fact independent of the choice of the section). For $h \in H_G(\sigma)$ one has ${}^h\sigma \sim \sigma$ and therefore there exists a unitary operator $\Psi(h)$, defined on $V = V_\sigma$, such that

$$ {}^h\sigma(n) = \sigma\left(t(h)^{-1}nt(h)\right) = \Psi(h)^{-1}\sigma(n)\Psi(h), \tag{7.1}$$

for all $n \in N$. By Schur's lemma, $\Psi(h)$ is uniquely determined up to a multiplicative constant in $\mathbb{T}$; we only assume that $\Psi(1_{H_G(\sigma)}) = I_V$ (see also Lemmas 7.2 and 7.6).

T. Ceccherini-Silberstein et al., *Representation Theory of Finite Group Extensions*, Springer Monographs in Mathematics,
https://doi.org/10.1007/978-3-031-13873-7_7

Then, for all $h, k \in H_G(\sigma)$ and $n \in N$, by means of a repeated application of (7.1) we get

$$\begin{aligned}
\Psi(k)^{-1}\Psi(h)^{-1}\sigma(n)\Psi(h)\Psi(k) &= \Psi(k)^{-1}\sigma\left(t(h)^{-1}nt(h)\right)\Psi(k)\\
&= \sigma\left(t(k)^{-1}t(h)^{-1}nt(h)t(k)\right)\\
\text{(by (1.56))} \qquad &= \sigma\left(n(h,k)^{-1}t(hk)^{-1}nt(hk)n(h,k)\right)\\
&= \sigma(n(h,k))^*\,\Psi(hk)^{-1}\sigma(n)\Psi(hk)\sigma(n(h,k)),
\end{aligned}$$

that is

$$\Psi(h)\Psi(k)\sigma(n(h,k))^*\,\Psi(hk)^{-1}\cdot\sigma(n) = \sigma(n)\cdot\Psi(h)\Psi(k)\sigma(n(h,k))^*\,\Psi(hk)^{-1}.$$

By Schur's lemma, there exists a constant $\eta(h,k) \in \mathbb{T}$ such that

$$\eta(h,k)I_V = \Psi(hk)\sigma(n(h,k))\Psi(k)^{-1}\Psi(h)^{-1}. \tag{7.2}$$

Proposition 7.1 *The function*

$$\eta\colon H_G(\sigma)\times H_G(\sigma) \to \mathbb{T},$$

defined in (7.2)*, is a (unitary) 2-cocycle. That is,* $\eta \in \mathscr{C}^2(H_G(\sigma), \mathbb{T})$.

Proof From the conditions $\Psi(1_H) = I_V$, (1.61) and (7.2) it follows that $\eta(h, 1_G) = \eta(1_G, k) = 1$, for all $h, k \in H_G(\sigma)$. Moreover, from (7.2) it also follows that, for all $h_1, h_2, h_3 \in H_G(\sigma)$,

$$\begin{aligned}
\eta(h_1h_2,h_3)\eta(h_1,h_2)I_V &= \Psi(h_1h_2h_3)\sigma(n(h_1h_2,h_3))\Psi(h_3)^{-1}\Psi(h_1h_2)^{-1}\\
&\quad\cdot\Psi(h_1h_2)\sigma(n(h_1,h_2))\Psi(h_2)^{-1}\Psi(h_1)^{-1}\\
&= \Psi(h_1h_2h_3)\sigma(n(h_1h_2,h_3))\Psi(h_3)^{-1}\sigma(n(h_1,h_2))\Psi(h_2)^{-1}\Psi(h_1)^{-1}\\
\text{(by (7.1))} \quad &= \Psi(h_1h_2h_3)\sigma\left(n(h_1h_2,h_3)n(h_1,h_2)^{h_3^\sharp}\right)\Psi(h_3)^{-1}\Psi(h_2)^{-1}\Psi(h_1)^{-1}\\
\text{(by (1.59))} \quad &= \Psi(h_1h_2h_3)\sigma(n(h_1,h_2h_3))\Psi(h_2h_3)^{-1}\cdot\\
&\quad\cdot\Psi(h_2h_3)\sigma(n(h_2,h_3))\Psi(h_3)^{-1}\Psi(h_2)^{-1}\Psi(h_1)^{-1}\\
\text{(by (7.2))} \quad &= \Psi(h_1h_2h_3)\sigma(n(h_1,h_2h_3))\Psi(h_2h_3)^{-1}\eta(h_2,h_3)\Psi(h_1)^{-1}\\
&= \eta(h_1,h_2h_3)\eta(h_2,h_3)I_V.
\end{aligned}$$

■

Recall that Ψ is defined up to a choice of a multiplicative constant in $\mathbb{T}$.

Lemma 7.2 *If we choose another Ψ' in (7.1), then the corresponding 2-cocycle η' is cohomologous to η.*

Proof We know that for each $h \in H_G(\sigma)$, the unitary operators $\Psi(h)$ and $\Psi'(h)$ differ by a multiplicative constant $\psi(h) \in \mathbb{T}$, that is, $\Psi'(h) = \psi(h)\Psi(h)$. Since $\Psi(1_H) = \Psi'(1_H) = \sigma(n(1_H)) = \sigma(1_N) = I_V$, we have $\psi(1_H) = 1$. Moreover, for $h, k \in H_G(\sigma)$, (7.2) yields

$$\begin{aligned} \eta'(h,k)I_V &= \Psi'(hk)\sigma(n(h,k))\Psi'(k)^{-1}\Psi'(h)^{-1} \\ &= \Psi(hk)\sigma(n(h,k))\Psi(k)^{-1}\Psi(h)^{-1}\psi(kh)\psi(h)^{-1}\psi(k)^{-1} \\ &= \eta(k,h)\psi(h)^{-1}\psi(k)^{-1}\psi(kh)I_V. \end{aligned} \tag{7.3}$$

That is, η and η' are cohomologous. ■

Conversely, given Ψ with associated 2-cocycle η, by Proposition 6.15.(2) we can find a map $\rho\colon H \to \mathbb{T}$ such that $\eta' := \eta\tau_\rho$ is equalized. Then, defining Ψ' by setting $\Psi'(h) := \rho(h)^{-1}\Psi(h)$ for all $h \in H$, the associated 2-cocycle η' is equalized.

Definition 7.3 The *Mackey obstruction* of $\sigma \in \widehat{N}$ is the element of $\mathscr{H}^2(H_G(\sigma), \mathbb{T})$ containing the cocyle η defined in (7.2). It is *trivial* if this cocycle is a coboundary.

Proposition 7.4 *The Mackey obstruction of σ is trivial if and only if σ may be extended (cf. Definition 2.3) to a representation of $I_G(\sigma)$.*

Proof Suppose that σ admits an extension $\widetilde{\sigma}$ of $I_G(\sigma)$. Then, for all $h \in H_G(\sigma)$ we can take $\Psi(h) = \widetilde{\sigma}(t(h))$ in (7.1), so that from (1.56) we get

$$\Psi(hk) = \widetilde{\sigma}(t(hk)) = \widetilde{\sigma}\left(t(h)t(k)n(h,k)^{-1}\right) = \Psi(k)\Psi(h)\sigma(n(h,k))^*$$

and (7.2) yields $\eta \equiv 1$.

Conversely, suppose that η is the identity of $\mathscr{H}^2(H_G(\sigma), \mathbb{T})$, so that (cf. (6.3)) there exists a map $\psi\colon H_G(\sigma) \to \mathbb{T}$ such that

$$\eta(h,k) = \psi(hk)\psi(h)^{-1}\psi(k)^{-1}, \tag{7.4}$$

for all $h, k \in H_G(\sigma)$. We then set

$$\widetilde{\sigma}(t(h)n) := \Psi(h)\sigma(n)\psi(h)^{-1}, \tag{7.5}$$

for all $h \in H_G(\sigma)$ and $n \in N$. For $h, k \in H_G(\sigma)$ and $m, \ell \in N$ we have

$$\begin{aligned} \widetilde{\sigma}(t(h)m \cdot t(k)\ell) &= \widetilde{\sigma}\left(t(hk)\left[n(h,k)m^{k^\sharp}\ell\right]\right) && \text{(by (1.60))} \\ &= \Psi(hk)\sigma\left(n(h,k)m^{k^\sharp}\ell\right)\psi(hk)^{-1} && \text{(by (7.5))} \end{aligned}$$

$$= \Psi(h)\Psi(k)\sigma\left(m^{k^\sharp}\ell\right)\eta(h,k)\psi(hk)^{-1} \qquad \text{(by (7.2))}$$

$$= \Psi(h)\sigma(m)\psi(h)^{-1}\cdot\Psi(k)\sigma(\ell)\psi(k)^{-1} \qquad \text{(by (1.57), (7.1) and (7.4))}$$

$$= \widetilde{\sigma}(t(h)m)\widetilde{\sigma}(t(k)\ell) \qquad \text{(by (7.5)).}$$

Therefore Ψ is an $I_G(\sigma)$-representation. Since $\Psi(1_H) = I_V$ and $\psi(1_H) = 1$, we have $\widetilde{\sigma}(n) = \sigma(n)$ for all $n \in N$. This shows that $\widetilde{\sigma}$ is an extension of σ. ■

Remark 7.5 The interpretation of the above theorem is the following: the cocycle η associated with $\sigma \in \widehat{N}$ is a measure of the obstruction that one finds when trying to extend σ from N to $I_G(\sigma)$. As a consequence, the little group method (cf. Theorem 2.26) can be applied exactly when the Mackey obstruction of σ is trivial for every $\sigma \in \widehat{N}$.

For some calculations, it is useful to lift Ψ to an operator valued function $\widetilde{\Psi}$ defined on $I_G(\sigma)$ by setting:

$$\widetilde{\Psi}(t(h)n) := \Psi(h)\sigma(n), \tag{7.6}$$

for all $h \in H_G(\sigma)$ and $n \in N$.

Moreover, in some places we explicitly use the quotient form $I_G(\sigma)/N$ to represent $H_G(\sigma)$; for instance, for $r, s \in I_G(\sigma)$, we write $\eta(rN, sN)$ to indicate the value of η on (rN, sN).

Lemma 7.6 *The operator valued function $\widetilde{\Psi}$ satisfies the following properties.*

$$\widetilde{\Psi}(rs) = \eta(rN, sN)\widetilde{\Psi}(r)\widetilde{\Psi}(s), \tag{7.7}$$

for all $r, s \in I_G(\sigma)$, where η is the 2-cocycle defined in (7.2). Moreover, choosing $\widetilde{\Psi}$ such that η is equalized, we have

$$\widetilde{\Psi}(s)^* \equiv \widetilde{\Psi}(s)^{-1} = \widetilde{\Psi}(s^{-1}), \tag{7.8}$$

for all $s \in I_G(\sigma)$. Finally, for all $s \in I_G(\sigma)$ and $m \in N$,

$$\sigma(s^{-1}ms) = \widetilde{\Psi}(s)^{-1}\sigma(m)\widetilde{\Psi}(s). \tag{7.9}$$

Proof If $r = t(h)n$ and $s = t(k)m$, with $h, k \in H_G(\sigma)$ and $n, m \in N$, then we have:

$$\widetilde{\Psi}(t(h)nt(k)m) = \widetilde{\Psi}\left(t(hk)n(h,k)n^{k^\sharp}m\right) \qquad \text{(by (1.60))}$$

$$= \Psi(hk)\sigma\left(n(h,k)\right)\sigma\left(n^{k^\sharp}m\right) \qquad \text{(by (7.6))}$$

$$= \eta(h,k)\Psi(h)\Psi(k)\sigma\left(n^{k^\sharp}\right)\sigma(m) \qquad \text{(by (7.2))}$$

$$= \eta(h,k)\Psi(h)\sigma(n)\Psi(k)\sigma(m) \qquad \text{(by (7.1))}$$
$$= \eta(h,k)\widetilde{\Psi}(t(h)n)\widetilde{\Psi}(t(k)m) \qquad \text{(by (7.6))},$$

and this proves (7.7). Then, by Proposition 6.15.(2) we may assume that $\eta(sN, s^{-1}N) = 1$ for all $s \in I_G(\sigma)$: in the notation therein, this corresponds to multiplying Ψ by the scalar function ρ^{-1}. Since $\widetilde{\Psi}(1_G) = I_V$, from (7.7) with $r = s^{-1}$ we deduce that

$$I_V = \widetilde{\Psi}(s)\widetilde{\Psi}(s^{-1}),$$

and (7.8) follows. Finally, we prove (7.9): for all $s = t(h)n \in I_G(\sigma)$ we have

$$\begin{aligned} \sigma(s^{-1}ms) &= \sigma(n^{-1}t(h)^{-1}mt(h)n) \\ &= \sigma(n^{-1}) \cdot \sigma(t(h)^{-1}mt(h)) \cdot \sigma(n) \\ \text{(by (7.1))} \qquad &= \sigma(n)^{-1}\Psi(h)^{-1} \cdot \sigma(m) \cdot \Psi(h)\sigma(n) \\ \text{(by (7.6))} \qquad &= \widetilde{\Psi}(s)^{-1}\sigma(m)\widetilde{\Psi}(s). \end{aligned}$$

■

Remark 7.7 We leave it to the reader, as an easy task, to deduce Proposition 7.4 from Lemma 7.6.

7.2 Unitary Projective Representations

Definition 7.8 (Unitary Projectived Representations) Let H be a finite group, let V be a unitary space, and denote by $\mathscr{U}(V)$ the group of unitary operators on V. Given a unitary 2-cocycle $\eta \in \mathscr{C}^2(H, \mathbb{T})$, a *(unitary) η-representation* (also called a *(unitary) projective representation* or *(unitary) cocycle representation*) on V is a map $\Phi \colon H \to \mathscr{U}(V)$ such that

$$\Phi(hk) = \eta(h,k)\Phi(h)\Phi(k), \tag{7.10}$$

for all $h, k \in H$, and $\Phi(1_G) = I_V$ (compare with (7.7)).

The notions of invariant subspaces, (unitary) equivalence (denoted by means of the symbol $\sim$), and irreducibility may be translated verbatim for (unitary) projective representations, and we leave it to the reader to fill in the details. For instance, if Φ_1, Φ_2, and Φ are projective representations of H, then Φ_1 and Φ_2 are equivalent if there exists a unitary operator $T \colon V_1 \to V_2$ such that $T\Phi_1 = \Phi_1 T$. Moreover, if $U \le V$ is a Φ-invariant subspace, then also $U^\perp \le V$ is Φ-invariant: it follows that

V decomposes into an orthogonal direct sum of Φ-irreducible subspaces. Finally, the Schur lemma also has the same formulation and proof as in the ordinary setting.

We denote by $\widehat{H}^\eta$ the set of all equivalence classes of irreducible unitary η-representations of H: as usual, we also regard $\widehat{H}^\eta$ as a fixed complete set of irreducible unitary η-representations of H, and we shall not make a distinction between these two points of view. Finally note that, if η is equalized, then we have $\Phi(h^{-1}) = \Phi(h)^*$ (cf. (7.8)). We denote by $d_\Phi := \dim V$ the *dimension* of Φ.

Let Φ be an η-representation of H. Then the following commutation relations hold: for all $h, k \in H$

$$\Phi(k)^{-1}\Phi(h)\Phi(k) = \eta(k, h^k)\overline{\eta(h, k)}\Phi(h^k), \tag{7.11}$$

where $h^k = k^{-1}hk$ (cf. (1.57)). Indeed, from (7.10) we immediately get

$$\eta(h, k)\Phi(h)\Phi(k) = \Phi(hk) = \Phi(kh^k) = \eta(k, h^k)\Phi(k)\Phi(h^k).$$

Remark 7.9

(1) Suppose that a map $\Phi\colon H \to \mathscr{U}(V)$ satisfies (7.10) for some map $\eta\colon H \times H \to \mathbb{T}$. Then necessarily η is a 2-cocycle (so that Φ is an η-representation). Indeed, from $(\Phi(h_1)\Phi(h_2))\Phi(h_3) = \Phi(h_1)(\Phi(h_2)\Phi(h_3))$ it follows easily that η satisfies (6.1), while from $\Phi(1_H h) = \Phi(h)$ one deduces (6.2).

(2) Let Φ_j be an η_j-representation on the space V_j, $j = 1, 2$, and suppose that Φ_1 and Φ_2 are equivalent. Then necessarily $\eta_1 = \eta_2$. Indeed, let $T \in \mathrm{Hom}(V_2, V_1)$ be a unitary operator such that $\Phi_1(h) = T\Phi_2(h)T^*$, for all $h \in H$. Then, for all $h, k \in H$,

$$\begin{aligned}\eta_1(h, k)\Phi_1(h)\Phi_1(k) &= \Phi_1(hk)\\ &= T\Phi_2(hk)T^*\\ &= \eta_2(h, k)T\Phi_2(h)\Phi_2(k)T^*\\ &= \eta_2(h, k)T\Phi_2(h)T^* \cdot T\Phi_2(k)T^*\\ &= \eta_2(h, k)\Phi_1(h)\Phi_1(k)\end{aligned}$$

and $\eta_1(h, k) = \eta_2(h, k)$ follows.

(3) With the notation in (2), suppose that $V_1 = V_2 =: V$ and there exists a map $\lambda\colon H \to \mathbb{T}$ such that $\Phi_1(h) = \lambda(h)\Phi_2(h)$, for all $h \in H$. Then η_1 and η_2 are cohomologous. Indeed, for all $h, k \in H$,

$$\Phi_1(hk) = \lambda(hk)\Phi_2(hk) = \lambda(hk)\eta_2(h, k)\Phi_2(h)\Phi_2(k) = \lambda(hk)\eta_2(h, k)\overline{\lambda}(h)\overline{\lambda}(k)\Phi_1(h)\Phi_1(k)$$

implies that

$$\eta_1(h, k) = \eta_2(h, k)\lambda(hk)\overline{\lambda}(h)\overline{\lambda}(k).$$

Conversely, suppose that η_1 and η_2 in $\mathscr{C}^2(H, \mathbb{T})$ are cohomologous, say $\eta_1 = \eta_2\tau_\lambda$, where $\lambda\colon H \to \mathbb{T}$, and let Φ_2 be an η_2-representation. Then, setting $\Phi_1(h) := \lambda(h)\Phi_2(h)$ for all $h \in H$ yields an η_1-representation. Moreover, the map $\Phi_2 \mapsto \Phi_1$ yields a bijection between the space of all η_2- and η_1-representations of H. This bijection preserves irreducibility and equivalence, so that it suffices to determine the dual $\widehat{H}^\eta$ for all $\eta \in \mathscr{E}$, where $\mathscr{E}$ is a complete set of representatives of the $\mathscr{B}^2(H, \mathbb{T})$ cosets in $\mathscr{C}^2(H, \mathbb{T})$.

(4) Finally, if η is not a coboundary then every η-representation has dimension greater than one. Indeed, if ξ is a one dimensional η-representation, then $\xi(hk) = \xi(h)\xi(k)\eta(h, k)$, forces η be a coboundary: $\eta(h, k) = \xi(hk)\overline{\xi}(h)\overline{\xi}(k)$.

Definition 7.10 Let Φ be a unitary projective representation of H. The *kernel* of Φ is the subgroup $K := \ker \Phi$ defined by

$$K := \{h \in H : \Phi(h) = \lambda I_V,\ \text{for some } \lambda \in \mathbb{T}\},$$

where V is the representation space of Φ.

Note that K uniquely determines a function $\lambda\colon K \to \mathbb{T}$ such that

$$\Phi(k) = \lambda(k) I_V \quad \text{for all} \quad k \in K. \tag{7.12}$$

Now we prove some general facts on the kernel of a unitary projective representation. They are taken form [16, Chap. 6], where more results on the kernels are discussed.

Lemma 7.11 *Let Φ be an η-representation of H and let K denote its kernel.*

(1) *K is a normal subgroup of H.*
(2) *Also, with $\lambda\colon K \to \mathbb{T}$ as in (7.12), we have*

$$\eta(k, k') = \lambda(kk')\overline{\lambda}(k)\overline{\lambda}(k') \quad \textit{for all} \quad k, k' \in K; \tag{7.13}$$

in particular, the restriction of η to K is a coboundary. Moreover,

$$\eta(k, h)\lambda(k) = \eta(h, k^h)\lambda(k^h) \quad \textit{for all} \quad k \in K \textit{ and } h \in H. \tag{7.14}$$

(3) *There exists $\eta' \in \mathscr{C}^2(H, \mathbb{T})$ which is cohomologous to η and, moreover, K-invariant:*

$$\eta'(h_1k_1, h_2k_2) = \eta'(h_1, h_2) \quad \textit{for all} \quad k_1, k_2 \in K,\ h_1, h_2 \in H, \tag{7.15}$$

that is, η' is constant on the cosets of K in H.

Proof

(1) The fact that K is a subgroup is clear. From a repeated application of (7.10) we deduce that for all $h \in H$ and $k \in K$:

$$\begin{aligned} \Phi(h^{-1}kh) &= \eta(h^{-1}k, h)\Phi(h^{-1}k)\Phi(h) = \eta(h^{-1}k, h)\eta(h^{-1}, k)\Phi(h^{-1})\Phi(k)\Phi(h) \\ \text{(by (7.12))} \quad &= \eta(h^{-1}k, h)\eta(h^{-1}, k)\eta(h^{-1}, h)^{-1}\lambda(k)I_V \in \mathbb{C}I_V, \end{aligned}$$

so that also $h^{-1}kh \in K$. This shows that K is normal in H.

(2) These identities are easy to prove: (7.13) is an immediate consequence of (7.10) and of (7.12), while (7.14) follows from (7.11).

(3) Now suppose that $t\colon H/K \to H$ is a section (cf. (1.54)) and write (1.56) and (1.60) in the simplified forms:

$$t(s)t(r) = t(sr)n(s,r) \quad \text{for all} \quad s, r \in H/K, \tag{7.16}$$

and

$$t(s)kt(r)k' = t(sr)n(s,r)k^{t(r)}k' \quad \text{for all} \quad s, r \in H/K \text{ and } k, k' \in K \tag{7.17}$$

and $n(r,s) \in K$. Now set

$$\rho(t(s)k) := \eta(t(s), k)\lambda(k) \tag{7.18}$$

and

$$\eta'(t(s)k, t(r)k') = \eta(t(s), t(r))\eta(t(sr), n(s,r))^{-1}\overline{\lambda(n(r,s))}, \tag{7.19}$$

for all $s, r \in H/K$ and $k, k' \in K$. It is obvious that η' satisfies (7.15). Now we prove the identity

$$\eta(t(s)k, t(r)k') = \eta'(t(s)k, t(r)k')\rho(t(s)kt(r)k')\overline{\rho(t(s)k)\rho(t(r)k')} \tag{7.20}$$

from which it follows that η' is a 2-cocycle and that it is cohomologous to η. First of all, from (7.17) and (7.18) we deduce that

$$\rho(t(s)kt(r)k') = \rho(t(sr)n(s,r)k^{t(r)}k') = \eta(t(sr), n(s,r)k^{t(r)}k')\lambda(n(s,r)k^{t(r)}k')$$

and therefore (7.20) is equivalent to the identity

$$\begin{aligned} &\eta(t(s)k, t(r)k')\eta(t(s), k)\eta(t(r), k')\eta(t(sr), n(s,r))\lambda(k)\lambda(k')\lambda(n(s,r)) \\ &\qquad = \eta(t(s), t(r))\eta(t(sr), n(s,r)k^{t(r)}k')\lambda(n(s,r)k^{t(r)}k'). \end{aligned}$$

By means of two applications of the cocycle identity (6.1) we get the following expression for the left hand side:

$$\begin{aligned}&\eta(t(s), t(r)k^{t(r)}k')\eta(k, t(r)k')\eta(t(r), k')\eta(t(sr), n(s,r))\lambda(k)\lambda(k')\lambda(n(s,r))\\&= \eta(t(s), t(r)k^{t(r)}k')\eta(kt(r), k')\eta(k, t(r))\eta(t(sr), n(s,r))\lambda(k)\lambda(k')\lambda(n(s,r)),\end{aligned}$$

while the right hand side is equal to:

$$\begin{aligned}&\eta(t(s), t(r))\eta(t(sr), n(s,r)k^{t(r)}k')\eta(n(s,r), k^{t(r)}k')\lambda(n(s,r))\lambda(k^{t(r)}k') && \text{(by (7.13))}\\&= \eta(t(s), t(r))\eta(t(sr)n(s,r), k^{t(r)}k')\eta(t(sr), n(s,r))\lambda(n(s,r))\lambda(k^{t(r)}k') && \text{(by (6.1))}\\&= \eta(t(s), t(r))\eta(t(s)t(r), k^{t(r)}k')\eta(t(sr), n(s,r))\lambda(n(s,r))\lambda(k^{t(r)}k') && \text{(by (7.16))}\\&= \eta(t(s), t(r)k^{t(r)}k')\eta(t(r), k^{t(r)}k')\eta(t(sr), n(s,r)) && \text{(by (6.1))}\\&\qquad \cdot \lambda(n(s,r))\eta(k^{t(r)}, k')\lambda(k^{t(r)})\lambda(k') && \text{(by (7.13))}\\&= \eta(t(s), t(r)k^{t(r)}k')\eta(kt(r), k')\eta(t(r), k^{t(r)})\lambda(k^{t(r)}) && \text{(by (6.1))}\\&\qquad \cdot \lambda(k')\eta(t(sr), n(s,r))\lambda(n(s,r))\\&= \eta(t(s), t(r)k^{t(r)}k')\eta(kt(r), k')\eta(k, t(r))\lambda(k)\lambda(k') && \text{(by (7.14))}\\&\qquad \cdot \eta(t(sr), n(s,r))\lambda(n(s,r)),\end{aligned}$$

so that the identity (7.20) is proved.

■

In what follows, we assume that $\eta \in \mathscr{C}^2(H, \mathbb{T})$ is equalized and, moreover, $\eta(h,k)^n = 1$ for all $h, k \in H$, where $n = |H|$ (see Theorem 6.19, Remarks 6.22, and 7.9.(3)), so that, in particular, $\eta \in \mathscr{C}^2(H, C_n)$. We show that unitary η-representations of H are in one-to-one correspondance with some particular unitary representations of the group extension $G(C_n, H, \eta)$ (cf. Sect. 6).

Proposition 7.12 *Let Φ be a unitary η-representation of H and set*

$$\widetilde{\Phi}(\lambda, h) := \lambda\Phi(h), \tag{7.21}$$

for all $(\lambda, h) \in C_n \times H$. Then $\widetilde{\Phi}$ is a unitary representation of $G(C_n, H, \eta)$. Morever, the map $\Phi \mapsto \widetilde{\Phi}$ is a bijection between the set of all unitary η-representations of H and the set of all unitary representations Θ of $G(C_n, H, \eta)$ such that

$$\Theta(\lambda, 1_H) = \lambda I_V, \tag{7.22}$$

for all $\lambda \in C_n$, where V is the space of Θ. The inverse map $\Theta \mapsto \Phi$ is obtained by setting $\Phi(h) := \Theta(1, h)$, for all $h \in H$. Finally, this bijection preserves irreducibility and equivalence in both directions.

Proof Let $\lambda_1, \lambda_2 \in C_n$ and $h_1, h_2 \in H$. We have:

$$\begin{aligned}
\widetilde{\Phi}\big((\lambda_1, h_1) * (\lambda_2, h_2)\big) &= \widetilde{\Phi}\big(\lambda_1\lambda_2\eta(h_1, h_2)^{-1}, h_1h_2\big) && \text{(by (6.8))}\\
&= \lambda_1\lambda_2\Phi(h_1h_2)\eta(h_1, h_2)^{-1} && \text{(by (7.21))}\\
&= \lambda_1\lambda_2\Phi(h_1)\Phi(h_2) && \text{(by (7.10))}\\
&= \widetilde{\Phi}(\lambda_1, h_2)\widetilde{\Phi}(\lambda_2, h_2), && \text{(by (7.21))}
\end{aligned}$$

so that $\widetilde{\Phi}$ is a unitary representation of $G(C_n, H, \eta)$ and clearly $\widetilde{\Phi}(\lambda, 1_H) = \lambda\Phi(1_H) = \lambda I_V$.

On the other hand, suppose that Θ is a unitary representation of $G(C_n, H, \eta)$ such that $\Theta(\lambda, 1_H) = \lambda I_V$ for all $\lambda \in C_n$. Then for $h \in H$ we have

$$\Theta(\lambda, h) = \Theta\big((\lambda, 1_H) * (1, h)\big) = \Theta(\lambda, 1_H)\Theta(1, h) = \lambda\Theta(1, h). \tag{7.23}$$

Therefore, setting $\Phi(h) := \Theta(1, h)$ we have

$$\begin{aligned}
\Phi(h_1)\Phi(h_2) = \Theta(1, h_1)\Theta(1, h_2) &= \Theta\big((1, h_1) * (1, h_2)\big) = \Theta(\eta(h_1, h_2)^{-1}, h_1h_2)\\
&= \eta(h_1, h_2)^{-1}\Theta(1, h_1h_2) = \eta(h_1, h_2)^{-1}\Phi(h_1h_2),
\end{aligned}$$

for all $h_1, h_2 \in H$, so that Φ is an η-representation of H. Moreover, (7.23) implies that $\widetilde{\Phi} = \Theta$. Finally, it is clear that Φ is irreducible if and only if $\widetilde{\Phi}$ is irreducible. ■

Remark 7.13 Setting $G := G(C_n, H, \eta)$, with the notation from the above proposition and from Definition 2.1, we have the following. Let $\xi \colon C_n \to \mathbb{T}$ such that $\xi(\lambda) = \lambda$ for all $\lambda \in C_n$ (viewed as a subgroup of $\mathbb{T}$). Then ξ is a character of C_n, and $I_G(\xi) = G$ since C_n is central in G and condition (7.22) is equivalent to $\Theta \in \widehat{G}(\xi)$. In other words, Proposition 7.12 establishes a bijective correspondence between $\widehat{G}(\xi)$ and the set of equivalence classes of irreducible η-representations of H.

Remark 7.14 The above proposition reduces the study of unitary projective representations of a group H to the study of ordinary representations of central extensions of H. This fact is useful from a theoretical point of view, but, in practice, as pointed out by Mackey [92, p. 270], this is a vicious circle. Indeed, in order to determine all ordinary representations of $G(C_n, H, \eta)$ one should apply the theory that will be developed in Sect. 7.3 which requires the knowledge of the unitary projective representations of certain intermediate quotients. However, Proposition 7.12 is useful to prove theoretical results on unitary projective representations starting from the corresponding reults holding for ordinary representations. See, for instance, Corollary 7.15 below.

The *character* of an η-representation Φ of H is the function $\chi^{\Phi} \in L(H)$ defined by setting

$$\chi^{\Phi}(h) := \operatorname{tr}[\Phi(h)]$$

for all $h \in H$.

Corollary 7.15

(1) *An η-representation Φ of H is irreducible if and only if*

$$\|\chi^{\Phi}\|_{L(H)}^2 = |H|.$$

(2) *If Φ, Ψ are irreducible η-representations of H then*

$$\langle \chi^{\Phi}, \chi^{\Psi} \rangle_{L(H)} = \begin{cases} |H| & \text{if } \Phi \sim \Psi, \\ 0 & \text{if } \Phi \nsim \Psi. \end{cases}$$

(3) *If Φ_j, $j = 1, 2, \ldots, r_\eta$, is a complete system of pairwise inequivalent η-representations then*

$$r_\eta = |\widehat{G}(\xi)| \tag{7.24}$$

(cf. Remark 7.13) and, denoting by d_j the dimension of Φ_j,

$$\sum_{j=1}^{r_\eta} d_j^2 = |H|. \tag{7.25}$$

(4) *Let Φ be an η-representation of H and let*

$$\Phi \sim \bigoplus_{i=1}^{k} m_i \Phi_i \tag{7.26}$$

be a decomposition of Φ into irreducible pairwise inequivalent η-representation. Then, for all $i = 1, 2, \ldots, k$,

$$m_i = \frac{1}{|H|} \langle \chi^{\Phi}, \chi^{\Phi_i} \rangle_{L(H)} \tag{7.27}$$

and therefore does not depend on the particular decomposition (7.26) of Φ.

Proof Just note that

$$\chi^{\widetilde{\Phi}}(\lambda, h) = \operatorname{tr}\left[\widetilde{\Phi}(\lambda, h)\right] = \lambda \operatorname{tr}[\Phi(h)] = \lambda \chi^{\Phi}(h)$$

and therefore, setting $G := G(C_n, H, \eta)$, we have (recall that $|G| = n|H|$)

$$\left\langle \chi^{\widetilde{\Phi}}, \chi^{\widetilde{\Psi}} \right\rangle_{L(G)} = \sum_{\lambda \in C_n} \sum_{h \in H} \lambda \overline{\lambda} \chi^{\Phi}(h) \overline{\chi^{\Psi}(h)} = n \langle \chi^{\Phi}, \chi^{\Psi} \rangle_{L(H)}. \tag{7.28}$$

(1) and (2) then follow from the classical results for ordinary characters; see [30, Proposition 10.2.17].

(3) The equality $r_\eta = |\widehat{G}(\xi)|$ follows from Remark 7.13. Also, Theorem 2.5.(4) (with $\sigma = \xi$ and $\theta = \Theta_j$) yields

$$d_j = d_{\Theta_j} = \ell_{\Theta_j}, \tag{7.29}$$

since $I_G(\xi) = G$ and $d_\xi = 1$. Then, (7.25) follows from Proposition 2.9.(1).

(4) This immediately follows from the bijection $\Phi \mapsto \Theta$ ($\Phi_i \mapsto \Theta_i$) which preserves irreducibility, (7.28), and from the classical results for ordinary characters: see, for instance, [30, Proposition 10.2.18]. ■

Let H_1 and H_2 be two finite groups, let $\eta_j \in \mathscr{C}^2(H_j, \mathbb{T})$, and let Φ_j be an η_j-representation of H_j, $j = 1, 2$. Define $\eta_1 \boxtimes \eta_2 \colon (H_1 \times H_2) \times (H_1 \times H_2) \to \mathbb{T}$ by setting

$$(\eta_1 \boxtimes \eta_2)[(h_1, h_2), (k_1, k_2)] := \eta_1(h_1, k_1)\eta_2(h_2, k_2)$$

and $\Phi_1 \boxtimes \Phi_2 \colon H_1 \times H_2 \to \mathscr{U}(V_1 \otimes V_2)$ by setting

$$(\Phi_1 \boxtimes \Phi_2)[h_1, h_2] := \Phi_1(h_1) \otimes \Phi_2(h_2)$$

for all $h_1, k_1 \in H_1$ and $h_2, k_2 \in H_2$.

Proposition 7.16 *With the above notation we have the following.*

(1) $\eta_1 \boxtimes \eta_2 \in \mathscr{C}^2(H_1 \times H_2, \mathbb{T})$.
(2) $\Phi_1 \boxtimes \Phi_2$ *is an* $(\eta_1 \boxtimes \eta_2)$*-representation of* $H_1 \times H_2$.
(3) *If* $H_1 = H_2 =: H$, *then denoting by* $\widetilde{H} := \{(h, h) : h \in H\} \cong H$ *the diagonal subgroup of* $H \times H$, *one has* $\eta_1\eta_2 := \mathrm{Res}^{H \times H}_{\widetilde{H}}(\eta_1 \boxtimes \eta_2) \in \mathscr{C}^2(H, \mathbb{T})$ *and*

$$\Phi_1 \otimes \Phi_2 := \mathrm{Res}^{H \times H}_{\widetilde{H}}[\Phi_1 \boxtimes \Phi_2]$$

is an $\eta_1\eta_2$*-representation of* H. ■

Algebraic characterizations of $r_\eta = |\widehat{H}^\eta|$ are in Corollary 7.38 below.

The cocycle $\eta_1 \boxtimes \eta_2$ is called the *outer product* of the cocycles η_1 and η_2, while the $(\eta_1 \boxtimes \eta_2)$-representation $\Phi_1 \boxtimes \Phi_2$ is called the *outer tensor product* of the η_1-representation Φ_1 and η_2-representation Φ_2. When $H_1 = H_2 =: H$, the cocycle $\eta_1\eta_2$ is just the ordinary product of the cocycles η_1 and η_2 in $\mathscr{C}^2(H, \mathbb{T})$, while the $\eta_1\eta_2$-representation $\Phi_1 \otimes \Phi_2$ is called the *internal tensor product* of the η_1-representation Φ_1 and η_2-representation Φ_2.

Let Φ be an η-representation of H. We define its *adjoint* as in (1.40), that is, by setting

$$[\Phi'(h)f](v) := f[\Phi(h^{-1})v] \tag{7.30}$$

for all $h \in H$, $v \in V$, and $f \in V'$. Then, arguing as in [30, Proposition 10.5.2], by using the Riesz map $\xi : V' \to V$ and (1.41), we have that

$$\Phi'(h) = \xi^{-1}\Phi(h)\xi,$$

for all $h \in H$, so that, keeping in mind antilinearity of ξ, we deduce that Φ' is an $\overline{\eta}$-representation of H. (Here $\overline{\eta}(h,k) := \overline{\eta(h,k)} = (\eta(h,k))^{-1}$ for all $h, k \in H$.)

It also follows that the inner tensor product $\Phi \otimes \Phi'$ is an ordinary representation of H.

Proposition 7.17 *Let (Ψ, W) and (Φ, U) be two η-representations of H and let*

$$W = W_1 \bigoplus W_2 \bigoplus \cdots \bigoplus W_n$$

$$U = U_1 \bigoplus U_2 \bigoplus \cdots \bigoplus U_m$$

be decompositions of W and U into H-invariant (not necessarily irreducible) subspaces. Then

$$\mathrm{Hom}_H(W, U) \cong \bigoplus_{i=1}^{n}\bigoplus_{j=1}^{m} \mathrm{Hom}_H(W_i, U_j).$$

Proof For $i = 1, 2, \ldots, n$, let $Q_i^W : W_i \to W$ denote the inclusion map and let $P_i^W : W \to W_i$ denote the projection along $W_1 \oplus \cdots \oplus W_{i-1} \oplus W_{i+1} \oplus \cdots \oplus W_n$ (that is, if $w \in W$, say $w = w_1 + w_2 + \cdots + w_n$, with $w_k \in W_k$, $k = 1, 2, \ldots, n$, then $P_i^W(w) := w_i$). Then $Q_i^W \in \mathrm{Hom}_H(W_i, W)$ and $P_i^W \in \mathrm{Hom}_H(W, W_i)$. Indeed, for $h \in H$ we have

$$Q_i^W(\Psi(h)w_i) = \Psi(h)w_i = \Psi(h)Q_i^W(w_i)$$

and $\Psi(h)w = \Psi(h)w_1 + \Psi(h)w_2 + \cdots + \Psi(h)w_n$, with $\Psi(h)w_k \in W_k$, $k = 1, 2, \ldots, n$, so that

$$P_i^W(\Psi(h)w) = \Psi(h)w_i = \Psi(h)P_i^W(w).$$

Analogously, one defines $Q_j^U \in \mathrm{Hom}_H(U_j, U)$ and $P_j^U \in \mathrm{Hom}_H(U, U_j)$.

For $T \in \mathrm{Hom}_H(W, U)$, $i = 1, 2, \ldots, n$, and $j = 1, 2, \ldots, m$, we then have

$$P_j^U T Q_i^W \in \mathrm{Hom}_H(W_i, U_j)$$

and the map

$$T \mapsto \bigoplus_{i,j=1}^{n,m} P_j^U T Q_i^W \tag{7.31}$$

yields a linear map from $\mathrm{Hom}_H(W, U)$ into $\bigoplus_{i,j=1}^{n,m} \mathrm{Hom}_H(W_i, U_j)$. Conversely, given $T_{j,i} \in \mathrm{Hom}_H(W_i, U_j)$ for $i = 1, 2, \ldots, n$ and $j = 1, 2, \ldots, m$, we define a linear map $T \colon W \to U$ by setting

$$T := \bigoplus_{i,j=1}^{n,m} Q_j^U T_{j,i} P_i^W$$

so that, if $w \in W$, say $w = w_1 + w_2 + \cdots + w_n$, with $w_i \in W_i$ for $i = 1, 2, \ldots, n$, then

$$Tw := \sum_{i,j=1}^{n,m} T_{j,i} w_i.$$

As, for $h \in H$ and $w \in W$,

$$\begin{aligned}
T\Psi(h)w &= \sum_{i,j=1}^{n,m} T_{j,i} P_i^W \Psi(h) w \\
(\text{since } P_i^W \in \mathrm{Hom}_H(W, W_i)) \quad &= \sum_{i,j=1}^{n,m} T_{j,i} \Psi(h) P_i^W w \\
(\text{since } T_{j,i} \in \mathrm{Hom}_H(W_i, U_j)) \quad &= \sum_{i,j=1}^{n,m} \Phi(h) T_{j,i} P_i^W w \\
&= \Phi(h) T w,
\end{aligned}$$

we deduce that $T \in \mathrm{Hom}_H(W, U)$. It is straightforward that the maps (7.31) and $(T_{j,i})_{i,j=1}^{n,m} \mapsto T$ are one inverse to the other, and this completes the proof. ■

Corollary 7.18

(a) *An η-representation Φ of H is irreducible if and only if* $\dim\mathrm{End}_H(\Phi) = 1$.
(b) *Two η-representations Φ and Θ have a common irreducible subrepresentation if an only if* $\dim\mathrm{Hom}_H(\Phi, \Theta) \geq 1$.

Proof

(a) The "only if" part follows from Schur's lemma. Conversely, denoting by V the representation space of Φ, a decomposition $V = V_1 \oplus V_2$ into nontrivial H-invariant subspaces leads to $\dim\mathrm{End}_H(V) \geq \dim\mathrm{End}_H(V_1) + \dim\mathrm{End}_H(V_1) \geq 2$.

(b) Use the decomposition in Proposition 7.17 and Schur's lemma; see [30, Sect. 10.6] for a similar, quite detailed analysis of the commutant of an ordinary representation.

■

The paper [11] (see also the recent monograph [12]) by Bekka and de la Harpe generalizes a number of known results about projective representations of finite groups to those of infinite countable groups.

7.3 The Dual of a Group Extension

We now state and prove a useful elementary fact on tensor products of representations of finite groups.

Lemma 7.19 *Let H be a finite group and let (σ, U), (θ, V), and (ρ, W) be three H-representations such that σ is irreducible. Suppose that $T \in \mathrm{Hom}(U \otimes V, U \otimes W)$ satisfies the condition*

$$T\,(\sigma(h) \otimes I_V) = (\sigma(h) \otimes I_W)\,T$$

for all $h \in H$. Then there exists a unique $L \in \mathrm{Hom}(V, W)$ such that $T = I_U \otimes L$.

Proof Express T as a sum of simple tensors, say $T = \sum_{j=1}^{m} S_j \otimes L_j$, where $S_j \in \mathrm{End}(U)$ and $L_j \in \mathrm{Hom}(V, W)$, for $i = 1, 2, \ldots, m$, with the L_js linearly independent. Then

$$\sum_{j=1}^{m} S_j\sigma(h) \otimes L_j = T\,(\sigma(h) \otimes I_V) = (\sigma(h) \otimes I_W)\,T = \sum_{j=1}^{m} \sigma(h)S_j \otimes L_j$$

and the linear independence of the L_js force $S_j\sigma(h) = \sigma(h)S_j$, $j = 1, 2, \ldots, m$. As σ is irreducible, by Schur's lemma we have $S_j = \alpha_j I_U$, with $\alpha_j \in \mathbb{C}$. Setting $L := \sum_{j=1}^{m} \alpha_j L_j$, the statement follows. ■

In what follows, we will use the notations in Sects. 2 and 7.1. In particular, we have fixed a normalized section $t\colon H := G/N \to G$. Note that if $\eta \in \mathscr{C}^2(H_G(\sigma), \mathbb{T})$ then also $\overline{\eta} \equiv \eta^{-1} \in \mathscr{C}^2(H_G(\sigma), \mathbb{T})$, where $\overline{\eta}(h, k) := \overline{\eta(h, k)} \equiv \eta(h, k)^{-1} = \eta^{-1}(h, k)$, for all $h, k \in H$.

Theorem 7.20 *Let $\sigma \in \widehat{N}$, let Ψ be as in (7.1), and let η be the Mackey obstruction of σ (cf. Definition 7.3). Let Φ be an irreducible $\overline{\eta}$-representation of $H_G(\sigma)$. Define $\Theta \colon I_G(\sigma) \to \mathscr{U}(V_\sigma)$ by setting*

$$\Theta(t(h)n) := \Psi(h)\sigma(n) \otimes \Phi(h), \tag{7.32}$$

for all $h \in H_G(\sigma)$ and $n \in N$. Then Θ is an irreducible $I_G(\sigma)$-representation acting on the tensor product $V \otimes U$, where V (resp. U) is the space of σ and Ψ (resp. of Φ). Moreover, the map

$$\begin{array}{ccc} \widehat{H_G(\sigma)}^{\overline{\eta}} & \longrightarrow & \widehat{I}_G(\sigma) \\ \Phi & \longmapsto & \Theta \end{array} \tag{7.33}$$

is a bijection.

Proof For $h, k \in H_G(\sigma)$ and $m, \ell \in N$ we have:

$$\begin{aligned}
\Theta(t(h)m \cdot t(k)\ell) &= \Theta\left(t(hk)\left[n(h,k)m^{k^\sharp}\ell\right]\right) && \text{(by (1.60))}\\
&= \Psi(hk)\,\sigma\left(n(h,k)m^{k^\sharp}\ell\right) \otimes \Phi(hk) && \text{(by (7.32))}\\
&= \Psi(h)\Psi(k)\,\sigma\left(m^{k^\sharp}\ell\right) \otimes \Phi(h)\Phi(k) && \text{(by (7.2) and (7.10))}\\
&= \Psi(h)\sigma(m)\Psi(k)\sigma(\ell) \otimes \Phi(h)\Phi(k) && \text{(by (1.57) and (7.1))}\\
&= [\Psi(h)\sigma(m) \otimes \Phi(h)]\,[\Psi(k)\sigma(\ell) \otimes \Phi(k)]\\
&= \Theta(t(h)m) \cdot \Theta(t(k)\ell).
\end{aligned}$$

Moreover, for $n \in N$,

$$\Theta(n) = \sigma(n) \otimes I_U \sim \underbrace{\sigma(n) \oplus \cdots \oplus \sigma(n)}_{\dim U-\text{times}}. \tag{7.34}$$

We now prove that Θ is irreducible. Let $F \in \mathrm{End}_{I_G(\sigma)}\Theta$. Then F commutes with $\sigma(n) \otimes I_U$ for each $n \in N$ and therefore, by Lemma 7.19, $F = I_V \otimes T$, with $T \in \mathrm{End}(U)$. It follows that, for all $h \in H_G(\sigma)$,

$$\Psi(h) \otimes [T\Phi(h)] = F\Theta(t(h)) = \Theta(t(h))F = \Psi(h) \otimes [\Phi(h)T],$$

which forces $T\Phi(h) = \Phi(h)T$. Since Φ is an irreducible η-representation, by Schur's lemma $T = \alpha I_U$ for some $\alpha \in \mathbb{C}$. Then $F = \alpha\,(I_V \otimes I_U)$ and, by using the converse to Schur's lemma, we conclude that Θ is irreducible.

We now show that the map $\Phi \mapsto \Theta$ is injective. Let Ξ be another irreducible $\overline{\eta}$-representation of $H_G(\sigma)$and denote by W its space. Define an irreducible $I_G(\sigma)$-representation Π by setting $\Pi(t(h)n) := \Psi(h)\sigma(n) \otimes \Xi(h)$, as in (7.32), that is,

$\Xi \mapsto \Pi$ in (7.33). Suppose that Θ and Π are equivalent, so that there exists a unitary isomorphism $P : V \otimes U \to V \otimes W$ such that:

$$P\,[\Psi(h)\sigma(n) \otimes \Phi(h)] = [\Psi(h)\sigma(n) \otimes \Xi(h)]\,P$$

for all $h \in H_G(\sigma)$ and $n \in N$. As in the first part of the proof, we have that $P\,[\sigma(n) \otimes I_U] = [\sigma(n) \otimes I_W]\,P$ forces $P = I_V \otimes T$, and T yields a unitary equivalence between Φ and Ξ.

We now prove that the map is surjective, that is, each $\Theta \in \widehat{I_G}(\sigma)$ is of the form (7.32). By virtue of Theorem 2.5.(3), we may assume that the representation space of Θ is of the form $V \otimes Y$ and that

$$\Theta(n) = \sigma(n) \otimes I_Y \sim \underbrace{\sigma(n) \oplus \cdots \oplus \sigma(n)}_{\dim Y-\text{times}} \tag{7.35}$$

for all $n \in N$ (compare with (7.34)). Therefore, for $h \in H_G(\sigma)$,

$$\begin{aligned}
\left[\Psi(h)^{-1} \otimes I_Y\right]\Theta(t(h))\,[\sigma(m) \otimes I_Y] &= \left[\Psi(h)^{-1} \otimes I_Y\right]\Theta(t(h))\Theta(m) && \text{(by (7.35))}\\
&= \left[\Psi(h)^{-1} \otimes I_Y\right]\Theta\left(t(h)mt(h)^{-1}\right)\Theta(t(h))\\
&= \left\{\left[\Psi(h)^{-1}\sigma(hmh^{-1})\right] \otimes I_Y\right\}\Theta(t(h)) && \text{(by (7.35))}\\
&= \left\{\left[\sigma(m)\Psi(h)^{-1}\right] \otimes I_Y\right\}\Theta(t(h)) && \text{(by (7.1))}\\
&= [\sigma(m) \otimes I_Y]\left[\Psi(h)^{-1} \otimes I_Y\right]\Theta(t(h)).
\end{aligned}$$

In other words, each operator $\left[\Psi(h)^{-1} \otimes I_Y\right]\Theta(t(h))$ commutes with $[\sigma(m) \otimes I_Y]$. From Lemma 7.19 it follows that there exists a unitary operator $\Xi(h)$ on Y such that

$$\left[\Psi(h)^{-1} \otimes I_Y\right]\Theta(t(h)) = I_V \otimes \Xi(h),$$

that is

$$\Theta(t(h)) = \Psi(h) \otimes \Xi(h). \tag{7.36}$$

If $h, k \in H_G(\sigma)$ then

$$\begin{aligned}
\Psi(hk) \otimes \Xi(hk) &= \Theta(t(hk)) && \text{by (7.36)}\\
&= \Theta(t(h))\Theta(t(k))\left[\sigma(n(h,k))^* \otimes I_Y\right] && \text{(by (7.35) and (1.56))}\\
&= \Psi(h)\Psi(k)\sigma(n(h,k))^* \otimes \Xi(h)\Xi(k) && \text{by (7.36)}\\
&= \Psi(hk) \otimes \Xi(h)\Xi(k)\eta(h,k)^{-1} && \text{by (7.2)}
\end{aligned}$$

which yields $\Xi(hk) = \Xi(h)\Xi(k)\eta(h,k)^{-1}$, that is, Ξ is an $\overline{\eta}$-representation of $H_G(\sigma)$. Finally, from (7.35) and (7.36) we get:

$$\Theta(t(h)n) = \Theta(t(h))[\sigma(n) \otimes I_Y] = \Psi(h)\sigma(n) \otimes \Xi(h),$$

that is, Θ may be written in the form (7.32) with $\Phi = \Xi \in \widehat{H_G(\sigma)}^{\overline{\eta}}$. This shows that the map (7.33) is surjective. ■

Theorem 7.21 *The map*

$$\Phi \longmapsto \mathrm{Ind}_{I_G(\sigma)}^{G}\Theta, \tag{7.37}$$

where Θ is as in (7.32), is a bijection between $\widehat{H_G(\sigma)}^{\overline{\eta}}$ and $\widehat{G}(\sigma)$ (cf. Definition 2.1).

Proof Combine Theorem 7.20 with Clifford correspondence (cf. Theorem 2.12). ■

Corollary 7.22 *The inertia index of $\mathrm{Ind}_{I_G(\sigma)}^{G}\Theta$ (= inertia index of Θ, cf. Theorem 2.12) is equal to the dimension of Φ.*

Proof From (7.32) and $\Phi(1_H) = I_U$ it follows that $\mathrm{Res}_N^{I_G(\sigma)}\Theta = \sigma \otimes I_U$; cf. Definition 2.7. ■

Recall that Theorem 2.26 may be applied only when the Mackey obstruction is trivial; cf. Remark 7.5. We are now in a position to give a more general result, which may be applied to all group extensions. It is the finite group case of a general theory of Mackey; see [48, 92]. In particular, [48] contains a discussion of the finite group case that was quite inspiring for our treatment.

Theorem 7.23 *Let Σ be as in Theorem 2.26. Then*

$$\widehat{G} = \left\{ \mathrm{Ind}_{I_G(\sigma)}^{G}\Theta : \sigma \in \Sigma, \Theta \text{ as in (7.32)}, \Phi \in \widehat{H_G(\sigma)}^{\overline{\eta}} \right\},$$

that is, the right hand side is a list of all irreducible G-representations and, for different values of σ and Φ, we get pairwise inequivalent representations.

Proof For a fixed σ they are pairwise inequivalent by Theorem 7.21. If Θ_1, Θ_2 correspond to $\sigma_1 \not\approx \sigma_2$ then they are inequivalent by Theorem 2.5(3).

Finally, if τ is an irreducible representation of G, from Theorem 2.5 we get that $\tau \in \widehat{G}(\sigma)$ for a $\sigma \in \Sigma$ and therefore, again by Clifford correspondence, τ is of the form $\mathrm{Ind}_{I_G(\sigma)}^{G}\Theta$ with $\Theta \in \widehat{I_G}(\sigma)$. But then Θ is the image of some $\Phi \in \widehat{H_G(\sigma)}^{\overline{\eta}}$ in (7.33). ■

7.4 Central Extensions and the Finite Heisenberg Group

We now suppose that $N = B$ is a central subgroup of G and we set $H := G/B$. We fix a normalized section $t : H \to G$ and denote by $b(h, k)$ the corresponding factor set as in (1.56) with N replaced by B. We then show how Theorems 7.20 and 7.23 simplify in this setting.

Proposition 7.24 *For every $\xi \in \widehat{B}$ we have $I_G(\xi) = G$ and its Mackey obstruction is just*

$$\eta(h, k) = \xi(b(h, k)). \tag{7.38}$$

Moreover, the map

$$\begin{array}{ccc} \widehat{H}^{\overline{\eta}} & \longrightarrow & \widehat{G}(\xi) \\ \Phi & \longmapsto & \Theta, \end{array} \tag{7.39}$$

where Θ is defined by setting

$$\Theta(t(h)b) := \xi(b)\Phi(h), \tag{7.40}$$

for all $h \in H(\xi), b \in B$, is a bijection. Finally,

$$\widehat{G} = \left\{ \Theta : \xi \in \widehat{B}, \Theta \text{ as in (7.40)}, \Phi \in \widehat{H_G(\xi)}^{\overline{\eta}} \right\}.$$

Proof From Remark 2.16 it follows that, for $\xi \in \widehat{B}$, we have $I_G(\xi) = G$, so that $H_G(\xi) = H$ and $\Sigma \equiv \widehat{B}$. In particular, in (7.37) induction disappears. Moreover, $\Psi(h) = 1$ for all $h \in H$, because in the right hand side of (7.1) we have: $\xi\left(t(h)^{-1}bt(h)\right) = \xi(b)$ for all $h \in H$ and $b \in B$. Finally, (7.2) yields (7.38) immediately. Then, taking all these simplifications into account, all the other statements follow from Theorems 7.20 and 7.23. ∎

It is worthwhile to give an elementary proof of Proposition 7.24. It does not involve induced representations, nor all the machinery of Clifford theory and Mackey obstructions.

Elementary and Direct Proof of Proposition 7.24 From (1.59) (that in this case reduces to the form (6.1)) it follows immediately that (7.38) defines a 2-cocycle. Suppose that Φ is a unitary $\overline{\eta}$-representation of H on a space V. Then Θ, defined as in (7.40), is a G-representation (this is a particular and simpler case of the corresponding calculation in the proof of Theorem 7.20). Indeed, for $h, k \in H$ and $a, b \in B$, taking into account that B is *central* in G, we have:

$$\begin{aligned} \Theta(t(h)a \cdot t(k)b) &= \Theta\left(t(hk)b(h, k)ab\right) && \text{(by (1.60))} \\ &= \Phi(hk)\xi\left(b(h, k)ab\right) && \text{(by (7.40))} \end{aligned}$$

$$= \Phi(hk)\eta(h,k)\xi(a)\xi(b) \qquad \text{(by (7.38))}$$
$$= \Phi(h)\Phi(k)\xi(a)\xi(b) \qquad \text{(by (7.10))}$$
$$= \Theta(t(h)m) \cdot \Theta(t(k)\ell) \qquad \text{(by (7.40))}.$$

Clearly, if Φ is irreducible then so is Θ. Moreover, if Θ_j is defined by means of Φ_j, $j = 1, 2$, then $\Theta_1 \sim \Theta_2$ if and only if $\Phi_1 \sim \Phi_2$: indeed, $\mathrm{Hom}_H(\Phi_1, \Phi_2) = \mathrm{Hom}_G(\Theta_1, \Theta_2)$ because for $T \in \mathrm{End}(V)$, $h \in H$ and $a \in B$, we have $T\Theta_1(t(h)b) = \xi(b)T\Phi_1(h)$ and $\Theta_2(t(h)b)T = \xi(b)\Phi_2(h)T$.

We have just proved that the map (7.39) is injective. Moreover, from (7.40) it also follows that

$$\mathrm{Res}^G_B\Theta \sim \underbrace{\xi \oplus \xi \oplus \cdots \oplus \xi}_{|B|-\text{times}}, \tag{7.41}$$

and therefore different characters of B yield inequivalent G-representations.

We now prove that *every* $\Theta \in \widehat{G}$ is of the form (7.40). In the general case (namely, Theorems 7.20 and 7.23) the proof of this fact depends on Clifford theory, but in the present setting we may use a shortcut.

Let $\Theta \in \widehat{G}$ with representation space V. Since B is central, for every fixed $b \in B$ we have $\Theta(b)\Theta(g) = \Theta(g)\Theta(b)$ for all $g \in G$, that is $\Theta(b) \in \mathrm{End}_G(\Theta)$. By Schur's lemma, there exists $\xi : B \to \mathbb{C}$ such that $\Theta(b) = \xi(b)I_V$. Therefore, $\xi \in \widehat{B}$ and the decomposition (7.41) follows. Setting $\Phi(h) := \Theta(t(h))$ for all $h \in H$, then

$$\Phi(hk) = \Theta(t(hk)) = \Theta\left(t(h)t(k)b(h,k)^{-1}\right) = \Phi(h)\Phi(k)\overline{\eta}(h,k),$$

so that Φ is an irreducible $\overline{\eta}$-representation and Θ is of the form (7.40). ■

Remark 7.25 Let G be a group and denote by B its center. If G does not factor as a direct product $G = B \times H$, then G is *not* even a semidirect product $B \rtimes H$. As a consequence, for the proof of Proposition 7.24 we cannot apply Theorem 2.54. See also Proposition 6.8.

Remark 7.26 It is immediate to check that, in the hypothesis of Proposition 7.24, the space of $\mathrm{Ind}^G_B\xi$ is a subalgebra of $L(G)$. See also Remark 7.31.

In the following, we illustrate Proposition 7.24 by developing a third approach to the representation theory of the finite Heisenberg group $G = H_q$, where $q = p^k$ (cf. Sect. 6.9). Let $B = Z(G) = \{(m, 0, 0) : m \in \mathbb{F}_q\} \cong (\mathbb{F}_q, +)$ and let $H = (\mathbb{F}_q, +) \oplus (\mathbb{F}_q, +) = \{(k, \ell) : k, \ell \in \mathbb{F}_q\} \cong G/B$. We choose as a section $t : H \longrightarrow G$ the function $t(k, \ell) = (0, k, \ell)$, so that

$$t(k_1,\ell_1)t(k_2,\ell_2) = (k_1\ell_2, k_1+k_2, \ell_1+\ell_2) = b\left((k_1,\ell_1),(k_2,\ell_2)\right)t\left((k_1,\ell_1)\cdot(k_2,\ell_2)\right)$$

where (cf. (1.56))

$$b\left((k_1,\ell_1),(k_2,\ell_2)\right) = (k_1\ell_2, 0, 0) \tag{7.42}$$

for all $k_1, k_2, \ell_1, \ell_2 \in \mathbb{F}_q$.

Now $\widehat{B} \cong \widehat{(\mathbb{F}_q,+)}$ consists of the one dimensional characters $\chi_\rho := \chi_{\rho,0,0}|_B$ with $\rho \in \mathbb{F}_q$ (cf. (6.74); that is, $\chi_\rho(m,0,0) = \exp((2\pi i/p)\mathrm{Tr}(\rho m))$ for all $m \in \mathbb{F}_q$). Moreover, (7.38) becomes:

$$\eta_\rho((k_1,\ell_1),(k_2,\ell_2)) = \chi_\rho(k_1\ell_2,0,0) = \exp((2\pi i/p)\mathrm{Tr}(\rho k_1\ell_2)), \tag{7.43}$$

while (7.40) yields $\Theta_\rho \in \widehat{G}$ given by

$$\Theta_\rho(m,k,\ell) = \chi_\rho(m,0,0)\Phi_\rho(0,k,\ell), \tag{7.44}$$

for all $k, \ell \in \mathbb{F}_q$, where $\Phi_\rho \in \widehat{H}^{\overline{\eta}_\rho}$.

If $\rho = 0$, then $\chi_0 \equiv 1$, $\eta_\rho = \eta_0 \equiv 1$ and $\widehat{H}^{\overline{\eta}_0} \equiv \widehat{H} = \widehat{(\mathbb{F}_q,+)\oplus(\mathbb{F}_q,+)} = \{\chi_{\alpha,\beta} : \alpha,\beta \in \mathbb{F}_q\}$ where

$$\chi_{\alpha,\beta}(k,\ell) = \exp((2\pi i/p)\mathrm{Tr}(\alpha k + \beta\ell))$$

for all $(k,\ell) \in H$. The G-representations resulting from (7.44) are exactly the one-dimensional representations $\chi_{0,\alpha,\beta}$, with $\alpha, \beta \in \mathbb{F}_q$.

Suppose now that $\rho \neq 0$. For $\ell, k \in \mathbb{F}_q$, define the $q \times q$ diagonal matrix

$$D_\rho(\ell) := \begin{pmatrix} \chi_\rho(0) & 0 & 0 & 0\cdots & 0 \\ 0 & \chi_\rho(\ell) & 0 & 0\cdots & 0 \\ 0 & 0 & \chi_\rho(2\ell) & 0\cdots & 0 \\ \vdots & \vdots & \vdots & \vdots\ddots & \vdots \\ 0 & 0 & 0 & 0\cdots & \chi_\rho((q-1)\ell) \end{pmatrix}$$

$$= \begin{pmatrix} 1 & 0 & 0 & 0\cdots & 0 \\ 0 & \exp(\frac{2\pi i}{p}\mathrm{Tr}(\rho\ell)) & 0 & 0\cdots & 0 \\ 0 & 0 & \exp(\frac{2\pi i}{p}\mathrm{Tr}(2\rho\ell)) & 0\cdots & 0 \\ \vdots & \vdots & \vdots & \vdots\ddots & \vdots \\ 0 & 0 & 0 & 0\cdots & \exp(\frac{2\pi i}{p}\mathrm{Tr}((q-1)\rho\ell)) \end{pmatrix}$$

and the $q \times q$ permutation matrix $W(k)$ where

$$W(k)_{i,j} := \delta_{i,j-k}$$

for all $i, j \in \mathbb{F}_q$. Observe that $D_\rho(\ell_1)D_\rho(\ell_2) = D_\rho(\ell_1+\ell_2)$ and $W(k_1)W(k_2) = W(k_1+k_2)$ for all $k_1, k_2, \ell_1, \ell_2 \in \mathbb{F}_q$, so that, in particular, $W(k)^{-1} = W(-k)$ for all $k \in \mathbb{F}_q$. We claim that

$$W(k)D_\rho(\ell)W(-k) = \chi_\rho(k\ell)D_\rho(\ell) \tag{7.45}$$

for all $k, \ell \in \mathbb{F}_q$. Indeed, for $i, j, k, \ell \in \mathbb{F}_q$ we have

$$\begin{aligned}
\big(W(k)D_\rho(\ell)W(-k)\big)_{i,j} &= \sum_{s,t\in\mathbb{F}_q} W(k)_{i,s}D_\rho(\ell)_{s,t}W(-k)_{t,j} \\
(D_\rho(\ell) \text{ is diagonal}) &= \sum_{s\in\mathbb{F}_q} W(k)_{i,s}D_\rho(\ell)_{s,s}W(-k)_{s,j} \\
&= \sum_{s\in\mathbb{F}_q} \delta_{i,s-k}\chi_\rho(s\ell)\delta_{s,j+k} \\
&= \begin{cases} \chi_\rho((i+k)\ell) = \chi_\rho(k\ell)D_\rho(\ell)_{i,i} & \text{if } i = j \\ 0 & \text{otherwise,} \end{cases}
\end{aligned}$$

and the claim follows. Setting

$$\Phi_\rho(k, \ell) := D_\rho(\ell)W(k)$$

for all $(k, \ell) \in H$ defines an $\overline{\eta}_\rho$-representation of H. Indeed,

$$\begin{aligned}
\Phi_\rho((k_1, \ell_1)(k_2, \ell_2)) &= \Phi_\rho((k_1 + k_2, \ell_1 + \ell_2)) \\
&= D_\rho(\ell_1 + \ell_2)W(k_1 + k_2) \\
&= D_\rho(\ell_1)D_\rho(\ell_2)W(k_1)W(k_2) \\
&= D_\rho(\ell_1)W(k_1)W(-k_1)D_\rho(\ell_2)W(k_1)W(k_2) \\
(\text{by } (7.45)) \quad &= \Phi_\rho(k_1, \ell_1)\chi_\rho(-k_1\ell_2)\Phi_\rho(k_2, \ell_2) \\
(\text{by } (7.43)) \quad &= \overline{\eta}_\rho((k_1, \ell_1), (k_2, \ell_2))\Phi_\rho(k_1, \ell_1)\Phi_\rho(k_2, \ell_2)
\end{aligned}$$

for all $k_1, k_2, \ell_1, \ell_2 \in \mathbb{F}_q$. We now compute the character of Φ_ρ:

$$\chi^{\Phi_\rho}(k, \ell) = \mathrm{tr}\big[D_\rho(\ell)W(k)\big] = \sum_{s,t\in\mathbb{F}_q} D_\rho(\ell)_{s,t}W(k)_{t,s} = \sum_{s\in\mathbb{F}_q} \chi_\rho(s\ell)\delta_{s,s-k}$$

$$= \begin{cases} 0 & \text{if } k \neq 0 \text{ or } \ell \neq 0, \\ q & \text{if } k = \ell = 0, \end{cases}$$

where, for $k = 0$ and $\ell \neq 0$ we used the orthogonality relations for characters. It follows that

$$\|\chi^{\Phi_\rho}\|^2_{L(H)} = \sum_{k,\ell\in\mathbb{F}_q} |\chi^{\Phi_\rho}(k, \ell)|^2 = q^2 = |H|,$$

so that, by virtue of Corollary 7.15.(7.15), Φ_ρ is irreducible. Then (7.44) yields the representations $U_\rho \in \widehat{G}$, $\rho \in \mathbb{F}_q^*$, defined by

$$U_\rho(m, k, \ell) := \chi_\rho(m)\Phi_\rho(k, \ell) = \chi_\rho(m)D_\rho(\ell)W(k)$$

for all $(m, k, \ell) \in G$.

Finally, recalling the one-dimensional representations described in the first part of this section, we get

$$\sum_{\alpha,\beta\in\mathbb{F}_q} (\dim\chi_{0,\alpha,\beta})^2 + \sum_{\rho\in\mathbb{F}_q^*} (\dim U_\rho)^2 = q^2 + (q-1)q^2 = q^3 = |G|,$$

so that, invoking (1.20), we deduce that the description of $\widehat{G}$ is complete. From Proposition 7.24 we also deduce that $\widehat{H}^{\eta_\rho} = \{\Phi_\rho\}$ for all $\rho \in \mathbb{F}_q^*$.

7.5 Analysis of the Commutant

In this section we present a more detailed analysis of the algebra $\mathrm{End}_G(\mathrm{Ind}_N^G\sigma)$, for $\sigma \in \widehat{N}$. From Theorem 2.5.(2) we know that $\dim\mathrm{End}_G(\mathrm{Ind}_N^G\sigma) = |H_G(\sigma)|$, where $H_G(\sigma) = I_G(\sigma)/N$. We now show that, indeed, $\mathrm{End}_G(\mathrm{Ind}_N^G\sigma)$ is isomorphic to the algebra $L(H_G(\sigma))$ endowed with a modified convolution, defined by means of a suitable cocycle.

Let H be a finite group and suppose that $\eta \in \mathscr{C}^2(H, \mathbb{T})$ is a unitary, equalized (cf. Proposition 6.15.(2)) cocycle. For $f_1, f_2 \in L(H)$ we define the *η-cocycle convolution* $f_1 *_\eta f_2 \in L(H)$ by setting

$$(f_1 *_\eta f_2)(k) := \sum_{h\in H} f_1(h^{-1}k)f_2(h)\eta(k^{-1}, h), \tag{7.46}$$

for all $k \in H$. We also define an involution $f \mapsto f^*$ on $L(H)$ by setting $f^*(h) = \overline{f(h^{-1})}$ for all $h \in H$.

Remark 7.27 Note that we choose (7.46) as a definition of η-cocycle convolution in order to get a $*$-isomorphism in Theorem 7.37 below.

Proposition 7.28 *The vector space $L(H)$ equipped with the η-*cocycle convolution *and the involution defined above is an involutive, unital, associative algebra, which will be denoted by $L_\eta(H)$.*

Proof We first prove that the η-cocycle convolution is associative. We need two equalities for the equalized cocycle η. Let $k, r, s \in H$. From (6.22) and (6.23), with $h = s^{-1}$, $k = r$, we immediately get

$$\eta(s, s^{-1}r) = \eta(r^{-1}, s). \tag{7.47}$$

We then have

$$\begin{aligned} \eta(k^{-1}s, s^{-1}r)\eta(k^{-1}, s) &= \eta(k^{-1}, r)\eta(s, s^{-1}r) \\ &= \eta(k^{-1}, r)\eta(r^{-1}, s), \end{aligned} \tag{7.48}$$

where the first equality may be obtained by setting $h_1 = k^{-1}$, $h_2 = s$ and $h_3 = s^{-1}r$ in (6.1), and the second equality follows from (7.47).

If $f_1, f_2, f_3 \in L(H)$ then for all $k \in H$ we have:

$$\begin{aligned} \left[(f_1 *_\eta f_2) *_\eta f_3\right](k) &= \sum_{s \in H} (f_1 *_\eta f_2)(s^{-1}k) f_3(s)\eta(k^{-1}, s) \\ &= \sum_{s\in H}\sum_{h \in H} f_1(h^{-1}s^{-1}k) f_2(h) f_3(s)\eta(k^{-1}s, h)\eta(k^{-1}, s) \\ (r := sh) \qquad &= \sum_{s\in H}\sum_{r \in H} f_1(r^{-1}k) f_2(s^{-1}r) f_3(s)\eta(k^{-1}s, s^{-1}r)\eta(k^{-1}, s) \\ (\text{by } (7.48)) \qquad &= \sum_{s\in H}\sum_{r \in H} f_1(r^{-1}k) f_2(s^{-1}r) f_3(s)\eta(k^{-1}, r)\eta(r^{-1}, s) \\ &= \sum_{r \in H} f_1(r^{-1}k)\eta(k^{-1}, r)\left[\sum_{s \in H} f_2(s^{-1}r) f_3(s)\eta(r^{-1}, s)\right] \\ &= \sum_{r \in H} f_1(r^{-1}k)\eta(k^{-1}, r)\left(f_2 *_\eta f_3\right)(r) \\ &= \left[f_1 *_\eta \left(f_2 *_\eta f_3\right)\right](k). \end{aligned}$$

Then we prove that $L_\eta(H)$ is involutive:

$$\begin{aligned}
(f_1^* *_\eta f_2^*)(k) &= \sum_{h\in H} f_1^*(h^{-1}k) f_2^*(h)\eta(k^{-1},h)\\
&= \sum_{h\in H} \overline{f_1}(k^{-1}h)\overline{f_2}(h^{-1})\eta(k^{-1},h)\\
\text{(by } s=k^{-1}h \text{ and (6.22))} \qquad &= \sum_{s\in H} \overline{f_1}(s)\overline{f_2}(s^{-1}k^{-1})\overline{\eta}(k,s)\\
&= (f_2 *_\eta f_1)^*(k).
\end{aligned}$$

The identity is δ_{1_H}. ∎

This algebra is a variation of the twisted group algebra in [16, 67, 128, 129].

Proposition 7.29 *Let $\eta, \alpha \in \mathscr{C}^2(H, \mathbb{T})$ and suppose that they are equalized and cohomologous. Then the algebras $L_\eta(H)$ and $L_\alpha(H)$ are $*$-isomorphic.*

Proof By the hypothesis, that there exists $\rho\colon H \to \mathbb{T}$ such that

$$\alpha(k,h) = \eta(k,h)\rho(kh)\rho(k)^{-1}\rho(h)^{-1}$$

for all $k, h \in H$, and $\rho(1_H) = 1$. Since η and α are equalized, we have

$$1 = \alpha(k,k^{-1}) = \rho(k)^{-1}\rho(k^{-1})^{-1}$$

so that

$$\rho(k^{-1}) = \rho(k)^{-1} \tag{7.49}$$

for all $k \in H$.

Setting

$$[Tf](h) := \rho(h^{-1})f(h),$$

for all $f \in L(H)$ and $h \in H$ defines $T \in \mathrm{End}(L(H))$, clearly a bijection, which also yields the seeked $*$-algebra isomorphism. Indeed, if $f_1, f_2 \in L(H)$ and $k \in H$, we have:

$$\begin{aligned}
\big([Tf_1] *_\eta [Tf_2]\big)(k) &= \sum_{h\in H} f_1(h^{-1}k)\rho(k^{-1}h) f_2(h)\rho(h^{-1})\eta(k^{-1},h)\\
\text{(by (7.49))} \qquad &= \sum_{h\in H} f_1(h^{-1}k) f_2(h)\rho(k^{-1}h)\rho(h)^{-1}\eta(k^{-1},h)\\
&= \sum_{h\in H} f_1(h^{-1}k) f_2(h)\alpha(k^{-1},h)\rho(k)^{-1}
\end{aligned}$$

$$= [f_1 *_\alpha f_2](k)\rho(k)^{-1}$$
$$= (T[f_1 *_\alpha f_2])(k).$$

Moreover, again from (7.49) we deduce that, for $h \in H$,

$$[Tf]^*(h) = \overline{[Tf]}(h^{-1}) = \overline{f}(h^{-1})\overline{\rho}(h) = f^*(h)\rho(h^{-1}) = [Tf^*](h).$$

In conclusion, $[Tf_1] *_\eta [Tf_2] = T[f_1 *_\alpha f_2]$ and $[Tf]^* = Tf^*$. ■

We now prove the analogue of Proposition 7.12.

Proposition 7.30 *Let $\eta \in \mathscr{C}^2(H, \mathbb{T})$ be an equalized cocycle and suppose that it satisfies the condition in Theorem 6.19, that is, $\eta^n = 1$, where $n = |H|$ (cf. Proposition 7.29). Then the map*

$$\varepsilon \colon L(H) \to L(G(C_n, H, \eta))$$

defined by setting

$$[\varepsilon(f)](\lambda, h) := \frac{1}{n}\overline{\lambda} f(h), \tag{7.50}$$

for all $f \in L(H)$ and $(\lambda, h) \in C_n \times H$, is a $*$-antiisomorphism between the algebra $L_\eta(H)$ and the subalgebra

$$\{F \in L(G(C_n, H, \eta)) : F(\lambda, h) = \lambda F(1, h) \text{ for all } (\lambda, h) \in C_n \times H\} \tag{7.51}$$

of the group algebra $L(G(C_n, H, \eta))$.

Proof It is obvious that ε is a bijection. Moreover, for all $f_1, f_2 \in L(H)$ and $(\alpha, k) \in C_n \times H$, we have:

$$\begin{aligned}
[\varepsilon(f_2) * \varepsilon(f_1)](\alpha, k) &= \sum_{(\beta,h)\in C_n \times H} [\varepsilon(f_2)](\beta, h) \cdot [\varepsilon(f_1)]\left((\beta, h)^{-1} * (\alpha, k)\right) \\
\text{(by (6.8)) and (6.9))} \quad &= \sum_{(\beta,h)\in C_n \times H} [\varepsilon(f_2)](\beta, h) \cdot [\varepsilon(f_1)]\left((\beta^{-1}\alpha\, \eta(h^{-1}, k)^{-1}, h^{-1}k)\right) \\
\text{(by (6.23))} \quad &= \sum_{(\beta,h)\in C_n \times H} [\varepsilon(f_2)](\beta, h) \cdot [\varepsilon(f_1)]\left(\beta^{-1}\alpha\, \eta(k^{-1}, h), h^{-1}k\right) \\
\text{(by (7.50))} \quad &= \frac{\overline{\alpha}}{n} \sum_{h \in H} f_2(h) f_1(h^{-1}k)\eta(k^{-1}, h) \\
\text{(by (7.46))} \quad &= \frac{\overline{\alpha}}{n}(f_1 *_\eta f_2)(k) \\
\text{(by (7.50))} \quad &= [\varepsilon(f_1 *_\eta f_2)](\alpha, k).
\end{aligned}$$

Finally, it is also easy to check that $\varepsilon(f^*) = \varepsilon(f)^*$ for all $f \in L(H)$. ■

Remark 7.31 Taking $B = C_n$ and $\xi(\lambda) = \lambda$ for all $\lambda \in C_n \subset \mathbb{T}$ as in Remark 7.13, then by Proposition 7.24, $\eta \equiv b$. Moreover, it is easy to check that the algebra (7.51) coincides with the representation space of $\mathrm{Ind}_{C_n}^G \xi$, so that the Proposition 7.30 may be expressed by saying that

$$L_\eta(H) \text{ is } * \text{-antiisomorphic to } \mathrm{Ind}_{C_n}^G \xi; \tag{7.52}$$

see also Remark 7.26.

We now characterize the center of the algebra $L_\eta(H)$. We follow [67, Exercises 11.4, 11.5, and 11.8], but with a slightly more direct approach than the one in the hints of Isaacs' book. We start with the following:

Definition 7.32 Let $\eta \in \mathscr{C}^2(H, \mathbb{T})$. One says that $h \in H$ is *η-special* provided that $\eta(h, k) = \eta(k, h)$ for all $k \in C_H(h)$, where $C_H(h) := \{s \in H : sh = hs\}$ is the centralizer of h in H.

For $s \in H$ denote by $\mathscr{C}(s) := \{tst^{-1} : t \in H\}$ its conjugacy class in H. By considering the conjugacy action of H on itself, we immediately get the classical equation: $|\mathscr{C}(s)| = [H : C_H(s)]$. Moreover, it is obvious that $C_H(tst^{-1}) = tC_H(s)t^{-1}$ for all $s, t \in H$.

Lemma 7.33 *Suppose that $h \in H$ is η-special. Then all elements in $\mathscr{C}(h)$ are η-special.*

Proof Suppose that $\eta(h, k) = \eta(k, h)$ for all $k \in C_H(h)$. We prove that $\eta(tht^{-1}, tkt^{-1}) = \eta(tkt^{-1}, tht^{-1})$, for all $t \in H$, by means of a repeated application of the cocycle identities (6.1) and (6.2), and of the commutation relations $hk = kh$ and $\eta(h, k) = \eta(k, h)$; we use the notation $\eta(h_1 \cdot h_2, h_3)$ or $\eta(h_1, h_2 \cdot h_3)$ to emphasize the factors h_1, h_2, h_3 when applying (6.2), always in one of the following forms:

$$\eta(h_1 \cdot h_2, h_3) = \eta(h_1, h_2h_3)\eta(h_1, h_2)^{-1}\eta(h_2, h_3) = \eta(h_1, h_2)^{-1}\eta(h_2, h_3)\eta(h_1, h_2h_3) \tag{7.53}$$

or

$$\eta(h_1, h_2 \cdot h_3) = \eta(h_1h_2, h_3)\eta(h_2, h_3)^{-1}\eta(h_1, h_2) = \eta(h_2, h_3)^{-1}\eta(h_1, h_2)\eta(h_1h_2, h_3). \tag{7.54}$$

Let $k \in \mathscr{C}(h)$ and $t \in H$. We have:

$$\begin{aligned}\eta(t \cdot ht^{-1}, tkt^{-1}) &= \eta(t, hkt^{-1})\eta(t, ht^{-1})^{-1}\eta(ht^{-1}, tkt^{-1})\\&= \eta(t, kt^{-1} \cdot tht^{-1})\eta(t, ht^{-1})^{-1}\eta(ht^{-1}, tkt^{-1})\\&= \eta(tkt^{-1}, tht^{-1})\eta(kt^{-1}, t \cdot ht^{-1})^{-1}\eta(t, kt^{-1})\eta(t, ht^{-1})^{-1}\eta(ht^{-1}, t \cdot kt^{-1})\\&= \eta(tkt^{-1}, tht^{-1})\eta(k, h \cdot t^{-1})^{-1}\eta(k \cdot t^{-1}, t)^{-1}\eta(h, k \cdot t^{-1})\eta(h \cdot t^{-1}, t)\\&= \eta(tkt^{-1}, tht^{-1}).\end{aligned}$$

■

By virtue of Lemma 7.33, we will say that a conjugacy class $\mathscr{C}$ of H is η-special if one, equivalently all, of its elements are η-special.

Lemma 7.34 *A conjugacy class $\mathscr{C}$ of H is η-special if and only if there exists a function $f : \mathscr{C} \to \mathbb{T}$ such that*

$$f(h)\eta(h, s^{-1}) = f(shs^{-1})\eta(s^{-1}, shs^{-1}), \tag{7.55}$$

for all $h \in \mathscr{C}$ and $s \in H$. Moreover, if such an f exists it is unique up to a multiplicative constant in $\mathbb{T}$.

Proof If such an f exists then, for $s \in C_H(h)$, (7.55) becomes $\eta(h, s) = \eta(s, h)$, so that h is η-special, and so is $\mathscr{C}$ by Lemma 7.33. Conversely, suppose that $\mathscr{C}$ is η-special, and let $h \in \mathscr{C}$, $k \in C_H(h)$, and $s \in H$. Then, using the cocycle identity (6.2) as in the proof of Lemma 7.33 as well as the fact that shs^{-1} is η-special and $sks^{-1} \in C_H(shs^{-1})$, we get:

$$\begin{aligned}\eta(ks^{-1}, shs^{-1}) &= \eta(s^{-1} \cdot sks^{-1}, shs^{-1})\\&= \eta(s^{-1}, sks^{-1})^{-1}\eta(s^{-1}, skhs^{-1})\eta(sks^{-1}, shs^{-1})\\&= \eta(s^{-1}, sks^{-1})^{-1}\eta(s^{-1}, shs^{-1} \cdot sks^{-1})\eta(shs^{-1}, sks^{-1})\\&= \eta(s^{-1}, sks^{-1})^{-1}\eta(h \cdot s^{-1}, sks^{-1})\eta(s^{-1}, shs^{-1})\\&= \eta(h, s^{-1})^{-1}\eta(h, ks^{-1})\eta(s^{-1}, shs^{-1}),\end{aligned}$$

so that

$$\frac{\eta(h, ks^{-1})}{\eta(ks^{-1}, shs^{-1})} = \frac{\eta(h, s^{-1})}{\eta(s^{-1}, shs^{-1})}. \tag{7.56}$$

We then define $f: \mathscr{C} \to \mathbb{T}$ by setting

$$f(shs^{-1}) := f_0\frac{\eta(h, s^{-1})}{\eta(s^{-1}, shs^{-1})}, \tag{7.57}$$

for all $s \in H$, where $f_0 = f(h) \in \mathbb{T}$ is arbitrary. This function is well defined: if $shs^{-1} = tht^{-1}$, for $s, t \in H$, then $t = sk$ with $k \in C_H(h)$, so that from (7.56) it follows that $f(shs^{-1}) = f(tht^{-1})$. Since (7.57) coincides with (7.55), this ends the proof. ■

Theorem 7.35 *A function $f \in L(H)$ belongs to the center of $L_\eta(H)$ if and only if it satisfies the identity* (7.55). *In particular, the dimension of the center of $L_\eta(H)$ is equal to the number of η-special conjugacy classes in H.*

Proof Let $s, k \in H$ and $f \in L(H)$. Then

$$\left[f *_\eta \delta_s\right](k) = f(s^{-1}k)\eta(k^{-1}, s) \quad \text{and} \quad \left[\delta_s *_\eta f\right](k) = f(ks^{-1})\eta(k^{-1}, ks^{-1}),$$

Thus, f belongs to the center of $L_\eta(H)$ if and only if

$$f(s^{-1}k)\eta(k^{-1}, s) = f(ks^{-1})\eta(k^{-1}, ks^{-1})$$

for all $s, k \in H$.

Setting $h := s^{-1}k \in H$ this identity becomes:

$$f(h)\eta(h^{-1}s^{-1}, s) = f(shs^{-1})\eta(h^{-1}s^{-1}, shs^{-1}). \tag{7.58}$$

Using the cocycle identities as in the proof of Lemma 7.33 as well as the fact that η is equalized, we get

$$\eta(h^{-1}s^{-1}, s) = \eta(h^{-1}, 1_H)\eta(s^{-1}, s)\eta(h^{-1}, s^{-1})^{-1} = \eta(h^{-1}, s^{-1})^{-1}$$

and

$$\begin{aligned}\eta(h^{-1} \cdot s^{-1}, shs^{-1}) &= \eta(h^{-1}, s^{-1})^{-1}\eta(h^{-1}, h \cdot s^{-1})\eta(s^{-1}, shs^{-1}) \\ &= \eta(h^{-1}, s^{-1})^{-1}\eta(h, s^{-1})^{-1}\eta(s^{-1}, shs^{-1}),\end{aligned}$$

which transforms (7.58) into (7.55).

As a consequence, f belongs to the center of $L_\eta(H)$ if and only if the following conditions are met: (i) f is supported in the η-special conjugacy classes; (ii) on each of such classes f is determined by (7.55) up to a multiplicative constant. ■

Corollary 7.36 *Let B be a finite Abelian group and let $\eta \in \mathscr{C}^2(B, \mathbb{T})$ be an equalized cocycle. Then*

$$\{\delta_b : b \in B \text{ such that } \eta(a, b) = \eta(b, a) \text{ for all } a \in B\},$$

is a basis for the center of $L_\eta(B)$.

Proof Just specialize Theorem 7.35 to the Abelian case, taking into account that $\mathscr{C}(b) = \{b\}$ and, equivalently, $C_B(b) = B$ for all $b \in B$, so that an element $b \in B$ is η-special exactly if $\eta(a, b) = \eta(b, a)$ for all $a \in B$. ■

In the following we assume again that N is normal in G, $\sigma \in \widehat{N}$, and keep all the notation in Sects. 2 and 7.1, in particular, in Proposition 6.15.(2) and (7.6).

Theorem 7.37 *Define*

$$\xi \colon L(H_G(\sigma)) \longrightarrow \mathrm{End}(\mathrm{Ind}_N^G V)$$

by setting

$$[\xi(f)F](g) := \frac{1}{|N|} \sum_{\substack{h \in G: \\ h^{-1}g \in I_G(\sigma)}} \widetilde{\Psi}(g^{-1}h) f(h^{-1}gN) F(h), \tag{7.59}$$

for all $f \in L(H_G(\sigma))$, $F \in \mathrm{Ind}_N^G V$, and $g \in G$. Then $\xi(f) \in \mathrm{End}_G(\mathrm{Ind}_N^G V)$ and the map ξ is an isometric, $*$-isomorphism between $L_\eta(H_G(\sigma))$ and $\mathrm{End}_G(\mathrm{Ind}_N^G V)$.

Proof Let $f \in L(H_G(\sigma))$, $F \in \mathrm{Ind}_N^G V$, and $g_0, g \in G$. We have:

$$\begin{aligned}
[\xi(f)\lambda(g_0)F](g) &= \frac{1}{|N|} \sum_{\substack{h \in G: \\ h^{-1}g \in I_G(\sigma)}} \widetilde{\Psi}(g^{-1}h) f(h^{-1}gN) F(g_0^{-1}h) \\
(s = g_0^{-1}h) \qquad &= \frac{1}{|N|} \sum_{\substack{s \in G: \\ s^{-1}g_0^{-1}g \in I_G(\sigma)}} \widetilde{\Psi}(g^{-1}g_0 s) f(s^{-1}g_0^{-1}gN) F(s) \\
&= [\xi(f)F](g_0^{-1}g) \\
&= [\lambda(g_0)\xi(f)F](g),
\end{aligned} \tag{7.60}$$

and this proves that $\xi(f) \in \mathrm{End}_G(\mathrm{Ind}_N^G V)$.

Then, for $f_1, f_2 \in L(H_G(\sigma))$ we have:

$$\begin{aligned}
[\xi(f_1 *_\eta f_2)F](g) &= \frac{1}{|N|} \sum_{\substack{h \in G: \\ h^{-1}g \in I_G(\sigma)}} \widetilde{\Psi}(g^{-1}h)[f_1 *_\eta f_2](h^{-1}gN) F(h) \\
&= \frac{1}{|N|} \sum_{\substack{h \in G: \\ h^{-1}g \in I_G(\sigma)}} \widetilde{\Psi}(g^{-1}h) F(h) \cdot
\end{aligned}$$

$$
\begin{aligned}
\text{(by (7.46))} \quad & \cdot \frac{1}{|N|} \sum_{s \in I_G(\sigma)} f_1(s^{-1}h^{-1}gN) f_2(sN) \eta(g^{-1}hN, sN) \\
\text{(by (7.7) and (7.8))} \quad &= \frac{1}{|N|^2} \sum_{\substack{h \in G: \\ h^{-1}g \in I_G(\sigma)}} \sum_{s \in I_G(\sigma)} \widetilde{\Psi}(g^{-1}hs)\widetilde{\Psi}(s^{-1}) \cdot \\
& \qquad \cdot F(h) f_1(s^{-1}h^{-1}gN) f_2(sN) \\
(s = h^{-1}t) \quad &= \frac{1}{|N|^2} \sum_{\substack{t \in G: \\ t^{-1}g \in I_G(\sigma)}} \widetilde{\Psi}(g^{-1}t) f_1(t^{-1}gN) \cdot \\
& \qquad \cdot \sum_{\substack{h \in G: \\ h^{-1}t \in I_G(\sigma)}} \widetilde{\Psi}(t^{-1}h) f_2(h^{-1}tN) F(h) \\
&= [\xi(f_1)\xi(f_2)F](g),
\end{aligned}
$$

that is,

$$
\xi(f_1 *_\eta f_2) = \xi(f_1)\xi(f_2). \tag{7.61}
$$

Moreover, for $F_1, F_2 \in \mathrm{Ind}_N^G V$,

$$
\begin{aligned}
\langle \xi(f)^* F_1, F_2 \rangle_{\mathrm{Ind}_N^G V} &= \langle F_1, \xi(f) F_2 \rangle_{\mathrm{Ind}_N^G V} \\
\text{(by (1.29))} \quad &= \frac{1}{|N|} \sum_{g \in G} \langle F_1(g), [\xi(f)F_2](g) \rangle_V \\
\text{(by (7.59))} \quad &= \frac{1}{|N|^2} \sum_{g \in G} \sum_{\substack{h \in G: \\ h^{-1}g \in I_G(\sigma)}} \langle F_1(g), \widetilde{\Psi}(g^{-1}h) f(h^{-1}gN) F_2(h) \rangle_V \\
\text{(by (7.8))} \quad &= \frac{1}{|N|^2} \sum_{h \in G} \sum_{\substack{g \in G: \\ g^{-1}h \in I_G(\sigma)}} \langle \widetilde{\Psi}(h^{-1}g) f^*(g^{-1}hN) F_1(g), F_2(h) \rangle_V \\
\text{(by (7.59))} \quad &= \langle \xi(f^*) F_1, F_2 \rangle_{\mathrm{Ind}_N^G V},
\end{aligned}
$$

that is

$$
\xi(f)^* = \xi(f^*). \tag{7.62}
$$

From (7.61) and (7.62) it follows that ξ is a $*$-homomorphism.

In order to prove that it is an isometry, we need a preliminary identity. Let $\{v_i : j = 1, 2, \ldots, d_\sigma\}$ be an orthonormal basis for V and consider the associated orthonormal basis for $\mathrm{Ind}_N^G V$ given by (1.30), where $\mathscr{T} = t(H)$. Then

$$\begin{aligned}
\mathrm{tr}[\xi(f)] &= \sum_{h\in H}\sum_{j=1}^{d_\sigma}\frac{1}{|N|}\sum_{g\in G}\langle[\xi(f)\lambda(t(h))f_{v_j}](g), [\lambda(t(h))f_{v_j}](g)\rangle_V \\
\text{(by (1.28) and (7.60))} \quad &= \frac{|G|}{|N|^2}\sum_{j=1}^{d_\sigma}\sum_{n\in N}\langle[\xi(f)f_{v_j}](n), \sigma(n^{-1})v_j\rangle_V \\
\text{(by (1.28) and (7.59))} \quad &= \frac{|G|}{|N|^3}\sum_{j=1}^{d_\sigma}\sum_{m,n\in N}\langle\widetilde{\Psi}(n^{-1}m)f(N)\sigma(m^{-1})v_j, \sigma(n^{-1})v_j\rangle_V \\
(\widetilde{\Psi}|_N = \sigma) \quad &= \frac{|G|}{|N|}d_\sigma f(N),
\end{aligned}$$

that is,

$$\mathrm{tr}[\xi(f)] = \frac{|G|}{|N|}d_\sigma f(N). \tag{7.63}$$

It follows that

$$\begin{aligned}
&\langle\xi(f_1), \xi(f_2)\rangle_{\mathrm{End}(\mathrm{Ind}_N^G V)} \\
&= \frac{1}{\dim V|G/N|}\mathrm{tr}\left[\xi(f_2)^*\xi(f_1)\right] && \text{by (1.1)} \\
&= \frac{1}{\dim V|G/N|}\mathrm{tr}\left[\xi(f_2^* *_\eta f_1)\right] && \text{by (7.61) and (7.62)} \\
&= (f_2^* *_\eta f_1)(N) && \text{by (7.63)} \\
&= \sum_{hN\in I_G(\sigma)/N} f_2^*(h^{-1}N)f_1(hN)\eta(N, hN) \\
&= \sum_{hN\in I_G(\sigma)/N} f_1(hN)\overline{f_2}(hN) \\
&= \langle f_1, f_2\rangle_{L(H_G(\sigma))},
\end{aligned}$$

and ξ is an isometry. This also proves that ξ is injective.

It remains to prove that ξ is surjective. Suppose that $T \in \mathrm{End}_G(\mathrm{Ind}_N^G V)$. We want to construct $f \in L(H_G(\sigma))$ such that $\xi(f) = T$. First of all, we define $\Xi : G \to \mathrm{End}(V)$ by setting:

$$\Xi(g)v := [Tf_v](g) \tag{7.64}$$

for all $g \in G$ and $v \in V$, where f_v is, again, as in (1.28). Then for $n, m \in N$ we have:

$$\begin{aligned} \Xi(ngm)v &= [Tf_v](ngm) \\ (\text{since } Tf_v \in \mathrm{Ind}_N^G V) \qquad &= \sigma(m^{-1})[\lambda(n^{-1})Tf_v](g) \\ (\text{since } T \in \mathrm{End}_G(\mathrm{Ind}_N^G V)) \qquad &= \sigma(m^{-1})[T\lambda(n^{-1})f_v](g) \\ (\text{since } \lambda(n^{-1})f_v = f_{\sigma(n^{-1})v}) \qquad &= \sigma(m^{-1})[Tf_{\sigma(n^{-1})v}](g) \\ &= \sigma(m^{-1})\Xi(g)\sigma(n^{-1})v. \end{aligned}$$

In particular,

$$\sigma(n)\Xi(g) = \Xi(gn^{-1}) = \Xi(gn^{-1}g^{-1} \cdot g) = \Xi(g)\sigma(gng^{-1}) = \Xi(g)\, {}^{g^{-1}}\sigma(n),$$

that is, $\Xi(g) \in \mathrm{Hom}_N({}^{g^{-1}}\sigma, \sigma)$, so that

$$\Xi(g) = 0 \quad \text{if} \quad g \notin I_G(\sigma), \tag{7.65}$$

(by Definition 2.1). On the other hand, from (7.8) and (7.9) it follows that for each $h \in I_G(\sigma)$ the operator $\widetilde{\Psi}(h)^*$ belongs to $\mathrm{Hom}_N({}^{h^{-1}}\sigma, \sigma)$. Since this space is one-dimensional (by Schur's lemma), for all $h \in I_G(\sigma)$ there exists $\widetilde{f}(h) \in \mathbb{C}$ such that $\Xi(h) = \widetilde{f}(h)\widetilde{\Psi}(h)^*$. From

$$\sigma(n^{-1})\widetilde{f}(h)\widetilde{\Psi}(h)^* = \sigma(n^{-1})\Xi(h) = \Xi(hn) = \widetilde{f}(hn)\widetilde{\Psi}(hn)^* = \sigma(n^{-1})\widetilde{f}(hn)\widetilde{\Psi}(h)^*$$

it follows that $\widetilde{f}$ in N-invariant; we set $f(hN) = \widetilde{f}(h)$ so that (recall (7.8))

$$\Xi(h) = f(hN)\widetilde{\Psi}(h)^* = f(hN)\widetilde{\Psi}(h^{-1}), \tag{7.66}$$

for all $h \in I_G(\sigma)$. Finally, by using again the basis in (1.30), for all $v \in V$, $g \in G$, and $k \in H_G(\sigma)$ we get:

$$\begin{aligned} [\xi(f)\lambda(t(k))f_v](g) &= \frac{1}{|N|} \sum_{\substack{h \in G: \\ h^{-1}g \in I_G(\sigma)}} \widetilde{\Psi}(g^{-1}h) f(h^{-1}gN) f_v(t(k)^{-1}h) \\ (t(k)^{-1}h = n \in N) \qquad &= \frac{1}{|N|} \sum_{\substack{n \in N: \\ n^{-1}t(k)^{-1}g \in I_G(\sigma)}} \widetilde{\Psi}(g^{-1}t(k)n) f(t(k)^{-1}gN)\sigma(n^{-1})v \\ (\text{by (7.6)}) \qquad &= \begin{cases} 0 & \text{if } t(k)^{-1}g \notin I_G(\sigma) \\ \widetilde{\Psi}(g^{-1}t(k)) f(t(k)^{-1}gN)v & \text{if } t(k)^{-1}g \in I_G(\sigma) \end{cases} \\ (\text{by (7.65) and (7.66)}) \qquad &= \Xi(t(k)^{-1}g)v \end{aligned}$$

$$
\begin{aligned}
&(\text{by } (7.64)) && = [Tf_v](t(k)^{-1}g) \\
& && = [\lambda(t(k))(Tf_v)](g)) \\
&(T \in \mathrm{End}_G(\mathrm{Ind}_N^G V)) && = [T\lambda(t(k)) f_v](g).
\end{aligned}
$$

Therefore $\xi(f) = T$ and this ends the proof. ■

Note that now Theorem 2.5.(3) becomes a corollary to Theorem 7.37.

Corollary 7.38 *The dimension of the center of $L_\eta(H)$ equals the number of all η-special conjugacy classes in H (cf. Lemma 7.33) as well as the number $r_\eta := |\widehat{H}^\eta|$ (cf. Corollary 7.15.(7.15)) of all irreducible, pairwise inequivalent η-representations of H.*

Proof The first equality is just the last statement in Theorem 7.35. The second equality follows from Theorem 7.37 applied in the setting of Remark 7.31. Indeed, if $n := |H|$, from (2.15) and (7.29) it follows that $\mathrm{Ind}_{C_n}^G \xi \sim \bigoplus_{\Theta \in \widehat{G}(\xi)} d_\Theta \Theta$ and therefore

$$
\begin{aligned}
L_\eta(H) &\cong \mathrm{End}_G\left(\mathrm{Ind}_{C_n}^G \xi\right) && (\text{by Theorem 7.37}) \\
&\cong \bigoplus_{j=1}^{r_\eta} M_{d_j}(\mathbb{C}) && (\text{by (1.38), (7.24), and (7.29)}).
\end{aligned}
$$

We deduce that the dimension of the center of $L_\eta(H)$ is equal to r_η. ■

Clearly, in the setting of Remark 7.31 and Corollary 7.38, we have that $\mathrm{End}_G\left(\mathrm{Ind}_{C_n}^G \xi\right) \cong \mathrm{Ind}_{C_n}^G \xi$.

7.6 The Hecke Algebra

Let $\sigma \in \widehat{N}$ with representation space V and associated 2-cocycle $\eta \in \mathscr{C}^2(H_G(\sigma), \mathbb{T})$. Let $\widetilde{\Psi}$ be as in (7.6) and fix a vector $v \in V$ with $\|v\| = 1$. We then define $\psi, \varphi \in L(I_G(\sigma))$ by setting

$$
\psi(s) := \langle v, \widetilde{\Psi}(s)v\rangle_V \tag{7.67}
$$

and

$$
\varphi(s) := \begin{cases} \frac{d_\sigma}{|N|}\psi(s) \equiv \frac{d_\sigma}{|N|}\langle v, \sigma(n)v\rangle_V & \text{if } s \in N \\ 0 & \text{otherwise} \end{cases} \tag{7.68}
$$

for all $s \in I_G(\sigma)$ (cf. (1.46)).

Theorem 7.39 *We have*

$$\psi = \psi * \varphi = \varphi * \psi = \varphi * \psi * \varphi \tag{7.69}$$

and, for all $r, s \in I_G(\sigma)$,

$$\sum_{n\in N} \psi(rns)\overline{\varphi}(n) = \overline{\eta}(rN, sN)\psi(r)\psi(s). \tag{7.70}$$

Proof Indeed, given $r \in I_G(\sigma)$, we have

$$\begin{aligned}
[\psi * \varphi](r) &= \sum_{s\in I_G(\sigma)} \psi(rs^{-1})\varphi(s) \\
&= \sum_{n\in N} \psi(rn^{-1})\varphi(n) \\
&= \sum_{n\in N} \langle v, \widetilde{\Psi}(rn^{-1})v\rangle \frac{d_\sigma}{|N|} \cdot \langle v, \sigma(n)v\rangle \\
\text{(by (7.6))} \quad &= \sum_{n\in N} \langle v, \widetilde{\Psi}(r)\sigma(n^{-1})v\rangle \frac{d_\sigma}{|N|} \cdot \langle v, \sigma(n)v\rangle \\
\text{(by (7.8))} \quad &= \sum_{n\in N} \langle \widetilde{\Psi}(r^{-1})v, \sigma(n^{-1})v\rangle \frac{d_\sigma}{|N|} \cdot \langle v, \sigma(n)v\rangle \\
&= \left\langle \widetilde{\Psi}(r^{-1})v, \frac{d_\sigma}{|N|} \sum_{n\in N} \langle \sigma(n)v, v\rangle \sigma(n^{-1})v \right\rangle \\
\text{(by (1.45))} \quad &= \langle \widetilde{\Psi}(r^{-1})v, v\rangle \\
&= \psi(r).
\end{aligned}$$

This shows the first equality in (7.69), the other ones follow after similar computations.

In order to show (7.70), let $r, s \in I_G(\sigma)$ and $n \in N$. We first note that

$$\begin{aligned}
\widetilde{\Psi}(rns) &= \widetilde{\Psi}(rs \cdot s^{-1}ns) \\
\text{(by (7.6))} \quad &= \widetilde{\Psi}(rs)\sigma(s^{-1}ns) \\
\text{(by (7.9))} \quad &= \widetilde{\Psi}(rs)\widetilde{\Psi}(s)^{-1}\sigma(n)\widetilde{\Psi}(s) \\
\text{(by (7.7))} \quad &= \eta(rN, sN)\widetilde{\Psi}(r)\sigma(n)\widetilde{\Psi}(s).
\end{aligned} \tag{7.71}$$

We deduce that

$$
\begin{aligned}
\sum_{n\in N} \psi(rns)\overline{\varphi}(n) &= \sum_{n\in N} \langle v, \widetilde{\Psi}(rns)v\rangle \frac{d_\sigma}{|N|}\langle \sigma(n)v, v\rangle \\
\text{(by (7.71) and (7.8))} \quad &= \overline{\eta}(rN, sN)\left\langle \widetilde{\Psi}(r^{-1})v, \frac{d_\sigma}{|N|}\sum_{n\in N}\langle \sigma(n^{-1})v, v\rangle \sigma(n)\widetilde{\Psi}(s)v \right\rangle \\
\text{(by (1.44))} \quad &= \overline{\eta}(rN, sN)\left\langle \widetilde{\Psi}(r^{-1})v, \langle \widetilde{\Psi}(s)v, v\rangle v\right\rangle \\
\text{(by (7.8))} \quad &= \overline{\eta}(rN, sN)\psi(r)\psi(s).
\end{aligned}
$$

■

Recall from Sect. 1.4 the linear operators $P\colon L(G) \to L(G)$ and $T_v\colon \mathrm{Ind}_N^G V \to L(G)$ defined in (1.47) and (1.49), respectively, the T_v-image, that is, the subspace

$$
\mathscr{I}(G, N, \varphi) := T_v\left(\mathrm{Ind}_N^G V\right) \subseteq L(G),
$$

as well as Theorem 1.5, in particular (1.51). In the following theorem we consider $L(I_G(\sigma))$ as a subalgebra of $L(G)$: a function $f \in L(I_G(\sigma))$ may be extended to the whole of G by setting $f(g) := 0$ if $g \in G \setminus I_G(\sigma)$; compare with Sect. 1.4. We complete the results in Theorems 1.5 and 7.37 by showing that $L_\eta(H_G(\sigma)) \cong \mathrm{End}_G(\mathrm{Ind}_N^G\sigma)$ is naturally anti-isomorphic to a subalgebra of $L(G)$ and that, with respect to this anti-isomorphism, the action of $\xi(f)$ on $F \in \mathrm{Ind}_N^G V$, where $f \in L(H_G(\sigma))$ and ξ is as in (7.59), becomes just right convolution. We will also use the following notation: if $f \in L(H_G(\sigma))$ we define $\overset{\vartriangle}{f} \in L(I_G(\sigma))$ by setting $\overset{\vartriangle}{f}(s) := f(sN)$ for all $s \in I_G(\sigma)$.

Theorem 7.40 *Define a map $\alpha\colon L(H_G(\sigma)) \to L(I_G(\sigma)) \subseteq L(G)$ by setting, for all $f \in L(H_G(\sigma))$,*

$$
\alpha(f) := \frac{d_\sigma}{|N|}\overset{\vartriangle}{f}\psi. \tag{7.72}
$$

Then

$$
\mathscr{H}(G, N, \sigma) := \alpha\left(L(H_G(\sigma))\right) = \{\overset{\vartriangle}{f}\psi : f \in L(H_G(\sigma))\}
$$

is contained in $\mathscr{I}(G, N, \varphi)$. Moreover,

(1) *$\alpha\colon L_\eta(H_G(\sigma)) \to L(I_G(\sigma)) \subseteq L(G)$ is an injective $*$-anti-homomorphism, so that $\mathscr{H}(G, N, \sigma)$ is a $*$-algebra;*

(2) *$\sqrt{\frac{|N|}{d_\sigma}}\,\alpha$ is an isometry;*

(3) *for all* $F \in \mathrm{Ind}_N^G V$ *and* $f \in L(H_G(\sigma))$ *we have*

$$T_v[\xi(f)F] = (T_v F) * \alpha(f).$$

Proof First of all, given $f \in L(H_G(\sigma))$ and $s \in I_G(\sigma)$ we have

$$\begin{aligned}
\left[\left(\overset{\Delta}{f\psi}\right) * \varphi\right](s) &= \sum_{n\in N} f(sn^{-1}N)\psi(sn^{-1})\varphi(n) \\
&= f(sN)(\psi * \varphi)(s) \\
\text{(by (7.69))} \quad &= f(sN)\psi(s) \\
&= \left(\overset{\Delta}{f\psi}\right)(s),
\end{aligned}$$

and this shows that $\mathscr{H}(G, N, \sigma) \leq \mathscr{I}(G, N, \varphi)$. Now we prove the properties of α.

(1) If $f_1, f_2 \in L(H_G(\sigma))$ and $s \in I_G(\sigma)$ then from (7.46) and (7.72) it follows that:

$$\begin{aligned}
\left[\alpha(f_1 *_\eta f_2)\right](s) &= \frac{d_\sigma}{|N|}\psi(s)(f_1 *_\eta f_2)(sN) \\
(|N| = |I_G(\sigma)|/|H_G(\sigma)|) \quad &= \frac{d_\sigma}{|N|^2}\sum_{t\in I_G(\sigma)} \psi(s) f_1(t^{-1}sN) f_2(tN)\eta(s^{-1}N, tN) \\
\text{(by (7.67) and (7.8))} \quad &= \frac{d_\sigma}{|N|^2}\sum_{t\in I_G(\sigma)} f_1(t^{-1}sN) f_2(tN) \\
&\quad \cdot \langle\widetilde{\Psi}(s^{-1})\widetilde{\Psi}(t)\widetilde{\Psi}(t^{-1})v, v\rangle\eta(s^{-1}N, tN) \\
\text{(by (7.7))} \quad &= \frac{d_\sigma}{|N|^2}\sum_{t\in I_G(\sigma)} f_1(t^{-1}sN) f_2(tN)\langle\widetilde{\Psi}(t^{-1})v, \widetilde{\Psi}(t^{-1}s)v\rangle \\
&= \frac{d_\sigma}{|N|^2}\sum_{t\in I_G(\sigma)} f_1(t^{-1}sN) f_2(tN)\sum_{i=1}^{d_\sigma}\langle\widetilde{\Psi}(t^{-1})v, v_i\rangle \\
&\quad \cdot \langle\widetilde{\Psi}(s^{-1}t)v_i, v\rangle \\
\text{(by (1.44))} \quad &= \frac{d_\sigma^2}{|N|^3}\sum_{t\in I_G(\sigma)} f_1(t^{-1}sN) f_2(tN) \\
&\quad \cdot \sum_{i=1}^{d_\sigma}\sum_{n\in N}\langle\widetilde{\Psi}(t^{-1})v, \sigma(n^{-1})v\rangle \cdot \langle v, \sigma(n)v_i\rangle
\end{aligned}$$

$$
\begin{aligned}
&\cdot \langle \widetilde{\Psi}(s^{-1}t)v_i, v\rangle \\
(t = rn) \quad &= \frac{d_\sigma^2}{|N|^3} \sum_{r \in I_G(\sigma)} f_1(r^{-1}sN) f_2(rN) \sum_{i=1}^{d_\sigma} \langle \widetilde{\Psi}(r^{-1})v, v\rangle \\
&\cdot \left\langle \widetilde{\Psi}(s^{-1}r) \sum_{n \in N} \langle v, \sigma(n)v_i\rangle \sigma(n)v_i, v \right\rangle \\
(\text{by } (1.45)) \quad &= \frac{d_\sigma^2}{|N|^2} \sum_{r \in I_G(\sigma)} f_1(r^{-1}sN) f_2(rN) \langle \widetilde{\Psi}(r^{-1})v, v\rangle \\
&\cdot \langle \widetilde{\Psi}(s^{-1}r)v, v\rangle \\
(\text{by } (7.67)) \quad &= \frac{d_\sigma^2}{|N|^2} \sum_{r \in I_G(\sigma)} f_2(rN) f_1(r^{-1}sN) \psi(r)\psi(r^{-1}s) \\
&= \frac{d_\sigma^2}{|N|^2} \left[\left(\overset{\triangle}{f_2}\psi \right) * \left(\overset{\triangle}{f_1}\psi \right) \right](s) \\
&= [\alpha(f_2) * \alpha(f_1)](s).
\end{aligned}
$$

Moreover,

$$
[\alpha(f^*)](s) = \frac{d_\sigma}{|N|} f^*(sN)\psi(s) = \frac{d_\sigma}{|N|} \overline{f}(s^{-1}N)\overline{\psi}(s^{-1}) = \overline{[\alpha(f)]}(s^{-1}) = [\alpha(f)^*](s).
$$

(2) We will use the notations in (7.6) For $f_1, f_2 \in L(H_G(\sigma))$ we have:

$$
\begin{aligned}
\langle \alpha(f_1), \alpha(f_2)\rangle_{L(G)} &= \frac{d_\sigma^2}{|N|^2} \left\langle \overset{\triangle}{f_1}\psi, \overset{\triangle}{f_2}\psi \right\rangle_{L(G)} \\
&= \frac{d_\sigma^2}{|N|^2} \sum_{s \in I_G(\sigma)} f_1(sN)\psi(s)\overline{f_2}(sN)\overline{\psi}(s) \\
(s = t(h)n) \quad &= \frac{d_\sigma^2}{|N|^2} \sum_{h \in H_G(\sigma)} f_1(h)\overline{f_2}(h) \sum_{n \in N} \langle v, \Psi(h)\sigma(n)v\rangle \\
&\cdot \langle \Psi(h)\sigma(n)v, v\rangle \\
&= \frac{d_\sigma^2}{|N|^2} \sum_{h \in H_G(\sigma)} f_1(h)\overline{f_2}(h) \sum_{n \in N} \langle v, \Psi(h)\sigma(n)v\rangle
\end{aligned}
$$

$$\cdot \sum_{j=1}^{d_\sigma} \langle \sigma(n)v, v_j\rangle \cdot \langle \Psi(h)v_j, v\rangle$$

$$= \frac{d_\sigma}{|N|} \sum_{h\in H_G(\sigma)} f_1(h)\overline{f_2}(h)$$

$$\cdot \sum_{j=1}^{d_\sigma} \left\langle \frac{d_\sigma}{|N|} \sum_{n\in N} \langle \sigma(n)v, v_j\rangle \sigma(n^{-1})\Psi(h)^* v, v \right\rangle \cdot \langle \Psi(h)v_j, v\rangle$$

$$\text{(by (1.44))} \quad = \frac{d_\sigma}{|N|} \sum_{h\in H_G(\sigma)} f_1(h)\overline{f_2}(h) \sum_{j=1}^{d_\sigma} \langle \Psi(h)^* v, v_j\rangle \cdot \langle \Psi(h)v_j, v\rangle$$

$$= \frac{d_\sigma}{|N|} \sum_{h\in H_G(\sigma)} f_1(h)\overline{f_2}(h) \langle \Psi(h)\Psi(h)^* v, v\rangle$$

$$= \langle f_1, f_2\rangle_{L(H_G(\sigma))}.$$

(3) For $f \in L(H_G(\sigma))$, $F \in \mathrm{Ind}_N^G V$, and $g \in G$, recalling (1.48) we have:

$$\{T_v[\xi(f)F]\}(g) = \sqrt{\frac{d_\sigma}{|N|}} \langle [\xi(f)F](g), v\rangle$$

$$\text{(by (7.59))} \quad = \frac{\sqrt{d_\sigma}}{|N|^{3/2}} \sum_{\substack{h\in G:\\ h^{-1}g\in I_G(\sigma)}} f(h^{-1}gN) \langle \widetilde{\Psi}(g^{-1}h)F(h), v\rangle$$

$$= \frac{\sqrt{d_\sigma}}{|N|^{3/2}} \sum_{\substack{h\in G:\\ h^{-1}g\in I_G(\sigma)}} f(h^{-1}gN) \sum_{i=1}^{d_\sigma} \langle F(h), v_i\rangle \cdot \langle \widetilde{\Psi}(g^{-1}h)v_i, v\rangle$$

$$\text{(by (1.45))} \quad = \frac{(d_\sigma)^{3/2}}{|N|^{5/2}} \sum_{\substack{h\in G:\\ h^{-1}g\in I_G(\sigma)}} f(h^{-1}gN) \sum_{i=1}^{d_\sigma} \sum_{n\in N} \langle F(h), \sigma(n^{-1})v\rangle$$

$$\cdot \langle v, \sigma(n)v_i\rangle \cdot \langle \widetilde{\Psi}(g^{-1}h)v_i, v\rangle$$

$$(h = kn) \quad = \frac{(d_\sigma)^{3/2}}{|N|^{5/2}} \sum_{\substack{k\in G:\\ k^{-1}g\in I_G(\sigma)}} f(k^{-1}gN) \langle F(k), v\rangle$$

$$\cdot \sum_{i=1}^{d_\sigma} \left\langle \widetilde{\Psi}(g^{-1}k) \sum_{n\in N} \langle \sigma(n^{-1})v, v_i\rangle \sigma(n)v_i, v \right\rangle$$

$$\text{(by (1.45))} \quad = \frac{(d_\sigma)^{3/2}}{|N|^{3/2}} \sum_{\substack{k\in G:\\ k^{-1}g\in I_G(\sigma)}} f(k^{-1}gN)\langle F(k), v\rangle \cdot \langle \widetilde{\Psi}(g^{-1}k)v, v\rangle$$

$$= \sum_{\substack{k\in G:\\ k^{-1}g\in I_G(\sigma)}} \sqrt{\frac{d_\sigma}{|N|}}\langle F(k), v\rangle \cdot \frac{d_\sigma}{|N|}\psi(k^{-1}g) f(k^{-1}gN)$$

$$= \sum_{k\in G} (T_v F)(k) \cdot [\alpha(f)](k^{-1}g)$$

$$= \{(T_v F) * [\alpha(f)]\}(g).$$

■

The algebra $\mathscr{H}(G, N, \sigma)$ is called the *Hecke algebra* associated with the triple (G, N, σ). An analysis of the multiplicity-free case is presented in [31, Sect. 4.6].

Chapter 8
Induced Projective Representations

The present and the subsequential chapters are devoted to the study of the induction of unitary projective representations. Our exposition is mainly based on the pioneering work of Mackey, [92, Sect. 4]. However, Mackey based most of his proofs on Proposition 7.12 (actually, he worked on locally compact groups). When possible, we prefer to give more direct proofs. See also Remark 8.8 below.

8.1 Basic Theory

Definition 8.1 Let K and H be finite groups, $K \leq H$, and let η be an equalized unitary 2-cocycle of H. We use the same symbol η to denote its restriction to K, which is still an equalized, unitary 2-cocycle. Let Ψ be a unitary η-representation of K on a space V. We set

$$\mathrm{Ind}_K^H V := \{f : H \to V \text{ such that } f(hk) = \overline{\eta}(h,k)\Psi(k^{-1})f(h), \ \forall h \in H, k \in K\} \tag{8.1}$$

we equip it with the same scalar product as in (1.29), and define $\Phi := \mathrm{Ind}_K^H \Psi : H \to \mathrm{Hom}(\mathrm{Ind}_K^H V, V^H)$ by setting

$$[\Phi(h)f](h_0) := \eta(h^{-1}, h_0) f(h^{-1}h_0) \tag{8.2}$$

for all $h, h_0 \in H$ and $f \in \mathrm{Ind}_K^H V$. One calls $(\Phi, \mathrm{Ind}_K^H V)$ the *induced η-representation* and writes $\mathrm{Ind}_K^H \Psi := \Phi$.

In the following we check that, indeed, $\Phi(h) \in \mathrm{End}(\mathrm{Ind}_K^H V)$ for all $h \in H$, and that Φ is a unitary η-representation.

T. Ceccherini-Silberstein et al., *Representation Theory of Finite Group Extensions*, Springer Monographs in Mathematics,
https://doi.org/10.1007/978-3-031-13873-7_8

Proposition 8.2 *Φ is a unitary η-representation of H on $\mathrm{Ind}_K^H V$.*

Proof Let $f \in \mathrm{Ind}_K^H V$ and let $h \in H$. We first check that $\Phi(h)f \in \mathrm{Ind}_K^H V$. For $h_0 \in H$ and $k \in K$, using (8.2) we have

$$\begin{aligned}[\Phi(h)f](h_0k) &= \eta(h^{-1}, h_0k)f(h^{-1}h_0k)\\ \text{(by (8.1))} \quad &= \eta(h^{-1}, h_0k)\overline{\eta}(h^{-1}h_0, k)\Psi(k^{-1})f(h^{-1}h_0)\\ \text{(by (6.1))} \quad &= \overline{\eta}(h_0, k)\eta(h^{-1}, h_0)\Psi(k^{-1})f(h^{-1}h_0)\\ \text{(by (8.2))} \quad &= \overline{\eta}(h_0, k)\Psi(k^{-1})[\Phi(h)f](h_0)\end{aligned}$$

so that $\Psi(h)f \in \mathrm{Ind}_K^H V$. Moreover, for $h_0, h_1, h_2 \in H$, using (8.2) we have

$$\begin{aligned}\{\Phi(h_1)[\Phi(h_2)f]\}(h_0) &= \eta(h_1^{-1}, h_0)[\Phi(h_2)f](h_1^{-1}h_0)\\ \text{(again by (8.2))} \quad &= \eta(h_1^{-1}, h_0)\eta(h_2^{-1}, h_1^{-1}h_0)f(h_2^{-1}h_1^{-1}h_0)\\ \text{(by (6.1))} \quad &= \eta(h_2^{-1}h_1^{-1}, h_0)\eta(h_2^{-1}, h_1^{-1})f(h_2^{-1}h_1^{-1}h_0)\\ \text{(by (6.23))} \quad &= \overline{\eta}(h_1, h_2)\eta(h_2^{-1}h_1^{-1}, h_0)f(h_2^{-1}h_1^{-1}h_0)\\ &= \overline{\eta}(h_1, h_2)[\Phi(h_1h_2)f](h_0).\end{aligned}$$

We deduce that

$$\Phi(h_1h_2) = \eta(h_1, h_2)\Phi(h_1)\Phi(h_2)$$

and this shows that Φ is an η-representation. It is immediate to check that Φ is unitary. ∎

Let $\mathcal{T}$ be a complete set of representatives for the left cosets of K in H, so that

$$H = \sqcup_{t \in T} tK, \tag{8.3}$$

such that $1_H \in \mathcal{T}$.

Lemma 8.3 *For $v \in V$ and $t \in \mathcal{T}$ define $f_{t,v} \colon H \to V$ by setting*

$$f_{t,v}(sk) := \begin{cases}\overline{\eta}(t, k)\Psi(k^{-1})v & \text{if } s = t\\ 0 & \text{otherwise}\end{cases}$$

for all $s \in \mathcal{T}$ and $k \in K$. Then $f_{t,v}$ belongs to $\mathrm{Ind}_K^H V$.

Proof For $k_1, k_2 \in K$ we have

$$\begin{aligned} f_{t,v}(tk_1k_2) &= \overline{\eta}(t, k_1k_2)\Psi(k_2^{-1}k_1^{-1})v \\ \text{(by (6.1))} \quad &= \overline{\eta}(tk_1, k_2)\overline{\eta}(t, k_1)\eta(k_1, k_2)\Psi(k_2^{-1}k_1^{-1})v \\ \text{(by (6.23) and (7.10))} \quad &= \overline{\eta}(tk_1, k_2)\overline{\eta}(t, k_1)\Psi(k_2^{-1})\Psi(k_1^{-1})v \\ &= \overline{\eta}(tk_1, k_2)\Psi(k_2^{-1})f_{t,v}(tk_1). \end{aligned}$$

■

Remark 8.4 Setting

$$Z_t := \{f: H \to V : f(sk) = \delta_{s,t}\overline{\eta}(t, k)\Psi(k^{-1})f(t), \ \forall s \in \mathcal{T}, k \in K\}$$

with the same proof of Lemma 8.3, we can check that Z_t is a subspace of $\mathrm{Ind}_K^H V$. Clearly, it is made up of all $f \in \mathrm{Ind}_K^H V$ that vanish outside tK and from (8.3) it follows that

$$\mathrm{Ind}_K^H V = \bigoplus_{t \in \mathcal{T}} Z_t. \tag{8.4}$$

Corollary 8.5 *For each $v \in V$, define $f_v := f_{1_H,v}$ so that*

$$f_v(h) = \begin{cases} \Psi(k^{-1})v & \text{if } h = k \in K \\ 0 & \text{otherwise} \end{cases} \tag{8.5}$$

for all $h \in H$. Then the functions $\Phi(t)f_v$, $v \in V$, and $t \in \mathcal{T}$, span the whole of $\mathrm{Ind}_K^H V$.

Proof For all $t \in \mathcal{T}$ and $h \in H$, by using (8.2) we have

$$\begin{aligned} [\Phi(t)f_v](h) &= \eta(t^{-1}, h)f_v(t^{-1}h) \\ &= \begin{cases} \eta(t^{-1}, h)f_v(k) \equiv \overline{\eta}(t, k)\Psi(k^{-1})v & \text{if } h = tk, k \in K \\ 0 & \text{if } h \notin tK \end{cases} \end{aligned}$$

because, by the cocycle identity,

$$\eta(t^{-1}, tk) = \eta(1, k)\overline{\eta}(t, k)\eta(t^{-1}, t) = \overline{\eta}(t, k),$$

so that $\Phi(t)f_v = f_{t,v}$. ■

Corollary 8.6

(1) *Let* $F \in \mathrm{Ind}_K^H V$. *Then* $\frac{1}{|K|}\sum_{h\in H}\Phi(h) f_{F(h)} = F$.

(2) *Let* $k \in K$ *and let* $v \in V$. *Then* $f_{\Psi(k)v} = \Phi(k) f_v$.

Proof

(1) If $h_0 \in H$ by (8.2) we have

$$\begin{aligned}
\frac{1}{|K|}\sum_{h\in H}\Phi(h)[f_{F(h)}](h_0) &= \frac{1}{|K|}\sum_{h\in H}\eta(h^{-1}, h_0) f_{F(h)}(h^{-1}h_0)\\
(h^{-1}h_0 = k \in K) \quad &= \frac{1}{|K|}\sum_{k\in K}\eta(kh_0^{-1}, h_0) f_{F(h_0k^{-1})}(k)\\
(\text{by } (8.5)) \quad &= \frac{1}{|K|}\sum_{k\in K}\eta(kh_0^{-1}, h_0)\Psi(k^{-1})F(h_0k^{-1})\\
(\text{by } (8.1)) \quad &= \frac{1}{|K|}\sum_{k\in K}\eta(kh_0^{-1}, h_0)\overline{\eta}(h_0, k^{-1})F(h_0)\\
&= F(h_0),
\end{aligned}$$

where the last equality follows from the cocycle identity

$$\eta(k \cdot h_0^{-1}, h_0)\overline{\eta}(h_0, k^{-1}) = \overline{\eta}(k, h_0^{-1}) \cdot \overline{\eta}(h_0, k^{-1}) = 1,$$

as η is equalized.

(2) For $k_0 \in K$ we have

$$\begin{aligned}
[f_{\Psi(k)v}](k_0) &= \Psi(k_0^{-1})\Psi(k)v\\
(\text{by } (7.10)) \quad &= \overline{\eta}(k_0^{-1}, k)\Psi(k_0^{-1}k)v\\
(\text{by } (6.23)) \quad &= \eta(k^{-1}, k_0) f_v(k^{-1}k_0)\\
(\text{by } (8.1)) \quad &= [\Phi(k) f_v](k_0).
\end{aligned}$$

■

We now relate the construction of an induced representation to Proposition 7.12 (in particular, the map $\Psi \mapsto \widetilde{\Psi}$ is as in (7.21)). First of all, note that, by (1.25), the space W of $\rho := \mathrm{Ind}_{G(C_n,K,\eta)}^{G(C_n,H,\eta)}\widetilde{\Psi}$ is made up of all functions $F\colon G(C_n, H, \eta) \to V$ such that

$$\begin{aligned}
F((\lambda, h) * (\mu, k)) &= \widetilde{\Psi}((\mu, k)^{-1})F(\lambda, h)\\
(\text{by } (6.21) \text{ and } (7.21)) \quad &= \overline{\mu}\, \widetilde{\Psi}(k^{-1})F(\lambda, h)
\end{aligned} \tag{8.6}$$

for all $\lambda, \mu \in C_n$, $h \in H$, and $k \in K$. In particular, since $(\lambda, h) = (1, h) * (\lambda, 1_H)$ (cf. 6.8) we have

$$F(\lambda, h) = \overline{\lambda} F(1, h). \tag{8.7}$$

Define $\gamma : \mathrm{Ind}_K^H V \to W$ by setting

$$[\gamma(f)](\lambda, h) := \overline{\lambda} f(h) \tag{8.8}$$

for all $f \in \mathrm{Ind}_K^H V$, $\lambda \in C_n$, and $h \in H$.

Theorem 8.7 *The map γ is an isometric isomorphism between the $G(C_n, H, \eta)$ ordinary representations $\widetilde{\Phi}$ and ρ.*

Proof First of all, we check that $\gamma(f) \in W$ for all $f \in \mathrm{Ind}_K^H V$. Indeed, for $\lambda, \mu \in C_n$, $k \in K$, and $h \in H$ by using (6.8) we have:

$$\begin{aligned}
[\gamma(f)]((\lambda, h) * (\mu, k)) &= [\gamma(f)](\lambda\mu\, \eta(h, k)^{-1}, hk) \\
\text{(by (8.8))} \quad &= \overline{\lambda\mu}\, \eta(h, k) f(hk) \\
\text{(by (8.1))} \quad &= \overline{\lambda\mu} \Psi(k^{-1}) f(h) \\
\text{(by (8.8))} \quad &= \overline{\mu} \Psi(k^{-1}) [\gamma(f)](\lambda, h) \\
\text{(by (7.21))} \quad &= \widetilde{\Psi}((\mu, k)^{-1}) [\gamma(f)](\lambda, h).
\end{aligned}$$

Moreover, for all $f \in \mathrm{Ind}_K^H V$, $h, h_0 \in H$, and $\lambda, \lambda_0 \in C_n$, we have by (8.8):

$$\begin{aligned}
\{\gamma[\widetilde{\Phi}(\lambda, h) f]\}(\lambda_0, h_0) &= \overline{\lambda_0}\, [\widetilde{\Phi}(\lambda, h) f](h_0) \\
\text{(by (7.21))} \quad &= \overline{\lambda_0} \lambda\, [\Phi(h) f](h_0) \\
\text{(by (8.2))} \quad &= \overline{\lambda_0} \lambda\, \eta(h^{-1}, h_0) f(h^{-1} h_0) \\
\text{(by (8.8))} \quad &= \gamma(f)[\overline{\lambda} \lambda_0\, \overline{\eta}(h^{-1}, h_0), h^{-1} h_0) \\
\text{(by (6.8))} \quad &= \gamma(f)[(\lambda^{-1}, h^{-1}) * (\lambda_0, h_0)] \\
\text{(by (1.26))} \quad &= [\rho(\lambda, h) \gamma(f)](\lambda_0, h_0),
\end{aligned}$$

that is,

$$\gamma[\widetilde{\Phi}(\lambda, h) f] = \rho(\lambda, h) \gamma(f).$$

It is easy to check that γ is an isometry. ■

Remark 8.8 In [92] Mackey uses systematically Theorem 8.7 to translate results from ordinary induced representations to projective induced representations (in the setting of locally compact groups). Here, in contrast, we adapt to the unitary projective representations the proofs of the ordinary case. We just give an example of Mackey's philosophy in the proof of Theorem 8.12. In any case, it is quite interesting that our methods lead to several elementary cocycle computations, where the assumption that η be equalized is quite useful. We believe that there must be some more general cohomological theory behind our computations.

Theorem 8.7 may be expressed as follows: for any η-representation Ψ of K, we have

$$\widetilde{\mathrm{Ind}_K^H \Psi} \sim \mathrm{Ind}_{G(C_n,K,\eta)}^{G(C_n,H,\eta)} \widetilde{\Psi}. \tag{8.9}$$

On the other hand, it is also obvious that, for any η-representation Φ of H, we have

$$\widetilde{\mathrm{Res}_K^H \Phi} \sim \mathrm{Res}_{G(C_n,K,\eta)}^{G(C_n,H,\eta)} \widetilde{\Phi}. \tag{8.10}$$

Theorem 8.9 (Transitivity of Induction) *Let L, K, H be three groups and suppose that $L \leq K \leq H$. Let η be an equalized unitary 2-cocycle of H and let Ψ be an η-representation of L. Then*

$$\mathrm{Ind}_L^H \Psi = \mathrm{Ind}_K^H \mathrm{Ind}_L^K \Psi.$$

Proof Let V be the space of Ψ, and define W to be the space of all maps $F\colon H \times K \to V$ such that

$$F(hk, k'\ell) = \eta(k,k')\overline{\eta}(h,k)\overline{\eta}(k',\ell)\Psi(\ell^{-1})F(h,kk') \tag{8.11}$$

for all $h \in H, k, k' \in K$, and $\ell \in L$. We define an η-representation Ξ of H on W by setting

$$[\Xi(h)F](h_0,k) := \eta(h^{-1},h_0)F(h^{-1}h_0,k) \tag{8.12}$$

for all $h, h_0 \in H$ and $k \in K$. The proof that

$$\eta(h_1,h_2)\Xi(h_1)\Xi(h_2) = \Xi(h_1h_2)$$

for all $h_1, h_2 \in H$ is as in the proof of Proposition 8.2 for Φ. In a similar manner we prove that $\Xi(h)F \in W$ for all $h \in H$ and $F \in W$. Indeed, for $h, h_0 \in H, k, k' \in K$, $\ell \in L$, and $F \in W$ we have

$$\begin{aligned}
[\Xi(h)F](h_0k, k'\ell) &= \eta(h^{-1},h_0k)F(h^{-1}h_0k,k'\ell)\\
\text{(by (8.11))}\quad &= \eta(k,k')\overline{\eta}(h^{-1}\cdot h_0,k)\overline{\eta}(k',\ell)\eta(h^{-1},h_0k)\Psi(\ell^{-1})F(h^{-1}h_0,kk')\\
\text{(by (7.53))}\quad &= \eta(k,k')\eta(h^{-1},h_0)\overline{\eta}(h_0,k)\overline{\eta}(k',\ell)\Psi(\ell^{-1})F(h^{-1}h_0,kk')\\
&= \eta(k,k')\overline{\eta}(h_0,k)\overline{\eta}(k',\ell)\Psi(\ell^{-1})[\Xi(h)F](h_0,kk'),
\end{aligned}$$

so that $\Xi(h)F$ satisfies the condition in (8.11).

Set now

$$\Theta := \mathrm{Ind}_L^K \Psi \quad \text{and} \quad \Phi := \mathrm{Ind}_K^H \Theta.$$

We prove that the map

$$\begin{array}{rcl} \mathrm{Ind}_K^H(\mathrm{Ind}_L^K V) & \to & W \\ f & \mapsto & \widetilde{f} \end{array}$$

where

$$\widetilde{f}(h,k) := [f(h)](k)$$

for all $h \in H$ and $k \in K$, is an isomorphism between Φ and Ξ. First of all, note that $\mathrm{Ind}_K^H(\mathrm{Ind}_L^K V)$ is made up of all $f\colon H \to \mathrm{Ind}_L^K V$ such that

$$f(hk) = \overline{\eta}(h,k)\Theta(k^{-1})f(h) \tag{8.13}$$

by (8.1),

$$[\Theta(k^{-1})f(h)](k') = \eta(k,k')[f(h)](kk') \tag{8.14}$$

by (8.2), and

$$[f(h)](k'\ell) = \overline{\eta}(k',\ell)\Psi(\ell^{-1})[f(h)](k'), \tag{8.15}$$

by (8.1).

Therefore, for all $h \in H$, $k, k' \in K$, and $\ell \in L$, we have

$$\begin{aligned} \widetilde{f}(hk, k'\ell) &= [f(hk)](k'\ell) \\ \text{(by (8.13) and (8.15))} \quad &= \overline{\eta}(k',\ell)\overline{\eta}(h,k)\Psi(\ell^{-1})[\Theta(k^{-1})f(h)](k') \\ \text{(by (8.14))} \quad &= \eta(k,k')\overline{\eta}(k',\ell)\overline{\eta}(h,k)\Psi(\ell^{-1})[f(h)](kk') \\ &= \eta(k,k')\overline{\eta}(k',\ell)\overline{\eta}(h,k)\Psi(\ell^{-1})\widetilde{f}(h,kk'), \end{aligned}$$

so that $\widetilde{f} \in W$.

Conversely, we define a map

$$\begin{array}{rcl} W & \to & \mathrm{Ind}_K^H(\mathrm{Ind}_L^K V) \\ F & \mapsto & \overset{\diamond}{F} \end{array}$$

by setting

$$[\overset{\diamond}{F}(h)](k) := F(h, k)$$

for all $h \in H, k \in K$. We first show that $\overset{\diamond}{F} \in \mathrm{Ind}_K^H(\mathrm{Ind}_L^K V)$. For all $h \in H$, $k, k' \in K$, and $\ell \in L$ we have:

$$\begin{aligned} [\overset{\diamond}{F}(h)](k'\ell) &= F(h, k'\ell) \\ \text{(by (8.11) with } k = 1_K) \quad &= \overline{\eta}(k', \ell)\Psi(\ell^{-1})F(h, k') \\ &= \overline{\eta}(k', \ell)\Psi(\ell^{-1})[\overset{\diamond}{F}(h)](k') \end{aligned}$$

so that (8.15) is satisfied and $\overset{\diamond}{F}(h) \in \mathrm{Ind}_L^K V$ for all $h \in H$. Moreover,

$$\begin{aligned} [\overset{\diamond}{F}(hk)](k') &= F(hk, k') \\ \text{(by (8.11) with } \ell = 1_K) \quad &= \eta(k, k')\overline{\eta}(h, k)F(h, kk') \\ &= \eta(k, k')\overline{\eta}(h, k)[\overset{\diamond}{F}(h)](kk') \\ \text{(by (8.14))} \quad &= \overline{\eta}(h, k)[\Theta(k^{-1})\overset{\diamond}{F}(h)](k') \end{aligned}$$

so that also (8.13) is satisfied and $\overset{\diamond}{F} \in \mathrm{Ind}_K^H \mathrm{Ind}_L^K V$ (just note that (8.14) is the *definition* of Θ and we may use it because, as we have already checked, $\overset{\diamond}{F}(h) \in \mathrm{Ind}_L^K V$). It is obvious that the maps $f \mapsto \widetilde{f}$ and $F \mapsto \overset{\diamond}{F}$ are inverses of each other. The next step consists in proving that $f \mapsto \widetilde{f}$ intertwines Ξ and Φ: for all $f \in \mathrm{Ind}_K^H \mathrm{Ind}_L^K V$, $h, h_0 \in H$, and $k \in K$, we have

$$\begin{aligned} [\widetilde{\Phi(h) f}](h_0, k) &= \{[\Phi(h) f](h_0)\}(k) \\ \text{(by (8.2))} \quad &= \eta(h^{-1}, h_0)[f(h^{-1}h_0)](k) \\ &= \eta(h^{-1}, h_0)\widetilde{f}(h^{-1}h_0, k) \\ \text{(by (8.12))} \quad &= [\Xi(h)\widetilde{f}](h_0, k). \end{aligned}$$

In conclusion we have proved that $\Phi \sim \Xi$.

We now define two maps

$$\begin{array}{rcl} \mathrm{Ind}_L^H V & \to & W \\ G & \mapsto & G^\sharp \end{array} \quad \text{and} \quad \begin{array}{rcl} W & \to & \mathrm{Ind}_L^H V \\ F & \mapsto & F^\flat \end{array}$$

by setting, for all $h \in H, k \in K, G \in \mathrm{Ind}_L^H V$, and $F \in W$

$$G^\sharp(h,k) := \eta(h,k)G(hk)$$
$$F^\flat(h) := F(h,1_K).$$

For $h \in H, k, k' \in K$, and $\ell \in L$ we have

$$\begin{aligned} G^\sharp(hk, k'\ell) &= \eta(hk, k'\ell)G(hkk'\ell) \\ \text{(by (8.1))} \quad &= \overline{\eta}(hkk', \ell)\eta(hk, k'\ell)\Psi(\ell^{-1})G(hkk') \\ &= \overline{\eta}(h, kk')\overline{\eta}(hkk', \ell)\eta(hk, k'\ell)\Psi(\ell^{-1})G^\sharp(h, kk') \\ &= \overline{\eta}(h,k)\overline{\eta}(k',\ell)\eta(k,k')\Psi(\ell^{-1})G^\sharp(h,kk'), \end{aligned}$$

where the last equality follows from the cocycle computations:

$$\begin{aligned} \overline{\eta}(h \cdot kk', \ell)\eta(hk, k'\ell)\overline{\eta}(h, kk') &= \overline{\eta}(h, k \cdot k'\ell)\overline{\eta}(k \cdot k', \ell)\eta(hk, k\ell') \\ &= \overline{\eta}(h,k)\overline{\eta}(k',\ell)\eta(k,k'). \end{aligned}$$

In conclusion, $G^\sharp \in W$. Similarly,

$$\begin{aligned} F^\flat(h\ell) &= F(h\ell, 1_K) \\ \text{(by (8.11))} \quad &= \overline{\eta}(h,\ell)F(h,\ell) \\ \text{(again by (8.11))} \quad &= \overline{\eta}(h,\ell)\Psi(\ell^{-1})F(h,1_K) \\ &= \overline{\eta}(h,\ell)\Psi(\ell^{-1})F^\flat(h) \end{aligned}$$

so that $F^\flat \in \mathrm{Ind}_L^H V$. Moreover,

$$\begin{aligned} (F^\flat)^\sharp(h,k) &= \eta(h,k)F^\flat(hk) \\ &= \eta(h,k)F(hk,1_K) \\ \text{(by (8.11))} \quad &= F(h,k) \end{aligned}$$

and

$$(G^\sharp)^\flat(h) = G^\sharp(h,1_K) = G(h),$$

so that $F \mapsto F^\flat$ and $G \mapsto G^\sharp$ are inverses of each other. Finally, we prove that $F \mapsto F^\flat$ intertwines Ξ and $\mathrm{Ind}_L^H \Psi =: \Omega$: for all $h, h_0 \in H$ and $F \in W$

$$\begin{aligned}[]
[\Xi(h)F]^\flat(h_0) &= [\Xi(h)F](h_0, 1_K) \\
\text{(by (8.12))} \quad &= \eta(h^{-1}, h_0)F(h^{-1}h_0, 1_K) \\
&= \eta(h^{-1}, h_0)F^\flat(h^{-1}h_0) \\
\text{(by (8.2))} \quad &= (\Omega(h)[F^\flat])(h_0).
\end{aligned}$$

In conclusion, $\Omega \sim \Xi \sim \Phi$, and this ends the proof. ∎

Theorem 8.10 *Let H be a group and let $K \leq H$ be a subgroup. Let η be an equalized unitary 2-cocycle of H. Let Θ be an η-representation of H on the space W and let Ψ be an η-representation of K on the space V. Then the map*

$$\phi\colon W \otimes \mathrm{Ind}_K^H V \to \mathrm{Ind}_K^H[(\mathrm{Res}_K^H W) \otimes V]$$

defined by setting

$$[\phi(w \otimes f)](h) := \Theta(h^{-1})w \otimes f(h)$$

for all $w \in W$, $f \in \mathrm{Ind}_K^H V$, and $h \in H$, is an isomorphism of η^2-representations of H (cf. Proposition 7.16.(1)).

Proof Set $\Phi := \mathrm{Ind}_K^H \Psi$ and $\lambda := \mathrm{Ind}_K^H[(\mathrm{Res}_K^H \Theta) \otimes \Psi]$. The space $W \otimes \mathrm{Ind}_K^H V$ is spanned by all simple tensors of the form $w \otimes f$, where $w \in W$ and $f\colon H \to V$ satisfies $f(hk) = \overline{\eta}(h, k)\Psi(k^{-1})f(h)$, for all $h \in H$ and $k \in K$. The space of λ is made up of all functions $F\colon H \to W \otimes V$ such that

$$F(hk) = \overline{\eta}^2(h, k)[\Theta(k^{-1}) \otimes \Psi(k^{-1})]F(h). \tag{8.16}$$

Moreover, λ is given by

$$[\lambda(h)F](h_0) = \eta^2(h^{-1}, h_0)F(h^{-1}h_0) \tag{8.17}$$

because it is an η^2-representation of H.

We now show that the image of ϕ is the space of λ. For $h \in H$ and $k \in K$ we have:

$$\begin{aligned}[]
[\phi(w \otimes f)](hk) &= [\Theta(k^{-1}h^{-1})w \otimes f(hk)] \\
\text{(by (7.10) and (8.1))} \quad &= \eta(k^{-1}, h^{-1})\overline{\eta}(h, k)[\Theta(k^{-1}) \otimes \Psi(k^{-1})]\left(\Theta(h^{-1})w \otimes f(h)\right) \\
(\eta \text{ is equalized}) \quad &= \overline{\eta}^2(h, k)[\Theta(k^{-1}) \otimes \Psi(k^{-1})]\left([\phi(w \otimes f)](h)\right)
\end{aligned}$$

so that, by (8.16), $\phi(w\otimes f) \in \mathrm{Ind}_K^H[(\mathrm{Res}_K^H W)\otimes V]$. We now show that ϕ intertwines $\Theta \otimes \Phi$ and λ. For all $h, h_0 \in H$, we have

$$\begin{aligned}
\{\phi[(\Theta(h) \otimes \Phi(h))(w \otimes f)]\}(h_0) &= (\phi\{[\Theta(h)w] \otimes [\Phi(h)f]\})(h_0) \\
&= [\Theta(h_0^{-1})\Theta(h)w] \otimes \{[\Phi(h)f](h_0)\} \\
\text{(by (7.10) and (8.2))} \quad &= \overline{\eta}(h_0^{-1}, h)\eta(h^{-1}, h_0)[\Theta(h_0^{-1}h)w \otimes f(h^{-1}h_0)] \\
(\eta \text{ is equalized}) \quad &= \eta^2(h^{-1}, h_0)[\phi(w \otimes f)](h^{-1}h_0) \\
\text{(by (8.17))} \quad &= \{\lambda(h)[\phi(w \otimes f)]\}(h_0).
\end{aligned}$$

By Corollary 8.5, $\mathrm{Ind}_K^H[(\mathrm{Res}_K^H W) \otimes V]$ is spanned by the functions $\lambda(h)F_{w\otimes v}$, $h \in H, w \in W$, and $v \in V$, where

$$F_{w\otimes v}(h) := \begin{cases} [\Theta(k^{-1})w] \otimes [\Psi(k^{-1})v] & \text{if } h = k \in K \\ 0 & \text{otherwise.} \end{cases}$$

Then,

$$[\phi(w \otimes f_v)](h) = [\Theta(h^{-1})w] \otimes [f_v(h)]$$

is zero if $h \notin K$ while if $h = k \in K$ it is equal to $F_{w\otimes v}(k)$, that is $\phi(w \otimes f_v) = F_{w\otimes v}$. This shows that ϕ is surjective. Finally, since

$$\dim[W \otimes \mathrm{Ind}_K^H V] = \dim W \cdot \dim V \cdot |H/K| = \dim \mathrm{Ind}_K^H[(\mathrm{Res}_K^H W) \otimes V]$$

we deduce that ϕ is also injective and therefore it is an isomorphism. ■

Proposition 8.11 *Let H be a group and let $K \leq H$ be a subgroup. Let η be an equalized unitary 2-cocycle of H. Let Ψ_j be an irreducible η-representation of K on the space V_j, $j = 1, 2,$. Then*

$$\mathrm{Ind}_K^H \Psi_1 \oplus \mathrm{Ind}_K^H \Psi_2 \sim \mathrm{Ind}_K^H (\Psi_1 \oplus \Psi_2).$$

Proof It is easy to check that the map

$$\alpha \colon \mathrm{Ind}_K^H V_1 \oplus \mathrm{Ind}_K^H V_2 \to \mathrm{Ind}_K^H (V_1 \oplus V_2)$$

given by

$$[\alpha(f_1 + f_2)](h) := f_1(h) + f_2(h)$$

for all $h \in H$, $f_j \in \mathrm{Ind}_K^H V_j$, and $j = 1, 2$, yields the required isomorphism. ■

Theorem 8.12 (Frobenius Reciprocity I) *Let H be a group and let $K \leq H$ be a subgroup. Let η be an equalized unitary 2-cocycle of H. Let Ψ_1 be an irreducible η-representation of K and let Φ_2 be an irreducible η-representation of H. Then the multiplicity of Φ_2 in $\mathrm{Ind}_K^H \Psi_1$ equals the multiplicity of Ψ_1 in $\mathrm{Res}_K^H \Phi_2$.*

Proof Setting $\Phi_1 := \mathrm{Ind}_K^H \Psi_1$ and $\Psi_2 := \mathrm{Res}_K^H \Phi_2$, we have

$$\widetilde{\Phi_1} = \mathrm{Ind}_{G(C_n,K,\eta)}^{G(C_n,H,\eta)} \widetilde{\Psi_1} \quad \text{and} \quad \widetilde{\Psi_2} = \mathrm{Res}_{G(C_n,K,\eta)}^{G(C_n,H,\eta)} \widetilde{\Phi_2} \tag{8.18}$$

by (8.9) and (8.10). Therefore if $n := |H|$ (see also the remark before the proof of Theorem 6.19) by virtue of (7.27) we have

$$\begin{aligned}
\text{multiplicity of } \Phi_2 \text{ in } \mathrm{Ind}_K^H \Psi_1 &= \frac{1}{|H|} \langle \chi^{\Phi_1}, \chi^{\Phi_2} \rangle_{L(H)} \\
\text{(by (7.28))} \quad &= \frac{1}{n|H|} \langle \chi^{\widetilde{\Phi_1}}, \chi^{\widetilde{\Phi_2}} \rangle_{L(G(C_n,H,\eta))} \\
\text{(by (8.18) and (1.34))} \quad &= \frac{1}{n|K|} \langle \chi^{\widetilde{\Psi_1}}, \chi^{\widetilde{\Psi_2}} \rangle_{L(G(C_n,K,\eta))} \\
&= \frac{1}{|K|} \langle \chi^{\Psi_1}, \chi^{\Psi_2} \rangle_{L(K)} \\
&= \text{multiplicity of } \Psi_1 \text{ in } \mathrm{Res}_K^H \Phi_2.
\end{aligned}$$

■

Example 8.13 Consider the case $K = \{1_H\}$ with the trivial representation $\iota_{\{1_H\}}$, which is clearly an η-representation of K. We then define the *η-regular representation* of H by setting $\Lambda_\eta := \mathrm{Ind}_{\{1_H\}}^H \iota_{\{1_H\}}$. Then by (8.2) we have

$$\left[\Lambda_\eta(h) f\right](h_0) = f(h^{-1}h_0)\eta(h^{-1}, h_0),$$

for all $h, h_0 \in H$ and $f \in L(H)$. Moreover, as $\mathrm{Res}_{\{1_H\}}^H \Theta \sim d_\Theta \iota_{\{1_H\}}$, from Frobenius reciprocity (Theorem 8.12) we deduce that

$$\Lambda_\eta \sim \bigoplus_{\Theta \in \widehat{H}^\eta} d_\Theta \Theta. \tag{8.19}$$

Remark 8.14 The interested reader may find a proof of formula (8.19) by constructing a basis for $\mathrm{Hom}_H(V, L(H))$.

Theorem 8.15 (Frobenius Reciprocity II) *Let Θ be an η-representation of H on the space W and let Ψ be an η-representation of K on the space V. Then the map*

$$\begin{array}{ccc} \mathrm{Hom}_H(W, \mathrm{Ind}_K^H) & \to & \mathrm{Hom}_K(\mathrm{Res}_K^H W, V) \\ T & \mapsto & \widehat{T} \end{array}$$

given by setting

$$\widehat{T}w := (Tw)(1_H)$$

for all $w \in W$ is a linear isomorphism. Its inverse is given by the map $L \mapsto \check{L}$, where

$$[\check{L}w](h) := L\Theta(h^{-1})w$$

for all $w \in W$ and $h \in H$.

Proof Set $\Phi := \mathrm{Ind}_K^H \Psi$. We first show that indeed $\widehat{T} \in \mathrm{Hom}_K(\mathrm{Res}_K^H W, V)$. For $w \in W$ and $k \in K$ we have

$$\begin{aligned} \widehat{T}\Theta(k)w &= \{T[\Theta(k)w]\}(1_H) \\ (\text{since } T \in \mathrm{Hom}_H(\Theta, \Phi)) \quad &= [\Phi(k)(Tw)](1_H) \\ (\text{by (8.2) and } \eta(k^{-1}, 1_H) = 1) \quad &= (Tw)(k^{-1}) \\ (\text{by (8.1)}) \quad &= \Psi(k)(Tw)(1_H) \\ &= \Psi(k)\widehat{T}w. \end{aligned}$$

We now show that, for $L \in \mathrm{Hom}_K(\mathrm{Res}_K^H W, V)$ and $w \in W$, one has $\check{L}w \in \mathrm{Ind}_K^H W$.

$$\begin{aligned} [\check{L}w](hk) &= L\Theta(k^{-1}h^{-1})w \\ (\text{by (7.10)}) \quad &= \eta(k^{-1}, h^{-1})L\Theta(k^{-1})\Theta(h^{-1})w \\ (\text{by (6.23)}) \quad &= \overline{\eta}(h, k)\Psi(k^{-1})[\check{L}w](h), \end{aligned}$$

so that (8.1) is satisfied. Moreover, $\check{L} \in \mathrm{Hom}_H(\Theta, \Phi)$. Indeed, for $h, h_0 \in H$ and $w \in W$ we have:

$$\begin{aligned} [\check{L}\Theta(h)w](h_0) &= L\Theta(h_0^{-1})\Theta(h)w \\ (\text{by (7.10)}) \quad &= \overline{\eta}(h_0^{-1}, h)L\Theta(h_0^{-1}h)w \\ (\text{by (6.23)}) \quad &= \eta(h^{-1}, h_0)[\check{L}w](h^{-1}h_0) \\ (\text{by (8.2)}) \quad &= [\Phi(h)\check{L}w](h_0), \end{aligned}$$

that is, $\check{L}\Theta(h) = \Phi(h)\check{L}$. Finally, we show that the maps $T \mapsto \widehat{T}$ and $L \mapsto \check{L}$ are inverses of each other: for all $w \in W$ and $h \in H$ we have, on the one hand for all $T \in \mathrm{Hom}_H(W, \mathrm{Ind}_K^H V)$

$$\begin{aligned}
[(\widehat{T})^\vee w](h) &= \widehat{T}\Theta(h^{-1})w \\
&= [T\Theta(h^{-1})w](1_H) \\
(T \in \mathrm{Hom}_H(\Theta, \Phi)) \quad &= \Phi(h^{-1})(Tw)(1_H) \\
(\text{by } (8.2)) \quad &= (Tw)(h)
\end{aligned}$$

and, on the other hand, $(\check{L})^\wedge w = [\check{L}w](1_H) = Lw$, for all $L \in \mathrm{Hom}_K(\mathrm{Res}_K^H W, V)$. ∎

Theorem 8.16 (The Other Side of Frobenius Reciprocity) *With the same notation as in Theorem 8.15, define the maps*

$$\begin{array}{ccc}
\mathrm{Hom}_H(\mathrm{Ind}_K^H V, W) & \to & \mathrm{Hom}_K(V, \mathrm{Res}_K^H W) \\
T & \mapsto & \mathring{T}
\end{array}$$

and

$$\begin{array}{ccc}
\mathrm{Hom}_K(V, \mathrm{Res}_K^H W) & \to & \mathrm{Hom}_H(\mathrm{Ind}_K^H V, W) \\
L & \mapsto & \overset{\diamond}{L}
\end{array}$$

by setting $\mathring{T}v := Tf_v$ for all $v \in V$ (see (8.5)) and

$$\overset{\diamond}{L}f := \frac{1}{|K|}\sum_{h \in H} \Theta(h)Lf(h)$$

for all $f \in \mathrm{Ind}_K^H V$. Then these maps are linear isomorphisms and inverses of each other.

Proof For $k \in K$ and $v \in V$ we have

$$\begin{aligned}
\mathring{T}\Psi(k)v &= Tf_{\Psi(k)v} \\
(\text{by Corollary } 8.6.(2)) \quad &= T\Phi(k)f_v \\
(T \in \mathrm{Hom}_H(\mathrm{Ind}_K^H V, W)) \quad &= \Theta(k)Tf_v \\
&= \Theta(k)\mathring{T}v,
\end{aligned}$$

so that $\overset{\circ}{T} \in \mathrm{Hom}_K(V, \mathrm{Res}^H_K W)$. Analogously, for $h \in H$ and $f \in \mathrm{Ind}^H_K V$ we have

$$\begin{aligned}
\overset{\diamond}{L}\Phi(h)f &= \frac{1}{|K|}\sum_{h_0 \in H} \Theta(h_0)L\left([\Phi(h)f](h_0)\right) \\
\text{(by (8.2))} \quad &= \frac{1}{|K|}\sum_{h_0 \in H} \eta(h^{-1}, h_0)\Theta(h_0)Lf(h^{-1}h_0) \\
(h_1 = h^{-1}h_0) \quad &= \frac{1}{|K|}\sum_{h_1 \in H} \eta(h^{-1}, hh_1)\Theta(hh_1)Lf(h_1) \\
\text{(by (7.10))} \quad &= \frac{1}{|K|}\Theta(h)\sum_{h_1 \in H} \eta(h^{-1}, hh_1)\eta(h, h_1)\Theta(h_1)Lf(h_1) \\
&= \Theta(h)\overset{\diamond}{L}f,
\end{aligned}$$

where the last equality follows from the cocycle identity

$$\eta(h^{-1}, h \cdot h_1)\eta(h, h_1) = \eta(1_H, h_1)\eta(h^{-1}, h)$$

and the latter equals 1 since η is equalized.

This shows that $\overset{\diamond}{L} \in \mathrm{Hom}_H(\mathrm{Ind}^H_K V, W)$.

We now show that the maps $T \mapsto \overset{\circ}{T}$ and $L \mapsto \overset{\diamond}{L}$ are inverses one of each other. For all $v \in V$ we have:

$$\begin{aligned}
(\overset{\diamond}{L})^\circ v &= \overset{\diamond}{L} f_v \\
&= \frac{1}{|K|}\sum_{k \in K} \Theta(k)Lf_v(h) \\
\text{(by (8.5))} \quad &= \frac{1}{|K|}\sum_{k \in K} \Theta(k)L\Psi(k^{-1})v \\
(L \in \mathrm{Hom}_K(V, \mathrm{Res}^H_K W)) \quad &= \frac{1}{|K|}\sum_{k \in K} \Theta(k)\Theta(k^{-1})Lv \\
\text{(by (7.10) and } \eta \text{ equalized)} \quad &= Lv.
\end{aligned}$$

Finally, for all $F \in \mathrm{Ind}_K^H V$ we have

$$
\begin{aligned}
(\overset{\circ}{T})^{\diamond} F &= \frac{1}{|K|} \sum_{h \in H} \Theta(h) \overset{\circ}{T} F(h) \\
&= \frac{1}{|K|} \sum_{h \in H} \Theta(h) T f_{F(h)} \\
(T \in \mathrm{Hom}_H(\mathrm{Ind}_K^H V, W)) \quad &= \frac{1}{|K|} \sum_{h \in H} T \Phi(h) f_{F(h)} \\
(\text{by Corollary 8.6.(1)}) \quad &= TF.
\end{aligned}
$$

■

Remark 8.17 Note that $\overset{\diamond}{L}$ may be defined for an arbitrary $L \in \mathrm{Hom}(V, W)$ and also in this case we have $\overset{\diamond}{L} \in \mathrm{Hom}_H(\mathrm{Ind}_K^H V, W)$. But the proof of the equality $(\overset{\diamond}{L})^{\circ} = L$ requires $L \in \mathrm{Hom}_K(V, \mathrm{Res}_K^H W)$.

One may also show that $(L^*)^{\diamond} = (\overset{\diamond}{L})^*$ for all $L \in \mathrm{Hom}_K(V, \mathrm{Res}_K^H W)$.

Proposition 8.18 *Let Ψ be an η-representation of K. Then*

$$(\mathrm{Ind}_K^H \Psi)' \sim \mathrm{Ind}_K^H (\Psi')$$

as η-representations of H.

Proof We consider the linear map

$$\alpha \colon \mathrm{Ind}_K^H (V') \to (\mathrm{Ind}_K^H V)'$$

defined by setting

$$[\alpha(f')](f) := \sum_{h \in H} f'(h)[f(h)] \tag{8.20}$$

for all $f' \in \mathrm{Ind}_K^H (V')$ and $f \in \mathrm{Ind}_K^H V$. Let $H = \sqcup_{t \in \mathcal{T}} tK$ be a decomposition of H into left K-cosets. We have

$$
\begin{aligned}
[\alpha(f')](f) &= \sum_{t \in \mathcal{T}} \sum_{k \in K} f'(tk)[f(tk)] \\
(\text{by (8.1)}) \quad &= \sum_{t \in \mathcal{T}} \sum_{k \in K} [\Psi'(k^{-1}) \eta(t, k) f'(t)] \left(\Psi(k^{-1}) \overline{\eta}(t, k) f(t) \right)
\end{aligned}
$$

$$(\text{by } (7.30)) \quad = \sum_{t\in\mathcal{T}} \sum_{k\in K} [\Psi'(k)\Psi'(k^{-1}) f'(t)]\,(f(t))$$

$$(\overline{\eta} \text{ is equalized}) \quad = |K| \sum_{t\in\mathcal{T}} f'(t)[f(t)].$$

Therefore, by Lemma 8.3, for all $t \in \mathcal{T}$ and $v \in V$,

$$[\alpha(f')](f_{t,v}) = |K| f'(t)v$$

so that $\alpha(f') = 0$ implies that $f' = 0$. This shows that α is a linear isomorphism. We now prove that it is also an intertwining operator. Set $\Phi := \mathrm{Ind}_K^H \Psi$ and $\Phi_1 := \mathrm{Ind}_K^H \Psi'$. For $f' \in \mathrm{Ind}_K^H(V')$, $f \in \mathrm{Ind}_K^H V$, and $h, h_0 \in H$, taking into account (7.30), we have:

$$\begin{aligned}
[\Phi'(h)\alpha(f')](f) &= \alpha(f')[\Phi(h^{-1})f] \\
(\text{by } (8.20)) \quad &= \sum_{h_0\in H} f'(h_0)\{[\Phi(h^{-1})f](h_0)\} \\
(\text{by } (8.2)) \quad &= \sum_{h_0\in H} \eta(h, h_0) f'(h_0)[f(hh_0)] \\
(h_1 = hh_0) \quad &= \sum_{h_1\in H} \eta(h, h^{-1}h_1) f'(h^{-1}h_1)[f(h_1)] \\
(\text{by } (8.2)) \quad &= \sum_{h_1\in H} \eta(h, h^{-1}h_1)\eta(h^{-1}, h_1)\{[\Phi_1(h)f'](h_1)\}[f(h_1)] \\
(\text{by } (6.1)) \quad &= \sum_{h_1\in H} \eta(1_H, h_1)\eta(h, h^{-1})\{[\Phi_1(h)f'](h_1)\}[f(h_1)] \\
(\eta \text{ is equalized}) \quad &= \alpha[\Phi_1(h)f'](f),
\end{aligned}$$

that is, $\Phi'(h) \circ \alpha = \alpha \circ \Phi_1(h)$ for all $h \in H$. This shows that $(\mathrm{Ind}_K^H \Psi)' = \Phi' \sim \Phi_1 = \mathrm{Ind}_K^H(\Psi')$. ■

Proposition 8.19 *Let K_1, K_2, H_1 and H_2 be four groups. Suppose that $K_j \leq H_j$ and let Ψ_j be an η_j-representation of K_j on V_j, for $j = 1, 2$. Then the map*

$$\epsilon\colon \mathrm{Ind}_{K_1}^{H_1} V_1 \otimes \mathrm{Ind}_{K_2}^{H_2} V_2 \to \mathrm{Ind}_{K_1\times K_2}^{H_1\times H_2}(V_1 \otimes V_2)$$

defined by setting

$$[\epsilon(f_1 \otimes f_2)](h_1, h_2) := f_1(h_1) \otimes f_2(h_2)$$

for all $h_j \in H_j$ *and* $f_j \in \mathrm{Ind}_{K_j}^{H_j} V_j$, $j = 1, 2$, *is a linear isomorphism of* $(\eta_1 \boxtimes \eta_2)$-*representations of* $H_1 \times H_2$ *(cf. Proposition 7.16) so that*

$$(\mathrm{Ind}_{K_1}^{H_1} \Psi_1) \boxtimes (\mathrm{Ind}_{K_2}^{H_2} \Psi_2) \sim \mathrm{Ind}_{K_1 \times K_2}^{H_1 \times H_2} (\Psi_1 \boxtimes \Psi_2).$$

Proof It is immediate to check that

- $\epsilon(f_1 \otimes f_2) \in \mathrm{Ind}_{K_1 \times K_2}^{H_1 \times H_2} (V_1 \otimes V_2)$ for all $f_j \in \mathrm{Ind}_{K_j}^{H_j} V_j$, $j = 1, 2$;
- setting $\Phi_j := \mathrm{Ind}_{K_j}^{H_j} \Psi_j$, $j = 1, 2$, and $\lambda := \mathrm{Ind}_{K_1 \times K_2}^{H_1 \times H_2} (\Psi_1 \boxtimes \Psi_2)$ we have

$$\epsilon[\Phi_1(h_1) \otimes \Phi_2(h_2)] = \lambda(h_1, h_2)\epsilon$$

  for all $h_1 \in H_1, h_2 \in H_2$;
- ϵ is injective, so that it is an isomorphism. We leave it to the reader to fill in the remaining details.

■

8.2 Mackey's Theory for Induced Projective Representations

Following [92], in order to develop Mackey's machinery on induced representation in the case of projective representations, we present some preliminary results.

Lemma 8.20 *Let* η *be an equalized, unitary 2-cocycle of the finite group* H *and let* $s \in H$. *Then* ${}^s\eta$, *defined by setting*

$${}^s\eta(h, k) := \eta(s^{-1}hs, s^{-1}ks)$$

for all $h, k \in H$, *is an equalized, unitary 2-cocycle, cohomologous to* η. *Moreover, we have:*

$${}^s\eta(h, k) = \eta(h, k)\rho_s(hk)\overline{\rho_s}(h)\overline{\rho_s}(k) \tag{8.21}$$

where

$$\rho_s(h) := \eta(s^{-1}h, s)\eta(s^{-1}, h) \tag{8.22}$$

for all $h, k \in H$.

Proof It is immediate to check that ${}^s\eta$ is an unitary, equalized 2-cocycle. Moreover, using repeatedly the cocycle identity (6.1) (a dot indicates, as usual, where we apply such an identity) we have

$$\begin{aligned}
{}^s\eta(h,k) &= \eta(s^{-1}h\cdot s, s^{-1}ks)\\
&= \eta(s^{-1}h, k\cdot s)\overline{\eta}(s^{-1}h, s)\eta(s, s^{-1}ks)\\
&= \eta(s^{-1}hk, s)\overline{\eta}(k,s)\eta(s^{-1}\cdot h, k)\overline{\eta}(s^{-1}h, s)\eta(s, s^{-1}ks)\\
&= \eta(s^{-1}hk, s)\overline{\eta}(k,s)\eta(s^{-1}, hk)\overline{\eta}(s^{-1}, h)\eta(h,k)\overline{\eta}(s^{-1}h, s)\eta(s, s^{-1}\cdot ks)\\
&= \eta(s^{-1}hk, s)\overline{\eta}(k,s)\eta(s^{-1}, hk)\overline{\eta}(s^{-1}, h)\eta(h,k)\overline{\eta}(s^{-1}h, s)\eta(1_H, ks)\overline{\eta}(s^{-1}, k\cdot s)\\
&= \eta(h,k)[\eta(s^{-1}hk, s)\eta(s^{-1}, hk)][\overline{\eta}(s^{-1}h, s)\overline{\eta}(s^{-1}, h)][\overline{\eta}(s^{-1}k, s)\overline{\eta}(s^{-1}, k)]\\
&= \eta(h,k)\rho_s(hk)\overline{\rho_s}(h)\overline{\rho_s}(k).
\end{aligned}$$

■

The following proposition shows what is the "right" conjugation for projective representations. It will play a fundamental role also in Chap. 9 where we develop a Clifford theory for induced projective representations. We use the notation in Lemma 8.20.

Proposition 8.21 *Let Ψ be an η-representation of $K \leq H$. Then ${}^s\Psi$, defined by setting*

$$ {}^s\Psi(r) := \Psi(s^{-1}rs)\overline{\rho_s}(r) \tag{8.23}$$

for all $r \in sKs^{-1}$, is an η-representation of sKs^{-1}.

Proof First of all, set $\Psi'(r) := \Psi(s^{-1}rs)$ for all $r \in sKs^{-1}$. Then, Ψ' is an ${}^s\eta$-representation of sKs^{-1}:

$$\begin{aligned}
\Psi'(r_1r_2) &= \Psi(s^{-1}r_1r_2s)\\
\text{(by (7.10))} \quad &= \eta(s^{-1}r_1s, s^{-1}r_2s)\Psi(s^{-1}r_1s)\Psi(s^{-1}r_2s)\\
&= {}^s\eta(r_1,r_2)\Psi'(r_1)\Psi'(r_2)
\end{aligned} \tag{8.24}$$

for all $r_1, r_2 \in sKs^{-1}$. Moreover, since ${}^s\Psi(r) = \Psi'(r)\overline{\rho_s}(r)$ for all $r \in sKs^{-1}$, we have, for all $r_1, r_2 \in sKs^{-1}$,

$$\begin{aligned}
{}^s\Psi(r_1r_2) &= \overline{\rho_s}(r_1r_2)\Psi'(r_1r_2)\\
\text{(by (8.24))} \quad &= {}^s\eta(r_1,r_2)\overline{\rho_s}(r_1r_2)\Psi'(r_1)\Psi'(r_2)\\
\text{(by (8.21))} \quad &= \overline{\rho_s}(r_1)\overline{\rho_s}(r_2)\eta(r_1,r_2)\Psi'(r_1)\Psi'(r_2)\\
&= \eta(r_1,r_2){}^s\Psi(r_1){}^s\Psi(r_2).
\end{aligned}$$

■

Remark 8.22 The function ρ_s has the following useful property:

$$\rho_s(h^{-1}) = \overline{\rho_s}(h) \tag{8.25}$$

for all $h \in H$. Indeed,

$$\begin{aligned} \rho_s(h^{-1}) &= \eta(s^{-1} \cdot h^{-1}, s)\eta(s^{-1}, h^{-1}) \\ &= \eta(s^{-1}, h^{-1}s)\eta(h^{-1}, s) \\ (\eta \text{ is equalized}) \quad &= \overline{\eta}(s^{-1}h, s)\overline{\eta}(s^{-1}, h) \\ (\text{by } (6.23)) \quad &= \overline{\rho_s}(h). \end{aligned}$$

We now expose the main results of this section, namely Mackey's machinery for projective representations. As already pointed out in Remark 8.8, our proofs are quite different from the original ones in [92]: Mackey makes a repeated use of Theorem 8.7 (note that he works on general locally compact groups); we give more direct proofs.

Let M be another subgroup of H and let $\mathcal{S}$ be a complete set of representatives for the M-K-cosets in H such that $1_H \in \mathcal{S}$. We thus have the decomposition

$$H = \sqcup_{s \in \mathcal{S}} MsK.$$

Let $s \in \mathcal{S}$. Then

$$H_s := M \cap sKs^{-1}$$

is a subgroup of M and $s^{-1}H_s s$ is a subgroup of K. It is easy to check that if $m, m_1 \in M$ and $k, k_1 \in K$ then

$$msk = m_1 s k_1 \Leftrightarrow \exists x \in H_s \text{ such that } m_1 = mx \text{ and } k_1 = s^{-1}x^{-1}sk. \tag{8.26}$$

(see [30, Lemma 11.3.1.]).

Let now Ψ be an η-representation of K on a space V. Then the restriction of (8.23) to H_s yields an η-representation on $V_s := V$.

Theorem 8.23 (Mackey's Lemma) *There exists an isomorphism*

$$\mathrm{Res}^H_M \mathrm{Ind}^H_K \Psi \sim \bigoplus_{s \in \mathcal{S}} \mathrm{Ind}^M_{H_s} {}^s\Psi$$

of η-representations of M.

Proof Let $s \in \mathcal{S}$ and set

$$Z_s := \{F \in \mathrm{Ind}^H_K V : F(ms'k) = \delta_{s,s'}\overline{\eta}(ms, k)\Psi(k^{-1})F(ms), \forall m \in M, s' \in \mathcal{S}, k \in K\}.$$

In other words, Z_s is made up of those $F \in \mathrm{Ind}_K^H V$ that vanish outside MsK. Clearly,

$$\mathrm{Ind}_K^H V = \bigoplus_{s\in\mathscr{S}} Z_s.$$

Consider the map

$$\begin{array}{rcl} Z_s & \to & \mathrm{Ind}_{H_s}^M V_s \\ F & \mapsto & F_s \end{array} \tag{8.27}$$

where $F_s\colon M \to V$ is defined by setting

$$F_s(m) := \eta(m,s)F(ms) \tag{8.28}$$

for all $m \in M$. First of all, we prove that indeed $F_s \in \mathrm{Ind}_{H_s}^M V_s$.

For $x \in H_s$ and $m \in M$, by (8.28) we have:

$$\begin{aligned} F_s(mx) &= \eta(mx,s)F(mxs) \\ &= \eta(mx,s)F(ms\cdot s^{-1}xs) \\ \text{(by (8.1))} \quad &= \overline{\eta}(ms,s^{-1}xs)\eta(mx,s)\Psi(s^{-1}x^{-1}s)F(ms) \\ \text{(by (8.28) and (8.23))} \quad &= \overline{\eta}(m,s)\rho_s(x^{-1})\overline{\eta}(ms,s^{-1}xs)\eta(mx,s)^s\Psi(x^{-1})F_s(m) \\ &= \overline{\eta}(m,x)^s\Psi(x^{-1})F_s(m), \end{aligned}$$

where the last equality follows from the cocycle computations (see (6.23) and (8.22)):

$$\begin{aligned} &\overline{\eta}(m,s)\rho_s(x^{-1})\overline{\eta}(ms,s^{-1}\cdot xs)\eta(mx,s) \\ &= \overline{\eta}(m,s)\overline{\eta}(m\cdot s,s^{-1})\overline{\eta}(m,x\cdot s)\overline{\eta}(x,s)\eta(mx,s) \\ &= \overline{\eta}(m,x). \end{aligned}$$

Conversely, given $f \in \mathrm{Ind}_{H_s}^M V_s$, define $f_s \in Z_s$ by setting

$$f_s(ms'k) := \delta_{s,s'}\overline{\eta}(ms,k)\overline{\eta}(m,s)\Psi(k^{-1})f(m) \tag{8.29}$$

for all $m \in M$, $s' \in \mathscr{S}$, and $k \in K$.

Note that f_s is well defined: indeed, if $msk = m_1sk_1$ then there exists $x \in H_s$ such that $k_1 = s^{-1}x^{-1}sk$ and $m_1 = mx$ (cf. 8.26) so that, by (7.10),

$$\begin{aligned}
\overline{\eta}(m_1 s, k_1)\overline{\eta}(m_1, s_1)\Psi(k_1^{-1})f(m_1) &= \eta(k^{-1}, s^{-1}xs)\overline{\eta}(m_1 s, k_1)\overline{\eta}(m_1, s_1) \\
&\quad \cdot \Psi(k^{-1})\Psi(s^{-1}xs)f(m_1) \\
\text{(by (8.23))} \quad &= \eta(k^{-1}, s^{-1}xs)\rho_s(x)\overline{\eta}(m_1 s, k_1)\overline{\eta}(m_1, s) \\
&\quad \cdot \Psi(k^{-1})^s\Psi(x)f(m_1) \\
(f \in \mathrm{Ind}_{H_s}^M {}^s\Psi) \quad &= \eta(k^{-1}, s^{-1}ks)\rho_s(x)\overline{\eta}(m_1 s, k_1)\overline{\eta}(m_1, s) \\
&\quad \cdot \eta(m_1, x^{-1})\Psi(k^{-1})f(m) \\
&= \overline{\eta}(ms, k)\overline{\eta}(m, s)\Psi(k^{-1})f(m),
\end{aligned}$$

where the last equality follows from the cocycle computations:

$$\begin{aligned}
&\eta(m_1, x^{-1})\eta(k^{-1}, s^{-1}xs)\rho_s(x)\overline{\eta}(m_1 s, k_1)\overline{\eta}(m_1, s) \\
&\qquad = \eta(mx, x^{-1})\eta(k^{-1}, s^{-1}x \cdot s)\eta(s^{-1}x, s) \\
&\qquad\quad \cdot \eta(s^{-1}, x)\overline{\eta}(mx \cdot s, s^{-1}x^{-1}sk)\overline{\eta}(mx, s) \\
&\qquad = \eta(mx, x^{-1})\eta(k^{-1}s^{-1}x, s)\eta(k^{-1}, s^{-1}x) \\
&\qquad\quad \cdot \eta(s^{-1}, x)\overline{\eta}(mx, x^{-1} \cdot sk)\overline{\eta}(s, s^{-1} \cdot x^{-1}sk) \\
&\qquad = \overline{\eta}(m, sk)\eta(x^{-1}, s \cdot k)\eta(k^{-1}, s^{-1}x)\eta(s^{-1}, x) \\
&\qquad = \overline{\eta}(m, sk)\overline{\eta}(s, k).
\end{aligned}$$

This shows that $f_s(msk) = f_s(m_1sk_1)$.

Moreover, $f_s \in Z_s$. Indeed, from (8.29) it follows that

$$\begin{aligned}
f_s(ms'k) &= \delta_{s,s'}\overline{\eta}(ms, k)\overline{\eta}(m, s)\Psi(k^{-1})f(m) \\
&= \delta_{s,s'}\overline{\eta}(ms, k)\Psi(k^{-1})f_s(ms')
\end{aligned}$$

for all $m \in M$, $s' \in \mathscr{S}$, and $k \in K$.

The maps

$$Z_s \ni F \mapsto F_s \in \mathrm{Ind}_{H_s}^M V_s \quad \text{and} \quad \mathrm{Ind}_{H_s}^M V_s \ni f \mapsto f_s \in Z_s$$

are inverses of each other: indeed, it is immediate to check that

$$F = f_s \Leftrightarrow f = F_s.$$

We deduce that (8.27) is a linear isomorphism.

Finally, we show that the map $F \mapsto F_s$ commutes with the action of M. Set $\Phi := \mathrm{Ind}_K^H \Psi$ and $\lambda_s := \mathrm{Ind}_{H_s}^M {}^s\Psi$. Then, for all $F \in W_s$ and $m, m_0 \in M$, by (8.28) we have:

$$\begin{aligned}
[\Phi(m)F]_s(m_0) &= \eta(m_0, s)[\Phi(m)F](m_0 s)\\
\text{(by (8.2))} \quad &= \eta(m^{-1}, m_0 \cdot s)\eta(m_0, s)F(m^{-1}m_0 s)\\
&= \eta(m^{-1}m_0, s)\eta(m^{-1}, m_0)F(m^{-1}m_0 s)\\
\text{(by (8.28))} \quad &= \eta(m^{-1}, m_0)F_s(m^{-1}m_0)\\
\text{(by (8.2))} \quad &= [\lambda_s(m)F_s](m_0).
\end{aligned}$$

■

Theorem 8.24 (Mackey's Tensor Product Theorem) *With the same notation as in Theorem 8.23 and with Θ an η-representation of M, there exists an isomorphism*

$$(\mathrm{Ind}_M^H \Theta) \otimes (\mathrm{Ind}_K^H \Psi) \sim \bigoplus_{s \in \mathscr{S}} \mathrm{Ind}_{H_s}^H \left[(\mathrm{Res}_{H_s}^M \Theta) \otimes {}^s\Psi\right].$$

of η^2-representations of H.

Proof Indeed, by virtue of Theorem 8.10, we have

$$\begin{aligned}
(\mathrm{Ind}_M^H \Theta) \otimes (\mathrm{Ind}_K^H \Psi) &\sim \mathrm{Ind}_M^H \left\{\Theta \otimes \left[\mathrm{Res}_M^H (\mathrm{Ind}_K^H \Psi)\right]\right\}\\
\text{(by Mackey's lemma)} \quad &\sim \mathrm{Ind}_M^H \left[\Theta \otimes \left(\bigoplus_{s \in \mathscr{S}} \mathrm{Ind}_{H_s}^M {}^s\Psi\right)\right]\\
\text{(again by Theorem 8.10)} \quad &\sim \mathrm{Ind}_M^H \left\{\bigoplus_{s \in \mathscr{S}} \mathrm{Ind}_{H_s}^M \left[(\mathrm{Res}_{H_s}^M \Theta) \otimes {}^s\Psi\right]\right\}\\
&= \bigoplus_{s \in \mathscr{S}} \mathrm{Ind}_{H_s}^H \left[(\mathrm{Res}_{H_s}^M \Theta) \otimes {}^s\Psi\right],
\end{aligned}$$

where the last equivalence follow from Proposition 8.11 and Theorem 8.9. ■

Theorem 8.25 (Mackey's Formula for Invariants) *With the same notation as in Theorem 8.24 we have the following isomorphism of vector spaces:*

$$\mathrm{Hom}_H(\mathrm{Ind}_M^H \Theta, \mathrm{Ind}_K^H \Psi) \cong \bigoplus_{s \in \mathscr{S}} \mathrm{Hom}_{H_s}(\mathrm{Res}_{H_s}^M \Theta, {}^s\Psi).$$

Proof By virtue of Theorem 8.16 (The other side of Frobenius reciprocity) we have

$$\begin{aligned}
\mathrm{Hom}_H(\mathrm{Ind}_M^H\Theta, \mathrm{Ind}_K^H\Psi) &\cong \mathrm{Hom}_M(\Theta, \mathrm{Res}_M^H\mathrm{Ind}_K^H\Psi)\\
\text{(by Mackey's lemma)} &\cong \mathrm{Hom}_M(\Theta, \bigoplus_{s\in\mathcal{S}} \mathrm{Ind}_{H_s}^M {}^s\Psi)\\
\text{(by Proposition 7.17)} &\cong \bigoplus_{s\in\mathcal{S}} \mathrm{Hom}_M(\Theta, \mathrm{Ind}_{H_s}^M {}^s\Psi)\\
\text{(by Theorem 8.15)} &\cong \bigoplus_{s\in\mathcal{S}} \mathrm{Hom}_{H_s}(\mathrm{Res}_{H_s}^M\Theta, {}^s\Psi).
\end{aligned}$$

■

By taking dimensions, we immediately deduce the following:

Corollary 8.26 (Mackey's Intertwining Number Theorem)

$$\dim\mathrm{Hom}_H(\mathrm{Ind}_M^H\Theta, \mathrm{Ind}_K^H\Psi) = \sum_{s\in\mathcal{S}} \dim\mathrm{Hom}_{H_s}(\mathrm{Res}_{H_s}^M\Theta, {}^s\Psi).$$

Finally, we have:

Corollary 8.27 (Mackey's Irreducibility Criterion) *With the above notation, suppose that $M = K$ and $\Theta = \Psi$. Then $\mathrm{Ind}_K^H\Psi$ is irreducible if and only if the following conditions are both met:*

(a) *Ψ is irreducible;*
(b) *for each $s \in \mathcal{S} \setminus \{1_H\}$, the η-representations $\mathrm{Res}_{H_s}^K\Psi$ and ${}^s\Psi$ contain no common irreducible subrepresentations.*

Proof First of all, note that $H_{1_H} = K$ and ${}^{1_H}\Psi = \Psi$, so that Mackey's intertwining number theorem (Corollary 8.26) yields

$$\dim\mathrm{End}_H(\mathrm{Ind}_K^H\Psi) = \dim\mathrm{End}_K(\Psi) + \sum_{s\in\mathcal{S}\setminus\{1_H\}} \dim\mathrm{Hom}_{H_s}(\mathrm{Res}_{H_s}^K\Psi, {}^s\Psi).$$

By Corollary 7.18, $\mathrm{Ind}_K^H\Psi$ is irreducible if and only if $\dim\mathrm{End}_K(\Psi) = 1$ (that is, Ψ is irreducible) and $\dim\mathrm{Hom}_{H_s}(\mathrm{Res}_{H_s}^K\Psi, \Psi_s) = 0$ for all $s \in \mathcal{S}\setminus 1_H$ (that is, $\mathrm{Res}_{H_s}^K\Psi$ and ${}^s\Psi$ have no common subrepresentations). ■

Chapter 9
Clifford Theory for Projective Representations

In this chapter, we establish some basic facts on Clifford theory for induced projective representations. We follow quite closely the exposition for the case of ordinary representations in Chap. 2 and we give the first applications of the theory. Another fundamental application is in Sect. 10.2 on irreducible projective representations of finite Abelian groups. This is not a standard topic and only some sporadic expositions may be found in the literature; see for instance [107].

9.1 Preliminaries and Notation

First of all, we establish all the notation used in this section. Let H be a finite group and let $N \trianglelefteq H$ be a normal subgroup of H (we will use the symbol B to denote the normal subgroup N when this is Abelian). Set $K := H/N$ (we will use the symbol A to denote this quotient when this is Abelian). Recall that if η is a unitary, equalized 2-cocycle of H, we denote by $\widehat{H}^\eta$ (resp. $\widehat{N}^\eta$) a complete set of pairwise-inequivalent irreducible, unitary η-representations of H (resp. N) (cf. Sect. 7.2).

We will use the notation on group extensions from Sect. 1.5 but we now think of N as a subgroup of H so that we avoid the use of the notation $\iota(n)$ or $\overline{n}$ to indicate $n \in N$ as an element of H.

Definition 9.1 Let $\Psi \in \widehat{N}^\eta$. We set (cf. Theorem 8.12)

$$\widehat{H}^\eta(\Psi) := \{\Phi \in \widehat{H}^\eta : \Psi \preceq \mathrm{Res}^H_N(\Phi)\} \equiv \{\Phi \in \widehat{H}^\eta : \Phi \preceq \mathrm{Ind}^H_N(\Psi)\}.$$

For $h \in H$, the *h-conjugate* of Ψ is the η-representation ${}^h\Psi \in \widehat{N}^\eta$ defined by

$$ {}^h\Psi(n) := \Psi(h^{-1}nh)\overline{\rho_h}(n), \tag{9.1}$$

T. Ceccherini-Silberstein et al., *Representation Theory of Finite Group Extensions*, Springer Monographs in Mathematics,
https://doi.org/10.1007/978-3-031-13873-7_9

where

$$\rho_h(n) := \eta(h^{-1}n, h)\eta(h^{-1}, n), \tag{9.2}$$

for all $n \in N$ (cf. Lemma 8.20 and Proposition 8.21).

The H-subgroup

$$I_H(\Psi) := \{h \in H : {}^h\Psi \sim \Psi\}$$

is called the *inertia group* of Ψ. Finally, we define

$$\widehat{I_H}^\eta(\Psi) := \{\Theta \in \widehat{I_H(\Psi)}^\eta : \Theta \preceq \mathrm{Ind}_N^{I_H(\Psi)}\Psi\}$$

and

$$K_H(\Psi) := I_H(\Psi)/N.$$

Lemma 9.2 *The η-conjugation (9.1) defines an action of H on $\widehat{N}^\eta$, the inertia group $I_H(\Psi)$ is the stabilizer of Ψ, and N is contained in $I_H(\Psi)$.*

Proof For $h_1, h_2 \in H, n \in N$ and $\Psi \in \widehat{N}^\eta$ we have:

$$\begin{aligned}\left[{}^{h_1}\left({}^{h_2}\Psi\right)\right](n) &= \left({}^{h_2}\Psi\right)\left(h_1^{-1}nh_1\right)\overline{\rho_{h_1}}(n)\\ &= \Psi\left((h_1h_2)^{-1}nh_1h_2\right)\overline{\rho_{h_1}}(n)\overline{\rho_{h_2}}\left(h_1^{-1}nh_1\right)\\ &= \Psi\left((h_1h_2)^{-1}nh_1h_2\right)\overline{\rho_{h_1h_2}}(n)\\ &= \left({}^{h_1h_2}\Psi\right)(n),\end{aligned} \tag{9.3}$$

by virtue of the cocycle identity

$$\begin{aligned}\rho_{h_1}(n)\rho_{h_2}\left(h_1^{-1}nh_1\right) &= \eta\left(h_1^{-1}n, h_1\right)\eta\left(h_1^{-1}, n\right)\eta\left(h_2^{-1}h_1^{-1}nh_1, h_2\right)\eta\left(h_2^{-1}, h_1^{-1}n\cdot h_1\right)\\ &= \eta\left(h_2^{-1}h_1^{-1}n, h_1\right)\eta\left(h_2^{-1}, h_1^{-1}\cdot n\right)\eta\left(h_1^{-1}, n\right)\eta\left(h_2^{-1}h_1^{-1}n\cdot h_1, h_2\right)\\ &= \eta\left(h_2^{-1}, h_1^{-1}\right)\eta\left(h_2^{-1}h_1^{-1}, n\right)\eta(h_1, h_2)\eta\left(h_2^{-1}h_1^{-1}n, h_1h_2\right)\\ \text{(by (6.23))} \quad &= \rho_{h_1h_2}(n).\end{aligned} \tag{9.4}$$

It is obvious that $I_H(\Psi)$ is the stabilizer of Ψ. Finally, we show that N is contained in $I_H(\Psi)$: for all $m, n \in N$ we have

$$\begin{aligned} {}^m\Psi(n) &= \Psi\left(m^{-1}nm\right)\overline{\rho_m}(n) \\ \text{(by (7.11))} \quad &= \Psi(m)^{-1}\Psi(n)\Psi(m)\overline{\rho_m}(n)\overline{\eta}(m, m^{-1}nm)\eta(n, m) \\ &= \Psi(m)^{-1}\Psi(n)\Psi(m), \end{aligned} \tag{9.5}$$

by virtue of the cocycle identity

$$\begin{aligned} \rho_m(n)\eta\left(m, m^{-1}nm\right)\overline{\eta}(n, m) &= \eta(m^{-1}n, m)\,\eta(m^{-1}, n)\,\eta(m, m^{-1}n \cdot m)\,\overline{\eta}(n, m) \\ &= \eta(m, m^{-1}\cdot n)\,\eta(m^{-1}, n) \\ &= \eta(1_H, n)\,\eta(m, m^{-1}) \\ \text{(by (6.23))} \quad &= 1. \end{aligned}$$

■

Then note that, when $N \equiv H$, we have a trivial conjugation action of H on $\widehat{H}^\eta$:

$$ {}^h\Phi \sim \Phi \qquad \text{for all } \Phi \in \widehat{H}^\eta \text{ and } h \in H. \tag{9.6}$$

Indeed, from Lemma 9.2 it follows that $I_H(\Phi) = H$.

From this remark, we deduce that in the general case (N is a possibly proper subgroup), $\widehat{H}^\eta({}^h\Psi) = \widehat{H}^\eta(\Psi)$ for all $\Psi \in \widehat{N}^\eta$ and $h \in H$.

We now fix a normalized section $t\colon K \to H$ and recall (1.54). As in (9.1), we may define an action of K on $\widehat{N}^\eta$ by setting ${}^k\Psi := {}^{t(k)}\Psi$, that is,

$$ {}^k\Psi(n) := \Psi(t(k)^{-1}nt(k))\overline{\rho_{t(k)}}(n) \equiv \Psi\left(n^{k^\sharp}\right)\overline{\rho_{t(k)}}(n), \tag{9.7}$$

for all $k \in K$ and $n \in N$.

Lemma 9.3 *The map $(k, \Psi) \mapsto {}^k\Psi$ defines an action of K on $\widehat{N}^\eta$. Moreover, it does not depend on the choice of the section t.*

Proof Given $k_1, k_2 \in K$ we have:

$$\begin{aligned} {}^{k_1}\left({}^{k_2}\Psi\right) &= {}^{t(k_1)t(k_2)}\Psi && \text{(by (9.7) and (9.3))} \\ &= {}^{t(k_1k_2)n(k_1,k_2)}\Psi && \text{(by (1.56))} \\ &= {}^{t(k_1k_2)}\left({}^{n(k_1,k_2)}\Psi\right) && \text{(by (9.3))} \\ &\sim {}^{t(k_1k_2)}\Psi && \text{(by (9.5))} \end{aligned}$$

so that, ${}^{k_1}\big({}^{k_2}\Psi\big) \sim {}^{k_1k_2}\Psi$. Moreover, since N is contained in the kernel of the action of H on $\widehat{N}^\eta$, we deduce that (9.7) is the induced action of $K = H/N$ on $\widehat{N}^\eta$.

We now show that it does not depend on the choice of t. Let $k \in K$ and $n \in N$ and set $h := t(k)n \in t(k)N$. For all $m \in N$ we have

$$\begin{aligned}
{}^h\Psi(m) &= \Psi(h^{-1}mh)\,\overline{\rho_h}(m) && \text{(by (9.1))}\\
&= \Psi\big(n^{-1}t(k)^{-1}mt(k)n\big)\overline{\rho_h}(m)\\
&= \Psi(n)^{-1}\Psi\big(t(k)^{-1}mt(k)\big)\Psi(n)\rho_n(t(k)^{-1}mt(k))\overline{\rho_h}(m) && \text{(by (9.5))}\\
&= \Psi(n)^{-1}\cdot\Psi\big(t(k)^{-1}mt(k)\big)\overline{\rho_{t(k)}}(m)\cdot\Psi(n) && \text{(by (9.4))}\\
&= \Psi(n)^{-1}\cdot{}^{t(k)}\Psi(m)\cdot\Psi(n) && \text{(by (9.1)).}
\end{aligned}$$

It follows that ${}^{t(k)}\Psi \sim {}^h\Psi$. ■

Suppose that $\mathscr{R} = \mathscr{R}(\Psi)$ is a set of representatives for the left $K_H(\Psi)$-cosets in K, i.e. $K = \sqcup_{r\in\mathscr{R}} rK_H(\Psi)$. Then $\{t(r) : r \in \mathscr{R}(\Psi)\}$ is a set of representatives for the left $I_H(\Psi)$-cosets in H:

$$H = \sqcup_{r\in\mathscr{R}} t(r)I_H(\Psi). \tag{9.8}$$

Moreover, $\{{}^h\Psi : h \in H\} \equiv \{{}^r\Psi : r \in \mathscr{R}\}$ and the representations ${}^r\Psi$'s are pairwise non–equivalent. In $\widehat{N}^\eta$ we define an equivalence relation $\approx$ by setting $\Psi_1 \approx \Psi_2$ if there exists $h \in H$ such that ${}^h\Psi_1 \sim \Psi_2$. Let Σ be a complete set of representatives of the corresponding quotient space $\widehat{N}^\eta/\approx$. Then

$$\widehat{N}^\eta = \sqcup_{\Psi\in\Sigma}\{{}^r\Psi : r \in \mathscr{R}(\Psi)\} \tag{9.9}$$

is the partition of $\widehat{N}^\eta$ into the orbits of H.

Proposition 9.4 *Let $N < L \lhd H$ be an intermediate subgroup. Then*

$$\mathrm{Ind}_N^L\left({}^h\Psi\right) \sim {}^h\!\left(\mathrm{Ind}_N^L\Psi\right), \tag{9.10}$$

for all $h \in H$.

Proof Let V_Ψ (resp. $V_{{}^h\Psi}$) denote the representation space of Ψ (resp. ${}^h\Psi$), as in (8.1), then the map $\mathrm{Ind}_N^L V_\Psi \ni f \mapsto {}^hf \in \mathrm{Ind}_N^L V_{{}^h\Psi}$, where

$${}^hf(\ell) := f(h^{-1}\ell h)\rho_h(\ell), \tag{9.11}$$

for all $\ell \in L$, is a vector space isomorphism. Indeed,

$$
\begin{aligned}
{}^h f(\ell n) &= f(h^{-1}\ell h \cdot h^{-1}nh)\rho_h(\ell n)\\
\text{(by (8.1))} \quad &= \Psi(h^{-1}n^{-1}h) f(h^{-1}\ell h)\overline{\eta}(h^{-1}\ell h, h^{-1}nh)\rho_h(\ell n)\\
\text{(by (9.1))} \quad &= {}^h\Psi(n^{-1})\rho_h(n^{-1}) f(h^{-1}\ell h)\overline{\eta}(h^{-1}\ell h, h^{-1}nh)\rho_h(\ell n)\\
\text{(by (9.11))} \quad &= {}^h\Psi(n^{-1})\, {}^h f(\ell)\overline{\rho_h}(\ell)\overline{\eta}(h^{-1}\ell h, h^{-1}nh)\rho_h(n^{-1})\rho_h(\ell n)\\
\text{(by (8.25))} \quad &= {}^h\Psi(n^{-1})\, {}^h f(\ell)\overline{{}^h\eta}(\ell, n)\overline{\rho_h}(\ell)\overline{\rho_h}(n)\rho_h(\ell n)\\
\text{(by (8.21))} \quad &= {}^h\Psi(n^{-1})\, {}^h f(\ell)\overline{\eta}(\ell, n)
\end{aligned}
$$

for all $n \in N$ and $\ell \in L$, so that ${}^h f \in \mathrm{Ind}_N^L V_{{}^h\Psi}$. It also establishes the equivalence (9.10): indeed, if $\Phi := \mathrm{Ind}_N^L \Psi$ and $\Phi_1 := \mathrm{Ind}_N^L \left({}^h\Psi\right)$ (cf. (8.2)) we have:

$$
\begin{aligned}
\left[\Phi_1(\ell)\, {}^h f\right](\ell_0) &= {}^h f(\ell^{-1}\ell_0) \cdot \eta(\ell^{-1}, \ell_0)\\
\text{(by (9.11))} \quad &= f(h^{-1}\ell^{-1}\ell_0 h)\eta(\ell^{-1}, \ell_0)\rho_h(\ell^{-1}\ell_0)\\
\text{(by (8.21))} \quad &= f(h^{-1}\ell^{-1}h \cdot h^{-1}\ell_0 h)\, {}^h\eta(\ell^{-1}, \ell_0)\rho_h(\ell^{-1})\rho_h(\ell_0)\overline{\rho_h}(\ell^{-1}\ell_0) \cdot \rho_h(\ell^{-1}\ell_0)\\
\text{(by (8.25))} \quad &= f(h^{-1}\ell^{-1}h \cdot h^{-1}\ell_0 h)\, {}^h\eta(\ell^{-1}, \ell_0)\overline{\rho_h}(\ell)\rho_h(\ell_0)\\
\text{(by (9.1) and (8.2))} \quad &= \left[{}^h\Phi(\ell) f\right](h^{-1}\ell_0 h)\rho_h(\ell_0)\\
\text{(by (9.11))} \quad &= {}^h\left[{}^h\Phi(\ell) f\right](\ell_0),
\end{aligned}
$$

for all $\ell, \ell_0 \in L$ and $f \in \mathrm{Ind}_N^L V_\Psi$. This shows that the unitary operator $f \mapsto {}^h f$ intertwines Φ_1 and ${}^h\Phi$. ■

We now give the analogues of (2.7) and (2.8) in the present setting:

$$I_H(\Psi) = \sqcup_{k \in K_H(\Psi)} t(k)N \tag{9.12}$$

is the coset decomposition of $I_H(\Psi)$ over N and we have

$$H = \sqcup_{r \in \mathscr{R}} t(r) I_H(\Psi) = \sqcup_{r \in \mathscr{R}} \sqcup_{k \in K_H(\Psi)} t(rk)N. \tag{9.13}$$

Also, we set

$$c = c_\Psi := [I_H(\Psi) : N]. \tag{9.14}$$

We end this section with a discussion on the notions of inflation and extension for cocycle representations.

Definition 9.5 Let τ be a unitary equalized 2-cocycle of K and let Θ be a τ-representation of K. The *inflations* of τ and Θ are the unitary equalized 2-cocycle of $\overset{\vartriangle}{\tau}$ of H and the $\overset{\vartriangle}{\tau}$-representation $\overset{\vartriangle}{\Theta}$ of H defined by setting

$$\overset{\vartriangle}{\tau}(h_1, h_2) := \tau(h_1 N, h_2 N) \quad \text{and} \quad \overset{\vartriangle}{\Theta}(h) := \Theta(hN) \tag{9.15}$$

for all $h_1, h_2, h \in H$, respectively.

Clearly, $\overset{\vartriangle}{\tau}(t(k_1)n, t(k_2)n) = \tau(k_1, k_2)$ and $\overset{\vartriangle}{\Theta}(t(k)n) = \Theta(k)$ for all $k, k_1, k_2 \in K$ and $n \in N$.

Moreover, if Θ is irreducible then $\overset{\vartriangle}{\Theta}$ is irreducible as well and if $\Theta \sim \Theta_1 \oplus \Theta_2$ then $\overset{\vartriangle}{\Theta} \sim \overset{\vartriangle}{\Theta}_1 \oplus \overset{\vartriangle}{\Theta}_2$.

Definition 9.6 Let η be a 2-cocycle of H, let L be a subgroup of H, and let $\Psi \in \widehat{L}^\eta$. An *extension* of Ψ to H is a representation $\widetilde{\Psi} \in \widehat{H}^\eta$ such that $\mathrm{Res}^H_L \widetilde{\Psi} = \Psi$.

In our setting, if $\Phi \in \widehat{H}^\eta$ then the condition $\mathrm{Res}^H_L \Phi = \Psi$ may be expressed by saying that Φ is an extension of Ψ. Clearly, an extension of a representation may fail to either exist or to be unique.

9.2 Basic Clifford Theory for Projective Representations

As in Sect. 2.2, we begin by specifying the classical Mackey's lemma (Theorem 8.23) in the present setting. We assume that η is a unitary equalized cocycle of H and N is a normal subgroup of H.

Theorem 9.7 *Let $\Psi \in \widehat{N}^\eta$ and $\Theta \in \widehat{H}^\eta(\Psi)$, and denote by ℓ the multiplicity of Ψ in* $\mathrm{Res}^H_N \Theta$. *Then we have:*

(1) $\mathrm{Res}^H_N(\mathrm{Ind}^H_N \Psi) \sim \bigoplus_{k \in K} {}^k\Psi \sim c \bigoplus_{r \in \mathcal{R}} {}^r\Psi$ *is the decomposition of* $\mathrm{Res}^H_N(\mathrm{Ind}^H_N \Psi)$ *into irreducible η-representations.*
(2) $\dim\mathrm{End}_H(\mathrm{Ind}^H_N \Psi) = c$.
(3) $\mathrm{Res}^H_N \Theta \sim \ell \bigoplus_{r \in \mathcal{R}} {}^r\Psi$.
(4) $d_\Theta = \ell |H/I_H(\Psi)| d_\Psi$ *and there is a $*$-algebra isomorphism*

$$\mathrm{End}_N(\mathrm{Res}^H_N \Theta) \cong \underbrace{M_c(\mathbb{C}) \oplus \cdots \oplus M_c(\mathbb{C})}_{\ell \text{ times}}.$$

Proof

(1) Let V denote the representation space of Ψ. For all $k \in K$, set (cf. (8.4))

$$Z_k := \{f : H \to V \text{ s.t. } f(t(k_0)n) = \delta_{k,k_0}\Psi(n^{-1})f(t(k))\overline{\eta}(t(k), n), \forall n \in N, k_0 \in K\},$$

that is, Z_k is made up of all $f \in \mathrm{Ind}_N^H V$ such that $f \equiv 0$ outside of $t(k)N$. Therefore, from (1.54) it follows that

$$\mathrm{Ind}_N^H V = \bigoplus_{k \in K} Z_k. \tag{9.16}$$

Moreover, the map $L_k\colon V \to Z_k$, given by

$$[L_k v](t(k_0)n) := \delta_{k,k_0}\overline{\eta}(t(k), n)\Psi(n^{-1})v, \tag{9.17}$$

for all $v \in V$, $k_0 \in K$, and $n \in N$, is a linear isomorphism of vector spaces, cf. Lemma 8.3.

Setting $\Phi := \mathrm{Ind}_N^H \Psi$, by (8.2) one has

$$\begin{aligned}
[\Phi(n)L_k v]\,(t(k_0)m) &= \eta(n^{-1}, t(k_0)m)\,[L_k v]\Big(n^{-1}t(k_0)m\Big) \\
\text{(by (1.57))} \quad &= \eta(n^{-1}, t(k_0)m)\,[L_k v]\left(t(k_0)\left(n^{-1}\right)^{k_0^\sharp} m\right) \\
\text{(by (9.17))} \quad &= \delta_{k,k_0}\,\eta(n^{-1}, t(k)m)\,\overline{\eta}\left(t(k), \left(n^{-1}\right)^{k^\sharp} m\right)\Psi\left(m^{-1}n^{k^\sharp}\right)v \\
\text{(by (6.23) and (7.10))} \quad &= \delta_{k,k_0}\,\eta(n^{-1}, t(k)m)\,\eta\left(m^{-1}n^{k^\sharp}, t(k)^{-1}\right)\eta\left(m^{-1}, n^{k^\sharp}\right)\cdot \\
&\quad \cdot \Psi\left(m^{-1}\right)\Psi\left(n^{k^\sharp}\right)v \\
\text{(by (9.7))} \quad &= \delta_{k,k_0}\,\eta\left(n^{-1}, t(k)m\right)\eta\left(m^{-1}n^{k^\sharp}, t(k)^{-1}\right)\eta\left(m^{-1}, n^{k^\sharp}\right)\cdot \\
&\quad \cdot \rho_{t(k)}(n)\Psi\left(m^{-1}\right)\,{}^{k}\Psi(n)v \\
&= \delta_{k,k_0}\,\overline{\eta}(t(k), m)\Psi(m^{-1})[\,{}^{k}\Psi(n)v] \\
&= \left[L_k\,{}^{k}\Psi(n)v\right](t(k_0)m)
\end{aligned} \tag{9.18}$$

for all $v \in V, k, k_0 \in K$ and $n, m \in N$. In the second to last identity we have used the following cocycle identity:

$$\begin{aligned}
&\eta\left(n^{-1}, t(k)m\right)\eta\left(m^{-1}\cdot n^{k^\sharp}, t(k)^{-1}\right)\eta\left(m^{-1}, n^{k^\sharp}\right)\rho_{t(k)}(n)\\
&= \eta\left(n^{-1}, t(k)m\right)\eta\left(m^{-1}, t(k)^{-1}\cdot n\right)\eta\left(t(k)^{-1}n\cdot t(k), t(k)^{-1}\right)\\
&\quad \times \eta\left(t(k)^{-1}n, t(k)\right)\eta\left(t(k)^{-1}, n\right)\\
&\qquad = \eta\left(n^{-1}, t(k)m\right)\eta\left(m^{-1}t(k)^{-1}, n\right)\eta\left(m^{-1}, t(k)^{-1}\right)\\
&\text{(by (6.23))} \qquad = \overline{\eta}(t(k), m)\,.
\end{aligned}$$

In other words, Z_k is Φ-invariant and $\Phi L_k = L_k{}^k\Psi$, that is, L_k is an intertwiner inducing an equivalence between $(\mathrm{Res}^H_N\Phi, Z_k)$ and $({}^k\Psi, V)$. Therefore, $\mathrm{Res}^H_N(\mathrm{Ind}^H_N\Psi)$ is equivalent to $\bigoplus_{k\in K}{}^k\Psi$ and the statement follows from (9.13):

$$\bigoplus_{k\in K}{}^k\Psi \sim \bigoplus_{r\in\mathscr{R}}\bigoplus_{k\in K_H(\Psi)}{}^{rk}\Psi \sim c\bigoplus_{r\in\mathscr{R}}{}^r\Psi.$$

(2) Since $\Psi \in \widehat{N}^\eta$ and by (1) its multiplicity in $\mathrm{Res}^H_N\mathrm{Ind}^H_N\Psi$ is equal to c, from Schur's lemma and Proposition 7.17 it follows that

$$\mathrm{Hom}_N(\mathrm{Res}^H_N\mathrm{Ind}^H_N\Psi, \Psi) \cong \mathbb{C}^c$$

and therefore, by the Frobenius reciprocity in Theorem 8.15,

$$\mathrm{dimEnd}_H(\mathrm{Ind}^H_N\Psi) = \mathrm{dimHom}_N(\mathrm{Res}^H_N\mathrm{Ind}^H_N\Psi, \Psi) = c.$$

(3) For all $h \in H$, the map

$$\begin{array}{ccc}
\mathrm{Hom}_N(\Psi, \mathrm{Res}^H_N\Theta) & \longrightarrow & \mathrm{Hom}_N({}^h\Psi, \mathrm{Res}^H_N\Theta)\\
T & \longmapsto & \Theta(h)T
\end{array}$$

is an isomorphism. Indeed, for all $n \in N$ we have $T\Psi(n) = \Theta(n)T$ and therefore:

$$\begin{aligned}
[\Theta(h)T]\,{}^h\Psi(n) &= \overline{\rho_h}(n)\Theta(h)\Theta(n^h)\Theta(h)^{-1}[\Theta(h)T] && \text{(by (9.1))}\\
&= \overline{\rho_h}(n)\overline{\eta}(h, n^h)\eta(n, h)\Theta(n)[\Theta(h)T] && \text{(by (7.11))}\\
&= \Theta(n)[\Theta(h)T],
\end{aligned}$$

where the last identity follows from the cocycle identity

$$\begin{aligned}\overline{\rho_h}(n)\,\overline{\eta}(h,n^h)\,\eta(n,h) &= \overline{\eta}(h^{-1}n,h)\,\overline{\eta}(h^{-1},n)\,\overline{\eta}(h,h^{-1}n\cdot h)\,\eta(n,h) && \text{(by (9.2))}\\ &= \overline{\eta}(h,h^{-1}\cdot n)\,\overline{\eta}(h^{-1},n) && \text{(by (6.23))}\\ &= 1.\end{aligned}$$

Therefore each ${}^h\Psi$, $h \in H$, has multiplicity ℓ in $\mathrm{Res}^H_N\Theta$. Moreover, by the Frobenius reciprocity in Theorem 8.12, we have that $\mathrm{Ind}^H_N\Psi$ contains exactly ℓ copies of Θ. Therefore, every irreducible sub-representation in $\mathrm{Res}^H_N\Theta$ is also a sub-representation of $\mathrm{Res}^H_N\mathrm{Ind}^H_N\Psi$. But the latter contains only irreducible sub-representations of the form ${}^r\Psi$, and this ends the proof.

(4) The first identity follows immediately from (3) and (9.8), while the isomorphism just requires the use of Proposition 7.17.

■

By Frobenius reciprocity, the multiplicity ℓ of Ψ in $\mathrm{Res}^H_N\Theta$ is also equal to the multiplicity of Θ in $\mathrm{Ind}^H_N\Psi$. Moreover, by Theorem 9.7.(3), it only depends on Θ and N, that is, $\ell = \ell_\Theta = \mathrm{dimHom}_N({}^r\Psi, \mathrm{Res}^H_N\Theta)$ for all $r \in \mathscr{R}$ (recall that $\{{}^r\Psi : r \in \mathscr{R}\}$ are the components of $\mathrm{Res}^H_N\Theta$). We thus give the following definition.

Definition 9.8 The number $\ell_\Theta := \mathrm{dimHom}_N(\Psi, \mathrm{Res}^H_N\Theta)$ (i.e., the multiplicity of Ψ in $\mathrm{Res}^H_N\Theta$) is called the *inertia index* of $\Theta \in \widehat{H}^\eta(\Psi)$ with respect to N.

Note that for all $r \in \mathscr{R}$

$$\mathrm{Ind}^H_N\Psi \sim \mathrm{Ind}^H_N\,{}^r\Psi \sim \bigoplus_{\Theta\in\widehat{H}^\eta(\Psi)} \ell_\Theta\Theta. \tag{9.19}$$

Remark 9.9 Clearly, $I_{I_H(\Psi)}(\Psi) = I_H(\Psi)$, and Theorem 9.7 with H replaced by $I_H(\Psi)$ yields:

$$\mathrm{Res}^{I_H(\Psi)}_N\mathrm{Ind}^{I_H(\Psi)}_N\Psi \sim \underbrace{\Psi\oplus\Psi\oplus\cdots\oplus\Psi}_{c\text{ times}} \quad\text{and}\quad \mathrm{Res}^{I_H(\Psi)}_N\Gamma \sim \underbrace{\Psi\oplus\Psi\oplus\cdots\oplus\Psi}_{\ell_\Gamma\text{ times}},$$

for all $\Gamma \in \widehat{I_H}^\eta(\Psi)$. In particular, $d_\Gamma = \ell_\Gamma d_\Psi$.

The following proposition contains some immediate but quite important consequences of Theorem 9.7.

Proposition 9.10 *With the above notation, we have the following.*

(1) $\sum_{\theta\in\widehat{H}^\eta(\Psi)}\ell_\Theta^2 = c$. *In particular,* $|\widehat{H}^\eta(\Psi)| = 1$ *if and only if* $\ell_\Theta^2 = c$ *for some* $\Theta \in \widehat{H}^\eta(\Psi)$.

(2) *Denoting by Σ a complete set of representatives of $\widehat{N}^\eta/\approx$ (cf. (9.9)),*

$$\widehat{H}^\eta = \sqcup_{\Psi\in\Sigma}\widehat{H}^\eta(\Psi) \tag{9.20}$$

is a partition of $\widehat{H}^\eta$ and the map

$$\{{}^r\Psi : r \in \mathscr{R}(\Psi)\} \longmapsto \widehat{H}^\eta(\Psi) \tag{9.21}$$

is a bijection between the parts of the partitions (9.9) and (9.20).

Proof (1) From (9.19) and Theorem 9.7.(3), setting $\ell := \sum_{\Theta\in\widehat{H}^\eta(\Psi)}\ell_\Theta^2$ we immediately get

$$\mathrm{Res}^H_N\mathrm{Ind}^H_N\Psi \sim \bigoplus_{\Theta\in\widehat{H}^\eta(\Psi)} \ell_\Theta\,\mathrm{Res}^H_N\Theta \sim \ell\left(\bigoplus_{r\in\mathscr{R}} {}^r\Psi\right),$$

and therefore Theorem 9.7.(1) forces $\ell = c$.

(2) This is an immediate consequence of Theorem 9.7.(3). Just note that if $\Xi \in \widehat{H}^\eta$, $\Omega \in \widehat{N}^\eta$ is contained in $\mathrm{Res}^G_N\Xi$, and $\Psi \in \Sigma$ is the representative of the G-orbit containing Ω, then $\Xi \in \widehat{H}^\eta(\Psi)$.

■

From (9.21) and (9.9) it follows that the left hand side of (7.25) may be written in the form:

$$\sum_{j=1}^{r_\eta} d_j^2 = \sum_{\Psi\in\widehat{N}^\eta}\frac{1}{|\mathscr{R}(\Psi)|}\sum_{\Theta\in\widehat{H}^\eta(\Psi)} d_\Theta^2 = \sum_{\Psi\in\Sigma}\sum_{\Theta\in\widehat{H}^\eta(\Psi)} d_\Theta^2. \tag{9.22}$$

Note also that, when $N = \{1_H\}$ is trivial, Proposition 9.10.(1) reduces to (7.25).

We now study the intermediate step of inducing from N to the inertia group $I_H(\Psi)$.

Lemma 9.11 *Let*

$$\mathrm{Ind}_N^{I_H(\Psi)}\Psi \sim \bigoplus_{\Gamma\in\widehat{I_H}^\eta(\Psi)} \ell_\Gamma\Gamma \tag{9.23}$$

be the decomposition of $\mathrm{Ind}_N^{I_H(\Psi)}\Psi$ into $I_H(\Psi)$-irreducible η-representations ($\ell_\Gamma > 0$ is the multiplicity of $\Gamma \in \widehat{I_H}^\eta(\Psi)$ and therefore it also equals its inertia index with respect to N). Then

$$\mathrm{Ind}^H_N\Psi \sim \bigoplus_{\Gamma\in\widehat{I_H}^\eta(\Psi)} \ell_\Gamma\,\mathrm{Ind}^H_{I_H(\Psi)}\Gamma \tag{9.24}$$

is the decomposition of $\mathrm{Ind}_N^H \Psi$ *into its* H*-irreducible components (i.e., the* $\mathrm{Ind}_{I_H(\Psi)}^H \Gamma$*'s, with* $\Gamma \in \widehat{I_H}^\eta(\Psi)$*, are pairwise inequivalent* H*-irreducible* η*-representations).*

Moreover, for $\Theta \in \widehat{H}^\eta(\Psi)$ *one has* $\Theta \sim \mathrm{Ind}_{I_H(\Psi)}^H \Gamma$ *for some (unique)* $\Gamma \in \widehat{I_H}^\eta(\Psi)$*.*

Proof First note that from Theorem 9.7.(2) and with the same notation therein we get

$$\mathrm{dimEnd}_{I_H(\Psi)}\left(\mathrm{Ind}_N^{I_H(\Psi)}\Psi\right) = c = \mathrm{dimEnd}_H\left(\mathrm{Ind}_N^H\Psi\right).$$

Indeed, $I_H(\Psi)$ coincides with the inertia group of Ψ with respect to $I_H(\Psi)$ (cf. (9.14) and Remark 9.9). Moreover, the decomposition (9.24) follows from (9.23) and transitivity of induction in Theorem 8.9. Then, by applying Proposition 7.17 to (9.23) and (9.24), we deduce that

$$\sum_{\Gamma\in\widehat{I_H}^\eta(\Psi)} \ell_\Gamma^2 = c = \sum_{\Omega,\Gamma\in\widehat{I_H}^\eta(\Psi)} \ell_\Omega \ell_\Gamma \; \mathrm{dimHom}_H(\mathrm{Ind}_{I_H(\Psi)}^H\Omega, \mathrm{Ind}_{I_H(\Psi)}^H\Gamma).$$

Therefore,

$$\mathrm{dimHom}_H(\mathrm{Ind}_{I_H(\Psi)}^H\Omega, \mathrm{Ind}_{I_H(\Psi)}^H\Gamma) = \delta_{\Omega,\Gamma},$$

and the irreducibility of the η-representations $\mathrm{Ind}_{I_H(\Psi)}^H\Gamma$ follows.

The remaining part is an immediate consequence of (9.24): each $\Theta \preceq \mathrm{Ind}_N^H\Psi$ must equal one of the $\mathrm{Ind}_{I_H(\Psi)}^H\Gamma$, with $\Gamma \in \widehat{I_H}^\eta(\Psi)$. ■

Theorem 9.12 (Clifford Correspondence for Projective Representations) *The map*

$$\begin{array}{ccc} \widehat{I_H}^\eta(\Psi) & \longrightarrow & \widehat{H}^\eta(\Psi) \\ \Gamma & \longmapsto & \mathrm{Ind}_{I_H(\Psi)}^H\Gamma \end{array}$$

is a bijection. Moreover, the inertia index of $\Gamma \in \widehat{I_H}^\eta(\Psi)$ *with respect to* N *coincides with the inertia index of* $\mathrm{Ind}_{I_H(\Psi)}^H\Gamma$ *with respect to* N*, that is,*

$$\ell_\Gamma = \ell_{\mathrm{Ind}_{I_H(\Psi)}^H\Gamma}. \tag{9.25}$$

Finally

$$\mathrm{Res}_N^{I_H(\Psi)}\Gamma \sim \ell_\Gamma\Psi. \tag{9.26}$$

Proof The first part is just a reformulation of Lemma 9.11. Moreover, from (9.23) and (9.24) it follows that the inertia index of Γ with respect to N coincides with the inertia index of $\mathrm{Ind}_{I_H(\Psi)}^H \Gamma$ with respect to N. Finally, an application of Theorem 9.7.(3), with H replaced by $I_H(\Psi)$ (cf. Remark 9.9), yields (9.26). ■

Corollary 9.13 *Suppose that* $I_H(\Psi) \trianglelefteq H$ *and let* $\Gamma \in \widehat{I_H}^\eta(\Psi)$ *and* $\Theta \sim \mathrm{Ind}_{I_H(\Psi)}^H \Gamma$. *Then*

$$\mathrm{Res}_{I_H(\Psi)}^G \Theta \sim \bigoplus_{r \in \mathcal{R}(\Psi)} {}^r\Gamma, \tag{9.27}$$

where the η*-representations of* $I_H(\Psi)$ *in the right hand side are (irreducible and) pairwise inequivalent.*

Proof First of all, from Clifford correspondence, Frobenius reciprocity (Theorem 8.12), and the equivalence ${}^r\Theta \sim \Theta$ (cf. (9.6)), it follows that $\mathrm{Res}_{I_H(\Psi)}^G \Theta$ contains each ${}^r\Gamma$ with multiplicity 1. Since $d_\Theta = [G : I_H(\Psi)] d_\Gamma$, the decomposition (9.27) follows. ■

Note that, conversely, from (9.27) and $\mathrm{Res}_N^{I_H(\Psi)} {}^r\Gamma \sim \ell_\Theta {}^r\Psi$ (cf. (9.26)), one recovers Theorem 9.7.(3).

Corollary 9.14 *We have* $\ell_\Theta = 1$ *if and only if there exists* $\Gamma \in \widehat{I_H}^\eta(\Psi)$ *such that*

$$\mathrm{Res}_N^{I_H(\Psi)} \Gamma \sim \Psi \qquad \textit{and} \qquad \mathrm{Ind}_{I_H(\Psi)}^H \Gamma \sim \Theta.$$

Proof Indeed, by Clifford's correspondence, each $\Theta \in \widehat{H}^\eta(\Psi)$ is equivalent to an induced η-representation of the form $\mathrm{Ind}_{I_H(\Psi)}^H \Gamma$, for some $\Gamma \in \widehat{I_H}^\eta(\Psi)$, with $\ell_\Theta = \ell_\Gamma$. We complete by using the second decomposition in Remark 9.9. ■

Remark 9.15 The Clifford correspondence provides a method for determining $\widehat{H}^\eta$, starting from $\widehat{N}^\eta$. These are the steps:

- Determine a set Σ as in (9.9).
- For each $\Psi \in \Sigma$ determine $I_H(\Psi)$.
- Compute the decomposition $\mathrm{Ind}_N^{I_H(\Psi)} \Psi = \oplus_{\Gamma \in \widehat{I_H}(\Psi)} \ell_\Gamma \Gamma$, in order to determine $\widehat{I_H}^\eta(\Psi)$.
- $\widehat{H}^\eta(\Psi) = \left\{ \mathrm{Ind}_{I_H(\Psi)}^H \Gamma : \Gamma \in \widehat{I_H}(\Psi) \right\}$.
- Finally, $\widehat{H}^\eta = \sqcup_{\Psi \in \Sigma} \widehat{H}^\eta(\Psi)$.

We end this section by examining the extremal cases $I_H(\Psi) = H$ and $I_H(\Psi) = N$. We begin with a simple remark.

Remark 9.16

(1) If $I_H(\Psi) = H$, that is, ${}^h\Psi \sim \Psi$ for all $h \in H$, then $K_H(\Psi) \equiv K, \mathscr{R} = \{1_H\}$, $c = |K|$, and we have:

$$\mathrm{Res}^H_N \mathrm{Ind}^H_N \Psi \sim \underbrace{\Psi \oplus \Psi \oplus \cdots \oplus \Psi}_{c \text{ times}} \quad \text{and} \quad \mathrm{Res}^H_N \Theta \sim \underbrace{\Psi \oplus \Psi \oplus \cdots \oplus \Psi}_{\ell_\Theta \text{ times}}.$$

Moreover, $d_\Theta = \ell_\Theta d_\Psi$, for all $\Theta \in \widehat{H}^\eta(\Psi)$, cf. Theorem 9.7 and Remark 9.9.

(2) In particular, if $I_H(\Psi) = H$ for all $\Psi \in \widehat{N}^\eta$, then $\Sigma \equiv \widehat{N}^\eta$ and, with Λ_η denoting the η-regular representation of H, (8.19) becomes

$$\Lambda_\eta \sim \bigoplus_{\Psi \in \widehat{N}^\eta} d_\Psi \bigoplus_{\Theta \in \widehat{H}^\eta(\Psi)} \ell_\Theta \Theta, \tag{9.28}$$

cf. (9.20). Alternatively, one may deduce (9.28) by starting from $\Lambda_\eta \sim \mathrm{Ind}^G_{\{1_H\}} I_{\{1_H\}}$ and using (9.19) and transitivity of induction in Theorem 8.9.

The following proposition is the normal subgroup version of Mackey's irreducibility criterion (cf. Corollary 8.27).

Proposition 9.17 *The induced representation* $\mathrm{Ind}^H_N \Psi$ *is irreducible if and only if* $I_H(\Psi) = N$.

Proof This follows from Theorem 9.7.(2) combined with Proposition 7.17 and Schur's lemma. ■

Proposition 9.18 *Suppose that* $I_H(\Psi) = N$. *Then the following holds.*

(1) $\Theta := \mathrm{Ind}^H_N \Psi$ *is the unique element in* $\widehat{H}^\eta(\Psi)$.
(2) $\mathrm{Res}^H_N \Theta \sim \oplus_{k \in K} {}^k\Psi$, *with the* ${}^k\Psi$ *pairwise inequivalent.*
(3) *The image of* $\{{}^k\Psi : k \in K\}$ *via the map* (9.21) *is the singleton* $\{\mathrm{Ind}^H_N \Psi\}$.

Proof

(1) This follows from Proposition 9.17 and Definition 9.1.
(2) This is just Theorem 9.7.(1), with $c = 1$ and $\mathscr{R} \equiv K$.
(3) This is obvious.

■

Corollary 9.19 *Let* $\Psi_1, \Psi_2 \in \widehat{N}^\eta$ *and suppose that* $I_H(\Psi_1) = I_H(\Psi_2) = N$. *Then* $\mathrm{Ind}^H_N(\Psi_1) \sim \mathrm{Ind}^H_N(\Psi_2)$ *if and only if* Ψ_1 *and* Ψ_2 *are conjugate (i.e., there exists* $k \in K$ *such that* $\Psi_1 = {}^k\Psi_2$*).*

Proof This follows from Frobenius reciprocity (Theorem 8.12) and Proposition 9.18.(2). ■

Corollary 9.20 *If* $I_H(\Psi) \trianglelefteq H$ *then* $I_H(\Gamma) = I_H(\Psi)$ *for all* $\Gamma \in \widehat{I_H}^{\eta}(\Psi)$.

Proof By Clifford correspondence (Theorem 9.12), $\mathrm{Ind}_{I_H(\Psi)}^H \Gamma$ is irreducible. ∎

9.3 Projective Unitary Representations of a Group Extension

In this section we extend the results in Sects. 7.1 and 7.3 to projective representations. Our exposition is based on [107]. Suppose that η is an equalized, unitary, 2-cocycle of H, N is normal in H, $\Psi \in \widehat{N}^{\eta}$, V is the space of Ψ and assume all the other notations in Sects. 9.1 and 9.2. For all $h \in I_H(\Psi)$ we have ${}^h\Psi \sim \Psi$ and therefore there exists a unitary operator $\Theta'(h)$ such that

$$ {}^h\Psi(n) \equiv \Psi(h^{-1}nh)\overline{\rho_h}(n) = \Theta'(h)^{-1}\Psi(n)\Theta'(h), \tag{9.29}$$

for all $n \in N$; we suppose that $\Theta'(1_H) = I_V$. From (9.3) it follows that

$$\Theta'(h_2)^{-1}\Theta'(h_1)^{-1}\Psi(n)\Theta'(h_1)\Theta'(h_2) = \Theta'(h_1h_2)^{-1}\Psi(n)\Theta'(h_1h_2)$$

and therefore

$$\Theta'(h_1h_2)\Theta'(h_2)^{-1}\Theta'(h_1)^{-1}\Psi(n) = \Psi(n)\Theta'(h_1h_2)\Theta'(h_2)^{-1}\Theta'(h_1)^{-1}.$$

By Schur's lemma there exists a constant $\gamma(k_1, k_2) \in \mathbb{T}$ such that

$$\Theta'(h_1h_2) = \gamma(h_1, h_2)\Theta'(h_1)\Theta'(h_2) \tag{9.30}$$

and from Remark 7.9.(1) it follows that γ is a unitary 2-cocycle and Θ' is a γ-representation of $I_H(\Psi)$. Now set

$$\widetilde{\Theta}(t(k)n) := \Theta'(t(k))\Psi(n)\eta(t(k), n), \tag{9.31}$$

for all $k \in K_H(\Psi), n \in N$, and

$$\mu(t(k_1)n_1, t(k_2)n_2) := \gamma(t(k_1), t(k_2))\eta\left(n_1^{k_2^\sharp}, n_2\right)\eta\left(n(k_1, k_2), n_1^{k_2^\sharp}n_2\right)\rho_{t(k_2)}(n_1), \tag{9.32}$$

for all $k_1, k_2 \in K_H(\Psi), n_1, n_2 \in N$.

Proposition 9.21

(1) μ *is a unitary* 2*-cocycle of* $I_H(\Psi)$ *and* $\widetilde{\Theta}$ *is a* μ*-representation of* $I_H(\Psi)$ *such that* $\mathrm{Res}_N^{I_H(\Psi)}\widetilde{\Theta} = \Psi$.

(2) *We have:*

$$\widetilde{\Theta}(hm) = \widetilde{\Theta}(h)\Psi(m)\eta(h, m),$$

for all $h \in I_H(\Psi), n \in N$.

(3) *For all* $h_1, h_2 \in I_H(\Psi), n_1, n_2 \in N$ *we have:*

$$\overline{\mu}(h_1n_1, h_2n_2)\eta(h_1n_1, h_2n_2) = \overline{\mu}(h_1, h_2)\eta(h_1, h_2). \tag{9.33}$$

Proof

(1) For all $k_1, k_2 \in K_H(\Psi), n_1, n_2 \in N$, we have:

$$\begin{aligned}
&\widetilde{\Theta}(t(k_1)n_1 \cdot t(k_2)n_2) \\
&= \widetilde{\Theta}\left(t(k_1k_2)n(k_1, k_2)n_1^{k_2^\sharp}n_2\right) && \text{(by (1.60))} \\
&= \Theta'(t(k_1k_2))\Psi\left(n(k_1, k_2)n_1^{k_2^\sharp}n_2\right) && \text{(by (9.31))} \\
&= \widetilde{\Theta}\left(t(k_1k_2)n(k_1, k_2)\right)\Psi\left(n_1^{k_2^\sharp}\right)\Psi(n_2)\cdot \\
&\quad \cdot\eta\left(n_1^{k_2^\sharp}, n_2\right)\eta\left(n(k_1, k_2), n_1^{k_2^\sharp}n_2\right) && \text{(by (9.31) and (7.10))} \\
&= \gamma(t(k_1), t(k_2))\Theta'(t(k_1))\Psi(n_1) && \text{(by (1.56), (9.29) and (9.30))} \\
&\quad \cdot\Theta'(t(k_2))\Psi(n_2)\cdot) \\
&\quad \cdot\rho_{t(k_2)}(n_1)\eta\left(n_1^{k_2^\sharp}, n_2\right) \\
&\quad \cdot\eta\left(n(k_1, k_2), n_1^{k_2^\sharp}n_2\right) \\
&= \widetilde{\Theta}(t(k_1)n_1)\widetilde{\Theta}(t(k_2)n_2)\mu(t(k_1)n_1, t(k_2)n_2). && \text{(by (9.31) and (9.32))}
\end{aligned}$$

From Remark 7.9.(1) it follows that μ is a unitary 2-cocycle and $\widetilde{\Theta}$ is a μ-representation of $I_H(\Psi)$:

$$\widetilde{\Theta}(h_1h_2) = \widetilde{\Theta}(h_1)\widetilde{\Theta}(h_2)\mu(h_1, h_2), \tag{9.34}$$

for all $h_1, h_2 \in I_H(\Psi)$. It is obvious that its restriction to N is equal to Ψ.

(2) By means of a repeated application of (9.31), (7.10) and the cocycle identity we easily get:

$$\begin{aligned}\widetilde{\Theta}(t(k)\cdot nm) =&\Theta'(t(k))\Psi(nm)\eta(t(k),nm)\\ =&\Theta'(t(k))\Psi(n)\Psi(m)\eta(n,m)\eta(t(k),nm)\\ =&\Theta'(t(k))\Psi(n)\Psi(m)\eta(t(k),n)\eta(t(k)n,m)\\ =&\widetilde{\Theta}(t(k)n)\Psi(m)\eta(t(k)n,m),\end{aligned}$$

for all $k \in K_H(\Psi)$ and $n, m \in N$.

(3) On the one hand we have:

$$\begin{aligned}\widetilde{\Theta}(h_1n_1)\widetilde{\Theta}(h_2n_2)=&\eta(h_1,n_1)\eta(h_2,n_2)\widetilde{\Theta}(h_1)\Psi(n_1)\widetilde{\Theta}(h_2)\Psi(n_2)\\ =&\eta(h_1,n_1)\eta(h_2,n_2)\overline{\rho_{h_2}}(n_1)\widetilde{\Theta}(h_1)\widetilde{\Theta}(h_2)\Psi(h_2^{-1}n_1h_2)\Psi(n_2)\\ =&\eta(h_1,n_1)\eta(h_2,n_2)\overline{\rho_{h_2}}(n_1)\overline{\mu}(h_1,h_2)\overline{\eta}(h_2^{-1}n_1h_2,n_2)\widetilde{\Theta}(h_1h_2)\\ &\times \Psi(h_2^{-1}n_1h_2n_2).\end{aligned}$$

On the other hand:

$$\begin{aligned}\widetilde{\Theta}(h_1n_1)\widetilde{\Theta}(h_2n_2) =&\overline{\mu}(h_1n_1,h_2n_2)\widetilde{\Theta}(h_1n_1h_2n_2)\\ =&\overline{\mu}(h_1n_1,h_2n_2)\eta(h_1h_2,h_2^{-1}n_1h_2n_2)\widetilde{\Theta}(h_1h_2)\widetilde{\Theta}(h_2^{-1}n_1h_2n_2).\end{aligned}$$

It follows that

$$\overline{\mu}(h_1n_1,h_2n_2)\eta(h_1h_2,h_2^{-1}n_1h_2\cdot n_2) = \eta(h_1,n_1)\eta(h_2,n_2)\overline{\rho_{h_2}}(n_1)\overline{\mu}(h_1,h_2)\overline{\eta}(h_2^{-1}n_1h_2,n_2).$$

By a repeated application of the cocycle identity we get the following equivalent forms:

$$\begin{aligned}&\overline{\mu}(h_1n_1,h_2n_2)\eta(h_1n_1\cdot h_2,n_2)\eta(h_1h_2,h_2^{-1}n_1\cdot h_2)\\ &\qquad = \eta(h_1,n_1)\eta(h_2,n_2)\overline{\eta}(h_2^{-1}n_1,h_2)\overline{\eta}(h_2^{-1},n_1)\overline{\mu}(h_1,h_2)\end{aligned}$$

$$\overline{\mu}(h_1n_1,h_2n_2)\eta(h_1n_1,h_2n_2)\eta(h_1h_2,h_2^{-1}\cdot n_1) = \eta(h_1,n_1)\overline{\eta}(h_2^{-1},n_1)\overline{\mu}(h_1,h_2)$$

$$\overline{\mu}(h_1n_1,h_2n_2)\eta(h_1n_1,h_2n_2)\eta(h_1\cdot h_2,h_2^{-1}) = \overline{\mu}(h_1,h_2)$$

and finally, invoking also the fact that η is equalized, one can get easily (9.33). ∎

From (9.33) it follows that we may consider $\overline{\mu}\eta$ as a cocycle of $K_H(\Psi)$ and therefore we set

$$\tau(k_1, k_2) := \overline{\mu}(t(k_1), t(k_2))\eta(t(k_1), t(k_2)), \tag{9.35}$$

for all $k_1, k_2 \in K_H(\Psi)$.

Theorem 9.22 *Suppose that $\Psi \in \widehat{N}^\eta$, $\widetilde{\Theta}$ is as in (9.31), τ is as in (9.35),*

$$\Phi \in \widehat{K_G(\Psi)}^\tau \tag{9.36}$$

and U is the space of Φ. Set

$$\Xi(t(k)n) := \widetilde{\Theta}(t(k)n) \otimes \Phi(k), \tag{9.37}$$

for all $k \in K_H(\Psi), n \in N$. Then Ξ is an irreducible τ-representation of $I_H(\Psi)$ and it acts on the tensor product $V \otimes U$ (V is the space of Ψ and $\widetilde{\Theta}$). Moreover, the map

$$\begin{array}{ccc} \widehat{K_H(\Psi)}^\tau & \longrightarrow & \widehat{I}_H(\Psi) \\ \Phi & \longmapsto & \Xi \end{array} \tag{9.38}$$

is a bijection.

Proof For all $k_1, k_2 \in K_H(\Psi), n_1, n_2 \in N$ we have:

$$\begin{aligned}
\Xi(t(k_1)n_1 \cdot t(k_2)n_2) =& \Xi\left(t(k_1k_2)n(k_1, k_2)n_1^{k_2^\sharp}n_2\right) && \text{(by (1.60))}\\
=& \widetilde{\Theta}(t(k_1)n_1 \cdot t(k_2)n_2) \otimes \Phi(k_1k_2) && \text{(by (9.37))}\\
=& \widetilde{\Theta}(t(k_1)n_1)\widetilde{\Theta}(t(k_2)n_2) \otimes \Phi(k_1)\Phi(k_2)\cdot && \text{(by (9.34) and (9.36))}\\
& \cdot \eta(t(k_1)n_1, t(k_2)n_2)\\
=& \left[\widetilde{\Theta}(t(k_1)n_1) \otimes \Phi(k_1)\right]\left[\widetilde{\Theta}(t(k_2)n_2) \otimes \Phi(k_2)\right]\cdot\\
& \cdot \eta(t(k_1)n_1, t(k_2)n_2)\\
=& \Xi(t(k_1)n_1)\Xi(t(k_2)n_2)\eta(t(k_1)n_1, t(k_2)n_2).
\end{aligned}$$

Therefore Ξ is an η-representation of $I_H(\Psi)$. Moreover, for $n \in N$,

$$\Xi(n) = \Psi(n) \otimes I_U \sim \underbrace{\Psi(n) \oplus \cdots \oplus \Psi(n)}_{\dim U-\text{times}}. \tag{9.39}$$

Now we prove that Ξ is irreducible. Suppose that $F \in \mathrm{End}_{I_H(\Psi)}\Xi$. Then F commutes with $\Psi(n) \otimes I_U$ for each $n \in N$ and therefore, by Lemma 7.19 (which may be easily adapted to projective representations), $F = I_V \otimes T$, with $T \in \mathrm{End}(U)$. It follows that, for all $k \in K_H(\Psi)$,

$$\widetilde{\Theta}(t(k)) \otimes [T\Phi(k)] = F\Xi(t(k)) = \Xi(t(k))F = \widetilde{\Theta}(t(k)) \otimes [\Phi(k)T],$$

which forces $T\Phi(k) = \Phi(k)T$ and therefore, by Schur's lemma, $T = \alpha I_U$ for some $\alpha \in \mathbb{C}$. Then $F = \alpha(I_V \otimes I_U)$ and by invoking the converse to Schur's lemma we may conclude that Ξ is irreducible.

Now we show that the map $\Phi \mapsto \Xi$ is injective. Suppose that Γ is another irreducible η-representation of $K_H(\Psi)$ and that it acts on a unitary space W. Denote by Π the irreducible $I_H(\Psi)$ representation obtained by setting $\Pi(t(k)n) = \widetilde{\Theta}(t(k)n) \otimes \Gamma(k)$, as in (7.32), that is, $\Gamma \mapsto \Pi$ in (9.38). Suppose that Ξ and Π are equivalent, so that there exists a unitary isomorphism $P : V \otimes U \to V \otimes W$ such that:

$$P\left[\widetilde{\Theta}(t(k)n) \otimes \Phi(k)\right] = \left[\widetilde{\Theta}(t(k)n) \otimes \Gamma(k)\right] P$$

for all $k \in K_H(\Psi), n \in N$. In particular, as in the first part of the proof, $P[\Psi(n) \otimes I_U] = [\Psi(n) \otimes I_W] P$ forces $P = I_V \otimes T$ and T is a unitary equivalence between Φ and Γ.

Now we prove that the map is surjective, that is every $\Xi \in \widehat{I_H(\Psi)}$ is of the form (7.32). From Theorem 9.7.(3) we may assume that the representation space of Ξ is of the form $V \otimes Y$ and that

$$\Xi(n) = \Psi(n) \otimes I_Y \sim \underbrace{\Psi(n) \oplus \cdots \oplus \Psi(n)}_{\dim Y - \text{times}} \tag{9.40}$$

for all $n \in N$ (compare with (9.39)). Therefore, for $k \in K_H(\Psi)$,

$$\begin{aligned}
&\left[\widetilde{\Theta}(t(k))^{-1} \otimes I_Y\right] \Xi(t(k)) [\Psi(n) \otimes I_Y] \\
\text{(by (9.40))} \quad &= \left[\widetilde{\Theta}(t(k))^{-1} \otimes I_Y\right] \Xi(t(k))\Xi(n) \\
&= \left[\widetilde{\Theta}(t(k))^{-1} \otimes I_Y\right] \Xi\left(t(k)nt(k)^{-1}\right) \Xi(t(k)) \\
\text{(by (9.40))} \quad &= \left\{\left[\widetilde{\Theta}(t(k))^{-1}\Psi(t(k)nt(k)^{-1})\right] \otimes I_Y\right\} \Xi(t(k)) \\
\text{(by (9.29))} \quad &= \left\{\left[\Psi(n)\widetilde{\Theta}(t(k))^{-1}\right] \otimes I_Y\right\} \Xi(t(k))\rho_{t(k)^{-1}}(n) \\
&= [\Psi(n) \otimes I_Y]\left[\widetilde{\Theta}(t(k))^{-1} \otimes I_Y\right] \Xi(t(k))\rho_{t(k)^{-1}}(n).
\end{aligned}$$

From Lemma 7.19 (the factor $\rho_{t(k)^{-1}}(n)$ is irrelevant) it follows that there exists a unitary operator $\Gamma(k)$ on Y such that

$$\left[\widetilde{\Theta}(t(k))^{-1} \otimes I_Y\right] \Xi(t(k)) = I_X \otimes \Gamma(k),$$

that is

$$\Xi(t(k)) = \widetilde{\Theta}(t(k)) \otimes \Gamma(k). \tag{9.41}$$

If $k_1, k_2 \in K_H(\Psi)$ then

$$\begin{aligned}
\widetilde{\Theta}(t(k_1k_2)) \otimes \Gamma(k_2k_2) &= \Xi(t(k_1k_2)) && \text{(by (9.41))}\\
&= \Xi(t(k_1))\Xi(t(k_2))\left[\Psi(n(k_1,k_2)^{-1}) \otimes I_Y\right]\cdot && \text{(by (9.40) and (1.56))}\\
&\quad \cdot \eta(t(k_1), t(k_2)\eta\left(t(k_1)t(k_2), n(k_1,k_2)^{-1}\right) \\
&= \widetilde{\Theta}(t(k_1))\widetilde{\Theta}(t(k_2))\Psi\left(n(k_1,k_2)^{-1}\right) \otimes \Gamma(k_1)\Gamma(k_2) && \text{by (9.41)}\\
&\quad \cdot \eta(t(k_1), t(k_2)\eta\left(t(k_1)t(k_2), n(k_1,k_2)^{-1}\right) \\
&= \widetilde{\Theta}(t(k_1k_2)) \otimes \Gamma(k_1)\Gamma(k_2)\tau(k_1,k_2) && \text{(by (9.34) and (9.35))}
\end{aligned}$$

which yields $\Gamma(k_2k_2) = \Gamma(k_1)\Gamma(k_2)\tau(k_1,k_2)$, that is, Γ is a τ-representation of $K_H(\Psi)$. Finally,

$$\begin{aligned}
\Xi(t(k)n) &= \Xi(t(k))[\Psi(n) \otimes I_Y]\eta(t(k),n) && \text{(by (9.40))}\\
&= \left[\widetilde{\Theta}(t(k))\Psi(n) \otimes \Gamma(k)\right]\eta(t(k),n) && \text{(by (9.41))}\\
&= \widetilde{\Theta}(t(k)n) \otimes \Gamma(k), && \text{(by (9.31))}
\end{aligned}$$

that is Ξ may be written in the form (9.37) with $\Phi = \Gamma \in \widehat{K_H(\Psi)}^{\tau}$, so that the map (9.38) is surjective. ■

Theorem 9.23 *The map*

$$\Phi \longmapsto \mathrm{Ind}_{I_H(\Psi)}^{H} \Xi, \tag{9.42}$$

where Ξ is as in (9.37), is a bijection between $\widehat{K_H(\Psi)}^{\tau}$ and $\widehat{H}(\Psi)$ (cf. Definition 9.1).

Proof Combine Theorem 9.22 with Clifford correspondence (cf. Theorem 9.12). ■

Corollary 9.24 *The inertia index of* $\mathrm{Ind}_{I_H(\Psi)}^H \Xi$ *(= inertia index of* Ξ*, cf. Theorem 9.12) is equal to the dimension of* Φ.

Proof From (9.31), (9.37) and $\Phi(1_H) = I_U$ it follows that $\mathrm{Res}_N^{I_H(\Psi)} \Xi = \Psi \otimes I_U$; cf. Definition 9.8. ■

Theorem 9.25 *Let* Σ *be as in* (9.9). *Then*

$$\widehat{H}^\eta = \left\{ \mathrm{Ind}_{I_H(\Psi)}^H \Xi : \Psi \in \Sigma, \Xi \text{ is as in (9.37)}, \Phi \in \widehat{K_H(\Psi)}^\tau \right\},$$

that is, the right hand side is a list of all irreducible, unitary η*-representations of* H *and, for different values of* σ *and* Φ*, we get pairwise inequivalent representations.*

Proof For a fixed Ψ they are pairwise inequivalent by Theorem 9.23. If Ξ_1, Ξ_2 correspond to $\Psi_1 \not\sim \Psi_2$ then they are inequivalent by Theorem 9.7.(9.7).

Finally, if Γ is an irreducible η-representation of H, from Theorem 9.7 we get that $\Gamma \in \widehat{H}(\Psi)$ for a $\Psi \in \Sigma$ and therefore, again by Clifford correspondence, Γ is of the form $\mathrm{Ind}_{I_G(\sigma)}^G \Xi$ with $\Xi \in \widehat{I}_H(\Psi)$. But then Ξ is the image of some $\Phi \in \widehat{K_H(\Psi)}^\tau$ in (9.38). ■

Chapter 10
Projective Representations of Finite Abelian Groups with Applications

In this chapter, we describe the irreducible projective representations of a finite Abelian group: we shall use the Clifford theory from the previous chapter. Then, by means of the machinery developed in Sect. 7.3, we describe the ordinary, irreducible representations of finite metabelian groups. As a particular case, we obtain an alternative description of the irreducible representations of finite 2-step nilpotent groups (cf. Chap. 6.6).

10.1 Bicharacters and 2-Cocycles on Finite Abelian Groups

The results in this and the subsequent section are taken from [16, Section 6, Chapter 6] and from [60]. Even if the former treatment shed light on the latter's computations, we have added more detailed arguments.

Let A be a finite Abelian group and denote by C_m the cyclic group of order m, both written in multiplicative notation. For $a \in A$ we denote by $\langle a \rangle$ (resp. $o(a)$) the cyclic subgroup generated by a (resp. the order of a). We first recall the *primary decomposition* of A [30, Theorem 1.3.7]: if $|A| = p_1^{k_1} p_2^{k_2} \cdots p_t^{k_t}$ is the decomposition of the order of $|A|$ into powers of distinct primes, then there exists a *unique* decomposition of A of the form

$$A \cong A_1 \times A_2 \times \cdots \times A_t, \tag{10.1}$$

where $|A_j| = p_j^{k_j}$, $j = 1, 2, \ldots, t$. Actually, A_j is made up of all $a \in A$ whose order is a power of p_j. Moreover, if $a \in A$, $a = a_1 a_2 \cdots a_t$, with $a_j \in A_j$, then $o(a) = o(a_1) o(a_2) \cdots o(a_t)$.

We will also use the *invariant factors decomposition* of A [30, Theorem 1.3.1]: there exists a *unique* set of positive integers $r_1, r_2, \ldots, r_k$ (the *invariant factors* of

T. Ceccherini-Silberstein et al., *Representation Theory of Finite Group Extensions*, Springer Monographs in Mathematics,
https://doi.org/10.1007/978-3-031-13873-7_10

A) such that:

$$|A| = r_1 r_2 \cdots r_k \quad \text{and} \quad r_j \text{ divides } r_{j-1}, \quad j = 2, 3, \ldots, k \tag{10.2}$$

and

$$A \cong \prod_{j=1}^{k} \langle a_j \rangle, \tag{10.3}$$

where $a_j \in A$ and $\langle a_j \rangle \cong C_{r_j}$ for all $j = 1, 2, \ldots, k$. Clearly, the a_j's are not uniquely determined; we will say that $\{a_1, a_2, \ldots, a_k\}$ is a *basis* of A. Combining (10.1) and (10.2) we get the *structure theorem for finite Abelian groups* [30, Corollary 1.3.10]: there exist unique positive integers t, h_i, $i = 1, 2, \ldots, t$, and m_{ij}, $i = 1, 2, \ldots, t$, $j = 1, 2, \ldots, h_i$, such that

$$h_i \le k_i \quad \text{and} \quad m_{i1} \ge m_{i2} \ge \cdots \ge m_{i,h_i} \quad \text{for } i = 1, 2, \ldots, t,$$

such that

$$A \cong \prod_{i=1}^{t} \prod_{j=1}^{h_i} C_{p_i^{m_{ij}}}. \tag{10.4}$$

Moreover,

$$A_i \cong \prod_{j=1}^{h_i} C_{p_i^{m_{ij}}} \quad \text{and} \quad C_{r_j} \cong \prod_{\substack{1 \le i \le t: \\ h_i \ge j}} C_{p_i^{m_{ij}}}$$

with

$$\sum_{j=1}^{h_i} m_{ij} = k_i \quad \text{and} \quad \prod_{\substack{1 \le i \le t: \\ h_i \ge j}} p_i^{m_{ij}} = r_j, \tag{10.5}$$

for all $i = 1, 2, \ldots, t$ and $j = 1, 2, \ldots, h_i$.

The positive integers m_{ij} are called the *elementary divisors* of A.

Definition 10.1 An *antisymmetric bicharacter* of A is a function $\omega \colon A \times A \to \mathbb{T}$ such that the maps $\omega(\cdot, a)$ and $\omega(a, \cdot)$ are characters of A and $\omega(a, a) = 1$ for all $a \in A$.

Remark 10.2 The theory of antisymmetric bicharacters on finite Abelian groups has a strict analogy with the theory of alternating bilinear forms on a vector space; for the latter we refer to the exposition in Grove's monograph [55].

The set $\Omega_a(A)$ of all antisymmetric bicharacters of A, endowed with pointwise multiplication is an Abelian group, as one may immediately check.

Lemma 10.3 *The following conditions are equivalent for a function* $\omega\colon A \times A \to \mathbb{T}$*:*

(a) $\omega \in \Omega_a(A)$*, that is,* ω *is an antisymmetric bicharacter of* A*;*
(b) ω *satisfies the following conditions:*

  (i) $\omega(a, \cdot)$ *(alternatively,* $\omega(\cdot, a)$*) is a character of* A *for all* $a \in A$*,*
  (ii) $\omega(b, a) = \overline{\omega}(a, b)$ *for all* $a, b \in A$*,*
  (iii) $\omega(a, a) = 1$ *for all* $a \in A$*.*

Proof Suppose first that $\omega \in \Omega_a(A)$. Conditions (i) and (iii) are trivially satisfied by definition. Moreover, given $a, b \in A$ one has

$$1 = \omega(ab, ab) = \omega(a, ab)\omega(b, ab) = \omega(a, a)\omega(a, b)\omega(b, a)\omega(b, b) = \omega(a, b)\omega(b, a)$$

so that

$$\omega(a, b) = \overline{\omega(b, a)} = \overline{\omega}(b, a) \tag{10.6}$$

and (ii) follows as well. This shows the implication (a) $\Longrightarrow$ (b). Conversely, suppose (b). In order to show that $\omega \in \Omega_a(A)$ we only need to check that $\omega(\cdot, a)$ is a character of A for all $a \in A$. This is easily shown: given $a \in A$, then, using (10.6), for all $b_1, b_2 \in A$, we have

$$\omega(b_1 b_2, a) = \overline{\omega}(a, b_1 b_2) = \overline{\omega}(a, b_1)\overline{\omega}(a, b_2) = \omega(b_1, a)\omega(b_2, a).$$

Since $\omega(\cdot, a) = \overline{\omega}(a, \cdot)$ for all $a \in A$, this completes the proof (covering both possibilities in (i)). ■

Definition 10.4 Let $\omega \in \Omega_a(A)$. The *kernel* of ω is the subgroup

$$\ker \omega := \{a \in A : \omega(a, b) = 1 \text{ for all } b \in A\} \leq A. \tag{10.7}$$

We say that ω is *nondegenerate* if $\ker \omega = \{1_A\}$.

We fix $\omega \in \Omega_a(A)$ and for each $a \in A$ we denote by $\omega_a\colon A \to \mathbb{T}$ the character defined by $\omega_a(b) := \omega(a, b)$ for all $b \in A$. Then the map

$$A \ni a \longmapsto \omega_a \in \widehat{A} \tag{10.8}$$

is a homomorphism (we recall that $\widehat{A}$ is a group under pointwise multiplication of characters and that $A \cong \widehat{A}$; see [30, Section 2.3]). Moreover, $\ker \omega$ coincides with the kernel of (10.8). In particular, ω is nondegenerate if and only if (10.8) is an isomorphism.

The *ω-annihilator* of a subgroup $B \le A$, is the subgroup $B_\omega^\perp \le A$ defined by setting

$$B_\omega^\perp := \{a \in A : \omega(b,a) = 1 \text{ for all } b \in B\} \equiv \bigcap_{b \in B} \ker \omega_b. \tag{10.9}$$

Note that, by virtue of (10.6), we also have

$$B_\omega^\perp = \{a \in A : \omega(a,b) = 1 \text{ for all } b \in B\} = \{a \in A : B \le \ker \omega_a\}. \tag{10.10}$$

Lemma 10.5 *Let $\omega \in \Omega_a(A)$ and suppose that it is nondegenerate. Let $B \le A$ be a subgroup. Then one has group isomorphisms*

$$B \cong A/B_\omega^\perp \quad \textit{and} \quad B_\omega^\perp \cong A/B.$$

Suppose that the restriction $\omega|_{B\times B} \in \Omega_a(B)$ is also nondegenerate. Then

$$A = B \times B_\omega^\perp. \tag{10.11}$$

Proof As already remarked, since ω is nondegenerate, (10.8) is an isomorphism. Moreover, the map

$$\widehat{A} \ni \xi \longmapsto \mathrm{Res}_B^A \xi \in \widehat{B}$$

is a surjective homomorphism. By composing, we get that the map $A \ni a \mapsto \omega_a|_B \in \widehat{B}$ is a surjective homomorphism whose kernel is $B_\omega^\perp$, by virtue of (10.10). We deduce that $A/B_\omega^\perp \cong \widehat{B} \cong B$.

Consider the subgroup $C_\omega := \{\omega_c : c \in B_\omega^\perp\} \le \widehat{A}$. Note that C_ω is the image of $B_\omega^\perp$ under the isomorphism (10.8) so that, in particular, $|C_\omega| = |B_\omega^\perp|$. Now, the image of the inflation map (cf. Definition 2.2)

$$\widehat{A/B} \ni \varphi \longmapsto \overset{\vartriangle}{\varphi} \in \widehat{A}$$

coincides with C_ω. Indeed, if $\omega_c \in C_\omega$ then $\omega_c = \overset{\vartriangle}{\varphi}$, where $\varphi(aB) := \omega_c(a)$ for all $a \in A$. This proves that C_ω is contained in the image of the inflation; the reverse inclusion follows from $|C_\omega| = |B_\omega^\perp| = |A|/|B| = |\widehat{A/B}|$. It follows that

$$B_\omega^\perp \cong C_\omega \cong \widehat{A/B} \cong A/B.$$

Finally, if $\omega|_{B\times B}$ is nondegenerate then

$$B \cap B_\omega^\perp = \{b \in B : \omega(c,b) = 1 \text{ for all } c \in B\} = \ker \omega|_{B\times B} = \{1_A\}.$$

Since $|B| \cdot |B_\omega^\perp| = |A|$, we deduce (10.11). ■

Now we analyze the behavior of ω with respect to the primary decomposition of A.

Lemma 10.6 *Let $\omega \in \Omega_a(A)$ and let $a, b \in A$. Set $s := o(a)$ and $t := o(b)$, and suppose that* $\gcd(s, t) = 1$. *Then $\omega(a, b) = 1$.*

Proof By the Bézout identity there exist $\ell, r \in \mathbb{Z}$ such that $\ell s + rt = 1$. Moreover

$$\omega(a, b)^s = \omega(a^s, b) = \omega(1_A, b) = \overline{\omega_b}(1_A) = 1$$

and, similarly, $\omega(a, b)^t = 1$, so that

$$\omega(a, b) = \omega(a, b)^{\ell s + rt} = \omega(a, b)^{\ell s}\omega(a, b)^{rt} = 1.$$

■

We say that a, b are *ω-orthogonal* if $\omega(a, b) = 1$, as in the previous lemma. Given two subgroups $A_1, A_2 \leq A$ such that $\omega(a_1, a_2) = 1$ for all $a_1 \in A_1$ and $a_2 \in A_2$, we say that A_1 and A_2 are ω-orthogonal.

Proposition 10.7 *Let $\omega \in \Omega_a(A)$. Then in the primary decomposition* (10.1), *one has that A_i and A_j are ω-orthogonal if $i \neq j$. Moreover, the map*

$$\begin{array}{ccc} \Omega_a(A) & \longrightarrow & \Omega_a(A_1) \times \Omega_a(A_2) \times \cdots \times \Omega_a(A_t) \\ \omega & \longmapsto & (\omega|_{A_1 \times A_1}, \omega|_{A_2 \times A_2}, \ldots, \omega|_{A_t \times A_t}) \end{array}$$

is an isomorphism.

Proof The first statement follows immediately from Lemma 10.6. For the second statement, just note that if $a = a_1 a_2 \cdots a_t, b = b_1 b_2 \cdots b_t \in A$, with $a_j, b_j \in A_j$, $j = 1, 2, \ldots, t$, then

$$\omega(a, b) = \prod_{i,j=1}^{t} \omega(a_i, b_j) = \prod_{j=1}^{t} \omega(a_j, b_j) \equiv \prod_{j=1}^{t} \omega|_{A_j \times A_j}(a_j, b_j).$$

■

We now analyze the behavior of ω with respect to the invariant factors decomposition (10.3). The proof of the following proposition is obvious.

Proposition 10.8 *Let $\omega \in \Omega_a(A)$. Let $a, b \in A$ and suppose that in the invariant factors decomposition* (10.3) *one has $a = \prod_{j=1}^{k} a_j^{\alpha_j}$ and $b = \prod_{j=1}^{k} a_j^{\beta_j}$, with α_j and β_j nonnegative integers. Then*

$$\omega(a, b) = \prod_{i,j=1}^{k} \omega(a_i, a_j)^{\alpha_i \beta_j}, \tag{10.12}$$

so that, in particular, ω is determined by the values $\omega(a_i, a_j)$, $1 \leq i < j \leq k$.

Moreover, with r_i and r_j as in (10.2), one has

$$\omega(a_i, a_j)^{r_i} = 1 = \omega(a_i, a_j)^{r_j} \quad \text{and} \quad \omega(a_i, a_j) = \overline{\omega}(a_j, a_i), \tag{10.13}$$

for all $i, j = 1, 2, \ldots, k$. ■

Theorem 10.9 *The invariant factors of the Abelian group $\Omega_{\mathrm{a}}(A)$ are the $\frac{k(k-1)}{2}$ numbers*

$$\underbrace{r_2}_{1}, \underbrace{r_3, r_3}_{2}, \ldots, \underbrace{r_{k-1}, r_{k-1}, \ldots, r_{k-1}}_{k-2}, \underbrace{r_k, r_k, \ldots, r_k}_{k-1}, \tag{10.14}$$

where the r_j are as in (10.2). In particular,

$$|\Omega_{\mathrm{a}}(A)| = \prod_{j=2}^{k} (r_j)^{j-1}.$$

Proof By virtue of Proposition 10.8, two bicharacters $\omega_1, \omega_2 \in \Omega_{\mathrm{a}}(A)$ are determined by the values $\big(\omega_1(a_i, a_j)\big)_{1 \le i < j \le k}$ and $\big(\omega_2(a_i, a_j)\big)_{1 \le i < j \le k}$. Moreover, their product is determined by the product of these values: $(\omega_1\omega_2)(a_i, a_j) = \omega_1(a_i, a_j) \cdot \omega_2(a_i, a_j)$ for all $1 \le i < j \le k$. Also, by virtue of (10.13), if $i < j$ one has that $\omega_1(a_i, a_j)$ is an r_j-th root of 1. Let $\mathcal{M}$ denote the set of all arrays $\big(\varepsilon_{ij}\big)_{1 \le i < j \le k}$, where each ε_{ij} is an r_j-th root of 1. Then $\mathcal{M}$ is an Abelian group under pointwise multiplication: it is the direct product of $\frac{k(k-1)}{2}$ cyclic groups and, by uniqueness, its invariant factors are precisely the numbers in (10.14). The map

$$\begin{aligned} \Omega_{\mathrm{a}}(A) &\longrightarrow \mathcal{M} \\ \omega &\longmapsto \big(\omega(a_i, a_j)\big)_{1 \le i < j \le k} \end{aligned}$$

is a group isomorphism. Indeed, it is injective by Proposition 10.8, and it is surjective because, if $\big(\varepsilon_{ij}\big)_{1 \le i < j \le k} \in \mathcal{M}$, then setting $\omega(a_i, a_i) := 1$, $\omega(a_i, a_j) := \varepsilon_{ij}$ for all $i < j$, and $\omega(a_i, a_j) := \overline{\varepsilon_{ij}}$, otherwise, and using (10.12), we get an element $\omega \in \Omega_{\mathrm{a}}(A)$ whose image is precisely the given $\big(\varepsilon_{ij}\big)_{1 \le i < j \le k}$. ■

Definition 10.10 An Abelian group A is *of symmetric type* if it admits a factorization $A = A' \times A''$, where A' and A'' are isomorphic. A finite Abelian group C is *of hyperbolic type* it admits a factorization $C = C_1 \times C_2$, where C_1 and C_2 are cyclic p-groups of the same order, i.e., C_1, C_2 are cyclic of order p^k for some prime p and some positive integer k.

The proof of the following proposition is obvious.

Proposition 10.11 *Let A be an Abelian group of symmetric type. Then the following holds:*

(1) *The invariant factor decomposition of A* (*cf.* (10.3)) *has the form*

$$A \cong \prod_{j=1}^{k} \left(\langle a_j \rangle \times \langle \overline{a}_j \rangle \right), \tag{10.15}$$

where $\langle a_j \rangle \cong \langle \overline{a}_j \rangle \cong C_{r_j}$ *for all* $j = 1, 2, \ldots, k$, *and*

$$|A| = r_1^2 r_2^2 \cdots r_k^2$$

where r_j *divides* r_{j-1} *for all* $j = 2, 3, \ldots, k$.

(2) *The structure theorem for A* (*cf.* (10.4)) *has the form*

$$A \cong \prod_{i=1}^{t} \prod_{j=1}^{h_i} \left(C_{p_i^{m_{ij}}} \times C_{p_i^{m_{ij}}} \right).$$

■

Corollary 10.12 *A finite Abelian group A is of symmetric type if and only if it is the product of groups of hyperbolic type.* ■

Theorem 10.13 (Yamazaki [134]) $\Omega_{\mathrm{a}}(A)$ *contains a nondegenerate element if and only if A is of symmetric type. Moreover, if this is the case, then* $\omega \in \Omega_{\mathrm{a}}(A)$ *is nondegenerate if and only if there exists a basis in the invariant factor decomposition* (10.15) *such that the matrix representation* (10.12) *of* ω *has the following "block diagonal form":*

$$\omega(a, b) = \prod_{j=1}^{k} \varepsilon_j^{h_j s_j - t_j m_j}, \tag{10.16}$$

for all $a, b \in A$ *with* $a = \prod_{j=1}^{k} \left(a_j^{h_j} \overline{a}_j^{t_j} \right)$ *and* $b = \prod_{j=1}^{k} \left(a_j^{m_j} \overline{a}_j^{s_j} \right)$, *with* $h_j, t_j, m_j, s_j \in \mathbb{Z}$, *where* ε_j *is a primitive* r_j*-th root of* 1, *and* $j = 1, 2, \ldots, k$.

Proof First of all, suppose that A is symmetric. Define ω by means of (10.16). Note that it is well defined: adding a multiple of r_j to one of the integers h_j, s_j, t_j, m_j does not affect (10.16). It is also easy to check that it is a bicharacter. Moreover, it is antisymmetric:

$$\omega(a, a) = \prod_{j=1}^{k} \varepsilon_j^{h_j t_j - t_j h_j} = 1.$$

If $a = \prod_{j=1}^{k} \left(a_j^{h_j} \overline{a}_j^{t_j}\right) \in \ker \omega$ then, for all $j = 1, 2, \ldots, k$,

$$1 = \omega(a, a_j) = \varepsilon_j^{-t_j} \text{ and } 1 = \omega(a, \overline{a}_j) = \varepsilon_j^{h_j}$$

implies that r_j divides both t_j and h_j, so that $a = 1_A$. In other words, ω is nondegenerate.

Conversely, suppose that $\Omega_a(A)$ contains a nondegenerate element, say ω. We prove, by induction on the number of factors in (10.3) that A is symmetric and that ω admits an expression as in (10.16). Let $a_1 \in A$ be an element of order r_1, where r_1 is the greatest integer in the invariant factors decomposition (10.2) of A. As (10.8) is an isomorphism, we have that the character ω_{a_1} is of order r_1 as well. We claim that there exists $\overline{a}_1 \in A$ such that $\varepsilon_1 := \omega_{a_1}(\overline{a}_1) \equiv \omega(a_1, \overline{a}_1)$ is a primitive r_1-th root of 1, that is, it generates a cyclic group of order r_1 in $\mathbb{T}$. Indeed, from (10.1) and (10.4) it follows that

$$o(\omega_{a_1}) = o\big(\omega_{a_1}|_{A_1}\big)\, o\big(\omega_{a_1}|_{A_2}\big) \cdots o\big(\omega_{a_1}|_{A_t}\big)$$

and, recalling (10.5), we deduce that

$$o(\omega_{a_1}) = r_1 \quad \Longrightarrow \quad o\big(\omega_{a_1}|_{A_i}\big) = p_i^{m_{i1}}, \; i = 1, 2, \ldots, t.$$

Take $c_i \in A_i$ such that $o(\omega_{a_1}(c_i))$ is maximal, namely, $p_i^{m_{i1}}$. Then $\overline{a}_1 := c_1 c_2 \cdots c_t$ is of order $p_1^{m_{11}} p_2^{m_{21}} \cdots p_t^{m_{t1}} = r_1$, and the claim follows. Now suppose that $\overline{a}_1^s = a_1^t$, for some integers s and t. Then

$$\varepsilon_1^t = \omega_{a_1}(\overline{a}_1)^t = \omega(a_1, \overline{a}_1)^t = \omega(a_1^t, \overline{a}_1) = \omega(\overline{a}_1^s, \overline{a}_1) = \omega(\overline{a}_1, \overline{a}_1)^s = 1.$$

This implies that r_1 divides t so that $a_1^t = 1_A$. We deduce that $\langle a_1 \rangle \cap \langle \overline{a}_1 \rangle = \{1_A\}$. Setting $C := \langle a_1 \rangle \cdot \langle \overline{a}_1 \rangle \cong \langle a_1 \rangle \times \langle \overline{a}_1 \rangle$, for $h, t, m, s \in \mathbb{Z}$ we have

$$\begin{aligned} \omega(a_1^h \overline{a}_1^t, a_1^m \overline{a}_1^s) &= \omega(a_1, a_1)^{hm} \omega(\overline{a}_1, \overline{a}_1)^{ts} \omega(a_1, \overline{a}_1)^{hs} \omega(\overline{a}_1, a_1)^{tm} \\ &= \omega(a_1, \overline{a}_1)^{hs} \omega(\overline{a}_1, a_1)^{tm} \\ &= \omega(a_1, \overline{a}_1)^{hs} \omega(a_1, \overline{a}_1)^{-tm} \\ &= \varepsilon_1^{hs-tm}. \end{aligned} \tag{10.17}$$

In particular, by the first part of the proof, $\omega|_{C \times C}$ is nondegenerate. From (10.11) in Lemma 10.5 we deduce that

$$A = C \times C^{\perp}.$$

Moreover, also $\omega|_{C^\perp\times C^\perp}$ is nondegenerate: if $d \in \ker(\omega|_{C^\perp\times C^\perp})$, then for all $a \in A$, say $a = cd'$ with $c \in C$ and $d' \in C^\perp$, we have

$$\omega(a,d) = \omega(cd',d) = \omega(c,d)\omega(d',d) = 1. \tag{10.18}$$

Since ω is nondegenerate, (10.18) forces $d = 1_A$. By the inductive hypothesis, we may conclude that $C^\perp$, and therefore also A, is of symmetric type. Finally, we deduce (10.16) from (10.17), (10.18) and the inductive hypothesis applied to $\omega|_{C^\perp\times C^\perp}$. ■

Corollary 10.14 *Suppose that A is of symmetric type and let $\omega \in \Omega_{\mathrm{a}}(A)$ be a nondegenerate bicharacter. Then there exists a factorization $A = A' \times A''$ such that $A' \cong A''$ and $\omega|_{A'\times A'} \equiv 1$, $\omega|_{A''\times A''} \equiv 1$.*

Proof Just set, in the notation in Theorem 10.13, $A' := \prod_{j=1}^k \langle a_j\rangle$ and $A'' := \prod_{j=1}^k \langle \overline{a}_j\rangle$, and use (10.16). ■

We now associate a bicharacter with every unitary 2-cocycle of A and study this correspondence in details. Let $\eta \in \mathscr{C}^2(A,\mathbb{T})$ (cf. Sect. 6) and define a map $\omega_\eta \colon A\times A \to \mathbb{T}$ by setting

$$\omega_\eta(a,b) := \eta(a,b)\overline{\eta}(b,a) \tag{10.19}$$

for all $a, b \in A$. See also the second formula in (6.37).

Proposition 10.15 $\omega_\eta \in \Omega_{\mathrm{a}}(A)$.

Proof Indeed, for all $a, b, c \in A$ we have:

$$\begin{aligned}
\omega_\eta(ab,c) &= \eta(ab,c)\overline{\eta}(c,ab)\\
\text{(by (6.1))}\quad &= \eta(a,cb)\eta(b,c)\overline{\eta}(a,b)\cdot\overline{\eta}(ca,b)\overline{\eta}(c,a)\eta(a,b)\\
\text{(again by (6.1))}\quad &= \overline{\eta}(c,b)\eta(ac,b)\eta(a,c)\eta(b,c)\overline{\eta}(ac,b)\overline{\eta}(c,a)\\
&= \eta(a,c)\overline{\eta}(c,a)\cdot\eta(b,c)\overline{\eta}(c,b)\\
&= \omega_\eta(a,c)\omega_\eta(b,c).
\end{aligned}$$

Moreover, $\omega_\eta(a,b) = \eta(a,b)\overline{\eta}(b,a) = \overline{\eta}(b,a)\eta(a,b) = \overline{\omega_\eta}(b,a)$ and $\omega_\eta(a,a) = \eta(a,a)\overline{\eta}(a,a) = 1$. By using Lemma 10.3, this shows that $\omega_\eta \in \Omega_{\mathrm{a}}(A)$. ■

In the following we use the notation in Definitions 6.3 and 6.20. Recall (cf. Sect. 6.2) that a cocycle $\eta \in \mathscr{C}^2(A,\mathbb{T})$ is *symmetric* if $\eta(a,b) = \eta(b,a)$, for all $a, b \in A$. We begin with a simple identity (where indeed $\mathbb{T}$ may be replaced by an arbitrary Abelian group (and $\overline{\eta}$ by η^{-1})).

Proposition 10.16 *Let $\eta \in \mathscr{C}^2(A, \mathbb{T})$ and suppose that it is symmetric. Then*

$$\eta(ab, cd) = \eta(ac, bd)\eta(a, c)\eta(b, d)\overline{\eta}(c, d)\overline{\eta}(a, b)$$

for all $a, b, c, d \in A$.

Proof Using the cocycle identity (6.1), the symmetry of η, and the fact that A is Abelian we get:

$$\begin{aligned}\eta(a \cdot b, cd) &= \eta(a, c \cdot bd)\overline{\eta}(a, b)\eta(cd, b)\\ &= \eta(ac, bd)\eta(a, c)\overline{\eta}(c, d \cdot b)\overline{\eta}(a, b)\eta(cd, b)\\ &= \eta(ac, bd)\eta(a, c)\eta(b, d)\overline{\eta}(c, d)\overline{\eta}(a, b).\end{aligned} \tag{10.20}$$

■

The following is a technical but fundamental result that deserves three different proofs. The first one involves representation theory (and is taken from [16]), the second one is purely group theoretic, while the third one is more direct and also gives an explicit formula for the coboundary. Actually, the third proof derives from the second one but avoids the construction of the group extension; it is useful in the applications, see Remark 10.30.(2) below and (10.65). Recall Proposition 6.2.4.

Theorem 10.17 *A cocycle $\eta \in \mathscr{C}^2(A, \mathbb{T})$ is symmetric if and only if η is a coboundary. Moreover, if this is the case, then $\eta = \tau_\rho$ (cf. (6.3)), where $\rho \colon A \to \mathbb{T}$, in the notation (10.3), is given by*

$$\rho\left(a_1^{h_1}a_2^{h_2}\cdots a_k^{h_k}\right) = \left[\prod_{j=1}^{k}\left(\prod_{i=1}^{h_j-1}\eta(a_j, a_j^i)\right)\overline{\xi}_j^{h_j}\right]\cdot\prod_{j=1}^{k-1}\eta\left(a_1^{h_1}a_2^{h_2}\cdots a_j^{h_j}, a_{j+1}^{h_{j+1}}\right), \tag{10.21}$$

for all $0 \le h_j \le r_j - 1$, and $\xi_j \in \mathbb{T}$ is an r_j-root of $\prod_{i=1}^{r_j-1}\eta\left(a_j, a_j^i\right)$, $j = 1, 2, \dots, k$, that is,

$$\xi_j^{r_j} = \prod_{i=1}^{r_j-1}\eta\left(a_j, a_j^i\right). \tag{10.22}$$

First proof of the first statement. Suppose first that $\eta \in \mathscr{B}^2(A, \mathbb{T})$ and let $\rho \colon A \to \mathbb{T}$ such that $\eta = \tau_\rho$. Then, since A is Abelian, for all $a, b \in A$ we have $\eta(a, b) = \overline{\rho}(a)\overline{\rho}(b)\rho(ab) = \overline{\rho}(b)\overline{\rho}(a)\rho(ba) = \eta(b, a)$, and this shows that η is symmetric. Conversely, suppose that η is symmetric. Let Φ be an irreducible η-representation of A (it exists: see Proposition 7.12 and Remark 7.13). From (7.10) and using symmetry of η and, again, commutativity of A, it follows that

$$\Phi(a)\Phi(b) = \Phi(ab)\overline{\eta}(a,b) = \Phi(ba)\overline{\eta}(b,a) = \Phi(b)\Phi(a), \tag{10.23}$$

for all $a, b \in A$. By Schur's lemma, there exists a function $\rho\colon A \to \mathbb{T}$ such that $\Phi(a) = \rho(a)I$ for all $a \in A$, where I is the identity operator on the space of Φ. Then, the first equality in (10.23) gives $\rho(a)\rho(b) = \rho(ab)\overline{\eta}(a,b)$, equivalently, $\eta = \tau_\rho$. This shows that $\eta \in \mathscr{B}^2(A, \mathbb{T})$. ■

Second Proof of the First Statement The "if" part is as in the first proof. Conversely, suppose that η is symmetric and consider the extension $G := G(\mathbb{T}, A, \eta)$ as in Proposition 6.5. The symmetry of η implies that G is Abelian. Moreover, $\mathbb{T}$ is a subgroup of G and $G/\mathbb{T} \cong A$. Denote by $\pi : G \to A$ the projection map, choose a section $t\colon A \to G$, decompose A as in (10.3), and consider the elements $t(a_j) \in G$, $j = 1, 2, \dots, k$. Then

$$a_j^{r_j} = 1_A \implies \pi\big[t(a_j)^{r_j}\big] = 1_A \implies t(a_j)^{r_j} \in \mathbb{T}$$

and we can choose $\xi_j \in \mathbb{T}$ such that $t(a_j)^{r_j} = \xi_j^{r_j}$ (namely, an r_j-th root of $t(a_j)^{r_j}$). Set $g_j := t(a_j)\overline{\xi_j} \in G$, $j =, 1, 2, \dots, k$. Then the elements $g_1, g_2, \dots, g_k$ generate a subgroup of G isomorphic to A. Indeed, if $0 \leq h_j \leq r_j - 1$, $j = 1, 2, \dots, k$, and $g_1^{h_1} g_2^{h_2} \cdots g_k^{h_k} = 1_G$, then

$$1_A = \pi\left(g_1^{h_1} g_2^{h_2} \cdots g_k^{h_k}\right) = a_1^{h_1} a_2^{h_2} \cdots a_k^{h_k} \tag{10.24}$$

implies that $h_1 = h_2 = \cdots = h_k = 0$. Moreover, the order of g_j is r_j. Now, the second identity in (10.24) implies that the map $s\colon A \to G$ defined by setting

$$s\left(a_1^{h_1} a_2^{h_2} \cdots a_k^{h_k}\right) = g_1^{h_1} g_2^{h_2} \cdots g_k^{h_k}$$

is a section. As G is Abelian, s is also a homomorphism. Thus, $G = \mathbb{T}s(A) \cong \mathbb{T} \times A$ and from Proposition 6.8 we deduce that η is a coboundary. ■

Third Proof The following calculations are similar to those in the computation of the second cohomology group of a finite cyclic group, cf. [90, Section 7.3.2]. Assume that $\eta \in \mathscr{C}^2(A, \mathbb{T})$ is symmetric. The proof that $\eta = \tau_\rho$, with ρ as in (10.21), is by induction on k (the number of invariant factors of A, cf. (10.2)). Let $k = 1$. We shall use the identity

$$\overline{\eta}\left(a_1, a_1^i\right) \eta\left(a_1, a_1^{h_1+i}\right) = \overline{\eta}\left(a_1^i, a_1^{h_1}\right) \eta\left(a_1^{i+1}, a_1^{h_1}\right), \tag{10.25}$$

with a_1 as in (10.3), $1 \le i \le h_1 - 1$ and $0 \le h_1 \le r_1 - 1$, which is an immediate consequence of (6.1). For $0 \le s_1 \le h_1 \le r_1 - 1$ and $s_1 + h_1 \le r_1 - 1$ we have

$$\begin{aligned}
\tau_\rho\left(a_1^{h_1}, a_1^{s_1}\right) &= \rho\left(a_1^{s_1+h_1}\right)\overline{\rho}(a_1^{s_1})\overline{\rho}\left(a_1^{h_1}\right) \\
\text{(by (10.21))} \quad &= \left(\prod_{i=1}^{s_1+h_1-1} \eta\left(a_1, a_1^i\right)\right)\overline{\xi}_1^{s_1+h_1} \cdot \left(\prod_{i=1}^{s_1-1}\overline{\eta}\left(a_1, a_1^i\right)\right)\xi_1^{s_1} \cdot \left(\prod_{i=1}^{h_1-1}\overline{\eta}\left(a_1, a_1^i\right)\right)\xi_1^{h_1} \\
&= \prod_{i=0}^{s_1-1} \overline{\eta}\left(a_1, a_1^i\right)\eta\left(a_1, a_1^{h_1+i}\right) \\
\text{(by (10.25))} \quad &= \prod_{i=0}^{s_1-1} \overline{\eta}\left(a_1^i, a_1^{h_1}\right)\eta\left(a_1^{i+1}, a_1^{h_1}\right) \\
&= \eta\left(a_1^{s_1}, a_1^{h_1}\right).
\end{aligned}$$

If $s_1 + h_1 \ge r_1$ we set

$$t_1 = s_1 + h_1 - r_1, \tag{10.26}$$

so that $0 \le t_1 \le r_1 - 1$, then

$$\begin{aligned}
\tau_\rho\left(a_1^{h_1}, a_1^{s_1}\right) &= \rho\left(a_1^{t_1}\right)\overline{\rho}\left(a_1^{s_1}\right)\overline{\rho}\left(a_1^{h_1}\right) \\
&= \prod_{i=1}^{t_1-1} \eta\left(a_1, a_1^i\right) \cdot \overline{\xi}_1^{t_1} \cdot \prod_{i=1}^{s_1-1} \overline{\eta}\left(a_1, a_1^i\right) \cdot \xi_1^{s_1} \cdot \prod_{i=1}^{h_1-1} \overline{\eta}\left(a_1, a_1^i\right) \cdot \xi_1^{h_1} \\
\text{(by (10.22) and (10.26))} \quad &= \prod_{i=1}^{s_1+h_1-1} \eta\left(a_1, a_1^i\right) \cdot \prod_{i=1}^{s_1-1} \overline{\eta}\left(a_1, a_1^i\right) \cdot \prod_{i=1}^{h_1-1} \overline{\eta}\left(a_1, a_1^i\right),
\end{aligned}$$

and we conclude as in the previous case. ■

Suppose now that $k > 1$. By isolating the terms with a_k in (10.21), we get

$$\begin{aligned}
\rho\left(a_1^{h_1}a_2^{h_2}\cdots a_k^{h_k}\right) &= \left[\prod_{j=1}^{k-1}\left(\prod_{i=1}^{h_j-1} \eta(a_j, a_j^i)\right)\overline{\xi}_j^{h_j}\right] \cdot \prod_{j=1}^{k-2} \eta\left(a_1^{h_1}a_2^{h_2}\cdots a_j^{h_j}, a_{j+1}^{h_{j+1}}\right) \cdot \\
&\quad \cdot \left(\prod_{i=1}^{h_k-1} \eta(a_k, a_k^i)\right)\overline{\xi}_k^{h_k} \cdot \eta\left(a_1^{h_1}a_2^{h_2}\cdots a_{k-1}^{h_{k-1}}, a_k^{h_k}\right) \\
&= \rho\left(a_1^{h_1}a_2^{h_2}\cdots a_{k-1}^{h_{k-1}}\right)\rho\left(a_k^{h_k}\right)\eta\left(a_1^{h_1}a_2^{h_2}\cdots a_{k-1}^{h_{k-1}}, a_k^{h_k}\right),
\end{aligned}$$

that is,

$$\eta\left(a_1^{h_1}a_2^{h_2}\cdots a_{k-1}^{h_{k-1}}, a_k^{h_k}\right) = \tau_\rho\left(a_1^{h_1}a_2^{h_2}\cdots a_{k-1}^{h_{k-1}}, a_k^{h_k}\right). \tag{10.27}$$

Finally, an application of (10.20) with

$$a = a_1^{h_1}a_2^{h_2}\cdots a_{k-1}^{h_{k-1}},\ b = a_k^{h_k},\ c = a_1^{s_1}a_2^{s_2}\cdots a_{k-1}^{s_{k-1}},\ \text{and } d = a_k^{s_k},$$

yields

$$\begin{aligned}
\eta\left(a_1^{h_1}a_2^{h_2}\cdots a_k^{h_k}, a_1^{s_1}a_2^{s_2}\cdots a_k^{s_k}\right) &\equiv \eta(ab, cd)\\
&= \eta(ac, bd)\eta(a, c)\eta(b, d)\overline{\eta}(c, d)\overline{\eta}(a, b)\\
\text{(by case } k = 1,\ \text{induction and (10.27))}\quad &= \tau_\rho(ac, bd)\tau_\rho(a, c)\tau_\rho(b, d)\overline{\tau_\rho}(c, d)\overline{\tau_\rho}(a, b)\\
\text{(by (10.20))}\quad &= \tau_\rho(ab, cd)\\
&= \tau_\rho\left(a_1^{h_1}a_2^{h_2}\cdots a_k^{h_k}, a_1^{s_1}a_2^{s_2}\cdots a_k^{s_k}\right).
\end{aligned}$$

This shows that $\eta = \tau_\rho$, and the proof is complete. □

Corollary 10.18 *$\mathscr{H}^2(C_m, \mathbb{T}) = \{0\}$ for all integers $m \geq 1$.*

Proof Let $\eta \in \mathscr{C}^2(C_m, \mathbb{T})$ and let us show that η is symmetric. Let $a \in C_m$ be a generator. Then, by using the cocycle identity, we have $\eta(a^2, a) = \overline{\eta}(a, a)\eta(a, a^2)\eta(a, a) = \eta(a, a^2)$. By induction on $r + s$ for $s > r \geq 1$ we have

$$\begin{aligned}
\eta(a^s, a^r) &= \eta(a^{r+s-r}, a^r)\\
\text{(by the cocycle identity)}\quad &= \overline{\eta}(a^r, a^{s-r})\eta(a^r, a^s)\eta(a^{s-r}, a^r)\\
\text{(by induction)}\quad &= \overline{\eta}(a^r, a^{s-r})\eta(a^r, a^s)\eta(a^r, a^{s-r})\\
&= \eta(a^r, a^s).
\end{aligned}$$

From Theorem 10.17 we deduce that $\eta \in \mathscr{B}^2(C_m, \mathbb{T})$. ■

Remark 10.19 From Corollary 10.18 we may deduce another proof of Proposition 2.23. Indeed, if G/N is cyclic, then the Mackey obstruction is trivial (by the corollary) and we can apply Proposition 7.4.

Theorem 10.20 *The map*

$$\begin{aligned}
\mathscr{C}^2(A, \mathbb{T}) &\longrightarrow \Omega_{\mathrm{a}}(A)\\
\eta &\longmapsto \omega_\eta
\end{aligned} \tag{10.28}$$

is a surjective homomorphism with kernel $\mathscr{B}^2(A, \mathbb{T})$. *Moreover, a pre-image of* $\omega \in \Omega_{\mathrm{a}}(A)$ *is the cocycle* $\eta_\omega \in \mathscr{C}^2(A, \mathbb{T})$ *defined by setting*

$$\eta_\omega(a, b) := \prod_{1 \le i < j \le k} \omega(a_i, a_j)^{\alpha_i \beta_j}, \tag{10.29}$$

for all $a, b \in A$, *where we use the notation in Proposition 10.8. In particular,*

$$\mathscr{M}(A) \cong \Omega_{\mathrm{a}}(A),$$

where $\mathscr{M}(A)$ *denote the Schur multiplier of* A *(cf. Definition 6.20).*

Proof It is clear that the map is a homomorphism. Also,

$$\omega_\eta \equiv 1 \iff \eta(a, b) = \eta(b, a) \quad \text{for all } a, b \in A. \tag{10.30}$$

By Theorem 10.17, η is in the kernel of (10.28) if and only if $\eta \in \mathscr{B}^2(A, \mathbb{T})$.

We now prove that the map (10.28) is surjective by showing that (10.29), in the notation in Proposition 10.8, defines an element in $\mathscr{C}^2(A, \mathbb{T})$ whose image is the given $\omega \in \Omega_{\mathrm{a}}(A)$. Indeed, if in the decomposition (10.3) one has $a = \prod_{j=1}^k a_j^{\alpha_j}$, $b = \prod_{j=1}^k a_j^{\beta_j}$, and $c = \prod_{j=1}^k a_j^{\gamma_j}$, with α_j, β_j, and γ_j nonnegative integers, then

$$\begin{aligned}\eta_\omega(ab, c)\eta_\omega(a, b)\overline{\eta}_\omega(a, bc\overline{\eta}_\omega(b, c) &= \prod_{1 \le i < j \le k} \omega(a_i, a_j)^{(\alpha_i+\beta_i)\gamma_j + \alpha_i\beta_j - \alpha_i(\beta_j+\gamma_j) - \beta_i\gamma_j} \\ &= \prod_{1 \le i < j \le k} \omega(a_i, a_j)^0 = 1,\end{aligned}$$

and the cocycle identity follows. Moreover,

$$\eta_\omega(1_A, a) = \prod_{1 \le i < j \le k} \omega(a_i, a_j)^{0\alpha_i} = 1.$$

This shows that $\eta_\omega \in \mathscr{C}^2(A, \mathbb{T})$. Finally, by Proposition 10.8,

$$\eta_\omega(a, b)\overline{\eta}_\omega(b, a) = \prod_{1 \le i < j \le k} \omega(a_i, a_j)^{\alpha_i\beta_j - \beta_i\alpha_j} = \prod_{1 \le i < j \le k} \omega(a_i, a_j)^{\alpha_i\beta_j}\omega(a_j, a_i)^{\alpha_j\beta_i} = \omega(a, b).$$

■

Assuming the invariant factor decomposition (10.2) of A, from the above theorem and from Theorem 10.9, we immediately deduce:

Corollary 10.21 *The invariant factors of* $\mathscr{M}(A)$ *are the numbers in* (10.14).

Corollary 10.22 *Every $\eta \in \mathscr{C}^2(A, \mathbb{T})$ is cohomologous to an $\widetilde{\eta} \in \mathscr{C}^2(A, \mathbb{T})$ which, in a suitable basis for the invariant factor decomposition* (10.2) *of A, has the form*

$$\widetilde{\eta}(a, b) = \prod_{1 \le i < j \le k} \varepsilon_{ij}^{\alpha_i \beta_j},$$

where $a = \prod_{j=1}^k a_j^{\alpha_j}$, $b = \prod_{j=1}^k a_j^{\beta_j}$, and ε_{ij} is an r_j-th root of 1 for $1 \le i < j \le k$.

Proof Let $\eta \in \mathscr{C}^2(A, \mathbb{T})$ and denote by $\omega_\eta \in \Omega_{\mathrm{a}}(A)$ the bicharacter associated via the map (10.28). Then, keeping in mind (10.13), the 2-cocycle $\widetilde{\eta}$ is the (canonical) preimage of ω_η given by (10.29). Since the kernel of (10.28) are exactly the coboundaries, we deduce that η and $\widetilde{\eta}$ are cohomologous. ■

We are now in a position to give a canonical form for 2-cocycles whose associated bicharacter is nondegenerate. By Theorem 10.13, we assume that A is of symmetric type. A further canonical form requires A be 2-divisible, that is, $|A|$ is odd. See also Example 6.16.

Proposition 10.23 *Let $\eta \in \mathscr{C}^2(A, \mathbb{T})$. Then ω_η is nondegenerate if and only if there exists a basis in the invariant factor decomposition* (10.15) *and $\widetilde{\eta} \in \mathscr{C}^2(A, \mathbb{T})$ cohomologous to η whose expression, in this basis, is given by*

$$\widetilde{\eta}(a, b) := \prod_{j=1}^{k} \varepsilon_j^{h_j s_j} \tag{10.31}$$

where $a = \prod_{j=1}^k \left(a_j^{h_j} b_j^{t_j}\right)$, $b = \prod_{j=1}^k \left(a_j^{m_j} b_j^{s_j}\right)$, and ε_j is an r_j-th primitive root of 1 for all $j = 1, 2, \ldots k$. Moreover, if $|A|$ is odd, setting

$$\widetilde{\eta}'(a, b) := \prod_{j=1}^{k} \varepsilon_j^{(h_j s_j - m_j t_j)/2}, \tag{10.32}$$

where $\sqrt{\varepsilon_j} \equiv \varepsilon_j^{1/2} := \varepsilon_j^{(r_j+1)/2}$, we get an equalized cocycle $\widetilde{\eta}' \in \mathscr{C}^2(A, \mathbb{T})$ cohomologous to η.

Proof The first statement follows from Theorem 10.13 and Corollary 10.22, noting that in the basis

$$a_1, b_1, a_2, b_2, \ldots, a_k, b_k$$

the expression (10.16) yields

$$\begin{aligned}
\omega(a_\ell, a_r) &= 1 = \omega(b_\ell, b_r) && \text{for all } 1 \le \ell, r \le k \\
\omega(a_\ell, b_r) &= 1 = \omega(b_r, a_\ell) && \text{for all } 1 \le \ell \neq r \le k \\
\omega(a_\ell, b_\ell) &= \varepsilon_\ell = \text{an } r_\ell\text{-th primitive root of } 1 && \text{for all } 1 \le \ell \le k.
\end{aligned}$$

Therefore (10.29) reduces to (10.31).

We now derive the expression (10.32). Actually, this is a slightly generalization of the procedure in the first part of Example 6.16. First of all, $\widetilde{\eta}(a, a^{-1}) = \prod_{j=1}^k \varepsilon_j^{-h_j t_j}$, so that the function $\rho \colon A \to \mathbb{T}$ in the proof of Proposition 6.15.(2) is given by

$$\rho(a) := \prod_{j=1}^k \varepsilon_j^{-h_j t_j/2} \equiv \prod_{j=1}^k \left(\varepsilon_j^{(r_j+1)/2}\right)^{-h_j t_j} \tag{10.33}$$

Then the construction in that same proof yields the equalized cocycle $\widetilde{\eta}' \in \mathscr{C}^2(A, \mathbb{T})$ defined by

$$\begin{aligned}
\widetilde{\eta}'(a, b) &:= \widetilde{\eta}(a, b) \cdot \tau_\rho(a, b) = \widetilde{\eta}(a, b) \cdot \rho(ab)\overline{\rho}(a)\overline{\rho}(b) \\
(\text{by (10.33)}) \quad &= \widetilde{\eta}(a, b) \cdot \prod_{j=1}^k \varepsilon_j^{-(h_j+m_j)(s_j+t_j)/2} \prod_{j=1}^k \varepsilon_j^{(h_j t_j + m_j s_j)/2} \\
(\text{by (10.31)}) \quad &= \prod_{j=1}^k \varepsilon_j^{h_j s_j} \cdot \prod_{j=1}^k \varepsilon_j^{-(h_j+m_j)(s_j+t_j)/2} \prod_{j=1}^k \varepsilon_j^{(h_j t_j + m_j s_j)/2} \\
&= \prod_{j=1}^k \varepsilon_j^{(h_j s_j - m_j t_j)/2}
\end{aligned}$$

clearly cohomologous to $\widetilde{\eta}$ and (cf. Corollary 10.22), in turn, to η. ∎

10.2 The Irreducible Projective Representations of Finite Abelian Groups

We are now in a position to describe all irreducible projective representations of a finite Abelian group A. We limit our exposition to the case of Abelian groups of odd order, where a simple expression for equalized cocycles is available; see Example 6.16 and Proposition 10.23.

Actually, Theorems 10.27 and 10.31 below do not require the hypothesis that A be of odd order because in their proofs the canonical form (10.32) is not relevant.

With respect to the original sources, we describe those representations by means of a suitable action of the group on function spaces defined on other finite Abelian groups. We assume and use all the notation and results in the previous section. In particular, recalling (10.15), we set

$$A' := \prod_{j=1}^{k} \langle a_j \rangle \quad \text{and} \quad A'' := \prod_{j=1}^{k} \langle \overline{a}_j \rangle,$$

so that $A = A' \times A''$. It is also convenient to establish the following notation:

$$a = \prod_{j=1}^{k} a_j^{h_j} \in A', \quad b = \prod_{j=1}^{k} \overline{a}_j^{t_j} \in A'', \quad c = \prod_{j=1}^{k} a_j^{m_j} \in A', \quad d = \prod_{j=1}^{k} \overline{a}_j^{s_j} \in A''. \tag{10.34}$$

We then parameterize the characters of A' (resp. A'') by A'' (resp. A'), by defining $\psi_b \in \widehat{A'}$ (resp. $\psi_a \in \widehat{A''}$) by setting

$$\psi_b(a) := \prod_{j=1}^{k} \varepsilon_j^{h_j t_j} \quad (\text{resp. } =: \psi_a(b)) \tag{10.35}$$

for $a \in A'$ and $b \in A''$ as in (10.34), where ε_j is a primitive r_j-root of 1, $j = 1, 2, \ldots, k$.

We first describe all irreducible projective representations (actually, just one!) of an Abelian group $A = A' \times A''$ (of symmetric type!) with respect to a 2-cocycle whose associated bicharacter is nondegenerate. We will use Clifford theory from Sect. 9.2.

Theorem 10.24 *Let* $\widetilde{\eta}' \in \mathscr{C}^2(A, \mathbb{T})$ *be as in* (10.32) *and let* $a, c \in A'$ *and* $b, d \in A''$ *be as in* (10.34).

(1) *The following identities hold:*

$$\begin{aligned}
&\widetilde{\eta}'(ab, cd) = \widetilde{\eta}'(a, d)\, \overline{\widetilde{\eta}'(c, b)} \\
&\widetilde{\eta}'(a, d) = \overline{\widetilde{\eta}'(d, a)} = \overline{\widetilde{\eta}'(a^{-1}, d)} = \overline{\widetilde{\eta}'(a, d^{-1})} \\
&\widetilde{\eta}'(ac, b) = \widetilde{\eta}'(a, b)\, \widetilde{\eta}'(c, b) \\
&\widetilde{\eta}'(a, bd) = \widetilde{\eta}'(a, b)\, \widetilde{\eta}'(a, d).
\end{aligned} \tag{10.36}$$

Moreover,

$$\widetilde{\eta}'(a,d)^2 = \psi_a(d). \tag{10.37}$$

(2) *Using $\widetilde{\eta}'$ to define the function ρ_h in (9.2), with H replaced by A and N by A', then*

$$\rho_{cd}(a) = \psi_d(a). \tag{10.38}$$

Moreover, $\widetilde{\eta}'|_{A'\times A'} \equiv 1$ and $\widehat{A'}^{\widetilde{\eta}'} \equiv \widehat{A'}$ (the ordinary characters).

(3) *For each ordinary character ψ_{b_0} of A', $b_0 \in A''$, we have:*

(i) $I_A(\psi_{b_0}) = A'$,
(ii) $\mathrm{Ind}_{A'}^{A}\psi_{b_0}$ *is the unique irreducible $\widetilde{\eta}'$-representation of A, and*
(iii) $\mathrm{Res}_{A'}^{A}\mathrm{Ind}_{A'}^{A}\psi_{b_0} \sim \oplus_{d\in A''}\psi_d$.

(4) *The representation $\mathrm{Ind}_{A'}^{A}\psi_{b_0}$ is equivalent to the projective representation Φ on $L(A'')$ defined by setting:*

$$[\Phi(cd)f](b) := \overline{\psi_c}(b)\,\widetilde{\eta}'(c,d)\,\psi_{b_0}(c)\,f(d^{-1}b). \tag{10.39}$$

for all $f \in L(A'')$.

Proof

(1) The identities in (10.36) and (10.37) follow immediately from (10.32). For instance, using the notation in (10.34) and (10.35), we get

$$\widetilde{\eta}'(ab,cd) = \prod_{j=1}^{k} \varepsilon_j^{(h_js_j - m_jt_j)/2} = \left(\prod_{j=1}^{k}\varepsilon_j^{h_js_j/2}\right)\left(\prod_{j=1}^{k}\varepsilon_j^{-m_jt_j/2}\right) = \widetilde{\eta}'(a,d)\,\overline{\widetilde{\eta}'}(c,b).$$

(2) We have

$$\begin{aligned}\rho_{cd}(a) &= \widetilde{\eta}'(c^{-1}d^{-1}a,cd)\,\widetilde{\eta}'(c^{-1}d^{-1},a) && \text{(by (9.2))}\\ &= \widetilde{\eta}'(c^{-1}a,d)\,\overline{\widetilde{\eta}'}(c,d^{-1})\,\overline{\widetilde{\eta}'}(a,d^{-1}) = \widetilde{\eta}'(a,d)^2 && \text{(by (10.36))}\\ &= \psi_d(a). && \text{(by (10.37))}\end{aligned}$$

(3) Now, (9.1) becomes

$$\begin{aligned}\left[{}^{cd}\psi_{b_0}\right](a) &= \psi_{b_0}(a)\overline{\rho}_{cd}(a)\\ \text{(by (10.38))}\quad &= \psi_{b_0}(a)\psi_{d^{-1}}(a) = \psi_{b_0d^{-1}}(a).\end{aligned}$$

It follows that ${}^{cd}\psi_{b_0} \equiv \psi_{b_0 d^{-1}} = \psi_{b_0}$ if and only if $d = 1_A$ showing that $I_A(\psi_{b_0}) = A'$, and (i) follows. Also, the $\widetilde{\eta}\,'$-conjugation action of A on $\widehat{A'}$ is transitive. Therefore, in the present situation, both the partitions (9.9) and (9.20) are trivial (that is, $|\Sigma| = 1$) and A has just one irreducible $\widetilde{\eta}\,'$-representation. From Proposition 9.18 we deduce (ii) and (iii).

(4) From (8.1) we deduce that the space W of $\mathrm{Ind}_{A'}^{A}\psi_{b_0}$ is made up of all functions $F\colon A \to \mathbb{C}$ such that

$$\begin{aligned} F(cda) &= \overline{\widetilde{\eta}\,'}(cd, a)\,\overline{\psi_{b_0}}(a)F(cd) \\ \text{(by (10.36))} \quad &= \widetilde{\eta}\,'(a, d)\overline{\psi_{b_0}}(a)F(cd). \end{aligned} \tag{10.40}$$

It follows that $F \in W$ is determined by its values on A'' and the map $T\colon L(A'') \to W$ given by

$$[Tf](cd) := \widetilde{\eta}\,'(c, d)\,\overline{\psi_{b_0}}(c)f(d), \tag{10.41}$$

for all $f \in L(A'')$, is a linear isomorphism; in particular, $[Tf](d) = f(d)$. From (8.2) we deduce the expression for the action of $\Phi' := \mathrm{Ind}_{A'}^{A}\psi_{b_0}$ on W:

$$\begin{aligned} \left[\Phi'(cd)F\right](ab) &= \widetilde{\eta}\,'(c^{-1}d^{-1}, ab)F(d^{-1}b \cdot c^{-1}a) \\ \text{(by (10.36) and (10.40))} \quad &= \widetilde{\eta}\,'(c^{-1}, b)\overline{\widetilde{\eta}\,'}(a, d^{-1})\widetilde{\eta}\,'(c^{-1}a, d^{-1}b)\psi_{b_0}(a^{-1}c)F(d^{-1}b) \\ \text{(by (10.36) and (10.37))} \quad &= \overline{\psi_c}(b)\widetilde{\eta}\,'(a, b)\widetilde{\eta}\,'(c, d)\psi_{b_0}(a^{-1}c)F(d^{-1}b). \end{aligned}$$

for all $F \in W$. In particular,

$$\left[\Phi'(cd)Tf\right](ab) = \overline{\psi_c}(b)\widetilde{\eta}\,'(a, b)\widetilde{\eta}\,'(c, d)\psi_{b_0}(a^{-1}c)f(d^{-1}b) \tag{10.42}$$

for all $f \in L(A'')$.

Now define $\Phi(cd)\colon L(A'') \to L(A'')$ as in (10.39). Then, for all $f \in L(A'')$,

$$\begin{aligned} [T\Phi(cd)f](ab) &= \widetilde{\eta}\,'(a, b)\overline{\psi_{b_0}}(a)\,[\Phi(cd)f](b) && \text{(by (10.41))} \\ &= \overline{\psi_c}(b)\widetilde{\eta}\,'(a, b)\widetilde{\eta}\,'(c, d)\psi_{b_0}(a^{-1}c)f(d^{-1}b) && \text{(by (10.39))} \end{aligned}$$

so that from (10.42) we deduce that $\Phi'(cd)T = T\Phi(cd)$. This shows that $\Phi \sim \mathrm{Ind}_{A'}^{A}\psi_{b_0}$. ■

Corollary 10.25 *Taking $\widehat{A''} := \{\psi_a : a \in A'\}$ as a basis for $L(A'')$, then the action of Φ (cf. (10.39)) is given by:*

$$\Phi(cd)\psi_a := \overline{\psi_a}(d)\psi_{b_0}(c)\widetilde{\eta}\,'(c, d)\psi_{c^{-1}a}, \tag{10.43}$$

for all $a, c \in A'$ and $d \in A''$.

Proof From (10.39) we immediately deduce that:

$$[\Phi(cd)\psi_a](b) = \overline{\psi_c}(b)\ \widetilde{\eta}'(c,d)\ \psi_{b_0}(c)\ \psi_a(d^{-1}b) = \overline{\psi_a}(d)\ \psi_{b_0}(c)\ \widetilde{\eta}'(c,d)\ \psi_{c^{-1}a}(b).$$

■

Remark 10.26 It is easy to directly check that (10.43) is indeed an irreducible unitary projective representation of A and that it does not depend on the choice of the element $b_0 \in A''$. For, given $a_1, a_2, c \in A'$ and $d \in A''$, we have:

$$\begin{aligned}\langle \Phi(cd)\psi_{a_1}, \Phi(cd)\psi_{a_2}\rangle_{L(A'')} &= \frac{1}{|A''|}\sum_{b\in A''}\left[\Phi(cd)\psi_{a_1}\right](b)\cdot\overline{\left[\Phi(cd)\psi_{a_2}\right]}(b)\\ &= \overline{\psi_{a_1}}(d)\psi_{a_2}(d)\frac{1}{|A''|}\sum_{b\in A''}\psi_{c^{-1}a_1}(b)\overline{\psi_{c^{-1}a_2}}(b)\\ &= \delta_{a_1,a_2}\end{aligned}$$

by the orthogonality relations in $\widehat{A''}$. This shows that $\Phi(cd)$ is unitary. We now check the cocycle identity (7.10). Let $a, c_1, c_2 \in A'$ and $d_1, d_2 \in A''$. We first observe that

$$\begin{aligned}\widetilde{\eta}'(c_1c_2, d_1d_2)\ \overline{\widetilde{\eta}'}(c_2,d_1)^2 &= \widetilde{\eta}'(c_1,d_1)\widetilde{\eta}'(c_1,d_2)\widetilde{\eta}'(c_2,d_1)\widetilde{\eta}'(c_2,d_2)\overline{\widetilde{\eta}'}(c_2,d_1)^2\\ &= \widetilde{\eta}'(c_1,d_1)\widetilde{\eta}'(c_1,d_2)\ \widetilde{\eta}'(c_2,d_2)\overline{\widetilde{\eta}'}(c_2,d_1)\\ &= \widetilde{\eta}'(c_1,d_1)\widetilde{\eta}'(c_2,d_2)\widetilde{\eta}'(c_1d_1,c_2d_2)\end{aligned} \tag{10.44}$$

and

$$\psi_a(d_1) = \psi_{c_2^{-1}a}(d_1)\psi_{c_2}(d_1) = \psi_{c_2^{-1}a}(d_1)\ \widetilde{\eta}'(c_2,d_1)^2. \tag{10.45}$$

We then have

$$\begin{aligned}\Phi(c_1d_1\cdot c_2d_2)\psi_a &= \Phi(c_1c_2\cdot d_1d_2)\psi_a\\ &= \overline{\psi_a}(d_1d_2)\ \psi_{b_0}(c_1c_2)\ \widetilde{\eta}'(c_1c_2,d_1d_2)\ \psi_{c_1^{-1}c_2^{-1}a}\\ &= \overline{\psi_a}(d_1)\overline{\psi_a}(d_2)\ \psi_{b_0}(c_1)\psi_{b_0}(c_2)\ \widetilde{\eta}'(c_1c_2,d_1d_2)\ \psi_{c_1^{-1}c_2^{-1}a}\\ \text{(by (10.45))}\quad &= \overline{\psi_{c_2^{-1}a}}(d_1)\overline{\widetilde{\eta}'}(c_2,d_1)^2\ \overline{\psi_a}(d_2)\ \psi_{b_0}(c_1)\psi_{b_0}(c_2)\ \widetilde{\eta}'(c_1c_2,d_1d_2)\ \psi_{c_1^{-1}c_2^{-1}a}\\ &= \overline{\psi_{c_2^{-1}a}}(d_1)\overline{\psi_a}(d_2)\ \psi_{b_0}(c_1)\psi_{b_0}(c_2)\ \widetilde{\eta}'(c_1c_2,d_1d_2)\ \overline{\widetilde{\eta}'}(c_2,d_1)^2\ \psi_{c_1^{-1}c_2^{-1}a}\\ \text{(by (10.44))}\quad &= \overline{\psi_{c_2^{-1}a}}(d_1)\overline{\psi_a}(d_2)\ \psi_{b_0}(c_1)\psi_{b_0}(c_2)\ \widetilde{\eta}'(c_1,d_1)\widetilde{\eta}'(c_2,d_2)\\ &\quad\times\widetilde{\eta}'(c_1d_1,c_2d_2)\ \psi_{c_1^{-1}c_2^{-1}a}\end{aligned}$$

$$
\begin{aligned}
&= \widetilde{\eta}'(c_1d_1, c_2d_2)\,\overline{\psi_a}(d_2)\psi_{b_0}(c_2)\widetilde{\eta}'(c_2, d_2)\\
&\quad\times\left(\overline{\psi_{c_2^{-1}a}}(d_1)\psi_{b_0}(c_1)\widetilde{\eta}'(c_1, d_1)\psi_{c_1^{-1}c_2^{-1}a}\right)\\
&= \widetilde{\eta}'(c_1d_1, c_2d_2)\left(\overline{\psi_a}(d_2)\psi_{b_0}(c_2)\widetilde{\eta}'(c_2, d_2)\right)\cdot\Phi(c_1d_1)\psi_{c_2^{-1}a}\\
\text{((by (10.43))}\quad &= \widetilde{\eta}'(c_1d_1, c_2d_2)\cdot\Phi(c_1d_1)\left(\overline{\psi_a}(d_2)\psi_{b_0}(c_2)\widetilde{\eta}'(c_2, d_2)\cdot\psi_{c_2^{-1}a}\right)\\
\text{((by (10.43))}\quad &= \widetilde{\eta}'(c_1d_1, c_2d_2)\Phi(c_1d_1)\Phi(c_2d_2)\psi_a.
\end{aligned}
$$

Finally, we compute the character of Φ:

$$
\begin{aligned}
\chi^{\Phi}(cd) &= \sum_{a\in A'}\langle\Phi(cd)\psi_a, \psi_a\rangle_{L(A'')}\\
&= \frac{1}{|A''|}\sum_{a\in A', b\in A''}[\Phi(cd)\psi_a](b)\,\overline{\psi_a}(b)\\
\text{((by (10.43))}\quad &= \frac{1}{|A''|}\sum_{a\in A', b\in A''}\overline{\psi_a}(d)\psi_{b_0}(c)\widetilde{\eta}'(c, d)\psi_{c^{-1}a}(b)\overline{\psi_a}(b)\\
&= \psi_{b_0}(c)\widetilde{\eta}'(c, d)\cdot\frac{1}{|A''|}\sum_{a\in A'}\overline{\psi_d}(a)\cdot\sum_{b\in A''}\overline{\psi_c}(b)\\
&= \begin{cases}|A'| & \text{if } c = 1_{A'} \text{ and } d = 1_{A''}\\ 0 & \text{otherwise,}\end{cases}
\end{aligned}
$$

where the last equality follows from the orthogonality relations for characters and the equalities $|A'| = |A''|$ and $\psi_{b_0}(1_{A'})\widetilde{\eta}'(1_{A'}, 1_{A''}) = \psi_{1_{A'}}(b_0)\widetilde{\eta}'(1_A, 1_A) = 1$. Therefore, $\|\chi^{\Phi}\|^2_{L(A)} = |A'|^2 = |A|$ and from Corollary 7.15.(1) we deduce that Φ is irreducible. Moreover, χ^{Φ} does not depend on the choice of the fixed element $b_0 \in A''$.

Theorem 10.27 *Let $\eta \in \mathscr{C}^2(A, \mathbb{T})$ and let ω_η be as in (10.19).*

(1) *The kernel $\ker\omega_\eta$ (cf. Definition 10.4) coincides with the set A_η of η-special elements of A (cf. (10.7) and Definition 7.32).*
(2) $|\widehat{A}^{\eta}| = |\ker\omega_\eta|$.
(3) *For each irreducible η-representation Φ we have (cf. Definition 7.10):*

$$\ker\Phi = \ker\omega_\eta \tag{10.46}$$

and

$$d_\Phi^2 = [A : \ker\omega_\eta]. \tag{10.47}$$

Suppose, moreover, that $\ker \omega_\eta \neq \{1_A\}$. *Set* $C := A/\ker \omega_\eta$ *and choose a section* $t: C \to A$ *such that* $t(1_C) = 1_A$, *and let* $b: C \times C \to \ker \omega_\eta$ *be given by*

$$t(c)t(d) = t(cd)b(c,d) \tag{10.48}$$

for all $c, d \in C$ *(cf. (1.56) and (7.16)). Then the function* $\mu: C \times C \to \mathbb{T}$ *defined by setting*

$$\mu(c,d) := \eta(t(c), t(d))\, \eta\Big(t(c)t(d), b(c,d)^{-1}\Big)\, \lambda\Big(b(c,d)^{-1}\Big), \tag{10.49}$$

for all $c, d \in C$, *where* $\lambda: \ker \omega_\eta \to \mathbb{T}$ *is as in (7.12), satisfies* $\mu \in \mathscr{C}^2(C, \mathbb{T})$. *Moreover,* ω_μ *is nondegenerate,* $|\widehat{C}^\mu| = 1$, *and* C *is of symmetric type.*

(4) *The restriction of* η *to* $\ker \omega_\eta$ *is a coboundary and there exists* $\eta' \in \mathscr{C}^2(A, \mathbb{T})$ *cohomologous to* η *such that* η' *restricted to* $\ker \omega_\eta$ *is identically 1 and satisfies*

$$\eta'(a_1 t(c_1), a_2 t(c_2)) = \mu(c_1, c_2) \tag{10.50}$$

for all $a_1, a_2 \in \ker \omega_\eta$ *and* $c_1, c_2 \in C$, *that is,* η' *is the inflation of* μ *to* A *(cf. Definition 9.5).*

Proof

(1) For $a \in A$ we have:

$$\begin{aligned} a \in \ker \omega_\eta &\iff \omega_\eta(a,b) = 1 \text{ for all } b \in A \\ &\iff \eta(a,b) = \eta(b,a) \text{ for all } b \in A \\ &\iff a \text{ is } \eta\text{-special.} \end{aligned}$$

(2) This follows immediately from (1) and Corollary 7.38, keeping in mind that, being A Abelian, the number of all η-special conjugacy classes in A (cf. Lemma 7.33) coincides with the number of all η-special elements in A.

(3) On one hand we have:

$$\begin{aligned} a \in \ker \Phi &\implies \Phi(a) = \lambda I_V \text{ for some } \lambda \in \mathbb{T} \\ \text{(by (7.10))} &\implies \Phi(ab)\overline{\eta}(a,b) = \Phi(a)\Phi(b) \\ &\qquad = \lambda\Phi(b) = \Phi(b)\Phi(a) = \Phi(ba)\overline{\eta}(b,a) \text{ for all } b \in A \\ &\implies \eta(a,b) = \eta(b,a) \text{ for all } b \in A \\ \text{(by (1))} &\implies a \in \ker \omega_\eta. \end{aligned}$$

On the other hand,

$$\begin{aligned} a \in \ker \omega_\eta &\Longrightarrow \eta(a, b) = \eta(b, a) \text{ for all } b \in A \\ &\Longrightarrow \Phi(a)\Phi(b) = \Phi(ab)\overline{\eta}(a, b) = \Phi(ba)\overline{\eta}(b, a) \\ &\qquad\qquad\qquad = \Phi(b)\Phi(a) \text{ for all } b \in A \qquad \text{(by (7.10) and (1))} \\ &\Longrightarrow \Phi(a) = \lambda I_V \text{ for some } \lambda \in \mathbb{T} \qquad \text{(by Schur's Lemma)} \\ &\Longrightarrow a \in \ker \Phi. \end{aligned}$$

This proves (10.46).

We now prove (10.47). First of all, assume that $\ker \omega_\eta = \{1_A\}$. From (2) it follows that $|\widehat{A}^\eta| = 1$, say $\widehat{A}^\eta = \{\Phi\}$, and therefore (7.25) becomes $d_\Phi^2 = |A| = [A : \ker \omega_\eta]$.

Now suppose that $\ker \omega_\eta$ is non trivial. Setting

$$\widetilde{\Phi}(c) := \Phi(t(c)), \tag{10.51}$$

for all $c \in C$, we have

$$\begin{aligned} \widetilde{\Phi}(cd) &= \Phi(t(cd)) \\ \text{(by (10.48))} \quad &= \Phi(t(c)t(d)b(c, d)^{-1}) \\ \text{(by (7.10))} \quad &= \Phi(t(c))\Phi(t(d))\eta(t(c), t(d))\Phi(b(c, d)^{-1})\eta(t(c)t(d), b(c, d)^{-1}) \\ \text{(by (7.12) and (10.49))} \quad &= \widetilde{\Phi}(c)\widetilde{\Phi}(d)\mu(c, d) \end{aligned}$$

for all $c, d \in C$. From Remark 7.9.(1) it follows that μ is a 2-cocycle and $\widetilde{\Phi}$ a unitary μ-representation. Moreover, every $b \in A$ may be written in the form $b = at(c)$, with $a \in \ker \omega_\eta$ and $c \in C$, so that

$$\begin{aligned} \Phi(b) = \Phi(at(c)) &= \Phi(a)\Phi(t(c))\eta(a, t(c)) \\ \text{(by (10.51))} \quad &= \lambda(a)\eta(a, t(c))\widetilde{\Phi}(c). \end{aligned} \tag{10.52}$$

In particular, a $\widetilde{\Phi}$-invariant subspace is also Φ-invariant, showing that $\widetilde{\Phi}$ is irreducible. Finally, by using (10.46) for both pairs (ω_η, Φ) and $(\omega_\mu, \widetilde{\Phi})$ and keeping in mind (10.51), we have

$$\ker \Phi = \ker \omega_\eta \Longrightarrow \ker \widetilde{\Phi} = \{1_C\} \Longrightarrow \ker \omega_\mu = \{1_C\}.$$

We then deduce that ω_μ is nondegenerate, so that, by Yamazaki's theorem (Theorem 10.13), C is of symmetric type, and, by the first step of the proof, on the one hand $\widetilde{\Phi}$ is the unique irreducible μ-representation of C, so that,

$|\widehat{C}^{\mu}| = 1$, and on the other hand,

$$d_{\Phi}^2 = d_{\widetilde{\Phi}}^2 = |C| = [A : \ker \omega_\eta].$$

(4) This is just Lemma 7.11 specialized in the present context. It suffices to prove (10.50). Let $\eta' \in \mathscr{C}^2(A, \mathbb{T})$ be as in (7.19) with $K = \ker \omega_\eta$. Then for all $a_1, a_2 \in \ker \omega_\eta$ and $c_1, c_2 \in C$ we have:

$$\begin{aligned}
\eta'(a_1 t(c_1), a_2 t(c_2)) &= \eta(t(c_1), t(c_2))\overline{\eta}(t(c_1c_2), b(c_1, c_2))\overline{\lambda}(b(c_1, c_2)) \\
\text{(by (10.48))} \quad &= \eta(t(c_1), t(c_2))\overline{\eta}\Big(t(c_1)t(c_2) \cdot b(c_1, c_2)^{-1}, b(c_1, c_2)\Big)\,\overline{\lambda}(b(c_1, c_2)) \\
\text{(by (6.1))} \quad &= \eta(t(c_1), t(c_2))\eta(t(c_1)t(c_2), b(c_1, c_2)^{-1}) \cdot \overline{\lambda}(b(c_1, c_2)) \\
\text{(by (7.12))} \quad &= \eta(t(c_1), t(c_2))\eta(t(c_1)t(c_2), b(c_1, c_2)^{-1})\lambda(b(c_1, c_2)^{-1}) \\
\text{(by (10.49))} \quad &= \mu(c_1, c_2).
\end{aligned}$$

■

Note that we have two distinct proofs of the following fact: if ω_η is non-degenerate then A has exactly one irreducible η-representation. The first one is in Theorem 10.24.(3), the second one is in Theorem 10.27.(2).

Remark 10.28 It is worthwhile to check directly that (10.49) is indeed a 2-cocycle. From the cocycle identity (6.1) we deduce the following alternative expression for μ:

$$\mu(c, d) = \eta(t(c), t(d)b(c, d)^{-1})\eta(t(d), b(c, d)^{-1})\lambda(b(c, d)^{-1})$$

After this preliminary result, by a repeated application of (10.48), of the fact that $b(c, d)$ is η-special and of (6.1), both to η and to $b(c, d)$, we get, for all $c_1, c_2, c_3 \in C$,

$$\begin{aligned}
\mu(c_1c_2, c_3)\mu(c_1, c_2) &= \eta(t(c_1c_2), t(c_3)b(c_1c_2, c_3)^{-1})\eta(t(c_3), b(c_1c_2, c_3)^{-1})\lambda(b(c_1c_2, c_3)^{-1}) \cdot \\
&\quad \cdot \eta(t(c_1), t(c_2)b(c_1, c_2)^{-1})\eta(t(c_2), b(c_1, c_2)^{-1})\lambda(b(c_1, c_2)^{-1}) \\
&= \eta(t(c_1)t(c_2)b(c_1, c_2)^{-1}, t(c_3)b(c_1c_2, c_3)^{-1})\eta(t(c_3), b(c_1c_2, c_3)^{-1}) \cdot \\
&\quad \cdot \eta(t(c_1), t(c_2)b(c_1, c_2)^{-1})\eta(t(c_2), b(c_1, c_2)^{-1}) \cdot \\
&\quad \cdot \lambda(b(c_1c_2, c_3)^{-1})\lambda(b(c_1, c_2)^{-1}) \\
&= \eta(t(c_1), t(c_2)t(c_3)b(c_1, c_2)^{-1}b(c_1c_2, c_3)^{-1})\eta(t(c_3), b(c_1c_2, c_3)^{-1}) \cdot \\
&\quad \cdot \eta(t(c_2)b(c_1, c_2)^{-1}, t(c_3)b(c_1c_2, c_3)^{-1})\eta(t(c_2), b(c_1, c_2)^{-1}) \cdot \\
&\quad \cdot \lambda(b(c_1c_2, c_3)^{-1})\lambda(b(c_1, c_2)^{-1})
\end{aligned}$$

$$= \eta(t(c_1), t(c_2)t(c_3)b(c_1, c_2)^{-1}b(c_1c_2, c_3)^{-1})\eta(t(c_3), b(c_1c_2, c_3)^{-1})\cdot$$
$$\cdot \eta(t(c_3)b(c_1c_2, c_3)^{-1}, b(c_1, c_2)^{-1})$$
$$\cdot \eta(t(c_2), t(c_3)b(c_1, c_2)^{-1}b(c_1c_2, c_3)^{-1})\cdot$$
$$\cdot \lambda(b(c_1c_2, c_3)^{-1})\lambda(b(c_1, c_2)^{-1})$$
$$= \eta(t(c_1), t(c_2)t(c_3)b(c_1, c_2)^{-1}b(c_1c_2, c_3)^{-1})$$
$$\cdot \eta(t(c_2), t(c_3)b(c_1, c_2)^{-1}b(c_1c_2, c_3)^{-1})\cdot$$
$$\cdot \eta(t(c_3), b(c_1c_2, c_3)^{-1}b(c_1, c_2)^{-1})\eta(b(c_1c_2, c_3)^{-1}, b(c_1, c_2)^{-1})$$
$$\cdot \lambda(b(c_1c_2, c_3)^{-1})\lambda(b(c_1, c_2)^{-1})$$
$$= \eta(t(c_1), t(c_2)t(c_3)b(c_1, c_2)^{-1}b(c_1c_2, c_3)^{-1})$$
$$\cdot \eta(t(c_2), t(c_3)b(c_1, c_2)^{-1}b(c_1c_2, c_3)^{-1})\cdot$$
(by (7.13))
$$\cdot \eta(t(c_3), b(c_1c_2, c_3)^{-1}b(c_1, c_2)^{-1})\lambda(b(c_1c_2, c_3)^{-1}b(c_1, c_2)^{-1})$$
$$= \eta(t(c_1), t(c_2)t(c_3)b(c_1, c_2)^{-1}b(c_1c_2, c_3)^{-1})$$
$$\cdot \eta(t(c_2), t(c_3)b(c_1, c_2)^{-1}b(c_1c_2, c_3)^{-1})\cdot$$
$$\cdot \eta(t(c_3), b(c_1c_2, c_3)^{-1}b(c_1, c_2)^{-1})\lambda(b(c_1, c_2c_3)^{-1}b(c_2, c_3)^{-1})$$

and similarly

$$\mu(c_1, c_2c_3)\mu(c_2, c_3) = \eta(t(c_1), t(c_2c_3)b(c_1, c_2c_3)^{-1})\eta(t(c_2c_3), b(c_1, c_2c_3)^{-1})\lambda(b(c_1, c_2c_3)^{-1})\cdot$$
$$\cdot \eta(t(c_2), t(c_3)b(c_2, c_3)^{-1})\eta(t(c_3), b(c_2, c_3)^{-1})\lambda(b(c_2, c_3)^{-1})$$
$$= \eta(t(c_1), t(c_2)t(c_3)b(c_2, c_3)^{-1}b(c_1, c_2c_3)^{-1})$$
$$\cdot \eta(t(c_2)t(c_3)b(c_2, c_3)^{-1}, b(c_1, c_2c_3)^{-1})\cdot$$
$$\cdot \eta(t(c_2), t(c_3)b(c_2, c_3)^{-1})\eta(t(c_3), b(c_2, c_3)^{-1})\cdot$$
$$\cdot \lambda(b(c_1, c_2c_3)^{-1})\lambda(b(c_2, c_3)^{-1})$$
$$= \eta(t(c_1), t(c_2)t(c_3)b(c_1, c_2)^{-1}b(c_1c_2, c_3)^{-1})$$
$$\cdot \eta(t(c_2), t(c_3)b(c_2, c_3)^{-1}b(c_1, c_2c_3)^{-1})\cdot$$
$$\cdot \eta(t(c_3)b(c_2, c_3)^{-1}, b(c_1, c_2c_3)^{-1})\eta(t(c_3), b(c_2, c_3)^{-1})\cdot$$
$$\cdot \lambda(b(c_1, c_2c_3)^{-1})\lambda(b(c_2, c_3)^{-1})$$
$$= \eta(t(c_1), t(c_2)t(c_3)b(c_1, c_2)^{-1}b(c_1c_2, c_3)^{-1})$$
$$\cdot \eta(t(c_2), t(c_3)b(c_1, c_2)^{-1}b(c_1c_2, c_3)^{-1})\cdot$$
$$\cdot \eta(t(c_3), b(c_2, c_3)^{-1}b(c_1, c_2c_3)^{-1},)\eta(b(c_2, c_3)^{-1}, b(c_1, c_2c_3)^{-1})\cdot$$
$$\cdot \lambda(b(c_1, c_2c_3)^{-1}\lambda(b(c_2, c_3)^{-1})$$

$$= \eta(t(c_1), t(c_2)t(c_3)b(c_1, c_2)^{-1}b(c_1c_2, c_3)^{-1})$$

$$\cdot \eta(t(c_2), t(c_3)b(c_1, c_2)^{-1}b(c_1c_2, c_3)^{-1})\cdot$$

$$\text{(by (7.13))} \qquad \cdot \eta(t(c_3), b(c_1c_2, c_3)^{-1}b(c_1, c_2)^{-1})\lambda(b(c_1, c_2c_3)^{-1}b(c_2, c_3)^{-1}).$$

This proves the cocycle identity. Clearly, $\mu(c, 1_C) = \mu(1_C, c) = 1$ for all $c \in C$ (assuming that t is normalized: $t(1_C) = 1_A$).

In the following we give a series of remarks in order to give a more algorithmic approach to the results in the present section. First we give the definition of inflation of a bicharacter.

Definition 10.29 Let A be an Abelian group and let $K \leq A$ be a subgroup. The *inflation* of a bicharacter $\omega \in \Omega_a(A/K)$ is the bicharacter $\overset{\vartriangle}{\omega} \in \Omega_a(A)$ defined by setting

$$\overset{\vartriangle}{\omega}(a, b) := \omega(aK, bK)$$

for all $a, b \in A$.

Note that if $\omega \in \Omega_a(A/K)$ and $a \in A$, then, as characters of A (cf. (10.8)),

$$(\overset{\vartriangle}{\omega})_a = (\overset{\vartriangle}{\omega_{aK}})$$

where the right hand side is the inflation of the character ω_{aK} (cf. Definition 2.2).

Remark 10.30

(1) First of all, we give a slightly alternative definition of the cocycle μ in (10.49). We define a bicharacter ω'_η of $C = A/\ker\omega_\eta$ by setting

$$\omega'_\eta(c_1 \ker\omega_\eta, c_2 \ker\omega_\eta) := \omega_\eta(t(c_1), t(c_2)), \tag{10.53}$$

for all $c_1, c_2 \in C$. Note that ω'_η is well defined. Indeed, $\omega_\eta(a_1t(c_1), a_2t(c_2)) = \omega_\eta(t(c_1), t(c_2))$ for all $a_1, a_2 \in \ker\omega_\eta$ (recall that ω_η is a bicharacter). Moreover, from (10.48) we deduce that ω'_η is indeed a bicharacter. This way, we may interpret ω_η as the inflation of ω'_η:

$$\omega_\eta = \overset{\vartriangle}{\omega'_\eta}.$$

Moreover, ω'_η is non degenerate and C is an Abelian group of symmetric type. We then may compute the canonical form (10.16) of ω'_η corresponding to a decomposition $C = C' \times C''$, with $C' \cong C''$, and then *define* $\mu \in \mathscr{C}^2(C, \mathbb{T})$ by means of (10.32); in particular, μ is equalized and satisfies all the properties of $\widetilde{\eta}'$ in Theorem 10.24.

(2) If μ is as in (1) above, then denoting by $\overset{\vartriangle}{\mu}$ its inflation to A (compare with (10.50)), we have $\omega'_\eta = \omega_\mu$ and

$$\omega_\eta = \overset{\vartriangle}{\omega'_\eta} = \omega_{\overset{\vartriangle}{\mu}}.$$

Moreover, $\omega_{\eta\overline{\overset{\vartriangle}{\mu}}} = 1$ so that, $\eta\overline{\overset{\vartriangle}{\mu}}$ is symmetric (cf. (10.30)) and by means of (10.21) we can construct $\rho : A \to \mathbb{T}$ such that $\eta\overline{\overset{\vartriangle}{\mu}} = \tau_\rho$, that is,

$$\eta = \overset{\vartriangle}{\mu}\tau_\rho.$$

Note also that $\overset{\vartriangle}{\mu}$ is equalized and that

$$\overset{\vartriangle}{\mu}(a, b) = 1 = \overset{\vartriangle}{\mu}(b, a) \tag{10.54}$$

for all $a \in \ker \omega_\eta$ and $b \in A$.

(3) By the last statement in Remark 7.9.(3), there is a bijective correspondence between the irreducible $\overset{\vartriangle}{\mu}$-representations and the irreducible η-representations: it is just given by multiplication by ρ (note that, in Remark 7.9.(3), our present map ρ is denoted by λ).

By Remark 10.30.(3), in view of determining the η-dual of A, we reduce our analysis to the $\overset{\vartriangle}{\mu}$-irreducible representations, where $\mu \in \mathscr{C}^2(C, \mathbb{T})$ is as in Remark 10.30.(1).

Let $\Phi \in \widehat{A}^{\overset{\vartriangle}{\mu}}$. Since $\overset{\vartriangle}{\mu} \equiv 1$ on $\ker \omega_\eta \equiv \ker \Phi$ (cf. (10.46)), formula (7.13) becomes $\lambda(ab) = \lambda(a)\lambda(b)$, for all $a, b \in \ker \omega_\eta$, that is, $\lambda_\Phi := \lambda|_{\ker \omega_\eta}$ is a character of $\ker \omega_\eta$. This way, we have a map

$$\begin{aligned} \widehat{A}^{\overset{\vartriangle}{\mu}} &\longrightarrow \widehat{\ker \omega_\eta} \\ \Phi &\longmapsto \lambda_\Phi. \end{aligned} \tag{10.55}$$

Clearly, this map is well defined: if Φ_1 and Φ_2 are equivalent irreducible $\overset{\vartriangle}{\mu}$-representations of A, then $\lambda_{\Phi_1} = \lambda_{\Phi_2}$. Indeed, if $T : V_1 \to V_2$, where V_1 and V_2 are the representation spaces of Φ_1 and Φ_2, is an intertwiner, for $a \in \ker \Phi_1$ we have

$$\Phi_2(a)T = T\Phi_1(a) = T\lambda_{\Phi_1}(a)I_{V_1} = \lambda_{\Phi_1}(a)I_{V_2}T$$

so that $\Phi_2(a) = \lambda_{\Phi_1}(a)I_{V_2}$, yielding $a \in \ker \Phi_2$. This gives $\ker \Phi_1 \subseteq \ker \Phi_2$. By symmetry, $\ker \Phi_1 = \ker \Phi_2$, and, moreover, $\lambda_{\Phi_2} = \lambda_{\Phi_1}$.

Theorem 10.31 *The map (10.55) is a bijection. Moreover, once fixed a section $t\colon C \to A$ and denoted by $\widetilde{\Phi}$ the unique (up to equivalence) irreducible $\overset{\Delta}{\mu}$-representation of $C = A/\ker\omega_\eta$ as in (10.51), then the inverse map of (10.55) is given by $\widehat{\ker\omega_\eta} \ni \lambda \mapsto \Phi_\lambda$, where*

$$\Phi_\lambda(at(c)) := \lambda(a)\widetilde{\Phi}(c) \tag{10.56}$$

for all $a \in \ker\omega_\eta$ and $c \in C$.

Proof By Theorem 10.27.(2), with η replaced by the cohomologous cocycle $\overset{\Delta}{\mu}$, it suffices to prove that the map (10.55) is injective. Let $\Phi_1, \Phi_2 \in \widehat{A}^{\overset{\Delta}{\mu}}$ and suppose that $\lambda_{\Phi_1} = \lambda_{\Phi_2}$. Denoting by $\widetilde{\Phi}_1$ and $\widetilde{\Phi}_2$ the corresponding (irreducible) $\overset{\Delta}{\mu}$-representations of C, as in (10.51), it follows from Theorem 10.27.(3) that $\widetilde{\Phi}_1 \sim \widetilde{\Phi} \sim \widetilde{\Phi}_2$ (the unique irreducible μ-representation of C). Finally, by virtue of (10.54), in the present setting (10.52) becomes (10.56), and this ends the proof of the Theorem. ■

We now apply the results in Proposition 10.23, Theorem 10.24, and Corollary 10.25 in order to get an explicit expression for $\widetilde{\Phi}$. We just have to adapt all the notation from A to its quotient C. Note also that we are using a cocycle μ of C that satisfies all the properties of $\widetilde{\eta}\,'$ in Theorem 10.24. We fix a decomposition of C as an Abelian group of symmetric type: $C = C' \times C''$, with $C' \cong C''$. Moreover, letting $\{\psi_c : c \in C'\}$ be a parameterization of the characters of C'' (equivalently, of C') as in (10.35), we use the trivial character $\iota_{C''}$ of C'' to define

$$\widetilde{\Phi} := \mathrm{Ind}_{C''}^{C}\iota_{C''}.$$

Then the analogue of the expression (10.39) (with $b_0 = 1_{C''}$) in Theorem 10.24, holds true. Finally, we use the correspondence in Remark 10.30.(3) and (10.56).

Theorem 10.32 *With the notation above, we have a natural bijection*

$$\widehat{A}^{\overset{\Delta}{\mu}} = \{\Phi_\lambda : \lambda \in \widehat{\ker\omega_\eta}\} \quad \longrightarrow \quad \widehat{A}^{\eta} = \{\rho\Phi_\lambda : \lambda \in \widehat{\ker\omega_\eta}\}$$

and a concrete realization of $\rho\Phi_\lambda$ as a representation on $L(C'')$ is given by

$$\{[(\rho\Phi_\lambda)\,(at(cd))]\,f\}\,(d_0) = \rho(at(cd))\lambda(a)\overline{\psi_c}(d_0)\mu(c,d)f(d^{-1}d_0),$$

for all $f \in L(C'')$, $a \in \ker \omega_\eta$, $c \in C'$, *and* $d, d_0 \in C''$. *Moreover, if we use* $\widehat{C''} = \{\psi_c : c \in C'\}$ *as a basis for* $L(C'')$ *we have, for a fixed* $c_0 \in C'$*:*

$$(\rho\Phi_\lambda)\,(at(cd))\psi_{c_0} = \rho(at(cd))\lambda(a)\overline{\psi_{c_0}}(d)\mu(c,d)\psi_{c^{-1}c_0},$$

for all $a \in \ker \omega_\eta$, $c \in C'$, *and* $d \in C''$. □

Remark 10.33 We end this section by summarizing the method we developed so far in order to construct a complete set of irreducible η-representations of an Abelian group A of odd order for any $\eta \in \mathscr{C}^2(A, \mathbb{T})$.

- Determine $\ker \omega_\eta$ (cf. (10.19) and (10.7)) and $C = A/\ker \omega_\eta$.
- Define the nondegenerate antisymmetric bicharacter ω'_η of C as in (10.53) and then use the algorithm in the second part of the proof of Theorem 10.13 to determine its canonical form (10.16) as well as the associated decomposition $C = C' \times C''$.
- Use the canonical form of ω'_η and (10.32) to define $\mu \in \mathscr{C}^2(C, \mathbb{T})$, as indicated in Remark 10.30.(1) and then its inflation $\overset{\triangle}{\mu} \in \mathscr{C}^2(A, \mathbb{T})$.
- Use (10.21) to compute $\rho \colon A \to \mathbb{T}$, as indicated in Remark 10.30.(2).
- Finally, apply the formulas in Theorem 10.32 and get an explicit description of $\widehat{A}^\eta$.

10.3 Representation Theory of Finite Metabelian Groups

In this section, G is a finite, metabelian (equivalently, step-2 solvable) group. In other words, G contains a normal Abelian subgroup B such that the quotient $A := G/B$ is Abelian as well. We apply the results in Sects. 7.3 and 10.2 in order to explicitly describe $\widehat{G}$. One of the key points is that, in the present setting, (7.32) simplifies since, with the notation therein,

$$N = B \text{ Abelian} \Rightarrow \dim \sigma = 1 \text{ and } \Psi \equiv 1$$

and

$$H = A \text{ Abelian} \Rightarrow \Phi \text{ has an explicit description (cf. Remark 10.33).}$$

Our final result is a description of the irreducible representations of G by means of suitable actions on function spaces defined on certain finite Abelian groups under the assumption that A is of odd order. Since the results are quite messy, we proceed in several steps. Some properties of the irreducible representations of metabelian groups were studied in [115].

Step 1: a description of G. First of all, we specify to the present setting, some of the notation in Sect. 1.5. We fix a normalized section $t : A \to G$ and we denote by $b\colon A \times A \to B$ the corresponding factor system, that is, for all $a_1, a_2 \in A$ we have:

$$t(a_1)t(a_2) = t(a_1a_2)b(a_1, a_2). \tag{10.57}$$

The conjugation action of A on B is denoted by $b^a := t(a)^{-1}bt(a)$, for $a \in A$ and $b \in B$. This way, (1.59) and (1.60) become respectively

$$b(a_1, a_2a_3)b(a_2, a_3) = b(a_1a_2, a_3)b(a_1, a_2)^{a_3} \tag{10.58}$$

and

$$t(a_1)b_1 \cdot t(a_2)b_2 = t(a_1a_2)\left[b(a_1, a_2)b_1^{a_2}b_2\right], \tag{10.59}$$

for all $a_1, a_2, a_3 \in A$ and $b_1, b_2 \in B$.

Step 2: a first description of $\widehat{G}$. We specify some of the notation and results in Sects. 2.1 and 7.3. The action of G on $\widehat{B}$ is given by:

$${}^g\xi(b) := \xi(g^{-1}bg)$$

for all $\xi \in \widehat{B}$, $b \in B$, and $g \in G$. We then define:

- $I_G(\xi) := \{g \in G : {}^g\xi = \xi\}$ is the stabilizer of ξ,
- $A_G(\xi) := I_G(\xi)/B$,
- $\widehat{G}(\xi) := \{\theta \in \widehat{G} : \xi \preceq \mathrm{Res}^G_B\theta\}$ and
- $\Xi \subseteq \widehat{B}$ is a complete set of representatives for the orbits of G on $\widehat{B}$.

Moreover, arguing as in the proof of Proposition 7.24, in (7.1) one has $\xi\left(t(a)^{-1}bt(a)\right) = \xi(b)$ for all $a \in A_G(\xi)$ and $b \in B$, and therefore $\Psi \equiv 1$. Then, from (7.2) we deduce that the Mackey obstruction of ξ is given by $\eta \in \mathscr{C}^2(A_G, \mathbb{T})$, where

$$\eta(a_1, a_2) := \xi(b(a_1, a_2)) \tag{10.60}$$

for all $a_1, a_2 \in A_G(\xi)$. Also, given $\Phi \in \widehat{A_G(\xi)}^{\overline{\eta}}$, the expression of $\Theta \in \widehat{I_G(\xi)}$ in (7.32) simplifies, yielding

$$\Theta(t(a)b) = \Phi(a)\xi(b), \tag{10.61}$$

for all $a \in A_G(\xi)$ and $b \in B$. Finally, from Theorem 7.23 we deduce a first description of all irreducible representations of G:

$$\widehat{G} = \left\{ \mathrm{Ind}_{I_G(\xi)}^G \Theta : \xi \in \Xi, \Theta \text{ as in (10.61), and } \Phi \in \widehat{A_G(\xi)}^{\overline{\eta}} \right\}.$$

From now on, we assume that A is of odd order.

Step 3: a description of $\widehat{A_G(\xi)}^{\overline{\eta}}$. We now fix $\xi \in \Xi$ and apply the theory in Sect. 10.2. The bicharacter $\omega_{\overline{\eta}} \in \Omega_{\mathrm{a}}(A_G(\xi))$ associated with $\overline{\eta}$ is given by

$$\omega_{\overline{\eta}}(a_1, a_2) := \overline{\xi}(b(a_1, a_2))\, \xi(b(a_2, a_1)) = \xi\left(b(a_1, a_2)^{-1} b(a_2, a_1)\right) \tag{10.62}$$

for all $a_1, a_2 \in A_G(\xi)$, and its kernel is

$$\ker \omega_{\overline{\eta}} = \left\{ a_1 \in A_G(\xi) : b(a_1, a_2)^{-1} b(a_2, a_1) \in \ker \xi \text{ for all } a_2 \in A_G(\xi) \right\}.$$

The quotient $C_{\overline{\eta}} := A_G(\xi)/\ker \omega_{\overline{\eta}}$ is an Abelian group of symmetric type. Let $s \colon C_{\overline{\eta}} \to A_G(\xi)$ be a normalized section, and denote by $a \colon C_{\overline{\eta}} \times C_{\overline{\eta}} \to \ker \omega_{\overline{\eta}}$ the associated factor set:

$$s(c_1)s(c_2) = a(c_1, c_2)s(c_1c_2), \tag{10.63}$$

for all $c_1, c_2 \in C_{\overline{\eta}}$. Define a bicharacter $\omega'_\eta \in \Omega_{\mathrm{a}}(C_{\overline{\eta}})$ as in (10.53):

$$\omega'_\eta(c_1 \ker \omega_{\overline{\eta}}, c_2 \ker \omega_{\overline{\eta}}) := \omega_{\overline{\eta}}(s(c_1), s(c_2)),$$

for all $c_1, c_2 \in C_{\overline{\eta}}$. Then use the algorithm in the second part of the proof of Theorem 10.13 to determine its canonical form (10.16) as well as the associated decomposition $C_{\overline{\eta}} = C' \times C''$ with $C' \cong C''$. Clearly,

$$A_G(\xi) = \bigsqcup_{\substack{c' \in C' \\ c'' \in C''}} \ker \omega_{\overline{\eta}} \cdot s(c'c''). \tag{10.64}$$

Now use the canonical form of ω'_η and (10.32) to define the unitary cocycle $\mu \in \mathscr{C}^2(C_{\overline{\eta}}, \mathbb{T})$ as in Remark 10.30.(1) and then consider its inflation $\overset{\vartriangle}{\mu} \in \mathscr{C}^2(A_G(\xi), \mathbb{T})$.
Next, as in Remark 10.30.(2), use (10.21) to compute $\rho \colon A_G(\xi) \to \mathbb{T}$ such that

$$\overline{\eta} = \overset{\vartriangle}{\mu} \tau_\rho. \tag{10.65}$$

Then from Theorem 10.32 we deduce that

$$\widehat{A_G(\xi)}^{\overline{\eta}} = \left\{\rho\Phi_\lambda : \lambda \in \widehat{\ker\omega_{\overline{\eta}}}\right\},$$

where a concrete realization of $\rho\Phi_\lambda$ as an η-representation on $L(C'')$ is given by

$$\left\{\left[(\rho\Phi_\lambda)\left(as(c'c'')\right)\right]f\right\}(c''_0) = \rho(as(c'c''))\lambda(a)\overline{\psi_{c'}(c''_0)}\mu(c',c'')f((c'')^{-1}c''_0), \tag{10.66}$$

for all $f \in L(C'')$, $a \in \ker\omega_{\overline{\eta}}$, $c' \in C'$, and $c'', c''_0 \in C''$. Moreover, using $\widehat{C''} = \{\psi_{c'} : c' \in C'\}$ as a basis for $L(C'')$ we have, for a fixed $c'_0 \in C'$:

$$(\rho\Phi_\lambda)\left(as(c'c'')\right)\psi_{c''_0} = \rho(as(c'c''))\lambda(a)\overline{\psi_{c'_0}(c'')}\mu(c',c'')\psi_{(c')^{-1}c''_0}$$

for all $a \in \ker\omega_{\overline{\eta}}$, $c' \in C'$, and $c'' \in C''$.

Step 4: a description of $\widehat{I_G}(\xi)$. From (10.66) it follows that (10.61) becomes $\Theta_\lambda(t(a)b) = \rho(a)\Phi_\lambda(a)\xi(b)$, and an explicit realization of Θ_λ as a representation on $L(C'')$ is given by

$$\left\{\left[\Theta_\lambda(t(as(c'c''))b)\right]f\right\}(c''_0) = \xi(b)\rho(as(c'c''))\lambda(a)\overline{\psi_{c'}(c''_0)}\mu(c',c'')f((c'')^{-1}c''_0), \tag{10.67}$$

for all $f \in L(C'')$, $a \in \ker\omega_{\overline{\eta}}$, $c' \in C'$, $c'', c''_0 \in C''$, and $b \in B$. Moreover, noting that b appears only in $\xi(b)$ and that ξ is invariant under conjugation by elements of $I_G(\xi)$, for $b_1, b_2 \in B$ we have:

$$\Theta_\lambda(t(as(c'c''))b_1b_2) = \Theta_\lambda(b_1t(as(c'c''))b_2). \tag{10.68}$$

Finally, from Theorem 7.20 it follows that

$$\widehat{I_G}(\xi) = \left\{\Theta_\lambda : \lambda \in \widehat{\ker\omega_{\overline{\eta}}}\right\}.$$

Clearly,

$$\dim\Theta_\lambda = \dim\Phi_\lambda = |C''| = \sqrt{|A_G(\xi)|/|\ker\omega_{\overline{\eta}}|}. \tag{10.69}$$

Step 5: other computations with sections and cocycles. Let $a_0 \in \ker\omega_{\overline{\eta}}$, $c' \in C'$, and $c'' \in C''$. By a repeated application of (1.62) we get:

$$s(c'c'')^{-1} = a\left((c'c'')^{-1}, c'c''\right)^{-1} s\left((c'c'')^{-1}\right)$$

and

$$t(a_0 s(c'c''))^{-1} = b\left(s(c'c'')^{-1}a_0^{-1}, a_0 s(c'c'')\right)^{-1} \cdot t(s(c'c'')^{-1}a_0^{-1})$$
$$= b\Big(s(c'c'')^{-1}a_0^{-1}, a_0 s(c'c'')\Big)^{-1} \cdot t\left(a_0^{-1}a\left((c'c'')^{-1}, c'c''\right)^{-1} s\Big((c'c'')^{-1}\Big)\right). \tag{10.70}$$

We then set $D := A/A_G(\xi)$ and consider a normalized section $\ell\colon D \to A$, so that $A = \sqcup_{d\in D}\ell(d)A_G(\xi)$. The associated cocycle $\alpha \in \mathscr{C}^2(D, A_G(\xi))$ then satisfies

$$\ell(d_1)\ell(d_2) = \alpha(d_1, d_2)\ell(d_1d_2) \tag{10.71}$$

for all $d_1, d_2 \in D$. Moreover,

$$G = \sqcup_{d\in D}t(\ell(d))\, I_G(\xi). \tag{10.72}$$

Indeed, if $g \in G$, say, $g = t(a)b_1$ with $b_1 \in B$ and $a = \ell(d)a_1 \in A$, with $d \in D$ and $a_1 \in A_G(\xi)$, then from (10.57) we deduce that

$$g = t(\ell(d)a_1)\, b_1 = t(\ell(d))\left[t(a_1)\, b(\ell(d), a_1)^{-1}b_1\right] \tag{10.73}$$

and $t(a_1)b(\ell(d), a_1)^{-1}\, b_1 \in I_G(\xi)$, so that G is the union of the cosets on the right side. Moreover, since

$$|D| = |A/A_G(\xi)| = |(G/B)/(I_G(\xi)/B)| = |G/I_G(\xi)|,$$

(10.72) is a disjoint union. See also (2.4).
In particular, for each $g \in G$ (written as in (10.73)) there is a unique choice of

$$b_0 \in B,\ d \in D,\ a_0 \in \ker\omega_{\overline{\eta}},\ c' \in C',\ \text{ and } c'' \in C''$$

such that

$$g = t(\ell(d)) \cdot t(a_0 s(c'c''))b_0. \tag{10.74}$$

Indeed, in (10.73) we can write a_1 in the form $a_1 = a_0 c(c'c'')$ and set $b_0 := b(\ell(d), a_1)^{-1}b_1$. Moreover, in (10.74) we have:

$$t(a_0 s(c'c''))b_0 \in I_G(\xi). \tag{10.75}$$

Finally, if also $d_0 \in D$, we need to write $g^{-1}t(\ell(d_0))$ in a form closely to that in (10.74): since

$$\begin{aligned} g^{-1}t(\ell(d_0)) &= b_0^{-1}t(a_0s(c'c''))^{-1}t(\ell(d))^{-1}t(\ell(d_0)) \\ \text{(by (1.64))} \quad &= b_0^{-1}t(a_0s(c'c''))^{-1}t\big(\ell(d)^{-1}\ell(d_0)\big)b\big(\ell(d),\ell(d)^{-1}\ell(d_0)\big)^{-1} \\ \text{(by (1.64) and (10.71))} \quad &= b_0^{-1}t\big(a_0^{-1}s(c'c'')^{-1}\ell(d^{-1}d_0)\alpha(d,d^{-1}d_0)^{-1}\big) \\ &\quad \cdot b\big(a_0s(c'c''),a_0^{-1}s(c'c'')^{-1}\ell(d)^{-1}\ell(d_0)\big)^{-1} \\ &\quad \cdot b\big(\ell(d),\ell(d)^{-1}\ell(d_0)\big)^{-1} \\ \text{(by (10.57))} \quad &= b_0^{-1}t\big(\ell(d^{-1}d_0)\big)t\big(a_0^{-1}s(c'c'')^{-1}\alpha(d,d^{-1}d_0)^{-1}\big)\cdot \\ &\quad \cdot b\big(\ell(d^{-1}d_0),a_0^{-1}s(c'c'')^{-1}\alpha(d,d^{-1}d_0)^{-1}\big)\cdot \\ &\quad \cdot b\big(a_0s(c'c''),a_0^{-1}s(c'c'')^{-1}\ell(d)^{-1}\ell(d_0)\big)^{-1} \\ &\quad \cdot b\big(\ell(d),\ell(d)^{-1}\ell(d_0)\big)^{-1} \end{aligned}$$

we have

$$g^{-1}t(\ell(d_0)) = t\big(\ell(d^{-1}d_0)\big)\cdot b_1t\big(a_1s(c_1'c_1'')\big)b_2, \tag{10.76}$$

where $a_1 \in \ker\omega_{\overline{\eta}}$, $c_1' \in C'$, $c_1'' \in C''$, and $b_1, b_2 \in B$ are given by:

$$\begin{aligned} &a_0^{-1}s(c'c'')^{-1}\alpha(d,d^{-1}d_0)^{-1} = a_1s(c_1',c_1'') \in A_G(\xi), \quad \text{(cf. (10.64))} \\ &b_1 = \Big[t\big(\ell(d^{-1}d_0)^{-1}b_0^{-1}t\big(\ell(d^{-1}d_0)\Big] \in B, \\ &b_2 = b\big(\ell(d^{-1}d_0),a_0^{-1}s(c'c'')^{-1}\alpha(d,d^{-1}d_0)^{-1}\big) \\ &\qquad \cdot b\big(a_0s(c'c''),a_0^{-1}s(c'c'')^{-1}\ell(d)^{-1}\ell(d_0)\big)^{-1}b\big(\ell(d),\ell(d)^{-1}\ell(d_0)\big)^{-1} \in B. \end{aligned} \tag{10.77}$$

Step 6: a description of $\theta_\lambda := \mathrm{Ind}_{I_G(\xi)}^G \Theta_\lambda$. From the definitions in Sect. 1.3 we deduce that the space W_λ of θ_λ is made up of all functions $F : G \to L(C'')$ such that

$$F(gh) = \Theta_\lambda(h^{-1})F(g), \tag{10.78}$$

for all $g, \in G$ and $h \in I_G(\xi)$, and the action of G on W_λ is given by

$$[\theta_\lambda(g)F](g_0) = F\left(g^{-1}g_0\right), \tag{10.79}$$

for all $g, g_0 \in G$ and $F \in W_\lambda$. In particular, from (10.68) it follows that, for $d \in D, a_1 \in A_G(\xi)$, and $b_0, b_1 \in B$, so that $t(a_1)b_0b_1 \in I_G(\xi)$ and $F \in W_\lambda$, we have:

$$F\left(t(\ell(d))t(a_1)b_0b_1\right) = F\left(t(\ell(d))b_0t(a_1)b_1\right). \tag{10.80}$$

If $F \in W_\lambda$, $c_0'' \in C''$, and g is as in (10.74) we then have:

$$\begin{aligned}
[F(g)](c_0'') &\equiv \left[F\big(t(\ell(d)) \cdot t(a_0 s(c'c''))b_0\big)\right](c_0'') \\
&= \left[\Theta_\lambda\left(t(a_0 s(c'c''))^{-1}b_0^{-1}\right) F\big(t(\ell(d))\big)\right](c_0'') \\
&\qquad\qquad \text{(by (10.68), (10.75) and (10.78))} \\
&= \left[\Theta_\lambda\Big(t\left(a_0^{-1}a\left((c'c'')^{-1}, c'c''\right)^{-1} s\left((c'c'')^{-1}\right)\right)\right. \\
&\qquad\qquad \text{(by (10.68) and (10.70))} \\
&\qquad \left.\cdot\, b(s(c'c'')^{-1}a_0^{-1}, a_0 s(c'c''))^{-1}b_0^{-1}\Big) F\big(t(\ell(d))\big)\right](c_0'') \\
&= \Big\{\xi\left(b(s(c'c'')^{-1}a_0^{-1}, a_0 s(c'c''))b_0\right) \rho\Big(a_0^{-1}a\left((c'c'')^{-1}, c'c''\right)^{-1} s\left((c'c'')^{-1}\right)\Big) \\
&\qquad\qquad \text{(by (10.67))} \\
&\qquad \cdot\, \overline{\lambda}\left(a_0 a\left((c'c'')^{-1}, c'c''\right)\right)\psi_{c'}(c_0'')\mu((c')^{-1}, (c'')^{-1})\Big\} \left[F\big(t(\ell(d))\big)\right](c''c_0'').
\end{aligned}$$

Therefore, the map

$$\begin{aligned}
T\colon W_\lambda &\longrightarrow L(D \times C'') \\
F &\longmapsto f
\end{aligned}$$

where $f := T(F)$ is defined by setting

$$f(d, c'') := \left[F\big(t(\ell(d))\big)\right](c''), \tag{10.81}$$

for all $d \in D$ and $c'' \in C''$, is a linear isomorphism. We then consider the G-representation $\widetilde{\theta}_\lambda$ on $L(D \times C'')$, equivalent to θ_λ, given by

$$T\theta_\lambda(g)F := \widetilde{\theta}_\lambda(g)f, \tag{10.82}$$

for all $g \in G$. It follows that, for $f \in L(D \times C'')$, $d_0 \in D$, $c_0'' \in C''$, and g as in (10.74), we have:

$$\begin{aligned}
\left[\widetilde{\theta}_\lambda(g)f\right](d_0, c_0'') &= \left[T\theta_\lambda(g)F\right](d_0, c_0'') && \text{(by (10.82))} \\
&= \left\{[\theta_\lambda(g)F]\big(t(\ell(d_0))\big)\right\}(c_0'') && \text{(by (10.81))}
\end{aligned}$$

$$= \left\{F\left[g^{-1}t(\ell(d_0))\right]\right\}(c_0'') \qquad \text{(by (10.79))}$$

$$= \left\{F\left[t\left(\ell(d^{-1}d_0)\right)\cdot b_1 t\left(a_1 s(c_1'c_1'')\right)b_2\right]\right\}(c_0'') \qquad \text{(by (10.76))}$$

$$= \left\{F\left[t\left(\ell(d^{-1}d_0)\right)\cdot t\left(a_1 s(c_1'c_1'')\right)b_1 b_2\right]\right\}(c_0'') \qquad \text{(by (10.80))}$$

so that from (10.3) and (10.81) we deduce that

$$\left[\widetilde{\theta}_\lambda(g)f\right](d_0, c_0'') = \left\{\overline{\xi}\left(b(s(c_1'c_1'')^{-1}a_1^{-1}, a_1 s(c_1'c_1''))b_1\right)\rho\left(a_1^{-1}a\left((c_1'c_1'')^{-1}, c_1'c_1''\right)^{-1}\right.\right.$$

$$\left.\cdot s\left((c_1'c_1'')^{-1}\right)\right)$$

$$\left.\cdot\overline{\lambda}\left(a_1 a\left((c_1'c_1'')^{-1}, c_1'c_1''\right)\right)\psi_{c_1'}(c_0'')\mu((c_1')^{-1}, (c_1'')^{-1})\right\} f(c_1''c_0'', d^{-1}d_0,),$$

where $a_1 \in \ker\omega_{\overline{\eta}}$, $c_1' \in C'$, $c_1'' \in C''$, and $b_1 \in B$ are given by (10.77).

Step 7: a final description of $\widehat{G}$. Recall that Ξ is a set of representatives for the orbits of G on $\widehat{B}$, η_ξ is the Mackey obstruction of ξ (cf. (10.60)), and $\omega_{\overline{\eta}_\xi}$ is the bicharacter of $A_G(\xi)$ associated with $\overline{\eta}_\xi$ (cf. (10.62)). Then, adding to $\widetilde{\theta}_\lambda$ the dependence also on ξ, we have:

$$\widehat{G} = \left\{\widetilde{\theta}_{\xi,\lambda} : \xi \in \Xi, \lambda \in \widehat{\ker\omega_{\overline{\eta}_\xi}}\right\}.$$

In particular,

$$|\widehat{G}| = \sum_{\xi\in\Xi}\left|\ker\omega_{\overline{\eta}_\xi}\right|$$

and, from (10.69) and the equality $|G/I_G(\xi)| = |A/A_G(\xi)|$, we get

$$\dim\widetilde{\theta}_{\xi,\lambda} = \left|\frac{A}{A_G(\xi)}\right|\sqrt{\frac{|A_G(\xi)|}{\left|\ker\omega_{\overline{\eta}_\xi}\right|}} = \frac{|A|}{\sqrt{|A_G(\xi)|\left|\ker\omega_{\overline{\eta}_\xi}\right|}}.$$

As a final check, it is convenient to verify (1.20) in the present setting:

$$\sum_{\xi\in\Xi}\sum_{\lambda\in\ker\omega_{\overline{\eta}_\xi}}\left(\dim\widetilde{\theta}_{\xi,\lambda}\right)^2=\sum_{\xi\in\Xi}\sum_{\lambda\in\ker\omega_{\overline{\eta}_\xi}}\frac{|A|^2}{|A_G(\xi)|\left|\ker\omega_{\overline{\eta}_\xi}\right|}$$
$$=|A|\sum_{\xi\in\Xi}\frac{|A|}{|A_G(\xi)|}$$
$$=|A|\left|\widehat{B}\right|=|G|,$$

because $\frac{|A|}{|A_G(\xi)|}$ is the cardinality of the A-orbit on $\widehat{B}$ containing ξ.

We end this section by describing some applications of Clifford theory to the representations Θ_λ of $I_G(\xi)$. Denote by $\pi : G \to A$ the natural projection and set

$$D_{\overline{\eta}}=\pi^{-1}\left(\ker\omega_{\overline{\eta}}\right).$$

Then the Mackey obstruction of ξ is trivial on $D_{\overline{\eta}}$ (cf. Step 3 above and Theorem 10.17) and therefore ξ has an extension $\widetilde{\xi}$ to $D_{\overline{\eta}}$. By setting $c' = c'' = 1_C$ and $\lambda = \mathbf{1}_{\ker\omega_{\overline{\eta}}}$ in (10.67) we deduce that

$$\widetilde{\xi}(t(a)b) := \xi(b)\rho(a), \text{ for all } b\in B, a\in\ker\omega_{\overline{\eta}}$$

is indeed such an extension. Moreover, since $B \trianglelefteq D_{\overline{\eta}} \trianglelefteq I_G(\xi)$, we may apply Theorem 2.18 and get

$$\widehat{D_{\overline{\eta}}}(\xi)=\left\{\widetilde{\xi}\overset{\triangle}{\lambda} : \lambda\in\widehat{\ker\omega_{\overline{\eta}}}\right\}.$$

Note that the right hand side in the above formula is the set of all extensions of ξ to $D_{\overline{\eta}}$. Finally, from (10.67) we also deduce that

$$\mathrm{Res}^{I_G(\xi)}_{D_{\overline{\eta}}}\Theta_\lambda\sim\underbrace{\widetilde{\xi}\overset{\triangle}{\lambda}\oplus\widetilde{\xi}\overset{\triangle}{\lambda}\oplus\cdots\oplus\widetilde{\xi}\overset{\triangle}{\lambda}}_{\dim\Theta_\lambda-\text{ times}}.$$

This implies that the inertia index of Θ_λ with respect to $D_{\overline{\eta}}$ is equal to

$$\dim\Theta_\lambda\equiv|C'|\equiv\sqrt{|I_G(\xi)/\dim\Theta_\lambda|}.$$

Therefore (2.23) becomes

$$\mathrm{Ind}^{I_G(\xi)}_{D_{\overline{\eta}}}\left(\widetilde{\xi}\overset{\triangle}{\lambda}\right)\sim\underbrace{\Theta_\lambda\oplus\Theta_\lambda\oplus\cdots\oplus\Theta_\lambda}_{\dim\Theta_\lambda-\text{ times}}. \tag{10.83}$$

This means that $\mathrm{Ind}_{D_{\overline{\eta}}}^{I_G(\xi)}\left(\widetilde{\xi}\,\overset{\Delta}{\lambda}\right)$ is an isotypic component of $L(I_G(\xi))$; since under the left action (cf. (1.26)) it decomposes as a sum of copies of Θ_λ then it is the isotypic component associated with the adjoint representation Θ'_λ; see [30, Theorem 10.5.9] which may be deduced from (1.18).

10.4 Representation Theory of Finite Step-2 Nilpotent Groups

Here we suppose that G is a finite step-2 nilpotent group, that is, G contains a central Abelian subgroup B such that the quotient $A := G/B$ is also Abelian. We show how the results in Sect. 10.3 simplify in the present setting.

- In Step 1, $b\colon A \times A \to B$ is a 2-cocycle (cf. Chap. 6).
- In Step 2, the action of G on $\widehat{B}$ is trivial,

$$\Sigma \equiv \widehat{B} \text{ and } A_G(\xi) = A \text{ for all } \xi \in \widehat{B};$$

  cf. Proposition 7.24. Moreover, the Mackey obstruction η is a unitary 2-cocycle of A and the first description of $\widehat{G}$ is:

$$\widehat{G} = \left\{\Theta : \Theta \text{ is as in (10.61) with } \xi \in \widehat{B} \text{ and } \Phi \in \widehat{A^{\overline{\eta}}}\right\}.$$

  From now on, we assume that A is of odd order.

- In Step 3, $\omega_{\overline{\eta}}$ is a bicharacter of A, $C_{\overline{\eta}} = A/\ker\omega_{\overline{\eta}}$, μ is a unitary cocycle of $C_{\overline{\eta}}$, with $C_{\overline{\eta}} = C' \times C''$ a decomposition of $C_{\overline{\eta}}$ as an Abelian group of symmetric type, and $\rho\colon A \to \mathbb{T}$. Finally,

$$\widehat{A^{\overline{\eta}}} = \left\{\rho\Phi_\lambda : \lambda \in \widehat{\ker\omega_{\overline{\eta}}}\right\},$$

  where a concrete realization of $\rho\Phi_\lambda$ as an $\overline{\eta}$-representation on $L(C'')$ is given again by (10.66).
- Step 4 now yields a description of $\widehat{G}(\xi)$; cf. again Proposition 7.24. That is,

$$\widehat{G}(\xi) = \left\{\Theta_\lambda : \lambda \in \widehat{\ker\omega_{\overline{\eta}}}\right\},$$

  where Θ_λ is as in (10.67).

- Step 5 and Step 6 disappear: there is not induction and, finally, Step 7 becomes:

$$\widehat{G} = \sqcup_{\xi \in \widehat{B}} \widehat{G}(\xi) = \left\{ \Theta_{\xi,\lambda} : \xi \in \widehat{B}, \lambda \in \widehat{\ker \omega_{\overline{\eta}_\xi}} \right\}.$$

In the present setting, (10.83) becomes:

$$\mathrm{Ind}_{D_{\overline{\eta}}}^{G} \left(\widetilde{\xi} \overset{\vartriangle}{\lambda} \right) \sim \underbrace{\Theta_\lambda \oplus \Theta_\lambda \oplus \cdots \oplus \Theta_\lambda}_{\dim \Theta_\lambda - \text{ times}}. \tag{10.84}$$

Now $\mathrm{Ind}_{D_{\overline{\eta}}}^{G}(\widetilde{\xi} \overset{\vartriangle}{\lambda})$ is the isotypic component of $L(G)$ associated with the adjoint representation Θ'_λ.

In Table 10.1, we compare the notation in the present section and the one in relation to the orbit method in Sect. 6.7 (the key points are (6.62) and (6.63)).

Table 10.1 Notational comparison between the methods in Sects. 6.7 and 10.4

Section 6.7	Section 10.4
A	B
$\xi \circ (\phi\vert_A)$	ξ
H_0	$\ker \omega_{\overline{\eta}}$
S_Ω	$D_{\overline{\eta}}$
$\chi \circ \phi\vert_{S_\Omega}$	$\widetilde{\xi} \overset{\vartriangle}{\lambda}$
$\mathrm{Ind}_{S_\Omega}^{G}(\chi \circ \phi)$	$\mathrm{Ind}_{D_{\overline{\eta}}}^{G}(\widetilde{\xi} \overset{\vartriangle}{\lambda})$

Appendix A
Notes

In this appendix, we review some historical aspects and discuss some additional topics on: group extensions, the Schur multiplier and the Hopf formula, cohomology of groups, Clifford theory, the Mackey-Wigner theory (the little group method and the Mackey obstruction) and applications, Lie rings and Kirillov's orbit method, and projective representations.

We do not treat the general history of Representation Theory which is too wide as a subject and it goes beyond our purposes. We rather refer the interested reader to the beautiful historical monograph by Charles W. Curtis [37] which, in the author's own words from his preface,

> contains an account of the history of the representation theory of finite groups, presented through an analysis of the published work of the four principal contributors to the theory in its formative stages: Ferdinand Georg Frobenius, William Burnside, Issai Schur, and Richard Brauer.

There is a plethora of excellent expositions of the theory of group representations. Here we just mention the treatments in the works of Yakov G. Berkovich and Emmanuel M. Zhmud [16], Daniel Bump [21], Curtis and Irving Reiner [38, 39], James M.G. Fell and Robert S. Doran [48], Bertram Huppert [65], Martin Isaacs [67], Mark A. Naimark and Alexander I. Stern [101], Jean-Pierre Serre [121], Ronald Shaw [123], Barry Simon [125], and Shlomo Sternberg [126].

Also, we shall not discuss induction and restriction of group representations. We limit ourselves to mentioning that induced representations were initially defined by Frobenius for characters of finite groups and that the central result is the *Frobenius reciprocity theorem* which expresses a duality between the process of restricting and inducting.

Our approach—as presented in our research-expository paper [25] (where different definitions of induction are shown to be equivalent) with an emphasis on the fundamental contributions of George W. Mackey, namely, the *Mackey lemma* (on the decomposition of $\mathrm{Res}^G_K \mathrm{Ind}^G_H$ for $H, K \leq G$), the *tensor product*

T. Ceccherini-Silberstein et al., *Representation Theory of Finite Group Extensions*, Springer Monographs in Mathematics,
https://doi.org/10.1007/978-3-031-13873-7

theorem, *Mackey's formula for invariants*, *Mackey's intertwining number theorem*, and *Mackey's irreducibility criterion*—has more of a functional flavor and therefore it is more suitable for applications in a harmonic analytical framework.

A.1 Group Extensions and Cohomology

Given two groups N and H, the problem of classifying the groups G such that we have a short exact sequence

$$1 \longrightarrow N \stackrel{\iota}{\longrightarrow} G \stackrel{\pi}{\longrightarrow} H \longrightarrow 1 \tag{A.1}$$

is called the *extension problem*.

In our monograph, following the standard terminology in Group Theory (as, e.g., in Bourbaki [17, Section 6], Isaacs [68], and Otto Schreier [117]), we refer to (A.1) as to an *extension of H by N* and we denote it by the triple $(G; \iota, \pi)$. Note that, however, other authors (e.g., Derek Robinson [108] and Joseph Rotman [114]) use the other terminology (as, e.g., in Field Theory), namely, regarding an extension of a mathematical structure K as a structure L of which K is a substructure.

The extension problem has been studied extensively since the late nineteenth century. One of its main motivations, is the study of the *composition series* of a finite group G: this is a finite sequence of subgroups $\{1_G\} = G_0 \leq G_1 \leq G_2 \leq \cdots \leq G_n = G$ which is (1) *subnormal*, that is, $G_i \lhd G_{i+1}$, (2) with strict inclusions, and (3) such that each G_i is a maximal proper normal subgroup of G_{i+1}, equivalently, the quotient G_{i+1}/G_i (called a *composition factor*) is a simple group, for all $i = 0, 1, \ldots, n-1$. The integer n is called the *length* of the composition series. Note that a subnormal series is a composition series if and only if it is of maximal length (there are no additional intermediate subgroups which can be inserted). It is clear that every finite group G admits a composition series: starting from the trivial subnormal series $\{1_G\} \leq G$, recursively refine any subnormal series of G by inserting intermediate subnormal subgroups into the series up to maximality (cf. [64, Theorem 8.4]). The celebrated *Jordan-Hölder theorem*, after Camille Jordan and Otto Hölder (cf. also the *Schreier refinement theorem*), states that any two composition series of a given group G, say $\{1_G\} = G_0 \leq G_1 \leq G_2 \leq \cdots \leq G_n = G$ and $\{1_G\} = H_0 \leq H_1 \leq H_2 \leq \cdots \leq H_m = G$ are *equivalent* in the sense that they have the same length ($n = m$) and the same composition factors (with their multiplicities) up to permutation and isomorphism, that is, there exists $\sigma \in \mathfrak{S}_n$ (the symmetric group of degree n) such that $G_{i+1}/G_i \cong H_{\sigma(i)+1}/H_{\sigma(i)}$ for all $i = 0, 1, \ldots, n-1$ (cf. [64, Theorem 8.11]).

The *classification of finite simple groups* (briefly *CFSG*), one of the milestones in Group Theory and in the whole Mathematics, states that every finite simple group is either cyclic ($\mathbb{Z}/p\mathbb{Z}$, for $p \in \mathbb{N}$ prime), or alternating ($\mathfrak{A}_n$, for $n = 3$ and $n \geq 5$), or it belongs to a broad infinite class, called the *groups of Lie type*, or else it is one of the finite exceptions, the so-called *sporadic* groups. CFSG thus provides a complete list of all finite simple groups and, in combination with a solution to the extension problem, gives a constructive general classification of all finite groups. As a consequence, the finite simple groups may be viewed as the "basic building blocks" of finite groups, thus playing the same role of the prime numbers vs. the natural numbers.

The program for CFSG essentially started with Évariste Galois (1832) who introduced the notions of a normal subgroup and proved the simplicity of the alternating groups and of $\mathrm{PSL}_2(\mathbb{F}_p)$, the projective special linear groups over prime finite fields ($p \geq 5$ prime), and has developed since in a monumental work consisting of a few thousands of pages (in several hundred journal articles and books) written by about 100 authors. The most relevant contributions to this classification theorem have been published between 1955 and 2004. Daniel Gorenstein, Richard Lyons, and Ronald Solomon, also in collaboration with Michael Aschbacher, Stephen Smith, and Inna Capdeboscq, have published a series of books (presently consisting of 10 volumes), where the project for a complete, simplified, and revised version of the whole proof is carried on.

Recall (cf. Definition 1.8) that, given two groups N and H, two extensions $(G_1; \iota_1, \pi_1)$ and $(G_2; \iota_2, \pi_2)$ of H by N are *equivalent* if there exists a group isomorphism $\eta\colon G_1 \to G_2$ such that $\iota_2 = \eta\iota_1$ and $\pi_1 = \pi_2\eta$.

The *short five lemma* (cf. [64, Lemma 1.17.(iii)]) states that given a commutative diagram of groups

$$\begin{array}{ccccccccc} 1 & \longrightarrow & N_1 & \xrightarrow{\iota_1} & G_1 & \xrightarrow{\pi_1} & H_1 & \longrightarrow & 1 \\ & & \varphi\downarrow & & \downarrow\eta & & \psi\downarrow & & \\ 1 & \longrightarrow & N_2 & \xrightarrow{\iota_2} & G_2 & \xrightarrow{\pi_2} & H_2 & \longrightarrow & 1, \end{array}$$

where φ and ψ are isomorphism and η is a homomorphism, then necessarily η is an isomorphism. As a consequence, in the definition of equivalence of two group extensions, one can relax the condition on η by only requiring that it be a homomorphism.

Recall (cf. Example 1.9, see also Example 1.19) that it may happen that two extensions $(G_1; \iota_1, \pi_1)$ and $(G_2; \iota_2, \pi_2)$ of H by N are not equivalent even if $G_1 \cong G_2$.

An extension (A.1) is *Abelian* (resp. *central*) if H is Abelian (resp. $\iota(N) \leq Z(G)$, so that, in particular, N is Abelian and one then usually denotes it by $\mathscr{A}$). Central extensions are classified by the group $\mathscr{H}^2(H, \mathscr{A})$ the second cohomology group of H with values in $\mathscr{A}$. Finally, if (A.1) admits a section $t\colon H \to G$ which is a homomorphism, the extension is called *split* and there exists a homomorphism $\alpha\colon H \to \mathrm{Aut}(N)$, such that $G \cong N \rtimes_\alpha H$, the *semi-direct product* of H acting on N.

The theory of Abelian group extensions is naturally and classically treated with tools of homological algebra, such as the theory of *Ext*-functors (see [97]). Schreier (1926) and, later, Samuel Eilenberg and Saunders MacLane (late 1940-s) [45, 46, 96], developed a theory of classification of nonabelian extensions of abstract groups leading to the low dimensional nonabelian group cohomology. This is sometimes called *Schreier's theory* (see [18, 20]).

Group cohomology is a powerful tool from algebraic topology which, in analogy to group representations, studies group actions of a given group G on G-modules. In particular, by treating any such G-module M as a kind of topological space with elements of $G^n = G \times G \times \cdots \times G$ representing n-simplices, one computes the cohomology groups $\mathscr{H}^n(G, M)$ which provide a useful method to investigate the structure of the group G and the G-module M themselves. The so-called *low-dimensional cohomology groups* are $\mathscr{H}^1(G, M)$ and $\mathscr{H}^2(G, M)$.

The first one, is the quotient of the Abelian group of the so-called 1-*cocycles* (or *crossed homomorphisms*), i.e., the (set theoretical) maps $f : G \to M$ such that $f(g_1 g_2) = f(g_1) + g_1 f(g_2)$ for all $g_1, g_2 \in G$, modulo the subgroup consisting of the so-called 1-*coboundaries* (or *principal crossed homomorphisms*), i.e., the maps $f_m : G \to M$ defined by $f_m(g) := gm - m$ for all $m \in M$ and $g \in G$. The first cohomology group $\mathscr{H}^1(G, M)$ classifies the *complements* of M in $E := M \rtimes G$ (i.e., the subgroups $H \leq E$ such that $HM = E = MH$ and $H \cap M = \{1_E\}$) modulo conjugation in E (i.e., $G \sim H$ if there exists $e \in E$ such that $G = eHe^{-1}$).

The second one is described as follows. If M is a trivial G-module then (cf. Definition 6.3) $\mathscr{H}^2(G, M)$ is the quotient of the Abelian group of the so-called 2-*cocycles*, i.e., the (set theoretical) maps $\tau : G \times G \to M$ such that $\tau(g_1 g_2, g_3) + \tau(g_1, g_2) = \tau(g_1, g_2 g_3) + \tau(g_2, g_3)$ (*cocycle identity*) for all $g_1, g_2, g_3 \in G$, and $\tau(1_G, g) = \tau(g, 1_G) = 0_M$ (*normalization*) for all $g \in G$, modulo the subgroup consisting of the so-called 2-*coboundaries*, namely, the 2 cocycles of the form τ_ρ defined by $\tau_\rho(g_1, g_2) := \rho(g_1 g_2) - \rho(g_1) - \rho(g_2)$ for all $g_1, g_2 \in G$, where $\rho : G \to M$ is any (set theoretical) map. The second cohomology group $\mathscr{H}^2(G, M)$ is in one-to-one correspondence with the set of all central extensions of G by M (the latter simply viewed as an Abelian group) up to the equivalence relation (cf. Theorem 6.6). More generally, if the action of G on M is nontrivial, $\mathscr{H}^2(G, M)$ classifies the isomorphism classes of all extensions $0 \longrightarrow M \overset{\iota}{\longrightarrow} E \overset{\pi}{\longrightarrow} G \longrightarrow 0$ of G by M, for which the action (by inner automorphisms) of G on $\iota(M)$ yields an isomorphism with the original G-module structure of M.

Historically, the first result in the spirit of cohomology of groups may be considered *Hilbert Satz 90* by David Hilbert [61] (on the trace and the norm associated with a cyclic Galois extension of fields: see [30, Section 6.7], [64, Chapter V, Section 7] and, for a more general treatment, [86]). In fact, a more general theorem, due to Emmy Noether (1933) states that if $\mathbb{F} \subset \mathbb{E}$ is a finite Galois extension of fields with Galois group $G = \mathrm{Gal}(\mathbb{E}/\mathbb{F})$, then $\mathscr{H}^1(G, \mathbb{E}^*) = \{0\}$ (here $\mathbb{E}^*$ denotes the multiplicative group $\mathbb{E} \setminus \{0_\mathbb{E}\}$).

In connection with the second cohomology group, the concept of a 2-cocycle (also known as a *factor set*) arose in the work of Hölder (1893), in Schur's study of projective representations (1904), in Schreier's 1926 treatment (cf. the *Schreier factor systems* in Definition 1.12), and in Brauer's study of simple algebras and the Brauer group (1928). We refer to Charles Weibel's contribution [130, pages 806–811] for a discussion of this history.

The *Schur multiplier* (or *Schur multiplicator*) of a finite group G is the Abelian group $\mathscr{M}(G) := \mathscr{H}^2(G, \mathbb{T}) = \mathscr{H}^2(G, C_n)$, where $C_n \leq \mathbb{T}$ is the group of n-th roots of unity and $n = |G|$ (cf. Definition 6.20 and Corollary 6.21). It was introduced by Schur in 1904 in his work on projective representations [118, 119]. With every finite group G, he associated a finite group C, called a *Schur cover* (also known as a *covering group* or *Darstellungsgruppe*) of G, with the property that every projective representation of G (these are parameterized by $\mathscr{M}(G)$) can be lifted to an ordinary representation of C (cf. Proposition 7.12 with a different notation: G is denoted by H). More precisely, he proved the following (cf. [90, Section 7.4]):

Theorem (Schur) *Let G be a finite group. Then there exists a covering group C of G, i.e., a (central) extension C of G by an Abelian group A such that:*

(i) $A \subset Z(C) \cap C'$;
(ii) $A \cong \mathscr{M}(G)$;
(iii) $C/A \cong G$.

Moreover, every projective representation of G lifts to an ordinary representation of C.

An entire monograph devoted to the Schur multiplier is Gregory Karpilovsky's [73] which offers a unified treatment of several important results related to the second cohomology group. In the Preface one reads:

> Schur multipliers have proved to be a powerful tool in other areas, such as algebraic number theory, block theory of group algebras, and classification of finite simple groups, to mention only a few. Historically, it was Fröhlich [50] who first pointed out a role of the Schur multiplier of the Galois groups. The role of the Schur multiplier in finite group theory was vividly demonstrated in a work of Gorenstein [53].

In 1941, Heinz Hopf discovered what is now called *Hopf's integral homology formula*, expressing the second homology group $\mathscr{H}_2(G, \mathbb{Z})$ of a finite group G (see [108, Section 11.2]), which is identical to the Schur multiplier of G (see [108, 11.4.15] and [90, Corollary 7.9]):

Hopf's Formula *Let G be a finite group. Let F be a finitely generated free group and let $N \trianglelefteq F$ such that $F/N \cong G$. Then*

$$\mathscr{M}(G) \cong \frac{N \cap F'}{[N, F]}.$$

This led Eilenberg and MacLane (in the United States) to the creation of the cohomology theory of groups [45, 46]. In fact, this was independently discovered by several other people in the mid 1940s, notably, Hopf and Beno Eckmann (in Switzerland), Hans Freudenthal (in the Netherlands), and Dmitry Faddeev (in USSR). The relation between Schur multipliers and the Eilenberg-MacLane cohomology groups corresponding to group extensions was remarked by Friederich Mautner in his review of Mackey's paper [92] and, in fact, Mackey himself had remarked on the relation between $\mathscr{M}(G)$ and $\mathscr{H}_2(G, \mathbb{Z})$ in his earlier paper [91].

The first person, however, to work out in detail this relationship (for locally compact groups) was Calvin Moore [99, 100].

Nowadays, a great deal is known about the cohomology of groups, including interpretations of low-dimensional cohomology, long exact sequence of cohomology, products (as in the *de Rham cohomology*), functoriality, and how to change groups. For instance, in relation to this last concept, we mention the fundamental results due to Roger Lyndon [89] and to Gerhard Hochschild and Serre [62] who introduced the now-called *Lyndon-Hochschild-Serre spectral sequence* which, given a short exact sequence $1 \to N \to G \to H \to 1$, relates the group cohomologies of N and H to the cohomology of G.

For a gentle introduction to extensions and cohomology theory of groups we refer to [90, Chapter 7], while classical references are the monographs by: Karl Gruenberg [56], Robinson [108, Chapter 11] (which is based on Gruenberg's), Serge Lang [85], Kenneth Brown [19], Alejandro Adem and James Milgram [2] (for finite groups), Rotman [113] (which includes a chapter on group extensions and one on cohomology of groups), Leonard Evens [47], and David John Benson [13, 14] (on representation theory and cohomology of finite groups).

Applications of cohomological methods in Applied and Computational Harmonic Analysis, and, in particular to the Fast Fourier Transform (cf. [30]) is witnessed by the paper by L. Auslander, J.R. Johnson, and R.W. Johnson [5], where the representation theory of Abelian groups—and, in particular, the study of a certain cocycle associated with a section for a short exact sequence of Abelian groups—is used to obtain an algebraic divide-and-conquer algorithm for computing the finite Fourier transform of a finite Abelian group in terms of the Fourier transforms of an arbitrary subgroup and its quotient. From this algebraic algorithm, a procedure is derived for obtaining a concrete factorization of the Fourier transform matrix in terms of smaller Fourier transform matrices, diagonal multiplications, and permutations. For instance, for cyclic groups this recovers, as special cases, the famous Cooley-Tukey and Good-Thomas algorithms, while, in the general case, this procedure gives a variety of multidimensional Cooley-Tukey type algorithms. More recently, a Fourier Analysis for Abelian group extensions was developed by D. Rockmore in [109], where the problem of computing the Fourier transforms (of complex-valued functions f on a finite group G) and performing Fourier inversion when G admits a normal subgroup $N \leq G$ such that G/N is Abelian, is considered. In particular, he showed that the number $T(G)$ of operations needed to compute such Fourier transforms is $O((|G|/|K|)T(K)+|G|\log(|G|/|K|))$. Further computational aspects of the Fast Fourier Transform are developed by Rockmore in [110] for wreath products of the form $G \wr \mathfrak{S}_n$.

With an extension (A.1) and a fixed section $H \to G$ one associates an embedding $\Phi\colon G \to N \wr H$, called a *Frobenius embedding* of G into the (regular) wreath product of N by H (this is the semi-direct product $N^H \rtimes H$, where H acts on the *base group* N^H by shifting coordinates) such that $\Phi(\iota(N)) = \Phi(G) \cap N^H$ and $\Phi(G)N^H = N \wr H$. This is the celebrated theorem of Leo Kaloujnine and Marc Krasner [82, 83] (cf. Theorem 1.16; see also [42, Theorem 2.6A], [72, Theorem 6.2.8 and Exercise 6.2.9], [108, Exercise 11.1.12], [114, Theorem 7.37], [127,

Theorem 10.9], and [28, Theorem 2.2.1]). The wreath product construction, which led Kaloujnine to the determination of the Sylow p-subgroups (and the structure of characteristic subgroups) of the symmetric groups, according to Adalbert Kerber [75, Chapter I, Section 2] was given by Cauchy in the third volume of his "Exercises d'analyse et de physique mathématique" which appeared in 1844.

Two embeddings $\Phi_i : G_i \to N \wr H$, $i = 1, 2$, are said to be equivalent if there exists a group isomorphism $\eta\colon G_1 \to G_2$ such that $\Phi_2 \circ \eta = \Phi_1$. In [72, Exercise 6.2.10] it is remarked that, up to conjugation, the image subgroup $\Phi(G) \leq N \wr H$ is independent of the section $s\colon H \to G$. In fact, two Frobenius embeddings associated with the same groups and distinct sections are equivalent (this was observed by Kaloujnine himself). All of this is (re)covered by Theorem 1.18, namely that two group extensions are equivalent if and only if the associated Frobenius embeddings are equivalent (see [32]).

A.2 Clifford Theory

Clifford theory, introduced by Alfred H. Clifford in [33] (1937), describes the relation between representations of a group G and those of its normal subgroups through the operations of restriction, induction, and extension.

This work was undoubtedly written under the influence of Herman Weyl. Indeed, from 1936 to 1938, Clifford was Weyl's assistant at the Institute for Advanced Study when the latter was writing his famous book [131]. In the Preface, Weyl writes:

> ... this relative accomplishment is to be ascribed solely to the devoted collaboration of my assistant, Dr. Alfred H. Clifford, and even more valuable to me than the linguistic, were his mathematical criticisms.

Roughly speaking, Clifford's theorem states as follows (we limit ourselves to the setting of finite groups: in general, one requires the normal subgroup having finite index in the ambient group).

Theorem (Clifford) *Let G be a group and let $N \leq G$ be a normal subgroup. Let $\theta \in \widehat{G}$. Then the restriction $\mathrm{Res}^G_N(\theta)$ of θ to N breaks up into a direct sum of irreducible representations of N of equal dimensions. These irreducible representations of N lie in one single orbit for the action of G by conjugation on $\widehat{N}$. In particular, the number of pairwise nonequivalent summands is no greater than the index of N in G.*

In our presentation (see Sect. 2.1 for the notation) the above result corresponds to Theorem 2.5.(3). Our proof, together with the surrounding related results, follows quite closely Huppert's monograph [65]. However, our approach, based on our research-expository paper [26], rests on an explicit analysis of the representation spaces and their decompositions, rather than on computations with characters (as in [65]).

An important corollary to Clifford's theorem is the *Clifford correspondence* which, given $\sigma \in \widehat{N}$ and denoting by $I = I_G(\sigma) := \{g \in G : {}^g\sigma \sim \sigma\} \leq G$ the *inertia subgroup* of σ (equivalently, the stabilizer of σ under the conjugation action of G on $\widehat{N}$) and setting $\widehat{I_G}(\sigma) := \{\psi \in \widehat{I} : \sigma \preceq \mathrm{Res}^I_N \psi\}$ (resp. $\widehat{G}(\sigma) := \{\theta \in \widehat{G} : \sigma \preceq \mathrm{Res}^G_N \theta\}$) establishes a bijection between $\widehat{I_G}(\sigma)$ and $\widehat{G}(\sigma)$. In fact, this bijection is provided by induction from $I = I_G(\sigma)$ to G (cf. Theorem 2.12).

Mackey [94], in the framework of the representation theory of locally compact groups, found a more precise version of the Clifford theorem in what has become known as the "Mackey machine" or "Mackey normal subgroup analysis".

Excellent expositions of Clifford theory are the classical monographs by Berkovich and Zhmud [16], Curtis and Reiner [38, 39], and Isaacs [67]. See also the books by Larry Dornhoff [43] and Larry C. Grove [54].

An account on Clifford theory may be found in our short monograph [28] (see also [26]) where applications to the representation theory of semi-direct products (e.g., the affine group and the *Heisenberg group* over a finite field) and of wreath products (e.g., the finite *lamplighter group* $C_2 \wr C_n$, the *hyperoctahedral group* $C_2 \wr \mathfrak{S}_n$, and $\mathfrak{S}_m \wr \mathfrak{S}_n$) are presented in full detail in Section 1.3 and Chapter 2 therein, respectively. A monograph entirely devoted to Clifford theory is Karpilovsky's [74] with an emphasis on group algebras. Among other things, group-graded algebras (especially crossed products) are introduced and used as a tool, and, in the spirit of our Sects. 7.5 and 7.6 and Chaps. 8 and 9, examples such as the endomorphism rings of the induced modules and the twisted group algebras corresponding to projective representations, are presented.

We end this section by mentioning that Clifford is also very well known for his important work on semigroups: his two-volumes book [34], written in collaboration with Gordon B. Preston, is a classic in the field.

A.3 The Little Group Method and Its Applications

Suppose that a group G acts on a set X. Let $g \in G$ and $x \in X$ such that $gx = x$. One then says that x is a *fixed point* of g. For every $x \in X$, the *stabilizer subgroup* of G with respect to x is the set $\mathrm{Stab}_G(x) := \{g \in G : gx = x\} \leq G$ of all elements in G that fix x. It is also called the *isotropy group* or the *little group*.

If $N \leq G$ is a normal subgroup of G and the latter acts on $X = \widehat{N}$ by conjugation—given $(\sigma, V) \in \widehat{N}$ and $g \in G$, one defines $({}^g\sigma, V) \in \widehat{N}$ by setting ${}^g\sigma(n) := \sigma(g^{-1}ng)$ for all $n \in N$—one calls the corresponding stabilizer of σ the *inertia group* of σ, denoted $I_G(\sigma)$ (cf. Definition 2.1). Note that, trivially, $N \leq I_G(\sigma)$, since we only look at (unitary) representations up to (unitary) equivalence.

By *the little group method* (cf. Theorem 2.26) one refers to a powerful tool—which goes back to Frobenius (for a particular case) and was developed by Eugene P. Wigner [132] for groups arising in physics such as the Poincaré group (see [126, Sections 3.9 and 3.20]) and Mackey who extended this method to topological groups

[95]—that provides a complete list of all irreducible representations of a (finite) group G under the following assumptions and conditions: there exists a normal subgroup $N \le G$, for which a complete knowledge of the representation theory is at hand, such that, additionally, (i) for every $\sigma \in \widehat{N}$ a complete knowledge of the representation theory of the quotient group $H_G(\sigma) := I_G(\sigma)/N$ is at hand, and (ii) there exists an extension $\widetilde{\sigma}$ to its inertia subgroup $I_G(\sigma)$ (i.e., $\mathrm{Res}_N^{I_G(\sigma)}\widetilde{\sigma} = \sigma$).

A solution to problem (ii) is provided by the following general result of Patrick X. Gallagher [52]:

Theorem (Gallagher) *Given a group G together with a normal subgroup $N \le G$, in order that every $\sigma \in \widehat{N}$ admits an extension $\widetilde{\sigma}$ to G it is necessary and sufficient that, for each intermediate subgroup $N \le H \le G$ such that H/N is a direct product of a cyclic group and a group of prime power order, σ can be extended to a representation σ^H of H in such a way that $\sigma^{g^{-1}Hg} = {}^g(\sigma^H)$ for all $g \in G$ and $\sigma^H|_K = \sigma^K$ for all $H \le K \le N$.*

As an application, one has that if $[G : N]$ is a prime number and $I_G(\sigma) = G$, then $\sigma \in \widehat{N}$ admits a unique extension to G (cf. Proposition 4.9).

Another solution to (ii) is due to Mackey for locally compact groups and particularized to finite groups by Fell and Doran (who then generalized it in an operator algebraic setting) [48] which we closely followed. This is in terms of the so-called *Mackey obstruction* (cf. Definition 7.3): $\sigma \in \widehat{N}$ admits an extension $\widetilde{\sigma}$ to its inertia subgroup $I_G(\sigma)$ exactly if the cocycle η, suitably associated with σ (cf. (7.2)), is a coboundary (cf. Proposition 7.4).

In Theorem 2.18 we discuss another theorem of Gallagher (also from [52]) which, once guaranteed the existence of an extension θ of $\sigma \in \widehat{N}$ to G (so that, necessarily, $I_G(\sigma) = G$), provides an explicit decomposition of $\mathrm{Ind}_N^G\sigma$ in terms of tensor products of θ by (the inflation of) all irreducible representations of $H_G(\sigma) = G/N$ (see (2.29)).

A beautiful and concise treatment of the Mackey-Wigner little group method may be found in Alan J. Coleman's lecture notes [35] where, together with basics on induced representation theory (including the Frobenius reciprocity theorem, the Mackey subgroup theorem, the tensor product theorem, and the intertwining number theorem, along with results on the tensor products of representations of $\mathfrak{S}_n$ and of the general linear group $\mathrm{GL}(n)$), in Chap. 7 a description of the little group method concludes with a discussion of its application to space groups (omitting, however, the concept of projective representations).

From a Mathematical- and Theoretical-Physics point of view, with his seminal 1939 paper [132] (see also the monograph [133]), Wigner—classifying the irreducible unitary representations of the Poincaré group—established a natural connection between particle physics and representation theory (different quantum states of an elementary particle give rise to irreducible representations of the Poincaré group), and started the so-called *Wigner classification*, linking together the properties of such elementary particles, including their spectra, to properties of the representations of Lie groups and Lie algebras (corresponding to "approximate symmetries" of the universe).

Judith Parker's [103] is a beautifully written survey of the theory of projective representations (of locally compact groups), the Mackey obstruction (and, more generally, of the "Mackey machine"), connections with Number Theory and the theory of Operator Algebras (C*- and von Neumann algebras as well as C*-dynamical systems), and the related applications to Theoretical Physics.

A.4 Lie Rings and the Orbit Method

A *Lie ring* is an Abelian group $\mathfrak{g}$ (written additively) equipped with a *bi-additive nilpotent* map, called the *Lie bracket*, $\mathfrak{g} \times \mathfrak{g} \to \mathfrak{g}$, denoted $(g_1, g_2) \mapsto [g_1, g_2]$, satisfying the Jacobi identity (cf. Definition 6.24).

For instance, an associative ring R can be turned into a Lie ring by defining the bracket by setting $[r, s] := rs - sr$ for all $r, s \in R$.

The *lower central series* of a Lie ring $\mathfrak{g}$ is the sequence $\gamma_1(\mathfrak{g}) \geq \gamma_2(\mathfrak{g}) \geq \cdots \geq \gamma_n(\mathfrak{g}) \geq \gamma_{n+1}(\mathfrak{g}) \geq \cdots$ of Lie subrings defined by setting $\gamma_1(\mathfrak{g}) := \mathfrak{g}$ and $\gamma_n(\mathfrak{g}) := [\gamma_1(\mathfrak{g}), \gamma_{n-1}(\mathfrak{g})]$, the Lie subring generated by all commutators $[g_1, g_{n-1}]$ with $g_1 \in \gamma_1(\mathfrak{g})$ and $g_{n-1} \in \gamma_{n-1}(\mathfrak{g})$ for all $n \geq 2$.

A Lie ring $\mathfrak{g}$ is said to be *nilpotent* (of class $\leq c$) provided that one of the following equivalent conditions is satisfied:

- $\gamma_{c+1}(\mathfrak{g}) = \{0\}$;
- $[g_1, g_2, \ldots, g_{c+1}] = 0$ for all $g_1, g_2, \ldots, g_{c+1} \in \mathfrak{g}$,

where, for $n \geq 3$, the *commutator of weight n* $[g_1, g_2, \ldots, g_n]$ of $g_1, g_2, \ldots, g_n \in \mathfrak{g}$ is recursively defined by setting $[g_1, g_2, \ldots, g_n] := [[g_1, g_2, \ldots, g_{n-1}], g_n]$.

Let G be a group. A sequence $G = G_1 \geq G_2 \geq \cdots \geq G_n \geq G_{n+1} \geq \cdots$ of subgroups of G is said to be *strongly central* provided that $[G_m, G_n] \leq G_{m+n}$ for all $m, n \geq 1$, where, given two subgroups $H, K \leq G$, $[H, K]$ denotes the subgroup generated by the commutators $[h, k] = h^{-1}k^{-1}hk$ with $h \in H$ and $k \in K$. Note that since $[G, G_n] \leq G_{n+1} \leq G_n$, the subgroups G_n are normal in G. For instance, the *lower central series* of G, given by setting $\gamma_1(G) := G$ and $\gamma_n(G) := [\gamma_1(G), \gamma_{n-1}(G)]$ for all $n \geq 2$, is strongly central. We recall that G is said to be *nilpotent* (of class $\leq c$) provided that $\gamma_{c+1}(G) = \{1_G\}$.

Given a strongly central series of G as above, the Abelian group

$$L(G) := \bigoplus_{n \geq 1} G_n/G_{n+1}$$

can be equipped with a stricture of a Lie ring with addition supplied by extending linearly the group operation in each homogeneous part

$$a_n G_{n+1} + b_n G_{n+1} := a_n b_n G_{n+1}$$

for all $a_n, b_n \in G_n$, and the bracket operation given by

$$[a_m G_{m+1}, b_n G_{n+1}] := [a_m, b_n] G_{m+n+1}$$

for all $a_m \in G_m$ and $b_n \in G_n$, $m, n \geq 1$. The centrality of the series ensures that the commutator gives to the bracket operation the appropriate Lie theoretic properties.

When the strong central series of G is the lower central series, one denotes the associated Lie ring $L(G)$ by $\mathfrak{g}$ and the following holds: G is nilpotent of class c if and only if $\mathfrak{g}$ is nilpotent of the same class c (cf. [76, Theorem 3.22]).

The construction of $L(G)$ goes back to Hans Zassenhaus [135] who introduced linear methods in the study of nilpotent groups. For an exposition of such methods, we refer to [66, Chapter VIII] by Huppert and Norman Blackburn. Under special conditions, namely, G being 2-step nilpotent and 2-divisible, in Theorem 6.27 we presented another way for associating a Lie ring with a group G (see also [66, Lemma 9.16, Chapter VIII] and there references therein).

The map which associates with a finite nilpotent group G its nilpotent Lie ring $\mathfrak{g}$ is called the *Malcev correspondence* after Anatoly Malcev. Such linear methods have been most successful in the study of the *restricted Burnside problem*, which was settled by Efim I. Zelmanov [136, 137].

The orbit method—also known as the *Kirillov method* (after Alexander Kirillov, the pioneer of the theory), or the *method of coadjoint orbits*—aims at establishing a correspondence between the irreducible unitary representations of a Lie group and its coadjoint orbits, namely, the orbits of the action of the group on the dual space of its Lie algebra. It serves as a unifying principle in the description of the unitary duals of real reductive Lie groups. The milestones of this theory are Kirillov's fundamental work on representations of nilpotent groups [77] and its extension to solvable groups due to Louis Auslander and Bertram Kostant [4] and, among others, Lajos Pukánszky [105] (see also the monograph [106, Chapter 4]) who found a geometric condition, known as the *Pukánszky condition*, that allowed an extension of the orbit method from nilpotent to solvable Lie groups and played a major role in the subsequent development of the theory, and Roger Howe [63] for compact p-adic Lie groups.

A flash-idea of the orbit method is presented in the first lines of David Vogan's clear and elegant review of Kirillov's monograph [79], the first systematic and reasonably self-contained exposition of the orbit method containing a very useful and interesting "user's guide" (for a previously written expository survey—including its history—see [78]):

> This book is about a wonderfully successful example of (if you will forgive some geometric language) circular reasoning. Here is the short version. Everywhere in mathematics, we find geometric objects M (like manifolds) that are too complicated for us to understand. One way to make progress is to introduce a vector space V of functions on M. The space V may be infinite-dimensional, but linear algebra is such a powerful tool that we can still say more about the function space V than about the original geometric space M. (With a liberal interpretation of "vector space of functions," one can include things like the de Rham cohomology of M in this class of ideas.) Often M comes equipped with a group G of symmetries, but G and M may be even less comprehensible together than separately.

> Nevertheless, G will act on our function space V by change of variables (giving linear transformations), and so we get a representation of G on V. Our original (and impossible) problem of understanding all actions of G on geometric spaces M is therefore at least related to the problem of understanding all representations of G. Because this is a problem about vector spaces, it sounds a bit less daunting. The goal of the orbit method is to say something about all representations of a Lie group G. What it says is that an irreducible representation should correspond (roughly) to a symplectic manifold X with an action of G. So the circle is complete: we understand geometric objects with group actions in terms of representations, and we understand representations in terms of geometric objects with group actions.

In our work, since the groups involved are finite, one looses the "differential" geometric interpretation of Vogan's description, but one rather focuses on a more "combinatorial" description. Also, we limit ourselves to 2-divisible 2-step nilpotent groups.

A.5 Projective Representations

Given a (complex) vector space V, the quotient group $\mathrm{PU}(V) := \mathrm{U}(V)/\mathbb{T}$—of the unitary group of V by its center—is called the *projective unitary group* of V. A unitary projective representation of a group G is a group homomorphism $\rho\colon G \to \mathrm{PU}(V)$. Clearly, given any ordinary unitary representation $\widetilde{\rho}\colon G \to \mathrm{U}(V)$, the composition $\rho := \pi \circ \widetilde{\rho}\colon G \to \mathrm{PU}(V)$, where $\pi : \mathrm{U}(V) \to \mathrm{PU}(V)$ denotes the canonical quotient homomorphism, is a unitary projective representation. However, there are unitary projective representations which do not come from this construction, in other words, they cannot be lifted to a ordinary unitary representation (or, cannot be *de-projectivized*): the obstruction to this lifting can be understood via group cohomology, as described below. Fix a section for π, that is, a map $s\colon \mathrm{PU}(V) \to \mathrm{U}(V)$ such that $\pi \circ s = \mathrm{id}_{\mathrm{PU}(V)}$. Then, given a unitary projective representation $\rho\colon G \to \mathrm{U}(V)$ and setting $\Phi(g) := s(\rho(g))$ yields maps $\Phi\colon G \to \mathrm{U}(V)$ and $\eta \in \mathscr{C}^2(G, \mathbb{T})$ such that $\Phi(gg') = \eta(g, g')\Phi(g)\Phi(g')$ for all $g, g' \in G$. One then refers to Φ as to a unitary η-representation. Then, one shows that different sections $s, s'\colon \mathrm{PU}(V) \to \mathrm{U}(V)$ give rise to homologous cocycles $\eta, \eta' \in \mathscr{C}^2(G, \mathbb{T})$, and such a lift is possible (i.e., there is no obstruction) exactly if the cocycle is a coboundary, that is, its class in $\mathscr{H}^2(G, \mathbb{T})$ is trivial. In any case, however, one can lift ρ to a ordinary unitary representation of a different group H which will be a central extension of G.

Projective representations (also called, especially in a Quantum Physics setting, *ray representations*) go back to Schur's seminal paper [120] on the now-called *spin characters* of the symmetric groups $\mathfrak{S}_n$. In general, spin representations are particular projective representations of the spin groups (i.e., double covers of the special orthogonal groups) in arbitrary dimension and signature. In his paper, Schur introduced a class of symmetric functions, now called *Schur's Q-functions*—in analogy to the earlier work by Frobenius [49] on the ordinary irreducible characters of the symmetric groups in terms of what are now called *Schur's S-functions*—and determined the conjugacy classes of the double covers of $\mathfrak{A}_n$ and $\mathfrak{S}_n$. He explicitly calculated the values of the basic spin characters of these groups, established that there is exactly one non-trivial class of Schur multiplier (thus yielding a unique η-representation), and completely determined all the corresponding irreducible representations.

Since the seminal article of Schur, very little attention seems to have been paid to projective representations until Wigner [132] used them in a Quantum Mechanical context, namely, in his treatment of the Lorentz group in 1939, and Valentine Bargmann [6] proved his famous theorem in 1954 in the setting of topological, and especially Lie, groups. This theorem gives a condition—namely, the vanishing of the two-dimensional Lie algebra cohomology $\mathscr{H}^2(\mathfrak{g}; \mathbb{R})$ of the Lie algebra $\mathfrak{g}$ of the Lie group G—for when every projective unitary representation of a Lie group comes from an ordinary unitary representation of its universal cover, or, as one says, can be de-projectivized. For instance, every projective unitary representation of the Poincaré group comes for an ordinary unitary representation of its universal cover, which is a double cover: Bargmann's theorem applies since the double cover of the Poincaré group admits no non-trivial one-dimensional central extensions. The relevance of this last result, is witnessed in Quantum Mechanics by the fact that two vectors in the quantum Hilbert space that differ by a scalar represent the same physical state and two unitary operators that differ by a multiple of the identity have the same action on physical states.

However, the first general treatment of projective unitary representations (of locally compact groups) with an emphasis on induced representations is due to Mackey [92]: this, together with the Fell and Doran monograph [48] (especially for the Little Group Method for projective representations), was the source for our treatment (only for finite groups!) in Chap. 8.

As mentioned above (cf. Sect. A.3), a nice survey on, among other things, unitary projective representations was written, as a tribute in memory of Mackey, by Parker [103].

References

1. R. Achilles, A. Bonfiglioli, The early proofs of the theorem of Campbell, Baker, Hausdorff, and Dynkin. Arch. Hist. Exact Sci. **66**(3), 295–358 (2012)
2. A. Adem, R.J. Milgram, *Cohomology of Finite Groups*, Grundlehren der mathematischen Wissenschaften, vol. 309 (Springer, Berlin, 1994)
3. J.L. Alperin, R.B. Bell, *Groups and Representations*, Graduate Texts in Mathematics, vol. 162 (Springer, New York, 1995)
4. L. Auslander, B. Kostant, Polarization and unitary representations of solvable Lie groups. Invent. Math. **14**, 255–354 (1971)
5. L. Auslander, J.R. Johnson, R.W. Johnson, Multidimensional Cooley-Tukey algorithms revisited, Adv. in Appl. Math. **17**(4), 477–519 (1996)
6. V. Bargmann, On unitary ray representations of continuous groups. Ann. Math. **59**, 1–46 (1954)
7. L. Bartholdi, R.I. Grigorchuk, On the spectrum of Hecke type operators related to some fractal groups. Proc. Steklov Inst. Math. **231**(4), 1–41 (2000)
8. L. Bartholdi, R.I. Grigorchuk, On parabolic subgroups and Hecke algebras of some fractal groups Serdica Math. J. **28**(1), 47–90 (2002)
9. L. Bartholdi, R.I. Grigorchuk, V. Nekrashevych, From fractal groups to fractal sets, in *Fractals in Graz 2001*, Trends Math. (Birkhäuser, Basel, 2003), pp. 25–118
10. H. Bass, M.V. Otero-Espinar, D. Rockmore, Ch. Tresser, *Cyclic Renormalization and Automorphism Groups of Rooted Trees*, Lecture Notes in Mathematics, vol. 1621 (Springer, Berlin, 1996)
11. B. Bekka, P. de la Harpe, Groups with faithful irreducible projective unitary representations. Forum Math. **25**(3), 505–531 (2013)
12. B. Bekka, P. de la Harpe, *Unitary Representations of Groups, Duals, and Characters*, Mathematical Surveys and Monographs, vol. 250 (American Mathematical Society, Providence, RI, 2020)
13. D.J. Benson, *Representations and Cohomology. I. Basic Representation Theory of Finite Groups and Associative Algebras*. Cambridge Studies in Advanced Mathematics, vol. 30 (Cambridge University Press, Cambridge, 1991)
14. D.J. Benson, *Representations and Cohomology. II. Cohomology of Groups and Modules*. Cambridge Studies in Advanced Mathematics, vol. 31 (Cambridge University Press, Cambridge, 1991)
15. C. Benson, G. Ratcliff, A family of finite Gelfand pairs associated with wreath products. Colloq. Math. **152**(1), 65–78 (2018)

T. Ceccherini-Silberstein et al., *Representation Theory of Finite Group Extensions*, Springer Monographs in Mathematics,
https://doi.org/10.1007/978-3-031-13873-7

16. Ya.G. Berkovich, E.M. Zhmud, *Characters of Finite Groups*. Part 1, Translations of Mathematical Monographs, vol. 172 (American Mathematical Society, Providence, RI, 1998)
17. N. Bourbaki, *Éléments de mathématique. Part I. Les structures fondamentales de l'analyse. Livre II. Algèbre. Chapitre I. Structures algbriques.* Actualités Scientifiques et Industrielles, No. 934 (Hermann & Cie, Paris, 1942)
18. L. Breen, Thórie de Schreier supfieure. Ann. Sci. École Norm. Sup. (4) **25**(5), 465–514 (1992)
19. K. Brown, *Cohomology of Groups*, Graduate Texts in Mathematics, vol. 87 (Springer, New York-Berlin, 1982)
20. R. Brown, T. Porter, On the Schreier theory of non-abelian extensions: generalisations and computations. Proc. Roy. Irish Acad. Sect. A **96**, 213–227 (1996)
21. D. Bump, *Lie Groups*, Graduate Texts in Mathematics, vol. 225 (Springer, New York, 2004)
22. D. Bump, P. Diaconis, A. Hicks, L. Miclo, H. Widom, An exercise(?) in Fourier analysis on the Heisenberg group. Ann. Fac. Sci. Toulouse Math. (6) **26**(2), 263–288 (2017)
23. T. Ceccherini-Silberstein, F. Scarabotti, F. Tolli, Trees, wreath products and finite Gelfand pairs. Adv. Math. **206**, 503–537 (2006)
24. T. Ceccherini-Silberstein, F. Scarabotti, F. Tolli, *Harmonic Analysis on Finite Groups. Representation Theory, Gelfand Pairs and Markov Chains*. Cambridge Studies in Advanced Mathematics, vol. 108 (Cambridge University Press, Cambridge, 2008)
25. T. Ceccherini-Silberstein, A. Machì, F. Scarabotti, F. Tolli, Induced representation and Mackey theory. J. Math. Sci. (N.Y.) **156**(1), 11–28 (2009)
26. T. Ceccherini-Silberstein, F. Scarabotti, F. Tolli, Clifford theory and applications. J. Math. Sci. (N.Y.) **156**(1), 29–43 (2009)
27. T. Ceccherini-Silberstein, F. Scarabotti, F. Tolli, *Representation Theory of the Symmetric Groups: The Okounkov-Vershik Approach, Character Formulas, and Partition Algebras*, Cambridge Studies in Advanced Mathematics, vol. 121 (Cambridge University Press, Cambridge, 2010)
28. T. Ceccherini-Silberstein, F. Scarabotti, F. Tolli: *Representation Theory and Harmonic Analysis of Wreath Products of Finite Groups*. London Mathematical Society Lecture Note Series, vol. 410 (Cambridge University Press, Cambridge, 2014)
29. T. Ceccherini-Silberstein, F. Scarabotti, F. Tolli, Mackey's theory of τ-conjugate representations for finite groups. Japan. J. Math. **10**, 1–54 (2015)
30. T. Ceccherini-Silberstein, F. Scarabotti, F. Tolli, *Discrete Harmonic Analysis. Representations, Number Theory, Expanders, and the Fourier Transform*, Cambridge Studies in Advanced Mathematics, vol. 172 (Cambridge University Press, Cambridge, 2018)
31. T. Ceccherini-Silberstein, F. Scarabotti, F. Tolli, *Gelfand Triples and their Hecke Algebras. Harmonic Analysis for Multiplicity-free Induced Representations of Finite Groups*. With a foreword by Eiichi Bannai, Lecture Notes in Mathematics 2267 (Springer, Cham, 2020)
32. T. Ceccherini-Silberstein, F. Scarabotti, F. Tolli, A note on the Kaloujnine-Krasner theorem. Preprint (2022)
33. A.H. Clifford, Representations induced in an invariant subgroup. Ann. Math. (2) **38**(3), 533–550 (1937)
34. A.H. Clifford, G.B. Preston, *The Algebraic Theory of Semigroups*. Vol. I and II, Mathematical Surveys, No. 7 (American Mathematical Society, Providence, 1961)
35. A.J. Coleman, *Induced Representations with Applications to S_n and* GL(n). *Lecture Notes Prepared by C.J. Bradley*. Queen's Papers in Pure and Applied Mathematics, No. 4 (Queen's University, Kingston, 1966)
36. K. Conrad, Groups of order p^3, in *Expository Papers*. https://kconrad.math.uconn.edu/blurbs/
37. Ch.W. Curtis, *Pioneers of Representation Theory: Frobenius, Burnside, Schur, and Brauer*. History of Mathematics, vol. 15 (American Mathematical Society/London Mathematical Society, Providence/London, 1999)
38. Ch.W. Curtis, I. Reiner, *Representation Theory of Finite Groups and Associative Algebras*. Reprint of the 1962 original, Wiley Classics Library. A Wiley-Interscience Publication (John Wiley & Sons, New York, 1988)
39. Ch.W. Curtis, I. Reiner, *Methods of Representation Theory. Vol. I. With Applications to Finite Groups and Orders*, Reprint of the 1981 original. Wiley Classics Library. A Wiley-

Interscience Publication (Wiley, New York, 1990)
40. D. D'Angeli, A. Donno, Self-similar groups and finite Gelfand pairs. Algebra Discrete Math. (2), 54–69 (2007)
41. D. D'Angeli, A. Donno, A group of automorphisms of the rooted dyadic tree and associated Gelfand pairs. Rend. Semin. Mat. Univ. Padova **121**, 73–92 (2009)
42. J.D. Dixon, B. Mortimer, *Permutation Groups*. Graduate Texts in Mathematics, vol. 163 (Springer, New York, 1996)
43. L. Dornhoff, *Group Representation Theory. Part A: Ordinary Representation Theory*, Pure and Applied Mathematics, vol. 7 (Marcel Dekker, New York, 1971)
44. M. Eichler, A new proof of the Baker-Campbell-Hausdorff formula. J. Math. Soc. Japan **20**, 23–25 (1968)
45. S. Eilenberg, S. MacLane, Group extensions and homology. Ann. Math. (2) **43**, 757–831 (1942)
46. S. Eilenberg, S. MacLane, Cohomology theory in abstract groups I and II. Ann. Math. (2) **48**, 51–78 and 326–341 (1947)
47. L. Evens, *The Cohomology of Groups*. Oxford Mathematical Monographs. Oxford Science Publications (The Clarendon Press/Oxford University Press, New York, 1991)
48. J.M.G. Fell, R.S. Doran, *Representations of *-algebras, Locally Compact Groups, and Banach *-algebraic Bundles*. Vol. 2, Banach *-Algebraic Bundles, Induced Representations, and the Generalized Mackey Analysis. Pure and Applied Mathematics, vol. 126 (Academic Press, Boston, 1988)
49. G.F. Frobenius, Über die Charaktere der symmetrischen Gruppe. *Sitzungsber. Königl. Preuss. Akad. Wiss.*, 516–534 (1900)
50. A. Fröhlich, On fields of class two. Proc. London Math. Soc. (3) **4**, 235–256 (1954)
51. W. Fulton, J. Harris, *Representation Theory. A First Course* (Springer, New York, 1991)
52. P.X. Gallagher, Group characters and normal Hall subgroups. Nagoya Math. J. **21**, 223–230 (1962)
53. D. Gorenstein, *Finite Simple Groups. An Introduction to their Classification*. University Series in Mathematics (Plenum Publishing Corp., New York, 1982)
54. L.C. Grove, *Groups and Characters*, Pure and Applied Mathematics (New York). A Wiley-Interscience Publication (Wiley, New York, 1997)
55. L.C. Grove, *Classical Groups and Geometric Algebra*, Graduate Studies in Mathematics, vol. 39 (American Mathematical Society, Providence, 2002)
56. K.W. Gruenberg, *Cohomological Topics in Group Theory*, Lecture Notes in Mathematics, vol. 143 (Springer, Berlin, 1970)
57. M. Hall, *The Theory of Groups*. Reprinting of the 1968 edition (Chelsea Publishing, New York, 1976)
58. C.E. Hempel, Metacyclic groups. Commun. Algebra **28**(8), 3865–3897 (2000)
59. I.N. Herstein, *Topics in Algebra*, 2nd edn. (Xerox College Publishing, Lexington, 1975)
60. R.J. Higgs, Projective representations of abelian groups. J. Algebra **242**(2), 769–781 (2001)
61. D. Hilbert, Die Theorie der algebraischen Zahlkörper. *Jahresbericht der Deutschen Mathematiker-Vereinigung* **4**, 175–546 (1897)
62. G. Hochschild, J.P. Serre, Cohomology of group extensions. Trans. Amer. Math. Soc. **74**, 110–134 (1953)
63. R. Howe, Kirillov theory for compact p-adic groups. Pacific J. Math. **73**(2), 365–381 (1977)
64. Th.W. Hungerford, *Algebra*, Reprint of the 1974 original. Graduate Texts in Mathematics, vol. 73 (Springer, New York, 1980)
65. B. Huppert, *Character Theory of Finite Groups*, De Gruyter Expositions in Mathematics, vol. 25 (Walter de Gruyter, 1998)
66. B. Huppert, N. Blackburn, *Finite Groups. II*. AMD, 44. Grundlehren der Mathematischen Wissenschaften, vol. 242 (Springer, Berlin, 1982)
67. I.M. Isaacs, *Character Theory of Finite Groups*, Corrected reprint of the 1976 original [Academic Press, New York] (Dover Publications, New York, 1994)

68. I.M. Isaacs, *Finite Group Theory*. Graduate Studies in Mathematics, vol. 92 (American Mathematical Society, Providence, 2008)
69. N. Iwahori, H. Matsumoto, Several remarks on projective representations of finite groups. J. Fac. Sci. Univ. Tokyo Sec. I **10**, 129–146 (1964)
70. G.D. James, A. Kerber, *The Representation Theory of the Symmetric Group*. Encyclopedia of Mathematics and its Applications, vol. 16 (Addison-Wesley, Reading, MA, 1981)
71. G.J. Janusz, Some remarks on Clifford theorem and the Schur index. Pacific J. Math. **32**, 119–129 (1970)
72. M.I. Kargapolov, Yu.I. Merzljakov, *Fundamentals of the Theory of Groups*. Translated from the second Russian edition by Robert G. Burns. Graduate Texts in Mathematics, vol. 62 (Springer, New York, 1979)
73. G. Karpilovsky, *The Schur Multiplier*, London Mathematical Society Monographs. New Series, 2 (The Clarendon Press/Oxford University Press, New York, 1987)
74. G. Karpilovsky, *Clifford Theory for Group Representations*, North-Holland Mathematics Studies, vol. 156 (North-Holland Publishing, Amsterdam, 1989)
75. A. Kerber, *Representations of Permutation Groups. I*. Lecture Notes in Mathematics, vol. 240 (Springer, Berlin, 1971)
76. E.I. Khukhro, *Nilpotent Groups and their Automorphisms*, De Gruyter Expositions in Mathematics, vol. 8 (Walter de Gruyter, Berlin, 1993)
77. A.A. Kirillov, Unitary representations of nilpotent Lie groups. Uspehi Mat. Nauk **17**(4) (106), 57–110 (1962)
78. A.A. Kirillov, Merits and demerits of the orbit method. Bull. Amer. Math. Soc. (N.S.) **36**(4), 433–488 (1999)
79. A.A. Kirillov, *Lectures on the Orbit Method*, Graduate Studies in Mathematics, vol. 64 (American Mathematical Society, Providence, 2004)
80. A.W. Knapp, *Basic Algebra*, Cornerstones (Birkhäuser Boston, Boston, 2006)
81. K.P. Kokhas, Finite factor representations of 2-step nilpotent groups, and orbit theory. J. Math. Sci. (N.Y.) **131**(2), 5508–5519 (2004)
82. M. Krasner, L. Kaloujnine, Le produit complet des groupes de permutations et le problème d'extension de groupes I. C. R. Acad. Sci. Paris **227**, 806–808 (1948)
83. M. Krasner, L. Kaloujnine, Produit complet des groupes de permutations et problème d'extension de groupes I, II, III. Acta Sci. Math. Szeged **13**, 208–230; **14** (1951), 39–66 and 69–82 (1950)
84. A.G. Kurosh, *The Theory of Groups*, vol. 2 (Chelsea Publishing, New York, 1960)
85. S. Lang, *Topics in Cohomology of Groups*, Lecture Notes in Mathematics, vol. 1625 (Springer, Berlin, 1996)
86. S. Lang, *Algebra*. Revised third edition, Graduate Texts in Mathematics, vol. 211 (Springer, New York, 2002)
87. M. Lazard, Sur les groupes nilpotents et les anneaux de Lie, Ann. Sci. École Norm. Sup. (3) **71**, 101–190 (1954)
88. F. Lazebnik, Groups of order p^3, in Teaching Materials, etc. http://www.math.udel.edu/~lazebnik/Info/teaching.html
89. R.C. Lyndon, The cohomology theory of group extensions. Duke Math. J. **15**, 271–292 (1948)
90. A. Machì, *Groups. An Introduction to Ideas and Methods of the Theory of Groups*, Unitext, vol. 58 (Springer, Milan, 2012)
91. G.W. Mackey, Les ensembles boréliens et les extensions des groupes, J. Math. Pures Appl. (9) **36**, 171–178 (1957)
92. G.W. Mackey, Unitary representations of group extensions. I. Acta Math. **99**, 265–311 (1958)
93. G.W. Mackey, Multiplicity-free representations of finite groups. Pacific J. Math. **8**, 503–510 (1958)
94. G.W. Mackey, *The Theory of Unitary Group Representations. Based on notes by James M.G. Fell and David B. Lowdenslager of Lectures given at the University of Chicago, Chicago, Ill., 1955*, Chicago Lectures in Mathematics (University of Chicago Press, Chicago, 1976)

95. G.W. Mackey, *Unitary Group Representations in Physics, Probability, and Number Theory*, 2nd edn. Advanced Book Classics, Advanced Book Program (Addison-Wesley Publishing Company, Redwood City, 1989)
96. S. MacLane, Cohomology theory in abstract groups. III. Operator homomorphisms of kernels. Ann. of Math. (2) **50**, 736–761 (1949)
97. S. MacLane, *Categories for the Working Mathematician*. Graduate Texts in Mathematics, vol. 5 (Springer, New York, 1971)
98. A. Mihailovs, The orbit method for finite groups of nilpotency class two of odd order, Preprint; arXiv.org: math.RT/0001092
99. C.C. Moore, Extensions and low dimensional cohomology theory of locally compact groups. I, II. Trans. Amer. Math. Soc. **113**, 40–63 (1964); ibid. **113** (1964), 64–86
100. C.C. Moore, Group extensions and cohomology for locally compact groups. III. Trans. Amer. Math. Soc. **221**(1), 1–33 (1976)
101. M.A. Naimark, A.I. Stern, *Theory of Group Representations* (Springer, New York, 1982)
102. R.C. Orellana, M.E. Orrison, D.N. Rockmore, Rooted trees and iterated wreath products of cyclic groups. Adv. Appl. Math. **33**(3), 531–547 (2004)
103. J.A. Packer, Projective representations and the Mackey obstruction – a survey, in *Group Representations, Ergodic Theory, and Mathematical Physics: A Tribute to George W. Mackey*, Contemp. Math., vol. 449 (Amer. Math. Soc., Providence, 2008), pp. 345–378
104. F. Pearson, A. Romanov, D. Soller, Finite Gelfand pairs and cracking points of the symmetric groups. Rocky Mountain J. Math. **50**(5), 1807–1812 (2020)
105. L. Pukánszky, On the theory of exponential groups. Trans. Amer. Math. Soc. **126**, 487–507 (1967)
106. L. Pukánszky, *Characters of Connected Lie Groups. With a Preface by J. Dixmier and M. Duflo*, Mathematical Surveys and Monographs, vol. 71 (American Mathematical Society, Providence, 1999)
107. E.W. Read, The projective representations of the generalized symmetric group. J. Algebra **46**(1), 102–133 (1977)
108. D.J.S. Robinson, *A course in the theory of groups*, Graduate Texts in Mathematics, 2nd edn., vol. 80 (Springer, New York, 1996)
109. D. Rockmore, Fast Fourier analysis for abelian group extensions. Adv. in Appl. Math. **11**(2), 164–204 (1990)
110. D. Rockmore, Fast Fourier transforms for wreath products. Appl. Comput. Harmon. Anal. **2**(3), 279–292 (1995)
111. R.L. Roth, A dual view of the Clifford theory of characters of finite groups. Canad. J. Math. **23**, 857–865 (1971)
112. R.L. Roth, A dual view of the Clifford theory of characters of finite groups. II. Canad. J. Math. **25**, 1113–1119 (1973)
113. J.J. Rotman, *An Introduction to Homological Algebra*. Pure and Applied Mathematics, vol. 85 (Academic Press, New York, 1979)
114. J.J. Rotman, *An Introduction to the Theory of Groups*. Graduate Texts in Mathematics, 4th edn., vol. 148 (Springer, New York, 1995)
115. D.J. Rusin, Representations of metabelian groups. J. Pure Appl. Algebra **18**(3), 283–291 (1980)
116. F. Scarabotti, F. Tolli, Induced representations and harmonic analysis on finite groups. Monatsh. Math. **181**(4), 937–965 (2016)
117. O. Schreier, Über die Erweiterung von Gruppen I. Monatsh. Math. Phys. **34**(1), 165–180 (1926)
118. I. Schur, Über die Darstellung der endlichen Gruppen durch gebrochene lineare Substitutionen. J. Reine Angew. Math. **127**, 20–50 (1904)
119. I. Schur, Untersuchungen über die Darstellung der endlichen Gruppen durch gebrochene lineare Substitutionen. J. Reine Angew. Math. **132**, 85–137 (1907)
120. I. Schur, On the representation of the symmetric and alternating groups by fractional linear substitutions. J. Reine Angew. Math. **139**, 155–250 (1911)

121. J.-P. Serre, *Linear Representations of Finite Groups*. Graduate Texts in Mathematics, vol. 42 (Springer, New York, 1977)
122. J.-P. Serre, *Finite Groups: An Introduction*. With assistance in translation provided by Garving K. Luli and Pin Yu. Surveys of Modern Mathematics, vol. 10 (International Press/Higher Education Press, Somerville/Beijing, 2016)
123. R. Shaw, *Linear Algebra and Group Representations. Vol. II. Multilinear Algebra and Group Representations* (Academic Press, London, 1983)
124. H.S. Sim, Metacyclic groups of odd order. Proc. Lond. Math. Soc. (3) **69**(1), 47–71 (1994)
125. B. Simon, *Representations of Finite and Compact Groups* (American Math. Soc., Providence, 1996)
126. S. Sternberg, *Group theory and Physics* (Cambridge University Press, Cambridge, 1994)
127. M. Suzuki, *Group Theory. I*. Grundlehren der Mathematischen Wissenschaften, vol. 247 (Springer, Berlin, 1982)
128. P.A. Tucker, On the reduction of induced representations of finite groups. Amer. J. Math. **84**, 400–420 (1962)
129. P.A. Tucker, Note on the reduction of induced representations. Amer. J. Math. **85**, 53–58 (1963)
130. Ch.A. Weibel, History of homological algebra, in *History of Topology* (North-Holland, Amsterdam, 1999), pp. 797–836
131. H. Weyl, *The Classical Groups. Their Invariants and Representations* (Princeton University Press, Princeton, 1939)
132. E. Wigner, On unitary representations of the inhomogeneous Lorentz group. Ann. Math. **40**, 149–204 (1939)
133. E.P. Wigner, *Group Theory and its Application to the Quantum Mechanics of Atomic Spectra. Expanded and Improved ed. Translated from the German by J. J. Griffin*, Pure and Applied Physics, vol. 5 (Academic Press, New York, 1959)
134. K. Yamazaki, A note on projective representations of finite groups. Sci. Papers College Gen. Ed. Univ. Tokyo **14**, 27–36 (1964)
135. H. Zassenhaus, Ein Verfahren, jeder endlichenp-Gruppe einen Lie-Ring mit der Charakteristikp zuzuordnen. Abh. Math. Sem. Univ. Hamburg **13**, 200–207 (1939)
136. E.I. Zelmanov, Solution of the restricted Burnside problem for groups of odd exponent. Izv. Akad. Nauk SSSR Ser. Mat. **54**(1), 42–59, 221 (1990); translation in *Math. USSR-Izv.* **36** (1991), no. 1, 41–60.
137. E.I. Zelmanov, Solution of the restricted Burnside problem for 2-groups. Mat. Sb. **182**(4), 568–592 (1991); translation in Math. USSR-Sb. **72**(2), 543–565 (1992)

Subject index

T. Ceccherini-Silberstein et al., *Representation Theory of Finite Group Extensions*, Springer Monographs in Mathematics,
https://doi.org/10.1007/978-3-031-13873-7

Index of authors

T. Ceccherini-Silberstein et al., *Representation Theory of Finite Group Extensions*, Springer Monographs in Mathematics,
https://doi.org/10.1007/978-3-031-13873-7